SOCIAL CHANGE IN THE
PACIFIC ISLANDS

SOCIAL CHANGE IN THE PACIFIC ISLANDS

Edited by

Albert B. Robillard

SOCIAL SCIENCE RESEARCH INSTITUTE
UNIVERSITY OF HAWAI'I

Routledge
Taylor & Francis Group

LONDON AND NEW YORK

First published 1992 by Kegan Paul International Ltd

Published 2019 by Routledge
2 Park Square, Milton Park, Abingdon, Oxon OX14 4RN
52 Vanderbilt Avenue, New York, NY 10017

First issued in paperback 2019

Routledge is an imprint of the Taylor & Francis Group,
an informa business

Distributed by
John Wiley & Sons Ltd

Set in Palatino 10/12pt by Intype.

British Library Cataloguing in Publication Data
Robillard, Albert Britton, *1943–*
Social change in the Pacific islands.
1. Pacific islands. Social change. History
I. Title
990

Library of Congress Cataloging-in-Publication Data
Social change in the Pacific islands / edited by Albert Britton
Robillard.
510 pp. 216 cm.
Includes bibliographical references and index.
ISBN 0–7103–0400–5
1. Oceania–Social conditions. 2. Social structure–Oceania.
3. Social change. I. Robillard, Albert B., 1943– .
HN930.7.A8S63 1990
306'.0995–dc20   90–38454
                                                   CIP

ISBN 13: 978-1-138-98215-4 (pbk)
ISBN 13: 978-0-7103-0400-1 (hbk)

Contents

Preface   vii
  Robert C. Kiste
Acknowledgements   x
Contributors   xiii
Maps   xv
Chapter 1: Introduction: Social Change as the Projection of
          Discourse   1
  Albert B. Robillard

MELANESIA   33

Chapter 2: Papua New Guinea: Changing Relations of
          Production   35
  Eugene Ogan and Terence Wesley-Smith

Chapter 3: New Caledonia: Social Change, Political Change and
          Tradition in a Settler Colony
  John Connell   65

Chapter 4: Social Change in Vanuatu   98
  Jean-Marc Philibert

Chapter 5: Social Change and the Survival of Neo-Tradition
          in Fiji   134
  Vijay Naidu

MICRONESIA   201

Chapter 6: The Expensive Taste for Modernity: Caroline and
          Marshall Islands   203
  Francis X. Hezel, S.J.

Chapter 7: The Militarization of Guamanian Society   220
  Larry W. Mayo

Chapter 8: Elements of Social Change in the Contemporary
             Northern Mariana Islands   241
       Samuel F. McPhetres

Chapter 9: Kiribati: Change and Context in an Atoll World   264
       Roger Lawrence

POLYNESIA   301

Chapter 10: Schooling and Transformations in Samoan Social
             Status in Hawai'i   303
       Robert W. Franco

Chapter 11: State Formation, Development, and Social Change
             in Tonga   322
       Christine Ward Gailey

Chapter 12: French Polynesia: A Nuclear Dependency   346
       Ben Finney

Chapter 13: Where is Social Change in Hawai'i? The Reyn's
             Aloha Shirt   372
             Albert B. Robillard

Chapter 14: Language and Social Change in the Pacific Islands 393
       Donald M. Topping

Chapter 15: Children's Survival in the Pacific Islands   414
       George Kent

Chapter 16: Some Concluding Thoughts on Social Change   428
       Cluny Macpherson

Notes   453
References   469
Index   501

Preface

The Pacific Ocean is the largest geographical feature on the face of the earth, covering about one third of its entire surface. Occupying part of that large expanse are the far flung islands of the Pacific.

In terms of overall human history, the peopling of the Pacific has been relatively recent; the region was the last global space to experience events common to a world history. Setting aside the hostile environment of Antarctica, the Pacific region was the last major world area to be penetrated by human beings. The archaeological record suggests that the first migrants began to explore the island of New Guinea around 50,000 to 40,000 years ago from their origins somewhere in Southeast Asia and went on to occupy other areas of Melanesia. Around 4,000 to 5,000 years before the present, it appears that yet other peoples with great sea-going and navigational skills also entered the region from Southeast Asia and eventually populated what are now known as the culture areas of Micronesia and Polynesia.

The islands and their peoples have always been dynamic and for much of the Pacific, there was considerable canoe travel and communication among island groups in pre-European times. It seems almost certain that Polynesians reached the western shores of South America and returned to their homelands with the sweet potato, other plants, and other unknown items. At the same time, some populations were isolated, were outside of events elsewhere, and developed unique cultures. Easter Island is perhaps the foremost example. Throughout the entire region, however, there was a general movement towards greater cultural and linguistic diversity. Change was the rule and not the exception.

Beginning in the early sixteenth century, a new chapter of Pacific history was opened. The explorer Magellan became the first European to cross the Pacific, making the journey from east to west without sighting an inhabited island until he came upon one of the southernmost of the Marianas (probably the island of Guam) in 1521. In the process, the Pacific became the last major world area to be penetrated

vii

by European explorers. Actual colonization did not occur, however, until 1668 when Spain established its rule over the Marianas and the Pacific became the last portion of the globe over which Europeans claimed hegemony. There was a lapse of decades before the partition of the regional by European powers began in earnest in the 1840s. France acquired Tahiti and other possessions in the Society Islands early in the decade. Thereafter, the process escalated, and by shortly after the turn of the century, the entire Pacific was under European control.

World War I witnessed major revisions in the colonial order of the Pacific; Germany's exit allowed Japan to become the region's first non-Western colonial power. During the interim between the two world wars, the Pacific lapsed into a somewhat sleepy backwater outside the mainstream of world events. The metropolitan powers enhanced their control over their island territories, and Christian missions became further entrenched and extended their spheres of influence.

World War II brought an abrupt end to the tranquil time, and changed perceptions of the world order. In many territories, the legitimacy and inequalities of the colonial structures were questioned. There was no turning back to the pre-war arrangements, but even then and in comparison with other world areas, decolonization came late to the Islands. Not until 1962 did Western Samoa became the first Pacific island nation to gain independence; others followed. Today, there are nine independent nations (five of which are members of the United Nations), and four are self-governing states in free association with their former colonial rulers. Decolonization is far from complete however; eight Island territories remain dependencies of one variety or another of three Western metropolitan nations.

As the papers of this volume clearly indicate, the postwar era and decolonization have brought unprecedented change, and the Pacific is now experiencing problems that were formerly associated with other Third World nations. Most Pacific countries have rapidly expanding populations, and over half of all Pacific Islanders are now in their teenage years or younger. Education and modern communications have served to increase aspirations, and attracted by hopes of employment and the distractions of urban life, islanders are gravitating to urban centers.

As Robillard points out in his introduction, the histories and analyses of cultural and social change in island societies may be cast in one or more of a variety of theoretical lenses. In earlier decades, Pacific history was imperialistic in orientation and focused on the actions and perspectives of the colonial powers, and islanders were portrayed as more or less passive actors in the process. A shift occurred in the 1950s with the development of the 'new Pacific history' and attention became focused on the active, participatory, and dynamic role that islanders played in shaping the transformations of their own societies and their interactions with the foreign interlopers.

New perspectives were gained from the reorientation of Pacific history, but the papers that make up this volume implicitly suggest that the influence of forces external to the islands has been neglected in many recent analyses of the dynamics of the region; the time may be ripe for yet another shift in orientation. The metropolitan powers of the Pacific rim, the members of the European Economic Community, and nations in yet other areas of the world are evidencing renewed interest in the islands, and they are increasingly influencing and intervening in island affairs. Transnational corporations are becoming more involved in island economies, parliamentary and congressional studies in Australia, New Zealand, and the United States debate their roles in the region, and strategic interests shape assistance programs. Japan is reasserting its interests in the islands in a variety of ways, and other Asian nations are becoming more involved. In short, the spectre of neo-colonialism is clearly on the horizon while island nations are determined to protect their recently gained statuses as independent states or self-governing nations in free association with a metropolitan power. New tensions between internal and external forces are in the making and will only add to the dynamics of the region. The papers of this volume address some of these issues and are a first step in what may be yet another reorientation of the analyses of change in and an examination of the history and current events of the island world.

Robert C. Kiste
Director, Center for Pacific Island Studies
University of Hawai'i
Honolulu, Hawai'i
June 1990.

Acknowledgements

This book would not be possible without the support of the University of Hawai'i. The University has long focused on the Pacific islands and Asia and has earned the reputation, along with the University of Hawai'i Press and the Pacific and Asian Collections of Hamilton Library, as a world center of scholarship on Oceania and East and Southeast Asia. The School of Hawaiian, Asian and Pacific Studies is the most recent institutionalization of the University's interest and commitment to research, teaching and service to the Pacific and Asia. The editor has benefited from membership in the faculty of the School's Center for Pacific Islands Studies. Bob Kiste, the Center's Director, has been a source of personal support and guidance since I first arrived in Hawai'i from Michigan. Kiste is largely responsible for establishing a highly productive and often heterodox atmosphere of serious Pacific islands study.

The Social Science Research Institute, headed by Donald Topping, has been a base of operations and a warm home for seven years. Topping, a Pacific linguist, has shared a keen interest in Micronesia and, more importantly, has provided the material resources to bring this book to fruition. Topping has also been instrumental in providing support for numerous other undertakings and for making life more comfortable.

Kiyoshi Ikeda, Chair of the Department of Sociology at Hawai'i, has been an inspiration and friend for ten years. He fully supported my often long term (six or seven months a year) wanderings in the Pacific. He has pushed for sociology's involvement in Pacific studies, an unusual position in a discipline dominated by interests in industrial societies.

In the College of Social Sciences at Hawai'i, I have had the benefit of discussion, readership, and criticism of the book's premises and the individual chapters. Deane Neubauer has provided a useful criticism and support at every step of the process. Alfred Fortin has read several chapters, some of them many times. Neal Milner has also read the initial outline and sample chapters, as did Thomas Maretzki, Stephen Yeh, Cullen Hayashida, Donald H. Rubinstein, and Geoffrey M. White. Karen Watson-Gegeo provided needed encouragement. Michael Shapiro shared his work and provided a helpful discussion on several main ideas of the book. When Ben Finney gave me the first complete chapter, I knew we had a book.

Wen-Shing Tseng, Professor of Psychiatry, gave me the first opportunity to travel in the Pacific islands. His conversations on development and social change are reflected in this volume. In the School of Public Health Jonathan Okamura has been a colleague in Pacific Island studies and also in our mutual interest in things Filipino.

Cluny Macpherson of the University of Auckland has been a sociological colleague in Pacific studies, a field with few sociologists. He deserves to be a co-editor of this work. He has read all of the chapters and worked with me in Hawai'i and from New Zealand.

At York University in Toronto, John O'Neill has read and critiqued several manifestations of the chapters. He was instrumental in bringing an idea of a book to an actual product.

Harold Garfinkel at UCLA read the outline and chapters by the editor and provided encouragement throughout the long process of making the book.

Francis X. Hezel, S.J. in Truk and Anthony Polloi in Palau have been friends and colleagues for many years. They read and reacted to multiple drafts of the book proposal. On Saipan, Sam McPhetres provided an early paper and kept up with the book's progress with visits to Honolulu and by a stream of letters.

Ron Crocombe, now living in the Cook Islands, provided a wide ranging discussion of the book as well as other Pacific topics from Washington, D.C.

Graduate students are always an occasion for thinking and a source of ideas. My seminar in social change in the Pacific islands has provided the critical participation in this volume of Mary Low O'Sullivan of Fiji, Dorothy Goldsborough of Alaska, and Ishmael Lebehn of Pingelap. Richard Chabot, Robert Anae, Deanna Chang, Jeffrey Kamakahi, Neghin Modavi, Moshe Rapaport, Harumi Sasaki-Karel, Kwok Kit Ng, Sungnam Cho, and Roger Reed have been subjected to various chapters and have thereby contributed to the manuscript.

Now to the really important people: the graduate student research assistants who put this volume together. Jan Shishido typed the book proposal. Kathleen Milled worked on the first drafts of several chapters. Nami-Anne Kosaka supervised the entire editing process, from relations with the publisher and the authors, to style, to print face, to references, to proof reading. Agatha Yap typed from my dictation the 'aloha shirt' chapter and much of the correspondence. Meredith Burns and Andrea Leung typed and re-typed, critiqued, and proofread many of the chapters. These people worked tirelessly and with a sense of humor.

Freda Hellinger copy edited the entire manuscript. She is the editor of the Social Science Research Institute, University of Hawai'i.

Peter Hopkins, chairman of Kegan Paul International, has an interest in the Pacific islands and is responsible for stimulating the editor and authors to complete what was merely a good idea in the original book proposal. Kaori O'Connor, editorial director at Kegan Paul, has provided patient, detailed direction and encouragement for an overdue manuscript. She knew the book before she saw the manuscript: she is from Hawai'i and she knows critical and postmodern social theory.

I must thank the authors of the chapters for consenting to write original papers at my request and have them published in a volume of edited works on critical political economy of social change in the Pacific islands. However much help we have received from the individuals named above, only the editor and the authors are responsible for the shortcomings of the book.

Albert B. Robillard
Honolulu, Hawai'i
April 1990

Contributors

John Connell
Department of Geography, University of Sydney, Australia

Ben Finney
Department of Anthropology, University of Hawai'i at Manoa

Robert W. Franco
Department of Anthropology, Kapiolani Community College

Christine Ward Gailey
Department of Sociology and Anthropology, Northeastern University

Francis Z. Hezel, S.J.
Micronesian Seminar, Truk

George Kent
Department of Political Science, University of Hawai'i

Robert C. Kiste
Center for Pacific Island Studies, University of Hawai'i

Roger Lawrence
Victoria University of Wellington, New Zealand

Cluny Macpherson
Department of Sociology, University of Auckland, New Zealand

Samuel F. McPhetres
Saipan

Larry W. Mayo
DePaul University

Vijay Naidu
Department of Sociology, University of the South Pacific

Eugene Ogan
Department of Anthropology, University of Minnesota

Jean-Marc Philibert
Department of Anthropology, University of Western Ontario, Canada

Albert B. Robillard
Social Science Research Institute, University of Hawai'i

Donald M. Topping
Social Science Research Institute, University of Hawai'i

Terence Wesley-Smith
Center for Pacific Island Studies, University of Hawai'i

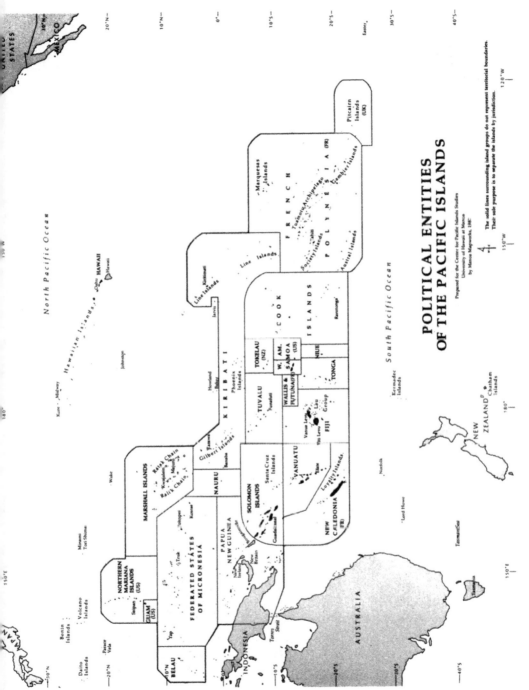

POLITICAL ENTITIES
OF THE PACIFIC ISLANDS

Maps published with permission of Professor Robert C. Kiste, Director, Center for
Pacific Islands, University of Hawaii, Manoa.

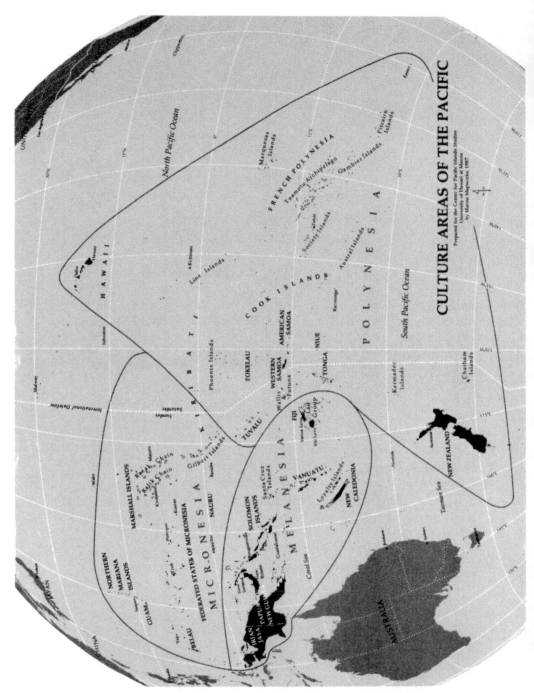

Maps published with permission of Professor Robert C. Kiste, Director, Center for Pacific Islands, University of Hawaii, Manoa.

1

Introduction: Social Change as the Projection of Discourse

Albert B. Robillard

Man is rendered incoherent by the coherence of his structural projection.
Baudrillard 1968, p. 69

This introduction is composed of two parts. The first is a postmodern formulation of the location of social change, where we should study it and how the formulation differs from other treatments of social change. The position is that social change is isomorphic with the discourse or language that makes social change a visible event, worthy of description, publication and circulation. This takes an expanded definition of discourse or language, one that includes every sign function imaginable. Discourse is not merely a neutral representational device but includes everything that can be represented – all the rationalities of interactional exchange. Discourse is the institutionalized world. This may be a strange part of the text for some readers, those not acquainted with the postmodern novel,[1] literary criticism, continental philosophy, and social theory. For the unacquainted, I have to ask for your indulgence for what will at first appear as a strange use of language.

In the first part of the introduction I am arguing for the possibility of an alternative analysis of social change. The possibility is an analysis of change based on the indigenous myths or cultures of the Pacific islands. Change would not be driven by the inner or teleological logic of capitalism interacting with precapitalist modes of production. Change would not be reduced to the anthropocentric notion of man creating a world by wresting his existence from nature. Change would be described as emanating from what secularized Europeans call gods, ghosts, spirits, magic, supernatural forces, rocks or powerful stones. The strangeness of the possibility of change as coming from other than man (as in Pacific island creation myths, Hawaiian Pele worship, or from the land, suffused with animistic powers, or from a claimed affinity with the land, as in the Taukei Movement in contemporary Fiji)[2] issues from our assumption that man is the fundamental datum.

1

The point of the argument in the first part is to identify that political economies of Pacific island social change are as socially produced as creation myth accounts.

Professor Ron Crocombe (now retired from the University of the South Pacific, Fiji) has correctly charged that this book is by Europeans. One Indo-Fijian is represented. However, he is writing from an entirely European political economy perspective, as does the part-Maori author of a chapter. To the extent that the language of political economy is the authorized (institutionalized) language for detecting and reporting social change, this book is an elaboration of the European sensibility of the Pacific islands. Alternative accounts will have to come from a different discourse, as is being developed by Albert Wendt (1987). An analogous new rhetoric of social historiography is being created by Eduardo Galeano in his three volume history of the Americas (1985; 1990).

However, from the fact that this is a European account a further point can be made about the purpose of a postmodern critique. The proposition that the Pacific islands are constructed by the particular discourse we inhabit is not put forward here merely to demystify or to claim some sort of nihilistic position (as in the claim that there is no 'ultimate' or 'true' Pacific islands). Hopefully, a postmodern critique of the language of theories of Pacific island social change will bring us, indigenous islander and Westerner alike, into a new and direct relationship with language, opening spaces for creating anew an even more diverse Pacific islands (Edmundson 1989, pp. 62–63). This should transform the hegemony of the Western languages.

This book is not comprehensive. The reader is referred to the *Pacific Islands Yearbook* (1991) for a comprehensive treatment of all Pacific island histories. The tour de force is, of course, Douglas Oliver's *The Pacific Islands*, now in its third edition (1989). There are many island groups not covered here: Nauru, the Solomons, the Cooks, Tuvalu, Tokelau, Niue, Pitcairn Islands, Wallis and Futuna, and Easter Island. New Zealand is also missing. The objective of this book is not to be a comprehensive treatment of Pacific island social history. Rather, the object of the book is to present theoretically recognizable political economies of social change in the Pacific islands. The essays are demonstrations of critical political economic exposition and, in various degrees, reflections on theory in Pacific island analysis.

The second part of the introduction discusses the sixteen chapters in the book. A thumbnail description of the substance and theory of each chapter is given. Some of the chapters are comprehensive, tracing social change usually from contact with the West to the present. Others focus on a theme of social change, and nutrition. Diverse as they are, all the chapters are written as political economies of social change.

The tension dealt with is the relationship between so-called traditional and capital-based political economies.

I. DISCOURSE OF SOCIAL CHANGE

This is a book about social change in the Pacific islands and theories of social change. This statement is not offered as a redundant statement of the title. I enter it here to fix before us, what for most of us, in our everyday attitude, are expressions taken for granted. Social change, the Pacific islands as a contiguous surface area of the globe, and theories of social change are not unfamiliar terms. They are used in academic discourse, the language of government and formal politics, the rhetoric of commerce and finance, the representations of non-government service organizations, and in the popular press, both print and electronic. We, writers and readers, are competent subjects in the referential exchange of these signs and their objects. We, who read this, are bound up in pragmatic activities, such as those listed above, wherein the use of these expressions produces and sustains our occupational and educated worlds. In short, the expressions are institutionalized features of our everyday pragmatic achievements, for example, our writing, reading, calculating, planning, educating, navigating, and correctly remembering the history of the social world. That the terms or expressions are part and parcel of a common world of subjects, that they account for and sustain a world of pragmatic achievements, is the phenomenon I will address in this part of the introduction.

This position, a genealogy of discourse, is not novel (Foucault 1977; Said 1979). At the same time, it is not the usual approach of Anglo-American social science, and it is not a common analytical frame in Pacific island studies. Certain lines of thought are roughly compatible. I am thinking of the recent work of Roger M. Keesing (1989) and also the theoretical work of Clifford Geertz (1983). Immanuel Wallerstein's idea of how the notions of the rise of industry and the bourgeoisie are elementary truisms, used to explain and form the modern world, when there is ample evidence for alternative explanations, are examples of linguistic categories sustaining a social structure (1989). Wallerstein, though, does not travel in the direction of discourse analysis beyond making this observation.

The usual social science approach regards language, discourse, as an a priori, a given. The consensus social science, as Anthony Giddens calls it (1981), proceeds from the assumption that language stands in a positive correspondence relationship with an external world of language independent objects. Language is a relatively neutral vehicle of reference for a positive world. Most research articles are written in the third person, indicating the passive role of language, the conveyor of objective facts. Referential disputes can be settled by inspection of a measurable extended world. For the reader steeped in this tradition of linguistic positivity, which includes all of us in the contemporary world of calculative rationality, the idea that social change is to be studied as the pragmatics of language, discourse, is strange. Social change is something that occurs 'out there', in this tradition, and not in the discourse of a theory of social change. A descriptive theoretical language is considered prior to the phenomenon of change, something distinct from the objects signified. This is the so-called bi-planar theory of language (Finlay 1988), language and signified objects each occupying a separate realm of reality.

I do not propose to review the tradition of linguistic positivism and its implications for social science and theories of change. I refer to the reader to a highly instructive book by Michael Shapiro, *The Politics of Representation* (1988) and an earlier book by the same author, *Language and Political Understanding* (1981). Shapiro's thesis is that language constructs and maintains a network of social institutions, including the institution of membership in the network of social institutions, including the institution of membership in the network. Also, the two-volume work by Jurgen Habermas, *The Theory of Communicative Action* (1981), provides a useful review of language and social science, as does his commentary on the methodology of the social sciences (1988). Habermas argues, following Alfred Schutz, Aaron Cicourel, and Melvin Pollner, that language contains common-sense knowledge which actors use to construct the intersubjective world. I merely want to acknowledge the origins of the difference between the orthodox consensus social science and what is being proposed here. However, I am not unaware of the need to illustrate the point that theories of social change are bound up with, are part of, the practices, pragmatics, of producing their object. I will illustrate this point with four examples.

Ships' Logs

Being that this book is about social change in the Pacific islands, it is perhaps appropriate, if not overly romantic, that the first illustration is from the age of 'discovery' in European sailing ships, an admittedly Eurocentric sense of exoticism and exploration. Let us consider the ship's log. A ship's log is the detailed record of the transit of the ship. It is written for the purpose of describing the procedures and conditions of sailing a particular route, from a point of embarkation to a destination. So detailed are the logs that they are often called rudders, providing instructions on how to sail to specific destinations. The rudders contain information on how to sail to the Pacific islands and

4

Asia from Europe, coming around Africa and South America. Some provided instructions on how to navigate up the west coast of the Americas and how to catch the northern Pacific prevailing trade winds to sail downwind across the Pacific – to Guam and Manila, for example – and how to sail the southern Pacific trade wind route back to Peru and up to Acapulco (Steinberg 1987, pp. 91–3).[3] This is the route of the Manila galleon trade. The ships' logs provided instructions based on the time of year, sea and weather conditions, observation points on land and ocean and how to measure the course of the ship, using celestial, sea conditions, flora and fauna, and the gauging of time. Ships' logs were precious possessions, guarded and fought over by the contending European explorer nations. They were regarded as maps to the material treasures and souls of Asia and Pacific.

The logs are written instructions. As such they span space and time, meaning that they are a continual repository or container of knowledge (Giddens 1981, p. 94). As a repository, even though located in objectivist time and space, they provide a kind of abstract cultural constancy of reference and review. Like books, reports, records, and computer files, the logs are a permanent description, something that is continuously available.

I do not want to argue that writing creates the constancy principle of objects. Much more is involved than writing, however important writing is to the maintenance of an objective world in correspondence to codified written knowledge. I want to submit that the logs, as usable representations, are embedded in the pragmatics of effectively reproducing the voyages that they describe. The ships' logs are productively a part of the adequate procedure of reproducing actual voyages and giving calculative accounts, the practices of commercial and state administration. The utility, relevance, truth, rationality, scientism and whatever other predications of the logs are to be found in the activity whereby the logs are a practice or practices for the material reproduction of a social life-world. In other words, the logs as representations achieve their entire character, whatever that might be, *'in and as of'* (Garfinkel 1967) the world they play a part in producing. The logs are a shared symbolic code, projecting and explaining the structure of sailing.

Ships' logs are not antiquated. They are artifacts. They are of interest mainly to historians. No one uses them to sail. They contain items of observation-like sea monsters – that seem to the contemporary reader to be incredible. The logs have been eclipsed by virtually more elaborate and more predictive methods of navigation, like computer driven LORAN and inertial navigation. But today, just as before, when the logs were being written and used as rudders, the logs receive their topicality as of the social world they produce and maintain. Today it is the world of the historian and archivist. Yesterday, it was the world-space of the expansionary European powers, looking for wealth and souls.

Pacific Islanders and Anthropological Literature

The second illustration is from an experience I had six years ago. This example presages the critique developed below of the hegemony of political economy in the representation of social change. I was involved in a teaching program for Pacific island mental health workers. Some of the students were Pacific island physicians and some were mental health counselors. they were asked to make presentations to the class about the social and cultural organization of everyday life in their home villages. The teaching faculty of the program was astonished to find that the students would only talk about their social and cultural structures in terms of the established and published ethnographies. The problem for us was that we wanted to have the trainees and ourselves think about contemporary social interaction in the village, how routine topics in everyday conversation were related – played a part in structuring events – to current mental health problems in the village population, for example, adolescent suicide, schizophrenia, depression, and violent behavior. Try as we might, we could not get the trainees, or the majority of the faculty, for that matter, to think of village social structure as sequentially produced, moving through time and space as of the symbolic interaction that made structure visible.

What we received were formal reports on kinship systems, land tenure, traditional political structure and religion, mythology, indigenous fishing, agriculture, and transportation technologies. The reports were entirely paraphrased versions of recognizable anthropological publications. The publications were often recited verbatim, frequently without attribution. The trainees and many of the faculty were astonished that we were insisting on descriptions of the mundane or ordinary (Pollner 1987); they could not see anything interesting or 'worth telling' (as in tell a story) in everyday life, focusing instead on the monumental, generalized descriptions of the past. These general or totalizing (Deleuze and Guattari 1972) descriptions of social and cultural organization were felt to be the descriptions of the essential social life of the village. They objectified the master cultural plan, as if the plan were the motor of interactional life. The accounts of published anthropology exercised a textual supremacy over experience. Current social life, especially everyday interaction, was approached as if it were a vulgar version of the indigenous culture, the present being corrupted by the West. Moreover, the real object of description, the influence of

6

the West aside, was contained in the totalizing or general descriptions of the thereby monumentalized or traditionalized social and cultural organization of the trainees' islands.

The issue here is not whether the trainees' use of the anthropological literature in making their descriptions produced false descriptions, as in the creation of neotraditional culture, discussed by Philibert in his chapter on Vanuatu in this volume. For the present, I want to point out that the trainees, when asked to provide a description, felt constrained to use a particular form of representation, the 'official' published anthropological literature. The storibility of their social world, with the professional Western audience, was contained in these monumentalized descriptions ignored and were incapable of describing the interactively mundane.

The use of the official language of anthropology, as the authorized voice, is analogous to the use of formal terms of address in the Pacific island cultures, for example, as in Pohnpei. As in the use of formal terms of address, even formal languages for each social caste, the use of the official language of anthropology reproduces and maintains a particular social organization. The general descriptions used by the trainees are part of the discourse of the pragmatics of producing, exchanging, teaching, presenting, publishing, library materials collecting and governmental administering the social and cultural domains reflexively produced and sustained in the scope of these descriptions. The trainees' choice of this discourse is the choice of being intelligible in a social world of prior experience and employment. The trainees were embedded in the bureaucratic-professional social organization produced in and through these descriptions, in their education, in their employment, in producing government-to-government relations, and in their general relations with the West.

Producing a National Culture

The third illustration is from a previously published article by one of the contributions to this book, Jean-Marc Philibert (1986). In this article, 'The Politics of Tradition: Toward a Generic Culture in Vanuatu', the argument is presented that the bureaucratic class, Western educated and cosmopolitan Vanuatuans, have formulated a national culture, drawing upon heterogeneous traditional practices, to form and control a civil policy. A single symbolic code was invented to assist in the creation of the state. Here we have the creation of *kastom*, a totalizing rhetoric reflexively producing the social cohesion requisite to the administered population of the modern nation-state. The theoretical discourse of *kastom* is identical to the national culture and social structure it invents, according to Philibert. This same phenomenon is to be seen in the history of newly independent former European colonies, such as Indonesia, Malaysia, Burma, and the Philippines. Where each had been composed of many cultures, often highly contrastive, they are now struggling to create a national culture.

An Account of Traditional Micronesian Navigation

Is is ironic that some Pacific islanders find themselves having to use the totalizing theoretical discourse of a single symbolic code or *kastom* in bureaucratic nation-building while other Pacific islanders are highly ambivalent about the use of this kind of language. An example can be found in the recent literary heralding of Micronesian canoe-sailing techniques (Thomas 1987; Brower 1983). The Satawalese reaction to a book on the master traditional navigator, Piailug, has been reportedly somewhere on the border between indifference and hostility. One account of the reaction of the population of this atoll in Yap State, in the Federated States of Micronesia, was that the traditional chiefs of the island were angry over the publicity given to Piailug. Piailug is not of chiefly status. The story goes on to say that Piailug was embarrassed, himself, over the attention brought to him and the disapproval of the traditional hierarchy of the island. In fact the book contains considerable detail on the perceived jealousy of the island chiefs. Another account of the Satawalese reaction is that people thought that the narrative by Thomas, *The Last Navigator*, expropriated something from their lives, the practices by which they lived. The feeling was that a Western writer was taking something from them and not giving anything in return, at least not in proportion to the benefits enjoyed by the author. This second account is increasingly typical of reactions of Pacific islanders to the work of social science research. This rhetoric can be heard from the Native Hawaiian community, particularly in the context of the study of social problems in Waianae on the west coast of Oahu.

Both kinds of reactions described point to the fact that Thomas's narrative entered the world of Satawal and those significant others with whom the Satawalese interacted. Whether or not the narrative was true, in a positivistic sense, it produced a certain kind of topicality, used over and over to find the Satawalese in Western conversation, publications, teaching, movies, television, and government administration. Chiefs do not live a life of consistent recognition of their chiefly status. Like the identity of everyone else, their leadership role is contextually produced. However, one of the unintended effects of the Thomas narrative was to produce the chiefs as consistently eclipsed by Piailug. The narrative entered the local and, more importantly, the international culture, fixing the identities of the chiefs and Piailug.

Perhaps the less than enthusiastic reception of the Thomas narrative can be explained as a reaction to a loss of control or freedom, even access, to the means of self-construction. Even traditional chiefs in the modern age experience being eclipsed – by government bureaucrats, by better-educated but lower-caste individuals, and by the prerogatives of children. But the effect of the Thomas book was to propel and fix the construction of the identity of the chiefs and Piailug into a social circuit, a symbolic code, far beyond the interactional grasp of the Satawalese.

The formulation of a national culture, *kastom*, a single written symbolic code, an ideology of national cohesion, operates in a fashion analogous to the effect of the Thomas work on Micronesian canoe sailing, or any of the other accounts of traditional navigation (Lewis 1972; Goodenough 1953). It makes abstract and general, beyond the scope of local interaction, the means and production of identity. The circuit of production of identity. The circuit of production of identity is embedded in what Anthony Giddens calls highly distanciated social relations, the hierarchically integrated modern division of labor made possible by the medium of writing (Giddens 1981; 1987). The written account is a circuit or realm beyond the high presence availability from traditional social structure, where almost everything is produced within face-to-face, mostly oral, interaction or physically close proximity.

Ships' logs, the use of the anthropological literature by Pacific islanders in making representations of village life, the issue of an 'invented' national culture in Vanuatu, and the reaction to the published account of traditional Micronesian navigation are all illustrations of the role of discourse in the pragmatic achievement of order. Each is an example of how discourse is constitutive of the human subjects and external objects of its use; how language (naming, describing, recording, predicting, storing, calculating, presenting, publishing, and displaying) reproduces not only the object of pragmatic reference but also the subjects of its use. In the case of the ships' logs, when the logs were actually rudders, not historical artifacts, the log was a form of representation that reproduced successful voyages and navigators. The logs as rudders, as adequate descriptions for making voyages, stand in the circuit of the practical achievement of seafaring, exploration, military and commercial affairs – including all of the personnel, from map makers, ship builders, ship provisioners, international traders and financiers, missionaries, and long distance fishermen, to the navy. In the case of the utilization of anthropological literature by Pacific islanders, the discourse reproduces the social structure of competent representation of indigenous culture, at least insofar as we think of adequate description to be identical with the instrumental practices of professional anthropology, the social sciences, and related administrative techniques. The voice of anthropology, with all the many occupational ways in which it is exchanged, has become the way of locating and exchanging a positivized indigenous culture with the West. Maybe we should think of this professional discourse as the cognitive-purposive rationalization of culture, the rendering of pretextual tradition social organization from absolute moral grounds into a specialized and separate, iteratively self-critical mode of representation (Habermas 1981). If this thought has any merit, it is no wonder that Pacific islanders have little to say to us outside of this type of discourse. I will return to this apparent Occidental reationalization of indigenous societies.

In the illustration of the creation of a national cultural ideology for Vanuatu, the discourse of representation, derived from the social sciences, produces and reproduces the integrated system of subjects and objects (here some of the objects, people, are subjects) of administrable territory, the modern nation-state. The idea of a secular state, the administration of a population by rational and independent means-ends calculation, is, I submit, selfsame with a liberal discourse of reciprocally oriented universe of individuals, persons ideally described as having internalized the same symbolic code. The secular bureaucrats who are building the nation of Vanuatu are using the discourse of social science in opening up and sustaining a social space of an exclusively and generally administered territory and population.

In the illustration of the reaction of the inhabitants of Satawal atoll to a description of traditional Micronesian way-finding, the protest was about an outsider's hypostatization of local social structure, taking away – expropriating, the freedom, authority – the practices by which the Satawalese could construct self. The Satawalese, being a remote and traditional atoll, are not wholly part of the government bureaucratic employment political economy, like the trainees in the mental health program. Their social structure is not reproduced – at least by and for themselves – by a standard professional account.

I expect that some readers will take the exception to these illustrations of the constitutive role of discourse in social theory. I anticipate that some, particularly social scientists, will argue that ships' logs, Pacific Islanders' use of the anthropological literature in making representations, the Vanuatuan elite bureaucrats' invention of a national culture, and the reaction of the Satawalese to a description of traditional Micronesian way-finding are not examples of contemporary social science research. Modern social science research is conducted by highly trained objective observers, using replicable methods. These people have spent many years in the field, in Pacific islands, 'working' with their subject populations. Furthermore, the body of researchers spend their entire professional lives interacting with colleagues, readers, and students who provide a continual source of criticism. The validity, reliability, and objectivity of professional social science of the Pacific islands are worked out through generations of multiple observers, each assessing the internal logic and empirical correspondence of the research. Indeed, there are disputes, but these are regarded as differences in theoretical perspective, for example, functionalist, cognitive, Marxian, neo-Marxian, world-system theory, conflict theory, modernization theory, postmodern theory, and so on. The different perspectives lead the researcher to make different interpretations of experience, albeit of one and the same world.

I have two responses to this criticism, admittedly self-constructed. First, I am not generically disputing the accuracy of social science accounts of the Pacific islands. Accuracy, objectivity, reliability, and validity, and whatever other truth correspondence issues, are practical social achievements, wherever they occur, in social science and in lay and other discourses. If I have an exception to social science accounts it is the naive and therefore unconscious political use of language. (Jameson 1981; Shapiro 1988). The prereflective use of language in social science, a failure to recognize that language is an already constituted social order of subjects and objects, is a limiting imposition on the real and potential possibilities for any people. My first response is that social science, one which takes itself as exempt from the constitution by the social order of its language, is not only reifying, maintaining, advancing the social order contained in discourse, but also fails to conceptualize social life, its object, as a practical achievement.

Second, the idea of a single world, a constancy hypothesis, with different theoretical interpretations, is a self-contradictory form of dualism, long in philosophical disrepute. A standard approach, using the assumed dualism of social science, would be to construct a taxonomy of theories of social change, a system of classification that would modify and contradict all of the preceding mutually contradictory taxonomies, putting all of the theoretical schools into their proper relation to the objective world. That social science insists on a methodological approach that is linguistically prereflective, that assumes a constant world, that it does not read philosophy or is not affected by it, is an illustration that social science is constitutive of a hypostatized world, which is basically unquestioned as to its social formation. This is a highly political position, albeit unrecognized. The political nature lies in the unconscious imposition of a social order contained in language.

How can I say that we should not assume a constant world? I have on the walls of my office four maps of the Pacific. One is a map of the cultural areas of the Pacific: Polynesia, Melanesia, and Micronesia. One is the political entities of the Pacific islands. Another is the 200-

11

mile exclusive economic zones of the Pacific islands. The other is a projection of Pacific islands, showing latitute and longitude. These are wonderful maps. They contain and remind me of useful information. There are many other useful maps and charts of the Pacific. I am particularly attracted to the navigational charts from the US Coast Guard. However, are the existence of such maps and charts and all of the successful activity guided by them a demonstration of a constant world? I merely refer the reader to contemporary Pacific islanders and Southeast Asians who have absolutely no idea of equal interval property space geography. Or is all of the successful activity guided by the maps a demonstration that the constancy of the world is achieved therein?

The position formulated here is a decentred approach, an analysis of how discourse is constitutive of subjects and objects, including authors. I do not want to propose a theory or theories of social change or even a 'new' history of Pacific islander active participation in social change after contact with the West (Howe 1984). But this is a book about social change. This may sound like a bit of Derrida (1973; 1982) 'madness' about writing. What I mean to do is examine the Pacific, the idea of the Pacific and social change, as produced in the political economy of the discourse of theoretical representation. This is far from saying that the chapters that do not take a decentered position, ones that propose a positivistic theory of social change, are not full-scale displays of the Pacific; that they are is the phenomenon to be explained and criticized for the purpose of opening other possible discourses. The papers written from a positivistic theoretical perspective *are* certainly the empirical or objective Pacific for the members of this type of discourse. I will propose that all members of the bourgeois social world of capitalism, particularly advanced capitalism, are members of this form of theoretical discourse. However, this is not the only actual or possible world. The writing-based world of theoretical discourse is a bourgeois-European world. The very conception of alternative worlds, domains with their own constitutive symbolic codes, is an eminently Eurocentric, post-traditional, achievement of an abstractive, objectifying and circulated discourse. Indeed, then, the idea of alternative worlds is not universally available, especially in preliterate and semiliterate societies, and is possible in the synoptic comparative survey of societies in literature.

What is Social Change?

Some Images

I will not take us on an excursus into some macrosociological formulations of social change in the Pacific. The first image I will figure is comparatively simple and unsophisticated; it is the familiar text of labor, production and distribution in Pacific subsistence societies being transformed by the commodification of Western capitalism. After laying out this textual figure, I will add the qualifications and repairs of the 'new' Pacific history (Howe 1977; 1984). Then I will argue that even the new Pacific history is a type of consciousness, acknowledging the active, rationalist participation of Pacific islanders in the transformation of their own postcontact circumstances, a theoretical object, produced and reproduced, as of a Western discourse of classical political economy. This is a limited discourse, reproducing the institutions of the West.

Before I present the first figure of social change, I want to consider the issue of where to begin: do we begin with prehistory or do we begin with history? We could study social change in the Pacific islands as a chain of migration from Asia into the Pacific (Bellwood 1978; Shutler and Shutler 1975; Allen, Golson, and Jones 1977). From mainland Southeast Asia to Indonesia and the Philippines into Melanesia, Polynesia, and Micronesia, using linguistic forms and pottery artifacts, DNA typing, for example, as tracers, we could formulate all kinds of 'push and pull' and random accretion theories to explain the entry and settlement of the Pacific by man. No doubt, the work in Pacific archaeology and physical anthropology provides fascinating and certainly truth adequate accounts, in the sense of having correspondent empirical referents; accounts that have legitimacy of validity and reliability in their general circulation among a society of private actors using a discourse of objectified nature. These accounts contrast with the creation myths of Oceania (Alkire 1977). The Vanuatu creation myth recounted by Philibert in his chapter, where large rocks descended from the heavens, rolled around the islands, giving the form, and otherwise acting like people, each rock having a personality and engaging in motivated action, sounds literally fantastic. For us, Western academic prehistory has replaced the creation myths.

We will not begin with prehistory. We will begin with history, thereby including, at least as a logical category, prehistory as a socially and historically located phenomenon of discourse. I will treat prehistory as a phenomenon of the discourse of Western social institutions and I will treat the creation myths on an equal footing with prehistory. This influences, of course, the treatment of the new Pacific history below.

A justification for the equal consideration of prehistory and creation myths, and by implication history and theories of social change, is methodological. If we are to take the analytic position that Pacific islanders were or now are part of an integrated system of subjects and objects in the context of creation myths, that the myths were objective reality, then we are obliged to treat our own discourse in a similar fashion: each should be analyzed as the production of the institutionalized discourse of subject-object predication. We cannot reify our own modern discourse, using it as an unanalyzed resource to describe, fault and recover, in terms of our own ontology, alien discourses.

Now I will turn to the simple model of social change. This model begins with the entry of the Europeans into the Pacific, finding the islands inhabited by people living in traditional societies. Some of these societies have been described as affluent subsistence societies. To be sure there were differences in the resources between atolls and high islands and between high rainfall and dry islands. Some of the island societies were stratified and complex, like Hawai'i and Tahiti, and some were less internally differentiated, like the Hall and Mortlock islands in Micronesia. Motivated by the search for commercial resources for the expanding world system of capitalism and for a sometimes related search for souls for Christianity, Europeans soon shattered the equilibrium of subsistence societies, incrementally superimposing cycles of expansion and contraction of commodified production and wage labor. The arrival of European and American ships, long anchorages, missionaries, beachcombers, barter, ascetic Christianity, plantation agriculture, work as seamen, and extractive early production (for example, sandalwood, whaling, sea slug, mother of pearl) and the introduction of money (the universal code of exchange and equivalency), all contributed to undermining the traditional order. Western education and writing (see chapter by Topping, p. 393) and the availability of Western consumer products further eroded the authority of tradition.

The population was alienated from an absolute relationship to the traditional interpretive codes, status and role, modes of production, and power hierarchies by the penetration, over time, of the bourgeois discourse of economy and religion. The attraction of Western commodities, commodified wage labor and money to pay for manufactured items and the religion of private salvation increasingly offered a system where people could buy prestige items and change the symbols of status and role, and caste – changes that released the population from static social positions.

Over time (we are characterizing a time frame from contact to the present) a discourse of individuation, of private individuals (MacPherson 1962), each seeking their own interests, began to dissolve adherence to the collective orientation of traditional political economy. This discourse of individuation was institutionalized in colonial governments based on the discourse of Natural Law, the positivization of the rights of men, property, and the labor contract and, later, the obligation to pay taxes. The rights of men included a congery of naturalizations of human behavior, such as the theory of Locke that the basic function of man is labor (Habermans 1973, pp. 82–120). Although the colonial administrations were paternal toward the indigenous islanders, often not extending them the rights of private individuals, this language permeated the islands, particularly in education and religion. After World War II, islanders began to write and talk of their own destiny in terms of positivized individual rights and the natural functions of government.

This highly characterized and compressed model of social change is one where Europeans imposed a political economy on the Pacific islanders. A more vulgar version of the model depicts the Europeans as overpowering the island cultures and social structures, stealing resources from innocent islanders. This is the story of rape, pillage, and corruption of the indigenous population by Europeans. As discredited as this version is, as a comprehensive description, it maintains a popular currency among nativistic political movements and among some of the bureaucratic elites who are organizing or creating national cultures. In any case, the model formulates a passive victim role for the Pacific islanders. The corrective to this passive view is the new history of the Pacific.

The new history of the Pacific postulates a more active participatory, if not collaborative, role of the indigenous islanders in the transformation of traditional subsistence political economies to at least mixed structures with a linear increase in commodification. The work by Howe (1984) is the most notable of this genre. His history displays the utilization of Western commodities, trade, money, and religion by the Pomare and Kamehameha families to create a central nation-state in Tahiti and Hawai'i. He argues that, at least on the big islands of Polynesia, where there was enough of a surplus to support a highly stratified society, with an elite and religious experts, that contact with the West provided the technology to hasten and elaborate the achievement of indigenous nation-state aspirations. Pomare and Kamehameha families used Western weapons and advisers to conquer their competitors among the chiefly elite. Both the Pomare and the Kamehameha families used Christianity and Christian education to pacify and create a single polity, almost in the fashion of the analysis of the French Physiocrats, a polity of public opinion.

The elites of Tahiti and the Society Islands and Hawai'i encouraged the penetration of Western commodities, techniques, religion, education, and the larger culture of the West. They were aggressive participants in the acquisition of Western ways, albeit turning them to their own objectives. To acquire Western goods they had to reorganize production and distribution. They soon engaged in petty commodity production for the provisioning of ships. Soon they were organizing

15

labor gangs for the China sandalwood trade. The participation of the elites and their minions in production for exchange, often through the medium of money, effected a profound change in social structure. These changes were deepened when islanders later became wage laborers and craftsmen (Beechert 1985).

The most fundamental change was from a collective identity to a society organized under the code of the production and exchange, with its division of labor, of commodities. The super-imposition of the general system of commodity production and exchange creates the individual as the fundamental locus of production and consumption. The collective identity is transformed into a society of privatized individuals (Baudrillard 1968, p. 147).

The pace of change has been different for the various island groups and within the regions of islands. In Hawai'i change came rapidly after contact. In Papua New Guinea, as described by Ogan and Wesley-Smith (p. 000), the rate of Western penetration is different for the coastal areas and the interior highlands. In every case, no doubt, change has not been a process of Western institutions suddenly and completely replacing tradition. Tradition exists to this day, mixed with and transformed by the cultures of the West. In Belau (or Palau), for example, wage labor permits increased participation in traditional communal house building through financial sponsorship. Relatives living on Guam, in Hawai'i, and in California are expected to and can, through the medium of money, contribute to the construction costs of home building in Palau.

Pauline King, in another tack of the new history, in her review of Hawai'i historiography shows how Western writers have portrayed Hawaiians as passive, almost grovelling, recipients of the Western normative order (King 1987). She makes the argument that Hawaiians, of every social status, had to be interactional collaborators in the imposition of occidental social structure. The very ability to interact between Westerner and Hawaiian assumed a communicative competence, a commonality of knowledge, motives, and procedures for claiming the validating of statements (Robillard 1987). Moreover, Hawaiians were essential to labor and consumption. Hawaiians, it is submitted, found it in their interest to enter the system of commodity production and exchange; it provided a freedom from the subjugation by the *alii*. The dominant historiography of Hawai'i has neglected the interactional partnership of Hawaiians and foreigners in social change in Hawai'i. This includes the work of Daws (1974) and Fuchs (1961). King calls for a new history of Hawai'i, one which details the everyday interaction between Hawaiians and foreigners.

There are differences between the two models, but I assert that the differences are minor and that both models are formulated in the productivist discourse of classical political economy (Baudrillard 1975). A justification of the logical superiority of the model of man in the new history, an active participant in the occidental penetration of Oceania, is futile. The model of man in the new history appears to be an improvement over the passive actor in the rape and pillage popular history (Moorehead 1971) of the Pacific. Only if we accept a priori a rationalist laboring actor, projecting him as the sign under which behavior must be recovered, assuming his objectivity, does the new history seem attractive. Why should we elevate one version of man (or even man instead of the anthropomorphic boulders of Vanuatu), above another to use it as an unanalyzed descriptive resource? Do we want to project naively an anthropocentric world, of *Homo economicus*, or are we interested in explaining how any worlds, anthropocentric or otherwise, even what appears to us as fantastic worlds of myths, are organized as integrated systems of subjects and objects?

It may have appeared that I was proposing an economic theory of social change, or at the very least making an equivalence between changes in production and the social relations of production and social change. In the first model of social change I highlighted the transition between a subsistence Pacific island political economy to a commodified economy, one increasingly dominated by production for exchange and the sale of labor. I formulated the supposed changed in the traditional order effected by the penetration of capitalism and Western religion. In the second model of social change, the new history, I tried to display the corrective to the passive image of Pacific island man in the first. The second model does not really propose a different relationship of man to nature. Both models assume that man wrests his existence from nature, that man moves through time and space by his own efforts, labor, thereby reproducing himself. Only in the first model, changes in the way man exerts himself upon nature to provide for his reproduction were wrought from the outside, the occident. In the second model, changes were the result of interactional collaboration, a situation where the Pacific islander calculatively selects Western practices to meet his own ends. Implicitly the ends, whether they are in conversation with Westerners or in formal employment, are the supposed necessity of human labor in producing and reproducing himself.

I am not interested in formulating an economic theory of social change. I will not propose an alternative positivism. I will not propose, either, that we do away with positive theories; there are inevitable discourse projections (O'Neill 1985; Levi-Strauss 1966; 1969) of any language system. What I am interested in doing is to inspect how the fiction of political economy, as the implicit assumption of labor as the reproduction of man, is a discourse domain, a system, within which

17

social change is produced. I am also interested in showing how this fictional code serves as an imperialistic ideology, effacing actual and possible alternative cultural worlds. I call political economy a fiction because it is a reductionary value used to replace contradictory explanations, like indigenous religious or animistic accounts.

I realize that social analysis is predicated upon the idea of man as a laboring individual and that this a priori sounds eminently reasonable in the light of 'biological reality'. Almost every social change analysis begins with the idea of a subsistence economy and how this economy, an equilibrium of needs and production, and changes from a production for use to a production for exchange. But I am suggesting that we treat this approach, this discourse, as the production and reproduction of a certain social order. How are we to treat, in this discourse, an incident like one I had in Pohnpei? I was working in Kolonia and one Saturday I visited the ruins of Nan Madol (Hanlon 1988; Athens 1980; Whiting 1954), a series of large structures of basalt columns constructed on islands made from coral land fill. The basalt columns are over thirty feet long and came from an area of the island of Pohnpei about ten miles away from Nan Madol. I asked a resident of the area how the columns were quarried and how they were transported to the site of Nan Madol. He told me that they were moved by gods. In this account the gods did the work, not man. If we are to use the occidental idea that man is the only one who creates his material existence, then this story is reinscribed; it doesn't mean what it says, it becomes ironic. How are we supposed to treat the work of agricultural sorcerers, for another example, in the Pacific islands (Rubinstein 1982)? In this case, the wizard intervenes in the natural order with his magic and entreats the gods to effect a harvest. The gods and the supernatural perform the work, not men.

There are many other theories of social change that we can consider. So far we have described the general approach of political economy. I have argued that the political economy approach, the macrosociological approach, assumes the unanalyzed discourse of man as having to reproduce himself in his labor. Whether we are talking about the subsistence economic ideal, the change from the utopia of subsistence to commodified exchange or the interactional competence of Pacific islanders interacting with Westerners, we are using the discourse of anthropocentricism; a world essentially built and maintained by human activity. This is the model of political economy as well as ethnography, cognitive anthropology, ethnomethodology, action theory or any perspective which places man at the origin of the sensible world. It can be argued, I think, that there are no professional theories except political economy theories of society. Where I have said above that there are other professional theories of social change we can consider, I mean that in the sense that other theories appear to be different but, in fact, the other theories are equally premised on man as a laborer, the producer of the world. Whether the theories are of technological diffusion (Volti 1988; Fried and Molnar 1978), or a history of material rationalism (Braudel 1967), or a teleological theory of the expansion and contraction of the world system of capitalism, in long waves of accumulation and crisis (Bergesen 1983; Wallerstein 1984), or the patently economistic action theory of Talcott Parsons (1937; 1954), or a critical theory of class relations, even in the form of literary criticism (Jameson 1981), or modernization theory (Bryant and White 1982; Sundrum 1983; Dogan and Pelassy 1984; World Bank 1985), with the idea of technocratic mastery over effecting the transition through epigenetic stages of development, or the theory that social change is effected by the creation of a central state (Mann 1986; Evans, Rueschemeyer, and Skocpol 1985; Giddens 1987), or the idea that social change is somewhat the same as changes in interpretive codes (Geertz 1983) – and the list can go on and on – all have man at the center, assuming a priori that man, and the medium of man, society, is in his action the producer of the world. This list of theories and my accusation that they are anthropocentric will have to stand here as a mere assertion. Much as this demands further substantiation, I do not have the room to develop this aside.

So what is social change? I began this section under the heading of 'What is Social Change? Some Images'. I adumbrated two images of social change; one was the ever-popular image of a passive transition from a subsistence economy to commodified production and individuation; the other was an image of the Pacific islander as an engaged actor, someone who was and is fully competent in a collaborative transformation of his own society. I then turned on the two models, calling them examples of the reifying discourse of political economy. The notion of political economy is that man is the producer of himself, extracting from nature his sustenance, and that society and social structure are a medium for this extraction. The idea of calling the discourse of political economy reifying is that it assumes and recovers everything as the effect of human labor upon nature. This is a rationalist and Western idea, from Moore, Hobbes, Locke, the Scottish Moral Philosophers and only later from Marx, however contrasting these writers are. I have contrasted the discourse of political economy with alternative discourses, creation myths, and personified boulders that made Vanuatu, a local explanation of the construction of Nan Madol and agricultural sorcery in Micronesia. The purpose of the contrast was to point to and take seriously other systems of combining subjects and predicates, other discourses, ways of thinking and making representations that do not assume that the knowable world is produced by man. This is not to say that the other discourses are any less ideological or any less reifying. All of them may be considered fictions, from the perspective of not belonging to any one of them. It is a point made for the purpose of maintaining the recognition, if not inhabitation, of truly alternative worlds; worlds which are not ironic or mystified expressions of people who have not yet learned that man is at the center of ontogenesis. Other discourses than our own are equally projective.

The discourse of political economy, an anthropocentric ontology, projects a cognitive-purposive rational actor, an individual who, given the recognized need to extract reproduction from nature, is able to evaluate means to achieve the necessary ends of life. This is a very different actor from one who is ordered or told by gods and spirits to do things. It is an actor very different from the non-theoretical, pre-literate actor. The rational actor of the discourse of political economy, as the primary mode of representation, is the institutionalization of a Western projection. It produces and maintains a systematic discipline of representation and corresponding social institutions (political, economic, aesthetic, axiological, intellectual, philological, ideological, scientifically and imaginatively) of Western intelligibility. The social change produced in this discourse is a Western, European, and American enterprise. Just as Edward Said has argued in *Orientalism* (1979) that the Orient, the so-called Middle East, is the creation of a Western or European discourse, the language of imperial hegemony, the Pacific and social change, as items, have to be inspected in the discourse that permits the combination of its elements.

So, what is social change? I am asking this question one more time. The temptation is to formulate a positive theory of social change; this would be a general theory. But I will submit that on methodological grounds there is no theory of social change nor is there a social change. There are many discourses of social change: sociological, anthropological, economic, political, in the popular novel, in the press, in government, in religion, strategic and military, and in financial and commercial enterprise. Social change exists in the discourse of its production. This is a little like making an identity statement (A=A). This does not mean that the social change is a figment of the imagination of the West. Social change, produced *in and as of* a discourse of representation and action, possesses the power and discipline of fact. I do not mean to deny social change; I only want to focus social science inquiry into social change on the discourses, the ontological relationship between subjects and predicates, which produce, maintain, and consolidate social change and the institutions of pointing to, writing about, planning, announcing or otherwise representing social change. It is in the reciprocal maintaining of institutions, in what Foucault (1980) calls the myriad capillaries of power that one is 'obligated' to say things (Barthes 1977, pp. 14, 25), that one is forced, as an intelligible member, to mirror a text. This obligated use of a language of representation, a discourse, is the reflexive maintenance of the system.

Social change then has no constancy or generality. Social change occurs wherever there is a discourse of its articulation. Social change occurs in whatever form and from whatever source discourse stipulates. Therefore, social change can be in any discourse, even discourse that attributes social change to the spirits. It is a mistake to reduce everything to production of an instrumental reflective ego, an event in bourgeois discourse. We should be able to have all types of discourses, social change from actors, for want of a better word, who do not look anything like the purposive laboring actor of our cultural system. To impose our ideology of rationalism, political economy, or to export it to different systems is a political act, even if unconscious. All acts of representation, working within a discourse of institutions, are political acts.

II. THE CHAPTERS

Two things have to be said right away about the chapters. First, they are organized by nation-state or colonial territorial designations. This is already to accept as natural a certain form of representation. The second thing is that all of the chapters are written from a political economy perspective. I will address the issue of nation-state organization of the book first.

A nation-state organization is problematic because we have seriously to question with whom do we locate with this kind of discourse. There are, for example, many people living within the Federated States of Micronesia, in the Central Caroline Islands, who do not find themselves within the discourse of a nation-state, even though they are, to our minds, geographically within one. These are individuals living on the outer islands of Pohnpei, Truk, and Yap states of the Federates States of Micronesia. That the outer islanders are living within a central state, even within a loose federation, is an outcome of colonialism, from the Spanish, Germans, Japanese, and Americans. These people are isolated and do not consider themselves as belonging to a common set of borders, a common national identity, and a central political administration with other islanders living far away. As bureaucrats who are involved in national development are all too well aware, many island groups and islanders are composed of heterogeneous populations and cultures; many of the inhabitants do not have a geographic projection of their island in relationship to other islands, the nation-state and other more distant land masses. Geography as we

21

know it, we in the West, is not universally distributed. Comprehensive representation of geographical empirical data is embedded in social institutions (for example, commercial shipping and air traffic, military and political administration, tourism, scholarly expertise, pedagogy, economic development and finance) of which the outer islanders are only peripheral members, if they are members at all.

Why are we proceeding with the nation-state perspective if it only locates and reproduces Western social institutions and individuals? The comprehensive treatment of the area afforded by the nation-state approach, projecting a property-space exclusively and totally divided up by political jurisdiction, does reproduce Western institutions. Howver, this volume recognizes that nation-state discourse increasingly encompasses more and more Pacific island populations and, at the same time, we recognize that this discourse does not locate everyone. The nation-state perspective is used here not for purposes of reifying political geography but for the purpose of finding our way to an objective of this book, in the ways we have of producing the referent field, and at the same time acknowledging that some of the people in these islands do not share the same locating institutions. The nation-state discourse is a part of the culture of the book and its readership but we can permit other discourses to occupy a logical space within the book.

The other problem is the use of the discourse of political economy to represent social change in the Pacific islands. Just as with the discourse of the nation-state, we do not want to reify this code and its social institutions. Nevertheless, the economistic discourse is a very real and palpable contemporary part of the social structure of the Pacific islands. This social structure does not include everyone, however. It does not include even the nation-building bureaucrats, in the Pacific or from the donor countries, all of the time. It does not include ancient Pacific societies. It is one discourse among many. The way it should be treated here is not that it locates any objective reality outside of social reproduction, but that it is a powerful discourse, although occidental, which is very much a part of the contemporary Pacific. It is the material, this schema of discourse, from which a raft of social institutions is constructed and maintained.

The hegemony of the discourse of political economy is the superimposition of an anthropocentric model of the world and the marginalization of magical and supernatural worlds or cultures. When the world becomes a construction of human labor, extracted from nature, discourses in which the world is projected as the creation of extra-anthropological natural forces become false consciousness, perceptions of an objective world mediated by cultural beliefs. The schematization, categorization, the exchange of accounts, and the discipline entailed in the reciprocity of exchange of the world as human labour ought to be looked at as the production and maintenance of the social institutions of political economy: the export of Occidental social institutions, particularly the liberal or bourgeois state. Of course, Pacific islanders are using this discourse to become self-governing states, coming to this discourse from their own history and situations. The discourse of the liberal anthropocentric world has been and continues to drive the anticolonial politics of the third world.

It is beyond doubt that the discourse of political economy has transformed the Pacific. What is at issue here is whether this discourse is aware of and analytically recovers its role in the creation and maintenance of the social institutions it describes. For the most part political economy obscures the ontological projection, its a priori assumptions, and its structuration of social institutions. Perhaps any code, any sign system, as an a priori, should be thought of as a positivization of the world, and political economy is just one such form. I suggest that all codes are political economies; all of them exercise a disciplinary power. The achievement of Marx and Marxism, including a neo-Marxism from which most of the following chapters are written, is the ability to make a critique of the object of the discourse of political economy. However, this critique is formulated totally within the ontological assumptions that the world is man-centered, a creation of labor, of political economy.

The purpose of this introduction is to bring us beyond the discourse of political economy, to permit an openness to radically different worlds, to permit an analytic recovery of different combinations of subjects and objects. To do this we have to be aware of the projective role of language, discourse, in producing and maintaining institutions of reality. Social change or stasis then, its source and destination, will be found in the discourse of its formulation. I am not arguing that we can escape discourse or that there is some unarticulatable experience under discourse such as desire (Deleuze and Guattari 1972), and the human reality always escapes depiction. That there are such positions is an assembly of code.

I have arranged the sequence of chapters in a conventional manner. The conventionality of the arrangement I have followed is the professional knowledge of physical anthropology, archaeology, and linguistics in placing the articles in three sections: Melanesia, Micronesia and Polynesia. Melanesia comes first because of the prehistoric movement of people from the Sahul or Southeast Asia into what is now Papua New Guinea, the Solomons, New Caledonia, Vanuatu and Fiji. This migration included Australia and Irian Jaya. Next is the section on Micronesia. Micronesia stretches from Palau (or Belau) to the edge of French Polynesia. Settlement was from the Philippines to the West and from Melanesia to the South. Polynesians also settled on the fringes of Micronesia and Melanesia. Other than the idea of a Micronesian culture and a Polynesian culture, there is no reason for having Micronesia before Polynesia. Movement into both was from the West, from Indonesia, the Philippines and other parts of Southeast Asia, and from Melanesia.

I must emphatically say there are other versions of how the Pacific islands were peopled. The alternative explanations are professional, like Thor Heyerdahl's (1950) idea that Polynesia was settled by South American Indians from Peru, and indigenous, like the common creation myth in Polynesia and Melanesia about the magical stones. Each version has its own social structure, no different from that which sustains the truth of the demographic expansion described above.

Papua New Guinea: Eugene Ogan and Terence Wesley-Smith in 'Papua New Guinea: Changing Relations of Production' (Chapter 2) take Marxist definitions of mode and relations of production, theoretically examine and modify them to conceptualize a mixed traditional-capitalist economy, and review the pre- and post-contact history of Papua New Guinea, focusing on the post-World War II period. The paper argues for a situational analysis, in contrast to what Ogan and Wesley-Smith take to be the over-general approach of the capitalist world economy of Wallerstein or the dependency analysis of Frank (1979). This position, situational analysis, is also taken by Christine Ward Gailey in her chapter on Tonga (Chapter 11).

Ogan and Wesley-Smith observe an uneven and late penetration of capitalism into the heterogeneous societies of Papua New Guinea, something the world system level of analysis would miss. Nevertheless, capitalism is changing the lives of people in the mining areas, in the coffee growing regions, in the towns, in the way land is held, in the role of women and children and in the level of participation by indigenes in the planning and production of wealth. Of particular interest is the impact of the United Nations agencies and the World Bank in the creation of a rhetoric and matching social structure of developmentalism. The post-independence government articulates itself in the same rhetoric, with what are no more than verbal feints in the direction of radical redistribution. However, capitalism and a market economy are the objectives of the government.

New Caledonia: Chapter 3 by John Connell gives the reader an overwhelming impression that history keeps elaborating itself in the same patterns, as in Nietzsche's law of eternal return. The Kanak rebellion of 1878 against the French colonizers has become a social institution in the French settler community and in the Melanesian community.

For the French, and perhaps for the allies Polynesian and Asian communities, the most destructive insurrection to take place in the Pacific islands is a retrospective and prospective vindication of the discriminatory practices which have alienated the Melanesians from their land, sequestered them physically, kept them from anything approaching equal participation in the developing capitalist economy, and excluded them from formal politics. For the Melanesians the bloody revolt has become a galvanizing event, projecting the symbolism of a unified Melanesian historical protest against French subordination. As a legendary heroic event, it collects and throws into relief the social organization, values and beliefs of a pre-colonized indigenous population. It creates the abstraction of national identity of the Melanesian opposition. The history of New Caledonia can be read as a demonstration of how historical events are not merely moments in time but social institutions in the memory of contending parties which are re-invoked over generations, maintaining and elaborating opposition.

Focusing on politics, Connell takes us from contact with Europeans, through early colonization, an agricultural economy, mining, World War II, French foreign and colonial policy, to the establishment of a tertiary level economy driven by government employment and subsidy. Of particular importance is how colonial administration and capitalism generated anti-systemic movements within the Melanesian population, culminating in the often violent call for an independent nation of Kanaky, and recent French attempts both to repress independence politics with the police and the army and to co-opt succession by increased Kanak political and economic participation. Integration in a consumer economy makes it unrealistic to think that the indigenous Melanesians could return to a traditional political economy. This problem, one of dependency born of commodity consumerism, is also addressed in the chapters on French Polynesia (chapter 12) and on the Caroline and Marshall islands (chapter 6).

Vanuatu: Chapter 4 by Jean-Marc Philibert on Vanuatu also takes the long view of the history of the country, from contact to the present. However, it is a very different form of analysis from the chapters by Ogan and Wesley-Smith or Connell. Philibert sets out to analyze social change by describing indigenous ideologies, from early creation myths, to the introduction of Christianity, to party politics, to the codes of contemporary consumerism, to the creation of *kastom* and a national culture. Philibert is trying to overcome the same abstract overdetermination of world system analysis that Ogan and Wesley-Smith find objectionable. Here the solution is different. Instead of an economic or material approach, as in the modes of production, Philibert restores the practical role of *ni-Vanuatu* (the people of Vanuatu) in the creative

25

construction and change of their own social structure. The metaphor here is not the extraction of sustenance from nature by human labor but rather the symbolic codes used to open and maintain social space. Symbolic analysis puts the actor in history, allowing him to confront foreign influences, making sense of them for himself, modifying them for his use from within his own horizon.

Fiji: Chapter 5 on Fiji begins with the recent military overthrow of the constitutional government. The author, Vijay Naidu, traces the two coups to the history of how Westerners institutionalized the chiefly mode of production so as to vertically integrate administrative and economic control for colonial and neo-colonial interests. At contact there were two modes of production, chiefly and communal. Administrators, missionaries, the military, and businessmen seized upon the chiefly directed and surplus appropriating mode of production, generalizing it and one Fijian dialect across the entire island chain, in an effort to rationalize the social structure for the bureaucratic state and capitalism. The Western penetration of Fiji created a neotraditional culture of chiefly dominance and a uniform language. This neotradition has served the interests of the chiefly class and the allied Anglo-economic oligarchy, even in the face of what has become an Indo-Fijian majority and an ethnic Fijian minority. Capital control of sugar, mining, timber, importing, retailing, and now the expansive tourist industry, has been through the chiefs' possession of land, and more recently the government. The coups were the chiefs' attempt to preserve their comprador status and capital's effort to expand the ability to generate surpluses (for example tax holidays, favorable labor laws and export processing zones).

The Fiji chapter is a structural analysis, using world system theory and the Marxian ideas of modes of production, labor, proletarianization, and class. The metaphor of structure is the describable organization of relations of production in the society. Naidu juxtaposes the communal and capitalist modes of production and exposes the teleology of capitalist development in terms of labor and surplus appropriation in Fiji.

Marshall and Caroline Islands: Chapter 6 by Francix X. Hezel, S. J., 'The Expensive Taste for Modernity: Caroline and Marshall Islands', traces the history of post-contact influences in this part of Micronesia. He takes us from subsistence affluence to the occasional Spanish ship, to the whaler trade, to the German copra industry, to the Japanese mandate and intensive settlement, to the American administration, and finally to the Compacts of Free Association with the United States. The impact of the Spanish, German, and Japanese administrations was minimal on the indigenous social structures. The Spanish governed from Manila. The Germans were primarily interested in copra, phosphate, and small scale trading. The Japanese imported their own labor from Japan, Okinawa, and Korea, became more numerous than Micronesians, and treated them as a subordinate class. It was the Americans who, in creating a pluralistic democracy, the liberal state of producing and consuming individuals, transformed Micronesian societies, structuring permanent dependency. Democracy, in the version of the liberal state, with the chief catalyst being wage employment of individuals in a public bureaucracy and the consumption of commodities, constructed a cash dependency on the United States in islands relatively bereft of the capacity to generate capital wealth.

Guam: Chapter 7 by Larry May, 'The Militarization of Guamanian Society', addresses how the American military has changed the indigenous Chamorro society, from the time of the Spanish-American War to the present. The US navy administered Guam for a long period after seizing it in 1898. At this time, most (90 percent) Chamorros were living in a subsistence agricultural economy. The navy again managed Guam after World War II. The military governments introduced an ecology and need for wage employment. The US military governments took control of all the land on Guam and bequeathed substantial amounts of acreage to the civilian government upon the demise of military rule. Between the Department of Defense and 'Gov-Guam' well over 50 percent of the land is in the public sector. The proletarianization of the Guamanian Chamorros has been through military and succeeding civilian governments. Government is the main business of Guam, as it is in much of the Pacific. This is reflected in the Guamanian occupational structure and the high rate of military enlistments.

Guam is addressed again by Donald Topping in his discussion of language change in the Pacific (chapter 14). Guam is used as an example of extreme linguistic change, the superimposition of English upon Chamorro.

Northern Marianas Islands: Chapter 8, by Samuel F. McPhetres, the archivist for the Trust Territory administration, telescopes history to the present social dilemmas of massive tourist industry development by Japanese capital and the expansion of a garment industry by Asian (mostly Chinese) capital. The number of Filipino and mainland Chinese workers is approaching equivalence to the number of Chamorros, the indigenous inhabitants. With major tracts of land leased to resort hotels, large numbers of tourists, mainly from Japan, hotel labor from the Philippines, construction and sewing factory workers from the

Peoples' Republic of China, what the author calls the neo-Chamorro has become restricted (or has restricted himself) to working for the Commonwealth government. The Chamorro has absorbed the influences of the Spanish, the Germans, the Japanese, and the Americans but now international capital and labor are relocating the Chamorro, alienating him from control of the Marianas.

The chapters on the Marshalls, Carolines, Guam and the Northern Marianas appear to eschew any theoretical perspective. They seem to go directly to the empirical structures in the islands. Appearances are deceptive in what can be identified as liberal functionalist social science. All three of the articles ride on deep assumptions, unidentified theories, about man, social organization in the reproduction of man, economic activity as the fundamental datum and the spatial relationship where one society imposes itself on another. In all three chapters outside societies overpower indigenous social orders, changing the ground of economic activity, to which the islanders adapt. Implicit in all the chapters is the natural opposition of the interests of the indigenous culture and the European colonizers, based on a posited contention over the means and social relations of production, between pre-capitalist and capitalist societies. Nascent class analysis drives the chapters. The chapter authors may not agree with the characterization.

Kiribati: Roger Lawrence in chapter 9 'Kiribati: Change and Content in an Atoll World', emphasizes the limited human ecology offered by low-lying, sandy atolls. The figure of the balance between human needs and the limited carrying capacity of atolls runs throughout the paper. The limited atoll resource base is a familiar explanatory theme in Pacific studies. Lawrence is not a geographical determinist, however. He devotes considerable attention to the impact of the early copra trade and of missionization by American and English Protestant churches. Though transformed in the nineteenth century by Christian cosmologies, an independent Kiribati has become a nation dependent on foreign aid, commercial fishing licenses, copra, government employment, and remittances from I-Kiribati (the name of the indigenous population) working as seamen on German ships and as unskilled laborers in other Pacific island countries (e.g., Nauru) and, for a few, in metropolitan centers. A small tourist industry has begun.

Tarawa has recently become an urban center, reflecting an integration into a world inter-state system. The metropolitan state bureaucracies replicate themselves in Kiribati in the very idea of a managed nation-state (i.e., in the centralized provision of education, health services, police, courts, transportation, sanitation, banking, and fiscal administration). The migration to Tarawa appears to be permanent, unlike the earlier circular labor migration to Australia, Hawai'i, Peru, and Nauru (Chapman and Prothero 1985). In an interesting twist, the chapter demonstrates that incorporation of the periphery, like Kiribati, into the world economy is driven by more than historical materialism. Kiribati has been integrated by religion and statecraft, as well as by copra, tuna and phosphate.

The Samoas: Chapter 10 is a biting critique of education in American Samoa and Western Samoa. Robert Franco argues that the American administration of education in American Samoa has been inconsistent, typified by endless cycles of new curricula and management. Western Samoa has had the New Zealand model of education. Franco calls it an elitist model. In both cases the educational systems have ill-equipped Samoans for employment in Honolulu. Moreover, the focus of Samoan education has been on the men. Women who migrate to Hawai'i experience a downward mobility in Samoaan social status. They no longer make wealth objects, like fine mats, and, with few employment skills, they have few avenues to contribute to family income and wealth.

Tonga: In a broad chapter, (chapter 11) Christine W. Gailey traces the changing practices of individual and social reproduction, from prehistory, to contact, through missionization, to the creation of the monarchy, the impact of the Pacific war, commodification, class forma-tion, labor migration and foreign aid.

The most theoretically notable things about her exposition are the integration of the macrosociological and microsociological levels of analysis in the description of the changes in gender relations in social system reproduction and the notion of a fluid and creative social structure, created and re-created through space and time through the invoked practices of culture.

It is not the only paper to focus on gender roles: Ogan and Wesley-Smith thematize the changing roles for women in Papua New Guknea and Robert Franco analyzes the downward status mobility of Samoan women in Hawai'i. The 'idiom' of a central state, in the form of a monarchy, and the incremental penetration of capitalism have changed the position of Tongan women in inheritance and the production and distribution of goods. Critical of teleological theories, Gailey sets out a position where social change and its meaning are historically contin-gent.

French Polynesia: When one looks at the map of the Pacific, the striking thing is the number of island chains and the vast distances included under the ribric of French Polynesia. Ben Finney (chapter 12) observes that before French colonizatjøn the islands shared no central

29

coordination. The islands and even the valleys on a single island constituted independent societies. The same observation covers what is now Kiribati and the Federated States of Micronesia. Finney describes the French colonization of the nineteenth century, the backwater status of the islands until the 1960s, and the rapid social change with the creation of the *Centre d'Experimentation du Pacifique*. The Centre is the atomic weapon testing effort on the French military.

Atomic testing brought a flood of French nationals, infrastructure projects and employment for Polynesians and the 'Demi' or mixed race people (European-Chinese, European-Polynesian). Wage employment made the Polynesians into urban proletarians. It crystallized nascent class divisions, making the French, Demi and the local Chinese into the petty bourgeoisie. Class consciousness manifested itself in nativistic politics, including calls for a return to subsistence lifestyles, and an initiative for an independent and nuclear free French Polynesia.

The Finney chapter is unique in that it describes not only the imposition of an external European social structure but also explicitly how this structure generates opposition and contradiction. French proletarization has transformed consciousness and culture, setting new social interests and generating new social groups. Finney ends by describing the dilemmas facing the Polynesian and part-Polynesian population in an effort to organize an independent state, without the current massive levels of French financial support.

Hawai'i: Chapter 13 on Hawai'i takes a different tack. It does not focus on the indigenous population, as do the rest of the chapters. While the indigenous populations of the Pacific islands are certainly substantively and ideologically important, we deprive outselves of the critical knowledge if we ignore other settler groups. There is a relatively small literature on the overseas Chinese in the Pacific islands (Wu 1982; Moench 1980) and on the mainland *Haole* in Hawai'i (Wittaker 1986). There are studies of Vietnamese labor in Vanuatu and New Caledonia and a popular literature on French settler communities. However, these literatures pale in size when compared to the literature on indigenous populations.

In any event, the Hawai'i chapter is not about an ethnic group. It is about a shirt. The aloha shirt, particularly the aloha shirt manufactured and marketed by Reyn's, is taken as a code for the production of a social order. The Reyn's aloha shirt produces one order among many in Hawai'i. The social order is that of professional management, whether the Reyn's shirt is worn by a Caucasian from the mainland, a Hawaiian trustee of the Bishop Estate, or a Japanese real estate investor. The shirt signifies not only membership in a group, but an entire production and consumption system based on the manipulation of signs.

Language in the Pacific: In a book that argues that discourse and social change are coterminous, it is fitting to have a chapter by a noted Pacific linguist (chapter 14). Donald Topping describes how the two world languages, English and French, have changed and continue to change social structures and thinking in the Pacific islands. Proceeding from the position that language is the embodiment of culture, the paper identifies cognitive and social changes that come with making the over eight hundred indigenous Pacific island languages into literate or written languages. Moreover, in the effort to fit into the world economy, the indigenous languages are rapidly losing ground to the languages of commerce, banking, navigation, law and international organizations. The world economic system articulates itself in English. Topping is not optimistic about the long term survival of indigenous Pacific languages.

Children's Survival in the Pacific Islands: Who has not been struck by the preference for Spam, vienna sausage, and canned corned beef among Pacific islanders? Both are high in fat and salt and yet command honored places of display in Pacific island households, stacked and placed where the visitor will not fail to notice. An abundance of processed food, most of them high in fat, salt, and sugar, is found at the many roadside market stands, in the urban center supermarkets, restaurants, hotels, bars, and in homes. These ingredients are cheap and yield substantial profits in food manufacture. The urbanized Pacific islander wage worker has become dependent on commodified food production, nutrition produced with the intent of generating surplus capital. Up to 70 per cent of nutritional requirements are imported in the Marshall Islands, probably an underestimate for the people living in the cramped space of Ebeye and Majuro.

Child mortality has been relatively high in the Pacific Islands. This mortality has been caused by chronic malnutrition in some areas. Kent (Chapter 15) reports a new form of malnutrition in the islands. The new form is caused by the powerlessness of the proletarian worker over the type of food he eats. He has neither the time nor space to grow or catch his food. The urban wage worker in the islands is dependent on capitalist (surplus generating) manufactured and marketed food. Without information about nutrition of commodified food, influenced by aggressive marketing and fashion, junk food has become a staple in many Pacific island families. This has led to malnutrition in adults and children. The change from subsistence to capital-market economies has altered the scale of life, segmenting the individual from access to and control of food production. The same centralized form of production that has affected food has skewed the distribution of medical care to urban-based acute care and has left public and preventative care without resources. Kent focuses on the Marshall Islands.

Conclusion: The conclusion (Chapter 16) addresses theoretical issues raised in part one of the introduction and illustrated in the various chapters. Cluny Macpherson, a sociologist, also brings in his own work on indigenous Samoan medicine in a detailed discussion of the construction from the present of descriptions of pre-contact Samoan medical practices. I will let the conclusion speak for itself.

MELANESIA

2

Papua New Guinea: Changing Relations of Production[1]

Eugene Ogan and Terence Wesley-Smith

Introduction

During the 1960s, American social scientists, especially anthropologists, took a renewed interest in Marxist concepts for analyzing sociocultural systems. A number of publications appeared in the next two decades (e.g., Bloch 1975; 1983; Wessman 1981; Seddon 1978) producing the predictable controversies over definition of terms, applicability to different kinds of societies, and the relevance of Marxism as a conceptual framework to issues of planned socioeconomic change.

While a considerable amount of this literature seems less interesting or useful today that it did on first appearance, one set of concepts continues to point analysts of social change toward key factors in understanding, in particular, the effects of Western capitalist practices and institutions on non-Western, 'traditional' cultures.

This is not to say that the notion of *mode of production*, comprising the factors of *material forces of production* and *social relations of production* is a matter of consensus among social scientists, of Marxist or other theoretical persuasions (see, for example, Wolpe 1980; Foster-Carter 1978). However, we suggest that, used as a heuristic device, rather than a Procrustean bed, these ideas may help bring into focus what might otherwise seem to be a bewildering array of changes in the economy and society – or, as some would have it, the economies and societies – of the modern nation-state of Papua New Guinea.

The *locus classicus* of the concepts organizing our discussion is, of course, Marx's *Preface to a Critique of Political Economy*:

> In the social production of their life, men enter into definite relations that are indispensable and independent of their will, relations of production which correspond to a definite stage of development of their material productive forces. The sum total of these relations of production constitute the economic structure of society, the real foundation, on which rises a legal and political superstructure and to which correspond definite forms of social

35

consciousness. The mode of production of material life conditions the social, political, and intellectual life process in general (McLellan 1977, p. 389).

Simple as this passage might seem, other writers have debated at length just what is meant by mode of production, especially in the application of the idea to non-capitalist systems. Problems arise primarily because Marx developed these concepts to explain the rise of capitalism in Europe, and paid little attention to their relevance in the rest of the world. In the classical account, only a limited number of modes of production exist, each having the capacity to 'reproduce' its particular forces and relations of production. Most important, a mode of production is associated with a particular stage of development of society, and modes are expected to succeed each other in a predetermined sequence.

Applying this type of stages theory in the periphery is fraught with difficulties. First, it is very difficult to characterize the rich variety of 'pre-capitalist' modes of production found there in terms of Marx's limited and very sketchy ideal types. Second, as Anthony Brewer (1980, p. 264) points out, in this perspective, 'one must either say that underdeveloped countries are pre-capitalist, that they are capitalist, or that they are in transition'. This is not very helpful in situations where 'pre-capitalist' forces and relations of production are as important as capitalist ones, even after a lengthy exposure to capitalism.

Third, there is the question of the source of the evolutionary dynamic. Marx described an internal evolution of capitalism in Europe, paying particular attention to how the capitalist mode of production arose phoenix-like out of the internal contradictions of the preceding feudal mode. But in most parts of the periphery, the dynamic for the transition to capitalism is clearly external, imposed from outside as capitalism expands from its homelands into the non-capitalist periphery. Hence a central problematic in analyses of change in regions that have been subjected to European imperialism is the relationship between an intrusive capitalism and the indigenous modes with which it comes in contact. The classical approach is not very helpful in this type of analysis.

Perhaps the best known attempt to overcome these conceptual problems is found in the works of Andre Gunder Frank (e.g., 1979) and Immanuel Wallerstein (e.g., 1979), who argue that capitalism must be analyzed as a world system. But by retreating to the global level to locate the mode of production, and to trace the stages of development of capitalism, these theorists provide us with few useful tools to investigate the internal mechanisms of change in third world societies. Furthermore, the emphasis on relations of production central to most definitions of mode of production is abandoned in favour of 'neo-Smithian' concern with exchange relations (Laclau 1971).

A more promising approach involves retaining Marx's emphasis on production, but modifying his conception of modes of production as successive stages of development. The seminal work in this vein is undoubtedly that of the French philosopher Louis Althusser (Althusser and Balibar 1970). He revived Marx's notion of the *social formation*, a conceptual construction whose boundaries correspond (more or less) with those of real societies, and which may contain more than one mode of production. His ideas have been embraced in one form or another by such writers as E. Laclau (1971) and P. P. Rey (1973), and have given rise to what can be loosely described as the 'articulation of modes of production' approach.

The novelty of this approach lies in the assumption that the relations of production associated with different modes can coexist in a social formation. The central focus is on their interaction because, rather than existing in isolation, each mode is seen to affect the workings of the other. Indeed, each is regarded as, in effect, needing and sustaining the other. However, this relationship is not strictly symbiotic, since one mode (usually the capitalist one, if that is involved) is assumed to be dominant. Nor is it static. What is involved is

the articulation of two modes of production, one of which establishes its dominance over the other . . . not as a static given, but as a process, that is to say a combat between two modes of production, with the confrontations and alliances which such a combat implies: confrontations essentially between the classes which these modes of production define (Rey in Foster-Carter 1978, p. 56).

This conceptualization incorporates external and internal factors, allows for change, and can accomodate the apparent complexities of 'concrete' situations in the periphery. However, it does not represent a set of laws governing change in the periphery, as the mysterious 'conceptual sculptures' (Brewer 1980, p. 265) of Hindness and Hirst (1975) and others might lead us to believe. Here the concepts and relationships suggested by modes of production theorists will be used as a loose framework to organize the material on Papua New Guinea, and to help make some sense of it.

Following Wolf's concern not to use concepts to create typologies but rather to underline strategic relationships (Wolf 1982, p. 76), we propose, in our attempt to trace social change in Papua New Guinea, to focus on relations of production. Thus, in what follows, we are concerned with the ways people organize themselves to produce the necessities of life, to include both what is usually called the division of labor and the rules for providing access to the material forces of production.

Setting our task in this way seems to us to have two major advantages: first, we believe the concept of relations of production thus broadly defined permits comparative discussions in both time and space, avoiding the common anthropological problem of describing in detail trees without ever mentioning the forest. Second, it permits us to go beneath abstractions like 'subsistence affluence,' 'colonialism,' 'neo-colonialism' or 'underdevelopment' to the level at which real people lead their daily lives. As other scholars have noted in regard to more advanced political economies, people do not experience such abstractions directly. Rather, in the case of Papua New Guinea, they face gardening, working on a plantation or at a mine, puzzling over changing prices for their crops, or resenting their fellows who inexplicably grow rich. To speak of relations of production seems to permit us to attempt, at least, the daunting task of describing, in a relatively few pages for an audience of diverse interests, more than a century of social change in a culturally diverse region.

As our bibliography shows, many other scholars (e.g., Connell 1982) have begun to deal with some of the issues treated here. What we hope to provide is a blend of theoretical and empirical discussion which will complement that of others, particularly by encompassing more Papua New Guinea societies than those in the highlands, which have received so much attention (e.g., Donaldson and Good 1988).

PRE-COLONIAL PAPUA NEW GUINEA

Pre-colonial societies receive short shrift in some of the most influential accounts of change in Papua New Guinea (e.g., Amarshi, Good, and Mortimer 1979; Griffin, Nelson, and Firth 1979). This leaves us with the unfortunate impression that the velocity and direction of change there have been determined wholly by intrusive forces. But, as every anthropologist is aware,

> global capitalism impacted on something. World capitalism came to a living, vibrant, changing social order, possessed of its own stresses, strains and motive forces (Donaldson 1982, p. 438).

There are several constructs developed elsewhere that might serve to capture the essential qualities of this dynamic social order. For example, the *lineage mode of production*, which derives from work by French anthropologists with African experience, has been suggested as particularly applicable to traditional life in Papua New Guinea. According to such writers as C. Meillassoux, P. P. Rey and E. Terray, agriculture provided the bulk of subsistence and, although some specialization (e.g., cloth production) existed, productive forces were at a relatively low level. What was especially noteworthy (and controversial) about the French portrayal of this mode of production was the exploitative relations which were said to exist between elder men, on the one hand, and junior men and women on the other. The nature of these relations became a matter of debate among the French anthropologists (Modjeska 1982, p. 60).

As some of the references cited below indicate, the notion of a lineage mode of production was particularly challenging to anthropologists who had worked in highlands regions of New Guinea where an ideology of patrilineal descent, emphasizing the lineage or clan as the basic unit of social organization, combined with concerns about control over women's productive and reproductive powers to produce distinctive cultural patterns. However, without yielding to a widespread tendency of anthropologists to overemphasize the cultural variety found in the entirety of Papua New Guinea, we suggest that the lineage mode of production concept is too narrowly defined to serve as a base point for the account of social change we seek to provide (Strathern 1982a, p. 43). In particular, the very name inclines one to give undue weight to descent as an organizing principle for the socio-economic life of Papua New Guineans.

Rather more promising is Wolf's (1982, pp. 88–100) category of a kin-ordered mode of production. As a 'specific, historically occurring set of social relations, through which labor is deployed to wrest energy from nature by means of tools, skill, organization and knowledge' (Wolf 1982, p. 75), this form of organization commits social labor through appeals to filiation and marriage, and to consanguinity and affinity (Wolf 1982, p. 91). This definition takes analysis beyond lineages or descent, but still treats kinship as rather too unproblematic (Gardner 1985, p. 86).

A number of other models, such as Marx's 'primitive-communal' and 'Asiatic' modes, or Rosa Luxemburg's 'natural economy', might be brought forward here, only to be rejected on a variety of grounds. To describe anything like life in the geographic region which came to be called Papua New Guinea, as lived before Europeans came upon the scene, is fraught with difficulty which no off-the-shelf conceptual package can completely dispel. Quite apart from the problem of doing justice to the cultural diversity apparent at contact, there is the challenge of explaining the diverse trajectories that individual societies followed during the 40,000 or so years before the European intrusion.

Most anthropologists have in recent decades come to realize the distortions created by the 'ethnographic present', the narrative convention that the anthropologist's account is exempt from historical considerations. No thoughtful scholar today would suggest that any human society is static and unchanging, although the pace of change may vary markedly from one group to the next. Change in precontact Papua New Guinean societies might have occasionally been profound, as in the shift from hunting-and-gathering to root crop agriculture, the rise of pig production, and the 'Ipomoean Revolution' postulated for the highlands region (Watson 1965; Golson 1982). Modjeska (1982) and others have attempted to add the historical dimension to their accounts of highlands societies, but much work of this sort remains to be done.

A comprehensive mode of production construct capable of adequately capturing the essence of pre-colonial Papua New Guinea will have to be tailor-made. Indeed, it is probably necessary to postulate several modes of production, suitably articulated, in order to incorporate especially the protean qualities of pre-contact Melanesia (Wesley-Smith 1988, Chapter 3). What we will attempt here is, of necessity, much more modest.

Despite the admitted diversity in detail which existed in pre-contact societies, it is still the case that, at the time of contact, virtually all were settled gardening communities (Chowning 1977).[2] Certainly these were the communities with which Europeans interacted first and most frequently. It is thus possible to sketch what we feel were the key relations of production with, again, a clear understanding that some features may be blurred which might be significant for the analysis of a single society.[3] Also, our account must perforce use a version of the 'ethnographic present' as a starting point, even though a convention is just that, not history.

The first of these keys to organizing people in relation to each other and to material forces in the productive process was an extraordinarily well-defined gender-based division of labor. Men's work – clearing ground for gardens, building houses and fences to keep out pigs, defending the community against attack – is contrasted with women's work – doing all the routine drudgery of daily garden work, cooking for her family and for pig herds, childrearing – with monotonous regularity in all accounts of traditional life. Although, as noted below, larger social units often encompassed significant relations of production, the household was generally the basic productive unit (Denoon 1987, p. 52; Modjeska 1982, p. 61). Hence the association of husband and wife (or, at least, adult male and adult female) in subsistence activities was both normative and statistically normal. Furthermore, gender governed access to such productive forces as tools: men were as closely identified with stone axes and weapons as women with digging sticks and netbags. This gender division of labor, together with all the varied ideological and political superstructure which arose upon it, can hardly be overemphasized in understanding the changes we seek to describe.

Second, relations of production were importantly affected by a descent principle. This principle was patrilineal in some societies (most notably those in the highlands), matrilineal in others equally well known in the literature (e.g., Trobriands, Siwai), and less often non-unilineal.[4] We know of no Papua New Guinea society described in the ethnographic literature in which descent was simply irrelevant to the recruitment of social labor or the establishment of rights to productive resources, especially land. On the other hand, for more than 25 years, an over-emphasis in earlier accounts on descent in relations of production has been recognized by anthropologists themselves (Crocombe and Hide 1971, p. 301; Gardner 1985, p. 83ff.; Strathern 1982a).

A third variable is locality, or residence. Part of the confusion in some earlier descriptions of highlands New Guinea societies stemmed from the failure to realize how significant this factor could be, since it often contradicted an ideology in which patrilineal descent was alleged to control where people might live or make gardens. It is now clear that, in Papua New Guinea generally, rights to productive resources could not always be maintained if those who possessed them by virtue of descent, or other attributes described below, were long absent from the area in which the resources were located. Conversely, long-term residence by individuals not otherwise related by descent or other social ties might result in incorporation – if not for the individual, then for his or her children – into a kin group or other resource-holding unit.

Third, ego-oriented kinship networks – that is, biological connections or social ties based on a biological model, linking an individual to any other – might operate in the recruitment of labor or the claim to productive resources. Given the small size of traditional Papua New Guinean communities and the social definitions of kinship which might prevail, it was often possible to find one's place in the production process by professing a kinship tie not easily understood by an outside observer.

Finally, relations of production could be established, transformed, or extinguished by a variety of 'gift' exchanges, whether of food, valuables or people. It is no accident that the anthropological literature on exchange finds some of its richest sources in descriptions of Papua New Guinean (and other Melanesian) societies (Gregory 1982). Under exchange, we include the various forms of marriage, since these were generally framed in terms of the exchange of people (not merely of women, though this is often stressed in the highlands) or of other items (e.g., shell valuables), the transfer of which was regarded as absolutely necessary to the establishment of an affinal tie.

We have thus argued that, despite variations in detail, the significant ways in which pre-colonial Papua New Guineans organized the productive process, to include the establishment and maintenance of access to land and other resources, can be summarized by attention to a limited number of factors. While some of these can be grouped as 'kinship',[5] the importance of gender and locality must ever be kept in mind.

By viewing these factors as forming a matrix in any individual Papua New Guinean society (Ogan 1971), it is possible to make comparisons without obliterating all the undeniable variety of detail that existed before European contact. A given society might emphasize one factor over another in regulating the productive process (although it would seem that gender took precedence everywhere), and such emphasis might change over time (for example, by shifting in response to changing demographic conditions). The notion of a matrix of factors governing relations of production also provides analytical room for individual agency: always excepting gender, an individual might utilize one factor rather than another in placing himself or herself in the community's economy.[6]

A relation of production, vital to Marxist analysis, which has not yet been mentioned for pre-contact Papua New Guinea is that of class. Not only is the Marxist literature itself unclear on this point; recent writing about Melanesia (including Papua New Guinea) has further muddied the waters. Partly in response to the Marxist challenge to pay greater attention to inequality in pre-capitalist social formations, some analysts have begun to attack earlier notions of Melanesian egalitarianism by finding something approximating to social class, especially in the highlands region. For example, Golson (1982, p. 111ff.) and Strathern (1982a, p. 46ff.) discuss the possibility that pre-contact class stratification, distinguishing 'big-men' and 'rubbish men' as separate classes, disappeared from the Mt Hagen area when Europeans broke the 'big-men''s monopoly on shell valuables. Still others, noting the obvious social disabilities women suffer in many highland communities, suggest that men and women should be regarded as separate classes.

Some of these debates seem flawed by inattention to the long tradition in social analysis which separates *class* from *status* from *power* (e.g., Runciman 1969). If one holds to the definition of class provided by that tradition – that is, class relates to a position in the production process whereby one group is able to control access to the forces of production, and thus to appropriate and accumulate surplus value produced by another group – then it is hard to find genuine social classes in pre-contact Papua New Guinea. Variations in status and power, especially separating women and men, can be found in abundance. But the same conditions that Denoon (1987, p. 54) suggests put limits on social change also operated to preclude the formation of real classes. These include: the impossibility of monopolizing weaponry, so that no group could subordinate large numbers to a single authority by force alone; the absence of long-term storable surplus from agriculture[7] (Modjeska 1982, p. 62, contrasting the New Guinea situation with those reported for Africa); and the high level of household self-sufficiency, impeding the development of a highly specialized division of labor.

Even more significant is the fact that access to land, the basic force of production, was guaranteed by the operation of some combination of descent, ego-oriented kinship, residence or exchange (Strathern 1982b, p. 138, in which he contrasts this guarantee with unequal access to means for achieving status). These factors could also protect women from becoming a subordinate class, whatever disabilities they faced in terms of status and power (Modjeska 1982, pp. 69, 85; Strathern 1985, p. 102). Therefore, we will argue that, to the extent that social class exists today in Papua New Guinea, it can only be an aftermath of European colonialism, to which we now turn.

EARLY STAGES OF COLONIAL PENETRATION

Although the southwestern Pacific was among the last parts of the world to feel the impact of European imperialism, the peoples of what is now Papua New Guinea have been significantly affected by Westerners for at least 150 years. Trading for iron took place in the Louisiade Archipelago by about 1849 (Griffin, et al 1979, p. 4), and the introduction of these tools had an instant impact on traditional relations of production wherever they appeared.

Just as stone axes and weapons had constituted a gender-defined prerogative of men, so too did metal implements. It was also men's traditional role to deal with outsiders, whether friendly or threatening. But while the way in which metal tools came into Papua New Guinea communities was thus shaped by indigenous gender roles, the effect was to increase men's domination within the community (Modjeska 1982, p. 74).

It has long been recognized (e.g., Salisbury 1962) that steel axes, in particular, greatly reduced the labor demands upon men in Papua New Guinea gardening. It was precisely the male tasks of clearing forest and building houses and fences which were facilitated by these new tools. The consequences varied in detail: men in some communities (e.g., Baruya; Godelier 1986, p. 192) increased pig herds dramatically, while those in other groups simply devoted more time to 'parish pump' political discussions. In any event, men had more time to spend on non-subsistence activities, while women's burdens did not decrease. Indeed, if men chose to use the greater efficiency steel axes provided to increase gardens and pig herds, the demands on women's labor became more onerous.

Thus, from the very beginning of sustained European contact with the peoples of what is now Papua New Guinea, women might become increasingly disadvantaged, a process which continued through what has been called a period of 'classic colonialism' (Amarshi et al 1979, p. 37) before World War II.[8]

The incorporation of what is now Papua New Guinea into European imperial schemes might be described as 'classic' in the sense that the motives for German claims made in 1884 stemmed from a concern for the 'survival of German commerce and plantation enterprise in Samoa' as part of a colonial policy aimed to satisfy 'the vocal body of expansionist business opinion in Germany' (McKillop and Firth 1981, p. 89). But Great Britain's annexation of what later was called Papua was less an economic expansion than a reluctant response to the strategic concerns of Australian colonists. As the Melbourne *Daily Telegraph* noted in May 1875, there was 'no promise of immediate wealth in New Guinea, and . . . every promise of an encounter with savage races'. But the editors supported annexation nonetheless: 'though we do not want the island ourselves, we do want very much that no one else shall have it' (in Thompson 1980, p. 39).

The subsequent fortunes of the two parts of colonial Papua New Guinea were undoubtedly influenced by the separate circumstances of their imperial incorporations. The northern colony was linked for thirty years to a rapidly-industrializing metropolitan power intent on carving out an overseas empire. British New Guinea to the south, in contrast, was only nominally an appendage of the British metropole. In fact its destiny was from the beginning tied to that of Australia, itself on the periphery of empire.

These different circumstances were reflected in the extent of foreign economic activity in each territory before World War I. In German New Guinea the volume of private capital investment was quite impressive, and the flow of government funds significant. British New Guinea, which became the Australian territory of Papua in 1906, knew no such largesse. By 1913, only about 32,000 acres of land were producing crops for export, and the metropolitan subsidy was less than half that provided by the Germans to their colony (McKillop and Firth 1981, p. 103; Papua Annual Report 1914/15).

Between the world wars, Australian economic energies were concentrated, not on Papua, nor on recently-acquired New Guinea, but on the vast untapped resources of South Australia, New South Wales, and Queensland. New agricultural investments in Papua practically ceased after 1917 (Gregory 1979, p. 392), and the expansion of plantation acreage in New Guinea was fueled by the reinvested profits of German planters, rather than by any influx of new capital. At the outbreak of World War II, the colonial administrations were poorly staffed, the metropolitan subsidy was minuscule, and modern infrastructure remained rudimentary.[9]

The pre-World War II capitalist assault on Papua New Guinea was not only weak, but the agents of the three major forms of capital involved (merchant capital, plantation capital, and mining capital) had little incentive to seek the radical transformation of indigenous modes of production. For merchants, the object was to stimulate trade, and particularly to extract agricultural commodities for export. Village-produced copra became a mainstay of their operations, mainly because it could be obtained cheaply. This 'trade copra' was cheap because it was produced in conjunction with subsistence production, rather than at its expense. In this sense, merchant capital not only needed village producers, but it needed village producers who were still essentially dependent on indigenous modes of production.

Much of the production of agricultural commodities for export, of course, occurred on expatriate-owned plantations. One of the essential characteristics of plantation production is its dependence on a small group of managers supervising the work of a large number of laborers (Lacey 1983, p. 16). Hence, by far the most significant demand that plantation capital placed upon the local economy was for constant supplies of unskilled labor at the lowest possible cost to the employer (Curtain 1984).

However, obtaining labor was a perennial problem. In Hess's (1983, p. 52) words, 'history has laid the precolonial foundations for economic exploitation so poorly' that techniques for providing necessary labor which had worked in Africa and other parts of the world could not be utilized. There were no political structures like chiefdoms through which workers could be impressed. Even more to the point, most Papua New Guineans lived in a state of 'subsistence affluence' (Fisk 1975) wherein they could provide all the necessities of life with relative ease. Colonizers found that the stick was impractical in recruiting labor, while such carrots as they could hold out worked only on the least sophisticated villagers.

Plantation activity was concentrated in just a few locations, many of them in the New Guinea islands. However, as the demand for labor grew and Papua New Guineans typically demonstrated their reluctance to sign up for more than one stint on the labor line, planters were forced to look further and further afield for new recruits. The result was a 'labor frontier', often associated with recent pacification efforts, that moved inexorably outwards from the plantation enclaves to

45

encompass most of Papua New Guinea by the 1970s (Brookfield with Hart 1971, p. 264; Gregory 1982, p. 118ff.).

Tight government regulation of all aspects of labor trade was a central feature of the colonial plantation economy. A major impetus for such rigorous regulation was the recognition (by state officials, if not always by planters) that the supplies of cheap labor on which the economy was based would continue to be forthcoming only if the village economy was allowed (or required) to maintain its integrity. If the creation of a 'landless proletariat' was to be avoided, it was necessary to protect areas from over-recruiting, and to ensure that the young men returned to their village duties after laboring for Europeans for strictly limited periods (Fitzpatrick 1978; Gregory 1979; Hess 1983; Curtain 1984).[10] Labor regulations, then, as an early report to the United Nations put it, were specifically 'designed to preserve the native economy' (New Guinea Annual Report 1946–7).

Unlike plantation capital, mining capital is not primarily concerned to exploit cheap labor power. Nevertheless, the miners who struggled to develop the nine goldfields (all but one located in Papua) discovered between 1889 and 1923 (Nelson 1976, pp. 172, 258), quickly found their mostly small-scale operations heavily dependent upon local resources such as food and labor. Indeed, the mining industry was the spur to the establishment of the whole system of migrant labor in Papua. Thus mining operations, like the plantations, were effectively dependent on and subsidized by the local economy through the provision of cheap indentured labor. Mining capital, as well as plantation and merchant capital, benefited from the preservation of the 'native economy'.

We have argued thus far that before World War II there was no concerted attempt by the capitalist in Papua New Guinea 'to clear out of his way by force, the modes of production and appropriation, based on the independent labor of the producer' (Marx 1987, p. 86). On the contrary, considerable efforts appear to have been devoted to preventing the separation of the producer from his means of production, and to preserving the essential integrity of traditional modes of production. However, it does not follow that pre-existing relations of production remained unsullied. A number of significant changes will be identified below.

Merchants may not have sought the destruction of the village economy, but they did encourage Papua New Guineans to produce commodities for the market, an activity that has significant destructive potential.[11] This potential was not realized in this or later periods, we suggest, because of the way individual Papua New Guineans responded to the opportunities and problems of commodity production. Men were always better placed to formulate a response of any sort than women, although cash-cropping typically increased women's workloads. Traditional leaders ('big-men') and those aspiring to leadership were usually among the first to be interested in the new opportunities for wealth and power, and were also well placed to mobilize the necessary resources of land and labor (Salisbury 1970; Epstein 1968).

But the traditional relations that allowed big-men to engage in production for the market also served to restrain their commercial ambitions. As they had for centuries, distributive demands, sometimes backed by threats of sorcery or physical assault, as well as the highly personalized nature of most transactions, continued to keep the big-man from rising too far above ordinary men, or from straying too far from traditional norms and values. Furthermore, many would-be entrepreneurs perceived 'modern' economic activity, not as a vehicle to escape from clan life, but as a means of achieving personal and group success within it (Finney 1987, p. 17). Such responses tended to confirm and even consolidate traditional structures and relations, not least those that served to keep big-men from becoming too big while ensuring that lesser men and women remained subordinate.

The planters' needs for land and labor also had their effects on traditional relations of production.[17] By the 1870s, land was being alienated in the New Guinea islands, a marked change in relations of production among people whose notion of land rights did not include permanent alienation beyond the local group by any means other than annihilation of the former landowners in warfare. Like other aspects of colonialism, land purchases varied widely in their impact. Connell (1979, p. 123) estimates that only about 2 percent of all land in Papua New Guinea was alienated during the colonial era, but for land with agricultural potential the figure was more like 10 percent. Gregory (1982, p. 162) gives a figure of 3.3 percent, the bulk of which was appropriated by the state. Both agree that the Tolai of East New Britain represent an extreme case, having the highest percentage of valuable land alienated by foreign companies.

The impact of the system of migrant labor on traditional structures was somewhat reduced by the planters' choice of crops. Although many products, such as vanilla, rubber, sisal, or cocoa, could be grown in the island environment, a chronic shortage of labor pushed planters toward copra, which made fewest demands on the unreliable supply of workers. For example, Richard Parkinson, the German scientist associated with one of the earliest plantation enterprises in the Bismarck Archipelago, experimented with a variety of crops (McKillop and Firth 1981, p. 90) but, like his peers and successors, bowed to 'the reluctant hegemony of copra' (Brookfield 1972, pp. 50–3) as the best solution to the intractable labor problem.

The fact that copra was the primary crop exported by colonizers between the World Wars had a special significance in changing relations of production during the 'classical colonial' period. Copra plantations are strangely old-fashioned, the antithesis of the 'factories in the field' which produce many other sorts of crops. There is no urgency to the production process, no need for careful time management, next to no modern technology required nor any incentive to make technological improvements to increase production, and no reason to improve the skills of the labor force.

From the point of view of the laborers who did the dull, routine tasks required, their lives were disrupted but not transformed. They saw planters enjoying a strange wealth (lanterns, metal tools, tinned foods) of inexplicable origin, but doing little of what could be called work. From the point of view of a planter who need pay scant attention to management in a modern sense, he could see himself as master of a little kingdom, behaving toward his subjects as paternalistically or in as brutally racist a manner as his individual personality led him.[13] From the point of view of women living in villages from which labor had been recruited, they might be deprived, not only of the companionship of spouses, siblings, or sons, but also of their help in maintaining subsistence production.

As Hank Nelson has described in detail, prospecting for gold in Papua and, later, what had become the Mandated Territory of New Guinea, often meant forays into areas whose inhabitants had little or no previous contact with Europeans. First encounters between prospectors and villagers could thus be dramatic, if not violent, while the dangers of mining itself, for example, the ravages wrought by dysentery in the Lakekamu goldfield (Nelson 1976, p. 199ff.), would seem to offer further contrast with plantation life and work.

Yet, from the perspective of their effects on Papua New Guinean socio-economic life, we suggest the differences are less than the similarities. Both miners and planters regularly complained about the difficulties of maintaining an adequate labor force. In both kinds of enterprise, an indenture system was utilized, removing men from their home areas while women and those men left behind maitained subsistence production. In neither mining nor plantation work did the laborer obtain new skills which could really transform village life toward patterns, especially of consumption, which workers saw Europeans enjoying.

Mining did differ from copra production in that the former demanded at least a minimum of modern technology. With a few notable exceptions, most individual mining operations before 1930 were small-scale and used simple technologies. But the opening up of the rich Morobe gold fields in the mid-1920s led to the establishment of companies, such as New Guinea Goldfields Limited and the Bulolo Gold Dredging Company Limited, that were large by any standards. These companies organized production on an unprecedented scale in order to exploit extensive but relatively low-grade mineral deposits. They were able to achieve this expansion in scale by applying advanced technologies to production and support services.

Despite this transformation of the mining industry, the role of Papua New Guineans in it did not alter radically. The companies saw clear economic advantages in training Papua New Guineans to replace expatriates in skilled and semi-skilled positions. But this initiative was effectively squashed by white mining company employees, who went on strike twice in 1935 to protest about labor substitution plans, and state officials, who feared a challenge to the whole system of native labor in New Guinea (Newbury 1975, pp. 30–3). As a result of these attitudes, it was not until the 1950s that formal apprenticeship programs for Papua New Guineans got underway on the Morobe goldfields.

Some changes did occur. Some Papua New Guineans acquired new skills, such as truck driving and machine operating. Periods of indenture in the mining industry were extended from two to three years in 1932, and rates of re-engagement appear to have increased (Healey 1967, p. 133). But the requirement that workers be returned to their home villages at the end of the contract was still strictly enforced, and wage rates remained very low. The segment of the workforce employed in large-scale mining may have been marginally more skilled that its counterparts in small-scale mining or on plantations, but it remained impermanent, impoverished, unorganized, and largely incapable of precipitating a radical transformation of traditional modes of production.

Therefore, on the eve of World War II, despite all differences of detail in the traditional societies of Papua and New Guinea, despite differences in the specific forms of European contact, despite separate administrations with allegedly different philosophies (but see Hess 1983, p. 53ff.), despite different emphases in colonial economy producing different degrees of success in the two territories, we suggest that the following similarities are more significant.

First, the population directly affected by colonial administration since the 1880s was relatively small. Most notably, the most densely populated parts of what would be Papua New Guinea had barely been contacted. Whether in the form of plantations, mines, trading stores or commercial shipping, European economic enterprise was likewise confined to a relatively small area.

Second, whatever individual success stories might be found in the history of the 'classic colonialism', by 1939 both mining and plantation industries had 'ended up in the hands of the big companies' (Griffin, et al 1979, p. 53). These 'big companies' (with the exception of the mining giants in Morobe) were themselves small operations by international standards, reflecting Australia's place on the periphery of a capitalist world system (Amarshi, et al 1979, p. 24).

Third, for those Papua New Guineans affected by pre-war colonialism, their traditional relations of production had been disrupted but hardly transformed. Steel tools had certainly provided a degree of male dominance where greater complementarity of gender-based productive roles had existed before their introduction. Men had been absorbed into a colonial economy to a degree, while women had continued their traditional work, but often at a more oppressive level. The standard experience of those men in the colonial economy was of indentured labor, at unskilled tasks, in situations in which racist attitudes were likely to be more notable than efforts at economic efficiency. Thus Denoon's (1987, p. 56) assessment of the early colonial state stands: 'The social consequences of this kind of development were few'.

POSTWAR COLONIALISM

Like the 'classic colonialism' that preceded it, World War II had a notably uneven impact on the peoples of Papua New Guinea. Griffin et al (1979, p. 91) claim that 'The war had almost no effect on about one-third' of the population. On the other hand, people in areas like Bougainville, East New Britain and the Sepik, which were occupied by the Japanese, endured considerable suffering, not least from Allied bombing.[14]

Even those villagers who were not directly involved in military action might have their lives seriously disrupted. More left their homes than ever before (Griffin et al. 1979, p. 96). By late 1944, the Australian New Guinea Administrative Unit (ANGAU) had impressed into service 28,000 laborers in Papua and 21,000 in the Territory of New Guinea, the two areas being administered as one by ANGAU (Ward n.d., p. 2). Villagers in the coastal or island regions were likely to be exposed to Europeans and their material wealth to a degree undreamed of before the war. Such experience was to have a lasting effect on attitudes in the years to come. Promises of postwar rewards for Papua New Guineans made by ANGAU and the Provisional Administration also created new expectations in many parts of the region (Griffin et al 1979, p. 106).

Australian attitudes towards Papua New Guinea had also changed, in part because of wartime experiences. The first postwar Minister for External Territories, E. J. Ward, gave classic expression to one of these new attitudes when he said 'The Government regards it as its bounden duty to further to the utmost the advancement of the native' (quoted in Hess 1983, p. 56). Furthermore, the Charter of the United Nations obliged Australia, a strong supporter of the organization, to take steps to prepare its dependent territories for eventual independence.[15] The resulting official emphasis on 'development' coincided with a renewed economic interest in Papua New Guinea by foreign investors, as well as a marked increase in the capacity of the Australian government to fund projects in the territory. While the pace of change was modest in the immediate postwar years, it accelerated significantly after about 1965.

One of the major changes which took place very early in the postwar period was the extension of full administrative authority to the densely populated highlands region. Although the area had been penetrated by miners, missionaries and administrative patrols in the 1920s and 1930s, it was only after World War II that the lives of highland villagers began to be transformed. But the pace of change was much greater than that experienced earlier by coastal and island people.

Planters in the lowlands attempting to rebuild their enterprises were even more desperate for labor than during the 'classic colonial' period, especially since local populations were reluctant to abandon their own reconstruction efforts in order to work for wages. All existing labor indentures had been cancelled in October 1945, and the paid labor force had dropped to 4,100 by February 1946 (Ward n.d., p. 2). In this situation, the newly pacified populations of the highlands represented a pool of workers that could hardly be ignored. The Highlands Labor Scheme organized recruitment under government supervision, and during its lifetime (1950–74) over 100,000 men served as agreement workers, mostly on coastal copra and cocoa plantations. The administration might have regarded this effort as 'advancing the native' by providing exposure to 'civilizing influences'. Certainly it helped keep labor costs down for expatriate planters and businessmen (Ward n.d.). Among other consequences, a new generation of men learned of the plantations' routine drudgery, while highland women followed their lowland sisters in shouldering increased burdens in maintaining village subsistence production.[16]

However, the highlands did not simply constitute a place to recruit labor. In 1951 the new Minister for Territories, Paul Hasluck (1976, p. 46), assured white residents that 'for some years to come, private enterprise will largely mean the enterprise of Australians', though he felt obliged to add a gentle warning, 'That will not be absolute for all time'. Development of private enterprise in the highlands, which accelerated when the region was opened for land lease applications in 1952 (Donaldson and Good 1988, p. 70), contrasted in a number of ways with the rise of copra plantations in the lowlands.

51

In the first place, the most important crop was coffee, production of which is significantly enhanced by processing equipment. Harvesting and processing must be done according to a definite timetable, quite in contrast to making copra. Many of the relevant tasks, especially harvesting, can be performed by women and children who, in fact, were early incorporated as casual labor when the first expatriate coffee came into production.

Second, and perhaps more important in changing relations of production for villagers, many of the first European coffee planters had an explicit philosophy for developing the eastern highlands. Influenced by ideas from Kenya, these men (a number of whom had been district officers) wanted to build a 'partnership' with the locals to produce a rural bourgeoisie as a buffer against anti-colonial resentment and unrest (Donaldson and Good 1988, p. 70ff.).

What this meant in practice was encouraging selected highlanders to plant their own coffee. Thus, in sharp contrast to prewar colonialism in the lowlands, the highland coffee industry managed to approach what might otherwise seem contradictory goals: facilitating indigenous 'private enterprise' while promoting expatriate business. By 1964, the World Bank was able to claim that 'Coffee production by indigenes is one of the top success stories in the Territory' (IBRD 1965, p. 101), a story celebrated in fulsome detail by Finney (1973). However, the story has been expanded by others to note the serious disadvantages created for highlands women, who have been, for the most part, included in the industry only as low-paid casual laborers (Donaldson and Good 1988, p. 155). Furthermore, it is the highlands coffee industry which first produced something approximating to social classes in rural Papua New Guinea, a point to which we will return.

Change in Papua New Guinea during the 1950s and 1960s was not, of course, confined to the highlands. One might argue that the swift pace of transformation there was actually in line with Hasluck's policy of gradualist, balanced social development (Ward and Ballard 1976, p. 442) in that these changes helped highlanders 'catch up' to lowland villagers with longer colonial experience. At any rate, Hasluck's policy involved much increased expenditure on health, education (Griffin et al 1979, pp. 124–8) and agricultural extension for the entire territory. Denoon has even claimed that 'Budgetary support allowed "welfare" to outstrip production by a wide margin . . . so that the colonial state almost was the cash economy during the 1950s' (1987, p. 57, emphasis in original).[17]

Despite official attitudes, these expanded administrative activities inevitably had differential impacts on different segments of the population. The faster pace of change in less developed areas, especially but not only in the highlands, has already been noted. Health programs were of particular benefit to those women they reached, but boys derived greater advantage from new opportunities for formal education.[18] Agricultural extension work, directed primarily toward increased production of cash crops, helped expatriates as least as much as the indigenous population.[19] The overwhelming emphasis on techniques for growing and marketing cash crops perforce neglected traditional subsistence cultivation, and thus further disadvantaged women in the emerging economy (Barnes 1982, p. 256).

Efforts to promote indigenous production of cash crops throughout the Territory increased greatly in the 1960s. The Hasluck approach to development continued to express its concern with 'the preservation of a stable social structure' maintaining the optimistic premise that 'a prosperous New Guinea society can be built on the foundations of native society' rather than at its expense (Downs 1980, p. 131). The colonial state did its best to ensure this by promoting cooperative rather than individual forms of cash-cropping, and by 'preserving the natives' rights to land' (Hasluck 1976, p. 161).

This comfortable policy of 'uniform and gradual' development was doomed to failure, not least because it did not build on the real 'foundations of native society'. Papua New Guinea societies were not actually rooted in communal forms of production, and some groups and individuals were clearly better placed than others to benefit from production for the market. Also, extension workers soon discovered the advantages of working in easily accessible areas and with the 'better growers'. Furthermore, this cautious approach appeared increasingly inappropriate as the global process of decolonization gathered momentum in the 1960s.

The turning point came with the release of the influential 1964 World Bank report which recommended, among other things, that resources be concentrated 'in areas and on activities where the prospective return is highest' (IBRD 1965, p. 35). In the period of 'accelerated development' that ensued, the emphasis shifted decisively to achieving rapid economic growth, and to the promotion of the 'progressive farmer' (Fitzpatrick 1980). The earlier goal of creating a relatively undifferentiated (and politically innocuous) mass of peasant farmers was effectively abandoned.

The colonial state orchestrated the rise of the 'big peasant' through the judicious use of agricultural extension services, credit facilities, and legislation (Fitzpatrick 1979; Connell 1979; Gerritsen 1975; MacWilliam 1988). The resources of the Department of Agriculture expanded rapidly after 1963 and were turned more explicitly to the needs of the individual farmer.[20] The Development Bank began operations in 1967 and dispensed more than nine million dollars for indigenous agricultural development over the next seven years (Development Bank

Annual Report 1974, p. 10). Their loans carried very low rates of interest and little security was required. The chances of failure were reduced significantly through the close supervision of projects by extension officers.[21]

In this heady atmosphere of change state officials finally 'had to face the fact that land retained in native ownership was not always available for the native farmers who were ready to work for a cash income' (Hasluck 1976, p. 319). But the instruments that were developed to implement a potentially far-reaching land reform policy were curiously inadequate for the task. In particular, the restrictions contained in the 1963 Land (Tenure Conversion) Act belied its explicit commitment to the individualization of land tenure (Fitzpatrick 1979, p. 104; James 1985, pp. 119–20). Furthermore, insufficient state resources were allocated to administer the reform, with the result that fewer than 600 conversion orders involving less than 17,000 acres of land were issued in the decade to 1973 (James 1985, pp. 36, 45).

Less radical alterations of customary tenure systems seem to have been preferred in practice by state officials. The Development Bank, for example, made extensive use of a Clan Land Usage Agreement to establish borrowers' rights to use clan land for commercial purposes. In effect, this instrument facilitated a strengthening of individual rights in land while preserving the wider context of customary tenure and group control (Fitzpatrick 1979, p. 98). Thus quite dramatic increases in rural commodity production occurred without the destruction of the traditional relations of production associated with land. Such attitudes towards land reform did much to preserve the essential integrity of traditional modes of production.

Papua New Guinean efforts to expand their role in the urban economy were generally much less successful. Even in the later years of colonial rule, towns remained forbidding places for Papua New Guineans. This was especially true for unskilled wage workers. Their wages were often insufficient to support a family in town, accommodation was provided by the employer and difficult to secure otherwise, and alternative means of support for the urban unemployed were practically nonexistent.[22] It is hardly surprising that many rural migrants returned home after only a short period in town.

Would-be businessmen also found the urban environment hostile. They generally lacked the capital, experience, and expertise to compete with the foreign companies and individuals that dominated modern enterprise to an extraordinary degree.[23] Furthermore, the promotional measures that were finally introduced by the state in the late 1960s and early 1970s were selective in operation and quite limited in effect. National entrepreneurs were officially encouraged to operate in areas of the economy not already claimed by foreign capital, such as small-scale rural retailing, or in enterprises where low capital requirements and labor intensity already gave them a definite comparative advantage.[24] Papua New Guineans who did achieve success in business or commerce often did so in close cooperation with metropolitan capital. Indeed, almost all national enterprise was ultimately and heavily dependent for its existence on foreign capital, whose control of the commanding heights of the economy remained virtually unchallenged as the colonial era drew to a close.[25]

Some skilled and semi-skilled Papua New Guinean workers, who had benefited from the postwar expansion of educational and training opportunities, did manage to become more-or-less permanently established in the urban economy during this period. Many were recruited to staff the mushrooming state bureaucracy where, for the first time, wages and conditions made a westernized urban existence a realistic long-term alternative to village life.[26]

According to colonial administrator and coffee planter Ian Downs (1980, p. 288), Australian policies in the twilight of colonial rule called for 'an economic revolution that would disrupt the life of the people in order to give them better opportunities'. The changes most Papua New Guineans actually experienced may have been disruptive but they were hardly revolutionary. In the first place, if census figures are to be believed, nearly 60 percent of Papua New Guineans remained completely outside the cash economy as late as 1971. Many of those who did participate in commodity production or wage labor continued to regard these activities as incidental to subsistence production, to be conducted in the broad context of traditional relationships and values. Only a very small minority of Papua New Guineans became fully committed to capitalist relations of production.

On the other hand, the rapid spread of commodity production in the rural areas, the continuing involvement of Papua New Guineans in wage labor, and the general influx of new goods and currencies, distorted traditional relations of production significantly. In particular, women found themselves shouldering an increasing share of the burden of 'development' while enjoying few of the benefits. State-sponsored agricultural schemes encouraged the emergence of a new commodity-producing rural elite, albeit one with strong ties to traditional modes of production. Finally, the spread of cash crops prompted usufruct land rights to become more individualized than before, and land itself moved ever closer toward commodification.

THE ENTRY OF TRANSNATIONAL MINING CAPITAL

The continued salience of traditional relations of production in Papua New Guinea is often explained with reference to the slight capitalist penetration of this part of the world, coupled with the benign nature of Australian colonialism. As the African political scientist, Ali Mazrui (1970), wryly observed after a visit in 1970, 'There is only one thing worse than exploitative colonialism – and that is indifferent colonialism'.

Certainly, much of the capital fueling postwar development in Papua New Guinea took the form either of direct grants from the Australian government, or of relatively small-scale private investments in agriculture, commerce, services, and some manufacturing and processing. As Denoon (1985, p. 133) has noted, these are not the forms of capital that one would expect to cut a swathe through existing social structures. But international capital has been anything but indifferent to the rich mineral resources of this part of the periphery since the mid–1960s, and governments (colonial and post-colonial) have proved eager to facilitate the resulting invasion of giant transnational mining corporations.[27]

This wave of mining capital is readily distinguishable from previous waves by its size. The mine at Panguna in Bougainville island, for example, cost nearly 450 million dollars to bring into production in 1972, and about 580 million dollars in additional capital expenditure between 1972 and 1984 (Mikesell 1975, p. 82; Bougainville Copper Annual Reports). In addition, this sort of mining is conducted on an extremely large scale, is capital- and technology-intensive, and generates huge amounts of wealth. Perhaps most important, it has taken Papua New Guinea out of Australia's shadow and connected it more directly to the dynamic centers of the global economy than at any time since the imperial links with Germany were severed in 1914.

This new type of capitalist activity had a major and immediate impact on the growth and structure of the modern economy in Papua New Guinea. Yet what is most striking about this massive influx of capital and technology is the relatively small impact it appears to have had on most Papua New Guineans and their social relations of production. In large part this is because modern mining places relatively few direct demands on indigenous modes of production. Thus the demand for local labor is relatively modest, the companies import a large proportion of the goods and services they require, and significant 'downstream' developments, such as processing facilities, have not eventuated. This means that the only way that most Papua New Guineans are affected by or can benefit from the exploitation of mineral resources is indirectly through the expenditure of state funds captured from the mining companies in the form of taxation.

However, this type of mining capital has had significant impacts at the local level. Individual landowners in Bougainville, for example, have been profoundly affected by exploration and mining activities since the early 1960s. When exploration work commenced, the Mining Ordinance did not require the landowners' permission, let alone the payment of any sort of compensation. Even when continued local opposition forced an amendment in 1966 to require compensation, work proceeded under police protection where necessary and, as a company official put it, 'unfortunately against the will of the landowners' (Vernon, in Bedford and Mamak 1977, p. 11).

When the agreement to develop the Bougainville mine was negotiated between the mining company and the state in 1966,[28] those speakers of the Nasioi language claiming primary land rights did not fare well. Initially, it appeared that compensation would be assessed at a minimum level, and only those directly affected by construction of the mine and associated facilities[29] would receive any payment at all. Only a remarkable effort by the Bougainville Member of the House of Assembly, Mr (later Sir) Paul Lapun, obtained for the landowners a royalty amounting to 5 percent of the government's royalty, i.e. 0.625 percent of the value of the minerals produced.[30]

Although the basic royalty provision remains the same, other mechanisms for compensating landowners affected by mining projects have become much more sophisticated and generous since the 1960s (Bedford and Mamak 1979; Connell 1989). Nevertheless, they can not possibly take into account the hierarchy of land rights which typically served traditional modes of production. For example, Nasioi villagers who had subsidiary claims under traditional tenure systems have received much less than those who are regarded, however correctly, as primary right holders.[31]

Large-scale mining projects inevitably cause tremendous social disruption locally. For example, when the Boungainville mine and associated facilities were being constructed between 1968 and 1972, and when the total population of Boungainville was less than 80,000, some 10,000 construction workers flooded into the Kieta area and inland to the mine site. Mostly unattached males, they changed what had been a quiet colonial port into something resembling a brawling frontier town (Momis and Ogan 1971). Peoples in the vicinity of other mining projects have experienced similar invasions of foreign personnel and equipment in more recent years.

But social disruption has been accompanied by other changes. Mining and construction companies have had good reason to treat Papua New Guineans very differently from the way they were treated by plantation managers in an earlier era. Modern mining may not be labor-intensive, but its efficiency is dependent upon a stable and relatively skilled workforce. Since profitability is enhanced by replacing expatriate with local employees at lower rates of pay, the mining companies have initiated extensive training programs, and established

57

university and technical school scholarships for Papua New Guineans. Indeed, mine workers now form a relatively skilled, privileged, and committed subset of wage workers in Papua New Guinea.[32]

These modern relations of production have been further enhanced by the provision of financial and managerial assistance to local entrepreneurs. The companies have a positive incentive to favor local suppliers even when this does not promise to reduce direct operating costs significantly. Efficient mine operation is critically dependent upon the cooperation of local people, and their goodwill may be fostered by involvement in satellite activities. With company support and encouragement, some quite sizeable indigenous commercial ventures based upon capitalist relations of production have sprung up around the major mining projects in Papua New Guinea.

The entry of transnational mining capital has served to accelerate the spread of capitalist relations of production in Papua New Guinea. But, because it influences relatively few people directly, it has exacerbated further the uneven nature of capitalist development there. Not only have the effects been felt unevenly throughout the country, but they have been quite selective in the vicinity of the mines. In Bougainville, for example, the combination of widely varying compensation payments, royalties, wage employment connected with the mine and related enterprises, as well as differential success in cash cropping, have produced extraordinary variation in income in an area that had previously displayed relatively egalitarian characteristics. Furthermore, Bougainville's traditional ideology of matriliny (Nash 1981) is under pressure as key positions in the mine, cooperative societies and private businesses are invariably occupied by men.

INDEPENDENCE AND 'THE MELANESIAN WAY'

Many commentators have noted the relative ease with which Papua New Guinea achieved political independence in September 1975 – no history of anti-colonial violence, no abrupt departure of the administering authority or foreign enterprise, no widespread memory of brutal colonial oppression.[33] On the other hand, neither was there any genuine consensus about an economic or political philosophy which would guide the future of the new nation-state. Indeed, despite some quite dramatic changes in the form of the post-colonial state and the aspirations of its operators, important inherited tendencies in economy and society persist in the late 1980s.

Papua New Guinea emerged from colonial rule with a market economy that was extraordinarily dependent on the outside world, and especially on Australia. Almost all commodity production was for export, rather than for domestic use. About half of all domestic expenditure went on imports of manufactured and capital goods, fuel, and foodstuffs. More important, the 'modern' economy was largely owned and controlled by foreigners, with an estimated 80 percent of gross monetary sector income accruing to foreign interests. Furthermore, Papua New Guinea entered independence with about 40 percent of the national government's budget financed directly by Australia.

Development of the Bougainville mine had not only created new, and sometimes unwelcome, distinctions among the local population, but further exacerbated the unevenness of social change throughout Papua New Guinea. Resources had been shifted from other areas to provide the infrastructure required by the original mining agreement, and now North Solomons Province[34] generated disproportionate amounts of export income and government revenue. At the time of independece, agricultural exports were produced in consequential amounts only in the islands (New Britain, New Ireland, and North Solomons) and in the coffee-growing highlands. Such areas as those around the Sepik and Fly Rivers had experienced little change.

The colonial state's growth-oriented, foreign investment-driven development strategy was vigorously challenged by members of the first national government that came to power in 1972, and an alternative charter for development began to emerge. Part of its inspiration came from younger politicians who evoked an image of an idealized past, in which egalitarian social relations (including relations of production) prevailed. They argued that this past might be recaptured in a form which could also encompass modern economic activities, and be free of the exploitation typical of European capitalism. Inspired by certain academics at the University of Papua New Guinea (established in the late 1960s), and by consultants from the University of East Anglia, who saw Tanzania as a possible model, some spoke of a 'Melanesian Socialism' analogous to Nyerere's 'African Socialism'. However, it was such educated and thoughtful men as Father John Momis and Bernard Narokobi (Narokobi 1980) who independently began to write about 'The Melanesian Way'.

The precise meaning of this concept, and the debates it produced cannot be treated here. What is clear is that 'The Melanesian Way' included the notion of greater equity in economic life than is usually associated with capitalist development. Even though all the introduced capitalist relations of production remained unchanged by independence in 1975, at least some influential leaders of the new nation were committed to modifying these toward the reduction of inequality. A concrete outcome of this commitment was the promulgation of a national development plan, in which four of the Eight Aims for development and one of the five National Goals and Directive Principles (enshrined in the Constitution) explicitly pointed toward attaining various forms of economic, social and spatial equality (Turner 1987, p. 25).

From the beginning, the Eight Aims were pursued selectively and, not surprisingly, the one called for a 'rapid increase in the proportion of the economy controlled by Papua New Guineans and in the proportion of personal and property income that goes to Papua New Guineans' received the most attention. However, no thoroughgoing anti-imperialist purge was attempted or even mooted. The state moved quickly to establish control of the major banking and airline networks, renegotiated the Bougainville Copper Agreement, and invested quite heavily in other resident foreign-owned enterprises. New foreign investment was excluded from the lower reaches of the economy, and foreign investors were permitted to operate in some other areas only in joint venture arrangements with Papua New Guineans. But the major regulating mechanism, the National Investment Development Authority (NIDA), also had a statutory requirement (until 1983) to promote metropolitan investment and, according to Fitzpatrick (1980, p. 221), this function was increasingly emphasized after about 1976.

Also about this time it became clear that the goal of economic self reliance (expressed in the fifth of the Eight Aims), at least insofar as it required less reliance on foreign investment, had effectively been abandoned. In its place was the objective of increased 'fiscal self-reliance' whose primary indicator was the proportion of government expenditure obtained from internal sources. This placed the emphasis on substituting internally-generated tax revenue for direct Australian budgetary support. This meant attracting more foreign enterprise, especially in the mining sector, and taxing it more effectively. Since 1977, the fiscal incentives offered foreign investors have gradually increased, and today the package of incentives is 'wide, especially for manufacturing enterprise' (Daniel and Sims 1986, pp. 40–47).

Garnaut and Baxter (1984, p. 66) estimate that the share of Gross Domestic Product accruing directly to Papua New Guineans had reached 50–60 percent by the mid–1980s. But these benefits have not been distributed evenly, and public servants have probably gained most from post-independence strategies. The localization of the public bureaucracy proceeded rapidly in the mid–1970s, and the number of state employees continued to grow steadily after independence.[35] The continuing involvement of the state in the economy would seem to be endorsed by the last of the Eight Aims, but whether this involvement has been 'necessary to achieve the desired kind of development' is debatable. Indeed, wages and salaries for public servants claim about 50 percent of total state expenditure, and a relatively small proportion of government spending has been allocated for capital improvements that might increase the wealth-generating capacity of the economy.

In contrast, those parts of the Eight Aims of obvious benefit to Papua New Guinean entrepreneurs, especially in the urban areas, have received relatively little attention. As noted above, the NIDA regulations cleared some space for urban businessmen, but foreigners continued to be allowed to operate relatively small businesses well into the 1980s.[36] The businessman's strongest advocate, the Development Bank, continued to lend money to expatriates until 1983. In more recent years it was renamed the Agricultural Bank and encouraged to focus more on the rural areas.[37]

The expansion of the crucial agricultural export sector poses a serious dilemma for planners, since too much 'development' too fast raises the same specter of a 'landless proletariat' that so concerned the colonialists. Like their Australian predecessors, the new leaders seek to increase rural commodity production without disturbing existing social relations of production. Thus the state continues to encourage the smallholder production of export commodities, but by quasi-traditional means. This approach receives strong ideological support from the Constitution, one part of which requires 'development to take place primarily through the use of Papua New Guinean forms of social and political organization'.

Strands of 'The Melanesian Way' also informed the work of the important Commission of Enquiry into Land Matters (Papua New Guinea 1973), which was to help the new national government formulate its policy towards land. The commission proposed that land tenure reforms be built 'on a customary base' and emphatically rejected the notion of 'a sweeping agrarian revolution or total transformation of society'. Nevertheless, at least some 'big-peasants", especially in the highlands, appear to have broken free of traditional entanglements in recent times to operate largely in the capitalist mode (Finney 1987).

Among those who have not benefited much from the uneven implementation of the Eight Aims are unskilled workers, who still enjoy little security, especially in the urban areas, and women, who continue to be underrepresented in all sectors of the modern economy.

The Eight Aims did not call for radical change. Indeed, as Fitzpatrick (1985, p. 26) has pointed out, they were premised on the then popular notion among development theorists that progress could be made within existing structures, in other words that 'It was going to be possible, at last, to make an omelette without breaking eggs'. Some progress has been made. Papua New Guineans have displaced foreigners in the state apparatus and made vigorous attempts to control the direction of change. Nationals have captured a greater share of the wealth generated by the exploitation of their resources. And they now produce most of the agricultural exports. But the goals of redistribution and equality that are at the heart of the ideology of 'The Melanesian Way' have proved more elusive.

Conclusion

Throughout this chapter, we have been at pains to emphasize the uneven pattern of social change in Papua New Guinea. Such grand theoretical schemes as those of Frank or Wallerstein cannot encompass such variety without the kind of modification we have sought to provide. Nevertheless, certain general points may be drawn from this variegated detail.

First, even where introduced, capitalist relations of production have had the greatest effect; they have yet to eliminate completely those which existed before colonialism. This is most obvious in the continuing conflict, increasingly violent, over land rights, especially in the highlands and, most recently, in Bougainville. Flexibility, a hallmark of traditional systems in Papua New Guinea, is incompatible with modern capitalist enterprise, which must depend on means to transfer property with absolute finality. One thread in the tangled skein of violence current (August 1989) in North Solomons is a system, still vital in the minds of local people, which permits continued claims on land which capitalist practice deems bought and paid for.

Nor have other traditional relations been destroyed in the manner theorists might have expected. Even the power of money has not always recreated the earlier practices in a capitalist mode (Gardner 1985, p. 82), since money is not always seen by Papua New Guineans as all-purpose currency in every context of exchange (Healey 1985, pp. 141–2). Traditional exchange systems have shown remarkable vitality, and Strathern (1982a, p. 49) claims for the people he knows best that as long as such ceremonial exchange persists, 'Hagen society will not become simply capitalist in structure'. Even after a century of exposure to capitalism, as Gregory (1982) points out, goods and labor are no longer always 'gifts', but neither are they always 'commodities'.[38]

However – and this is our second general point – one widespread effect of those changes which have taken place in Papua New Guinean economy is the increased advantage men possess over women in access to productive forces. A few statistics: in 1985 women accounted for only 16 percent of full-time university students (Turner 1987, p. 30); in June 1981, the overall participation of women in public service was 19.1 percent, almost all in the four lowest grades of clerks (Barnes 1982, p. 262); 17 female candidates out of a total of 1,124 contested the national elections of 1982, and one was elected (Turner 1987, p. 31). Women continue to work in subsistence production but, in the export-oriented economy which has continued since independence, this work enjoys little status as well as providing almost no access to cash (Ibid; Barnes 1982, p. 263). The degree to which women have resisted this growing inequality and fought back is, as Denoon (1987, p. 60) has pointed out, one of the most fertile fields for further analysis.

Third, and most important, changed relations of production in Papua New Guinea have produced true social classes where none existed before colonialism. The literature on this subject is complex and controversial; for example a decade ago Connell (1979, p. 132) could assert that 'Classes are not yet clearly defined in PNG'. However, today at least one scholar who doubted the utility of class analysis of change in Papua New Guinea (Turner 1984) now says

> there can be little doubt that a working class, a white collar class and a bourgeoisie do exist although the name chosen to label a class and the precise delineation and role of that class will vary according to theoretical and ideological preference (Turner 1987, p. 29).

Indeed, arguments over 'rich peasants' versus 'bureaucratic elites' or other categorizations continue. MacWilliam (1988, p. 96), for example, speaks of the unsuccessful struggle of an indigenous bourgeoisie to enlarge its strength against opposition from 'upper peasants, traders, salariat, and a radical intelligentsia'. It seems clear that ethnic ties still provide a strong countervailing force to class as an organizing principle for group action (Nash and Ogan 1990), and genuine 'class consciousness' may be hard to discern. Nevertheless, the detailed study of class formation and action seems to us the most crucial task for the future study of social change in Papua New Guinea.

Although social change is an open-ended process, any single discussion must have an end. In closing, we can only point to what many observers (Malik 1989) regard today (August 1989) as a critical juncture in Papua New Guinea's recent history. Violence produced by a complicated mixture of conflicting claims to land and compensation, 'cargo cult' beliefs, and inter-ethnic fighting have kept the Bougainville mine closed for three months, depriving Papua New Guinea of 45 percent of its exports and 17 percent of its government revenue. This situation has inevitably created doubts about the feasibility of exploiting the mineral wealth which might bring unprecedented prosperity to the country. Prices for the 'big three' agricultural exports – coffee, cocoa, and copra – are seriously depressed. Groups of armed ruffians create fear of robbery, rape and murder in towns and along a vital transportation link, the Highlands Highway. The future of a nation which had seemed on its way to becoming by far the most economically successful among the Pacific islands now appears problematic.

Just as we have attempted to show how a certain theoretical stance could illuminate the changes of the past, so we would suggest that further study of transitions and transformations of relations of production will provide the best guide to understanding that future.

3

New Caledonia: Social Change, Political Change and Tradition in a Settler Colony

John Connell

Melanesia can be crudely distinguished from other regions of the South Pacific by its long history of settlement, extending over many millennia in the larger northern islands. Societies are small, fragmented over small islands and mountain chains, separated by language, historical experience and distinctions in cultural practice. Melanesian societies have also been loosely characterized by their achieved (big-man) leadership structures, though in New Caledonia (and especially in the Loyalty Islands) more obviously inherited, chiefly systems are common. Above all Melanesia is typified by the extent of social differentiation and fragmentation across small areas and by such late contact with the global economy that these distinctions remain considerable, with tribal and regional affiliations more significant in most contexts than new national identity. More than any other region of the world the construction of national policies is extraordinarily difficult; the problem of creating modern states has been frustrated by the great extent to which their peoples' sense of self remains bound up in the gross actualities of blood, race, language, locality, religion or tradition (Geertz 1973, p. 258) and hence limited the extent to which individuals might identify with a wider society in which all these separate elements were fused. In New Caledonia this situation had been made even more complex because of substantial migration in the past century, from Europe, Asia, and Polynesia, and the fact that New Caledonia is part of France, and thus change has occurred in a different context from that of the remainder of Melanesia, and much of the rest of the South Pacific.

New Caledonia is one of nine French overseas departments and territories, the last 'confetti' of France's empire (Guillebaud 1976). New Caledonia and French Polynesia are the oldest colonies in the South Pacific. New Caledonia became the only settler colony and penal settlement in Melanesia, resulting in a structure of double domination – the role of European colonial power in the colony as a whole and the dominance of the indigenous population by the settlers (Houbert 1985, p. 217). By the 1870s, three crucial themes in New Caledonia's history were already present: (1) the first nickel rush (and foreign investment); (2) Melanesian opposition in the struggle for land; (3) the growth of a European population at the expense of the Melanesian population, who were displaced principally into east-coast reservations, becoming a minority in their own land. Underlying almost every aspect of the struggle for independence and identity is land: the restoration of authority to land and the ties between Melanesians and land; society was written on the ground in the past; a new generation seeks to restore that relationship. The extent of land alienation, the bitterness of the dispossessed, and the mutual incomprehension between Melanesians and Europeans provoked in 1878 a sustained and bloody revolt, the longest and most violent reaction to European colonization in the island Pacific. The eventual triumph of the French emphasized the withdrawal of Melanesians and their demographic decline, and resulted in Atai, the assumed leader of the revolt, becoming a seemingly legendary figure and a rallying point in the contemporary struggle for independence. Nowhere else in Melanesia has so much land been alienated, or so many Melanesians killed or displaced. That nineteenth-century legacy lives on.

Formal French colonization of New Caledonia after 1853 transformed the landscape of New Caledonia as European settlers colonized the West coast, moving northwards from Noumea, replacing Melanesian gardens with European cattle. By the 1860s New Caledonia had the largest white population in the Pacific islands. The establishment of the penal colony emphasized white settlement and the deterioration of racial relations; a land of exile for French criminals was rapidly becoming a land of exile for the Melanesians, whose land it had been, but who were forcibly removed into reservations in the mountains or on the east coast. Settlers and Melanesians feared and distrusted each other. Melanesians were governed by a code of laws, the *indigénat*, that turned them into cheap labor but otherwise confined them to the reservations. Only in the Loyalty Islands was no land alienated and the most oppressive forms of colonialism absent.

Though the extent of the initial decline of the Melanesian population has probably been exaggerated, there is little doubt that European contact had some fatal impact through the diffusion of new diseases, guns, and the intermittent violence of occupation. More important, however, was the slow decline of the Melanesian population, in the confines of the reservations, where helplessness and despair, and the absence of medical care, contributed to alcoholism and the degradation of Melanesian life. This decline continued into the 1930s, when modern medical care and distancing from the past encouraged a slow revival. At the same time the European population had grown, albeit slowly in the twentieth century, and because of the difficulties of obtaining Melanesian workers they sought Asian migrants as cheap labor. Though there had been Asian migrants from the 1860s, the 1920s witnessed a massive increase, to the extent that the Asian population briefly passed that of the European. Together they made the Melanesian a minority in New Caledonia.

The Second World War began the transformation of the history of New Caledonia. Melanesians earned substantial wages from occupying troops, mainly Americans, witnessed black American soldiers working in similar conditions to Europeans, and participated in the first commercial boom. After the war New Caledonia became a territory rather than a colony, the *indigénat* was repealed, a slow process of rural-urban migration and urbanization began, and Melanesians were incorporated into the modern world. Population began to grow rapidly, and Melanesians again became a majority in New Caledonia in the mid–1950s. However massive politically-inspired migration in the nickel boom from 1970–72 comprehensively ended this short-lived superiority. The French Prime Minister sought to encourage migration from France or other overseas possessions, such as Réunion, and to discourage migration from within the Pacific to ensure that indigenous nationalist movements would be discouraged and swamped, and that a docile labor force would be created. The strategy worked, though in a rather different manner, in that most migrants were from the two other French Pacific territories, Wallis and Futuna and French Polynesia, rather than more distant possessions, but the subsequent population growth of Wallisians has emphasized the new 'black colonization' of New Caledonia. Political decisions had again ensured that New Caledonia remain a settler colony. Melanesians are thus a minority; moreover, a minority that is concentrated in the outer islands and in the more remote rural areas, in contrast to the strongly urban European, Polynesian, and Asian populations.

In the census of 1989 New Caledonia had a population of over 164,000 and the Melanesian population of 73,600 (45 percent of the total) was only marginally a greater proportion of the total than a decade earlier. New Caledonia is one of the largest states in the island Pacific. The Melanesian birthrate is almost twice that of the European, though the Melanesian infant mortality rate is 40 compared with the European rate of 8 per thousand, and Melanesians have a life expectancy of 59 compared with Europeans of 72.5. In the (improbable) absence of migration the continuation of present population growth rates would mean that Melanesians would not become a majority in New Caledonia again before the 2030s decade. A quarter of the population of New Caledonia was born elsewhere, though two-thirds of Europeans were born in New Caledonia. Most of the overseas-born live in Noumea. Almost all those born outside were born in France or other French Pacific territories, particularly Wallis and Futuna (half of whose ethnic population live in New Caledonia) and French Polynesia. Salary and tax structures favor residence in New Caledonia and the majority of migrants, especially those from Asia and the Pacific, have vastly greater economic opportunities in New Caledonia than in their home countries. A small proportion of migrants have come from former French colonies, including the *pieds-noir* from Algeria, and are highly conservative and strongly, sometimes violently, opposed to independence. Virtually all support the retention of New Caledonia as a French territory to ensure that their minority status, slowed their movement into skilled employment and contributed to denying independence for New Caledonia. This has contributed to tension between ethnic groups and emphasized the role of population migration in every aspect of contemporary change.[1]

CONTEMPORARY ECONOMY: FROM SUBSISTENCE TO SUBSIDY

The nineteenth-century economy of New Caledonia was agricultural, with Melanesians growing subsistence root crops and European settlers struggling to produce and market cattle and a diversity of cash crops. Although New Caledonia experienced a brief gold and nickel rush before the end of the century, the economy remained agriculturally based until the Second World War, which was a watershed in New Caledonia development. The war ensured rising demand for nickel and chrome, and a mining boom was matched by a commercial boom, high levels of consumption, and unprecedented prosperity. In the postwar years, as the mining industry grew, the New Caledonian economy became increasingly urban, the agricultural sector declined, and the tertiary sector – both public and private – absorbed the bulk of the wage labor force. Economic deversification favored the establishment of tourism rather than the revitalization of agriculture. Continued and increased financial subsidies, especially for the bureaucracy, emphasized the desire of France to retain control of its distant territory, as New Caledonia became transformed into a 'consumer colony' much like French possessions elsewhere (Ormerod 1981, pp. 1–12; see also Connell and Aldrich 1988).

In all these transformations, the gulf between urban prosperity and rural poverty has tended to widen. Melanesians have played their part in the expansion of the urban economy, but essentially as wage laborers rather than owners or employers and, as the New Caledonian economy has become more firmly incorporated into the world economy, Melanesian ability to determine the economic destiny of New Caledonia has scarcely increased. For all the importance of land, the economic history of postwar New Caledonia is essentially that of an externally oriented economy, increasingly looking elsewhere for markets, finance, and even labor – quite unlike that of other parts of Melanesia, where the production of agricultural commodities dominates economic life. It is an economy structured by European interests in which the Melanesian economy is peripheral, and Melanesians largely remain bystanders in the political economy that has shaped their destiny.

Nickel mining remains the basis of the New Caledonia productive economy, and French emphasis on this industry has resulted in the neglect of other sectors, especially agriculture and fisheries. Nickel has absolutely dominated the economic history of New Caledonia. The mining industry, from its nineteenth-century antecedants, became, in the twentieth century, an alliance of overseas (French) capital and a local (mainly landowning) rural elite, with a significant part of its labor force imported into New Caledonia. The mining industry has increasingly become dominated by Sociétal Le Nickel (SLN) which owns two-thirds of the island's richest concessions, between Houailou and Thio, and has the sole smelter at Doniambo in Noumea. SLN is now entirely owned and subsidized by the French government. Depressed nickel prices have resulted in the industry being heavily subsidized throughout the 1980s. Contraction of mining operations since the 'boom' turned to 'bust' in the early 1970s resulted in the total mining workforce falling to around 2,600 at the end of 1983, and at the start of 1986 it was probably not much more than 2,000, as nickel went through one of its worst depression in two decades. In the past few years SLN has closed mining operations at Poro and Nepoui, but maintained output by concentrating operations at Thio, the main center of the mining industry. High nickel prices in 1988 resulted in new optimism over a revival in nickel mining, with the nickel mines reaching close to full production for the first time in several years. In New Caledonia there is now a firm belief that a new 'nickel boom' has already begun. Mining has however directly generated limited employment, and then primarily for skilled workers rather than mainly unskilled Melanesians. The future of the mining industry is important to the future of New Caledonia and there is some possibility of diversification, since New Caledonia contains about 20 percent of global cobalt resources.

Agriculture plays a minor role in the commercial economy, far behind mining, tourism, and commerce. Since 1956 the proportion of the workforce engaged in agriculture has fallen from 50 percent to 20 percent, a quite spectacular decline, producing proportionately one of the smallest agricultural workforces in the island Pacific. Extensive migration out of the agricultural sector has resulted in an increasingly urban economy and population distribution. High labor costs have reduced the viability of intensive agriculture, and European agriculture now principally involves commercial cattle ranching, cereals, and vegetable production. The great European cattle ranches (stations), such as that of Jacques Lafleur at Ouaco, are all on the West coast, but the cattle economy is static, investment has been limited because of concern over land tenure, much land is no more than rough grazing, and the agricultural workforce is steadily aging. Contemporary attempts to diversify the agricultural economy have had limited success. Melanesian agriculture combines the subsistence production of mainly root crops (yams and taro) and the commercial production of coffee and vegetables. Agriculture is as far now as it has ever been from recognizing its potential. Forestry and fisheries are also poorly developed though both sectors have some development potential.

Tourism developed rapidly in the 1980s after the end of the nickel boom, drawing a clientele initially from Australia and increasingly from Japan. Club Med began operations in 1980, and until late 1984 tourism was not only New Caledonia's second major industry, after nickel, but was steadily increasing its significance. Tourism is almost wholly confined to Noumea, though there are a number of small hotels in the Isle of Pines. Political unrest, especially away from Noumea, associated with the independence issue, has severely affected the tourist industry (Connell 1987b, pp. 54–65), substantially reduced the number of visitors, resulted in the destruction of three hotels, and increased the relative importance of Noumea to the industry. Since 1986 there has been a steady but slow growth in the tourist industry, following more active promotion in Japan and Australia, though tourist numbers have yet to regain 1984 levels.

France closely controls public finance, including the money supply, public investment, and private foreign investment. Much of the most important elements of the budget of New Caledonia are direct contributions and grants from France; in 1987 the total territorial revenue amounted to CFP36.969 million ($A385 million) of which 28.5 percent was a direct payment from France (Christnacht 1987). There is also widespread support for special development programs such as public works and health, so that France contributes more than half New Caledonia's financial needs. An independent state could never obtain such lavish overseas financial support, and this situation has weakened demands for independence as the economy of New Caledonia has become structured much like that of smaller Pacific micro-states (see Bertram 1986, pp. 808–822). The bureaucracy has doubled in size since 1970, employs a quarter of the wage labor force, and contributes half the wages and salaries in New Caledonia. This steady growth of employment and incomes in the bureaucracy, relative to the rest of the economy, and especially to the agricultural sector, has emphasized what has become an increasingly artificial and dependent economy, tied to France and isolated from the rest of the South Pacific. Much of the income of New Caledonia is repatriated to France, or diverted to Australia and elsewhere, and in the past decade a substantial trade deficit has developed.

Noumea, the most industrialized city in the island Pacific, dominates the island economy, as employment has shifted from the primary sector to the tertiary, and wage levels have been effectively subsidized by France. Relative regional and ethnic economic inequalities have worsened in the past decade. During the nickel boom Noumea expanded rapidly, along with suburban Mont Dore and Dumbea, resulting in more than half the population of New Caledonia being concentrated in the metropolitan area. Noumea, with 80,000 people, and its outlying suburbs now claim about 60 percent of the population; the next largest town, Bourail, has barely 2,000 people. Noumea is essentially a European city, and Melanesians constitute 20 percent of the population. The major concentration of Melanesians is at Montravel, a grim suburb of tenement buildings in the shadows of the Doniambo nickel works. Many are unemployed but have chosen to remain in Noumea rather than return to reserves where their ties have been poorly maintained. Outside Noumea, Melanesians are almost invariably a majority, especially on the east coast of the main island (the Grande Terre), and in the islands where less than 2 percent of the population are Europeans. The contrast between a European city and a Melanesian countryside remains, as opposition to European settlement continues in the countryside (Guiart 1988),[2] and this division is critical for politics and economic development.

Melanesians are incorporated into the periphery of the New Caledonia economy through wages, taxes, pensions, medical assistance, and a variety of legal and institutional means, and there is no longer a 'traditional' self-reliant Melanesian economy. Melanesians may now represent about a third of the wage labor force, though mainly in unskilled and poorly paid sectors, even relative to the recent migrants from other French Pacific territories. Few Melanesians are in skilled employment; in 1981 there was one Melanesian doctor (who worked in France), one architect, one journalist, but a number of public servants. However, since then Melanesians have made further inroads into this particular part of the modern sector. There have been tentative attempts to encourage Melanesian participation in formal sector employment; in contrast to the independent states of Melanesia there has been no 'localization' policy in New Caledonia since no statutes

71

differentiate Melanesians from other races, all being French citizens. Explanations for the limited participation of Melanesians are numerous; they include a history of job and racial discrimination, less adequate education and skill training, geographically remote location, the costs of urban life, and some preference for rural life. Young Melanesians aspire to formal sector employment, yet the unequal results of education and uneven access to prestigious or well-paid employment have blunted the hopes of many and encouraged greater adherence to traditional Melanesian values.

Village agricultural development is much like that elsewhere in Melanesia, though the extent of cash crop production is often less substantial. The land is the only resource unequivocally owned by Melanesians, and attitudes to the retention, expansion, exploitation, and alienation of land areas underlie economic and political history. Subsistence cultivation of root crops remains the basis of the agricultural system, though these are generally combined with cash crops and, where possible, small-scale cattle ranching. Melanesian coffee cultivation has existed in many places since the early decades of the century. There is widespread acceptance of the commercial economy. A small minority of Melanesians, many politically active, favor withdrawal from the urban capitalist economy, to a more self-reliant cooperative, agricultural system. However, there is heavy Melanesian dependence on consumer goods, hence such withdrawal is not practical. Prominent Melanesian supporters of independence have been enthusiastic about the revival of Melanesian commerce, and its potential for encouraging and enabling more self-reliant development in rural areas.

Most Melanesians accept that New Caledonia has irrevocably moved into the modern world, recognize many of the virtues of a monetized economy, and are willing to accept some of the risks of disruption and dissolution of some traditional customs that new economic enterprises make inevitable. More conservative attitudes, that place a strong emphasis on the conservation of land and customs, are most strongly held on the main island, the Grande Terre, where the greater part of the land has been alienated and the virtues and certainties of the past exercise the strongest influence. Many urban workers retain rural cash crops and, where possible, circulate between urban wage employment and rural commerce. This pattern of short-term employment has given Melanesians a reputation for being work-shy and lazy when, for many, wage employment is marginal to the principal goal of consolidating a rural livelihood. Increasingly, however, greater demands for material goods, and the difficulty of obtaining urban unskilled employment, have resulted in more permanent urban migration and employment, though few Melanesians deny their rural roots. The relative success

72

of Loyalty Islanders in the urban economy, and the dependence of the outer islands on cash incomes, has meant that the Melanesians are most in support of commercial development and least likely to be in favor of independence are from the Loyalty Islands and the Isle of Pines.

THE GENESIS OF POLITICS

After the Second World War, when the status of New Caledonia officially changed from 'colony' to 'overseas territory', Melanesians were gradually enfranchized, were no longer confined to reservations, and were free to be employed at wage rates comparable to those of Europeans. The first significant multiracial political party, Union Caldédonienne (UC), was founded in 1951, and its mild reformist policies attracted substantial Melanesian support. However, Melanesian frustrations with the slow pace of reform, racial discrimination, economic recession, and continued opposition to their aspirations towards greater autonomy produced a radicalization of politics in the 1970s.

A few Melanesians returned from the turbulent French universities to stimulate radical demands, primarily through attempts to rediscover and assert Kanak (Melanesian) identity and culture. Confrontations between militant Kanaks and the administration focused on land rights, and a number of wholly Melanesian parties broke away from UC or were spontaneously created. The new parties were based on regional and religious differences rather than on ideological issues. To prevent further fission, UC became more radical, resulting in the loss of most European support. In opposition to the genesis of more radical Melanesian parties, fragmented conservative parties consolidated into the Rassemblement pour la Calédonie dans la République (RPCR), a primarily European party. Despite its losses, UC retained considerable power within New Caledonia, but not compared with those of the High Commissioner, and therefore, through him, with the French state that he represented.

By the mid-1970s the Melanesian parties were becoming more radical, in part through frustration with the conservatism of other parties and catalyzed by the murder of a Melanesian in 1975. Politics primarily focused on land. In many Melanesian reserves land pressures were and are considerable, strengthened by natural population increase and return migration which have limited the potential of cash cropping and cattle ranching and stimulated demands for land reform. Though the speed of restoring land to Melanesians increased in the 1960s and the 1970s, it was still far short of Melanesian expectations and needs. In the second half of the 1970s Melanesians mounted direct action to regain land by occupying it. Increased amounts of land were also purchased and returned to Melanesians, but it was invariably too little and too late to defuse tension and political pressure. Indeed, Melanesians had no more land per capita in 1980 than they had at the start of the century (Ward 1982). Vast inequalities in land ownership remain a source of tension; land issues pervade every aspect of Melanesian life and land is both symbol and substance in the struggle for independence.

By the late 1970s each of the predominantly Melanesian parties had come out in favor of independence, though the strength of their demands varied considerably, from rhetoric to realistic expectation. In large part militant Melanesians (Kanaks) were forced into more radical positions because of the unwillingness of the French government to adopt reformist policies. More radical Melanesian demands antagonized the majority European electorate which consolidated its strength in the RPCR and, at the same time, all political debate became subsumed into support for or opposition to independence, rather than discussion over domestic policies. The RPCR lost what little zeal it had for reformism as the UC and the other Melanesian parties shifted to more radical nationalistic positions. The population composition of New Caledonia has ensured that as long as Europeans broadly vote for retention of ties with France, and are supported by the Asian and Polynesian electorate, militant Melanesian (Kanak) demands for independence are wholly unlikely to be satisfied through the ballot box. As Kanak nationalists increased their pressure for independence and conservatives strengthened their opposition, the population of New Caledonia became still more polarized in ethnic terms. The sole central reformist party, the Fédération pour une Nouovelle Société Calédonienne (FNSC), lost most of its support, and any lingering European support for the UC disappeared, leaving demands for independence almost entirely in the hands of Kanaks. On the east coast and in the islands Kanak frustrations increased and electoral geography combined with population and economic geography to contrast Noumea and the west coast with the east coast and islands, the latter regions being Melanesian and hence substantially and increasingly pro-independence.

Despite the amalgamation of the pro-independence parties into a loose Front Indépendantiste coalition, the pro-independence vote in elections never reached more than 40 percent. Kanak electoral frustrations led to more direct action, and there was confrontation between Kanaks and extremist right-wing groups. As other parts of Melanesia (especially nearby Vanuatu) became independent and a Socialist government took power in France in 1981, there was renewed Melanesian hope that independence would no longer be denied to New

Caledonia. However, success through the ballot-box was no more likely, there was strong opposition to Kanak independence, and no sign that France intended to move towards independence for New Caledonia. For a brief period the Front Indépendantiste, in coalition with the reformist FNSC, gained power in the Territorial Assembly, promoting right-wing reaction, alarmed at a Socialist government in France and a minority government in New Caledonia. Tension and violence mounted and there were further confrontations in several areas. France attempted to devise a new statute for New Caledonia; the statute satisfied few and Kanaks, angry that no electoral reform was proposed (to disenfranchize recent arrivals and enable them to achieve a majority) and that there was no firm timetable for independence, came together in 1984 in a new coalition, the Front de Libération Nationale Kanake et Socialiste (FLNKS), to demand independence for the state of Kanaky. As the Kanak position hardened into confrontation, the conservative position also became increasingly extreme, and new right-wing parties, the Front National, allied to Jean-Marie Le Pen's French party, and the Front Calédoniene, emerged to oppose the FLNKS. The stage was set for further confrontation.

THE STRUGGLE FOR KANAKY

The emergence of the FLNKS heralded an escalation of conflict as Kanaks abandoned the unbalanced struggle for constitutional change, ignored the French government and embarked on direct and violent action to secure Kanak independence. The FLNKS boycotted the November 1984 territorial elections, and through roadblocks and the destruction of ballot boxes ensured that the boycott was 'active'. The RPCR inevitably won the election. The FLNKS undertook more direct action (barricades and the occupation of town halls and gendarmeries), briefly held the small town of Thio, and declared a provisional government of the Republic of Kanaky. Jean-Marie Tjibaou (of UC) was declared President. Violent conservative reaction followed. Ten Kanaks were killed in an ambush and ten more Kanaks and Europeans were killed in various incidents. At the end of 1984 the French government sent Edgard Pisani as a new High Commissioner to devise a peace strategy. The essence of Pisani's proposals was for independence in association with France, with France retaining control of defense and foreign affairs, French citizens having special status, and New Caledonia moving to independence in 1986 if a referendum was approved. The proposals attempted to reconcile three conflicting interests: Melanesian claims to independence, the rights of French settlers (*Caldoches*), and French strategic interests. The FLNKS were unenthusiastic about a 'neo-colonial' solution, with the franchise barely changed, and withdrew from negotiation after the death of their most militant leader, Eloi Machoro, Minister of Security in the provisional Kanaky government, killed by military sharpshooters on a raid on a farmhouse near La Foa. The military presence was strengthened, right-wing opposition to Kanak militancy grew and, without French or urban support, Kanak militants were unable to gain power. Pisani's proposals were ignored; he was later recalled to Paris, and the French Prime Minister, Laurent Fabius, devised new proposals which divided New Caledonia into four regions, each with its own council responsible for a range of development planning issues. The FLNKS eventually accepted the basis of this plan, though they continued to distrust the French government, believing that they would win two and perhaps three of the four regions. Conservatives, increasingly well-armed, were concerned about these developments but confident that in March 1986 a more conservative government would gain power in France and abandon the 'conciliatory' mood of its socialist predecessor.

As the French government stepped up its military presence and the RPCR organized private militias, so the FLNKS withdrew from violent action and sought to develop a more self-reliant Melanesian society and economy in rural areas, in preparation for regional councils and eventual independence. Kanak people's schools (EPK) were established and cooperative agriculture was encouraged in a futile bid to destabilize the economy of Noumea. In the elections for the regional councils in September 1985, FLNKS won three of the four regions, though RPCR won so comprehensively and predictably in the Noumea region that it retained control of the Territorial Congress. The election was fought solely over the issue of independence and, in the highest turnout in history, some 38 percent of the voters, mainly on the east coast and in the islands, supported independence, again demonstrating the improbability of independence being gained through the ballot-box without some drastic restriction to universal suffrage.

Throughout the past decade a significant proportion of Melanesians, variously estimated at between 15 and 25 percent, have opposed independence and supported the RPCR or other conservative parties. These Melanesians were opposed to a Kanak independence that could be construed as racial, opposed socialism and communism (which they feared independence would degenerate into), and were concerned over their jobs (often in the bureaucracy) and the privileges that might be lost after independence, when overseas aid dried up, and so preferred stability, even if it was not wholly satisfactory. Such Melanesians were predominantly from the Loyalty Islands (especially Lifou), the Isle of Pines, and Noumea, and were more dependent on economic achievement and modern sector employment than many of the Melanesians of the Grande Terre. Others were older people who simply sought stability. In many places, especially Lifou, political divisions followed long-standing social divisions that had been maintained by different religions, Catholic and Protestant, and ensured continuous tension and occasional friction.

In New Caledonia there was concern within the FLNKS that the minimal gains of the regional councils would be obliterated and that the possibility of independence in association with France, which tenuously remained on the Socialist government's agenda, would be canceled out. In the run-up to the 1986 French elections the constituents parties of the FLNKS were divided over whether to compete, but eventually boycotted the elections on the grounds that 'the French elections are the concern of France, not Kanaky'. Differences of opinion maintained existing divisions, especially between FULK and the more conservative UC, which had first surfaced over ties with Libya. Yann Celene Uregei, leader of FULK and Foreign Affairs spokesman of FLNKS, visited Libya in April 1986 to attend a congress of world liberation movements. The FLNKS censured the visit and temporarily suspended Uregei from his official position within the FLNKS. Agreement to support the boycott smoothed over the divisions within the FLNKS but could not disguise the differences of opinion, in relation to policy and practice, born of frustration over the inability to achieve power. For more than a decade Melanesian attitudes had become more radical. Though based on regional and religious variations the different policies of the pro-independence parties also reflect the different speeds at which different groups have reached radical nationalist positions and the ideologies and attitudes of their leaders.

The March 1986 national elections were boycotted by the FLNKS and the conservative RPCR-Front National coalition swept the territory, gaining 89 percent of the vote, though the participation rate was no more than 50 percent. The detailed election results demonstrated an almost identical pattern to that of all earlier elections over the past decade, with a massive conservative vote around Noumea and in most other parts of the west coast, and very low participation rates on the east coast, where Melanesians and support for the FLNKS are concentrated. The effect of the boycott was not only that Jacques Lafleur was re-elected but the RPCR gained a second seat in Paris, that of Maurice Nenou, a Melanesian from Poindimie, at the expense of losing its seat in Paris, and thus its sole constitutional representation outside New Caledonia. In France Jacques Chirac and the conservative parties triumphed and a new era now dawned.

Though it had been hitherto assumed that Chirac would not immediately attempt to revert to the situation before the socialist government's limited reforms, for fear of provoking extreme Kanak reaction, the proposals that Pons announced in Noumea in April 1986 demonstrated that he had done exactly that. Although the regional councils remained in place, their funds were effectively frozen by the new concentration of power in the hands of the Territorial Congress (controlled by the RPCR) and the French High Commissioner. The shift of authority necessarily removed any possibility of the FLNKS regional government adopting radical initiatives towards developing viable regional economies. A further package of economic measures was proposed that, whilst formally directed at reducing unemployment by encouraging economic growth, effectively tied New Caledonia yet more firmly to France. Pons made no reference to any Melanesian demands, either political or economic, specifically abolished the Lands Office (although a land reform program officially continued), merely commenting that 'New Caledonia is French because its inhabitants wish it to be', a clear denigration of the historic basis of Kanak nationalism.

The prospect of further migration to New Caledonia was also raised. Troop numbers were increased and some of the 6,000 troops in New Caledonia were dispersed throughout the countryside, in a policy of nomadization previously practiced before independence in Algeria, a situation which augered badly both for peace and for future Kanak independence. Regiments of marines patrolled rural areras, maintaining a military presence, providing such technical assistance as road construction, and contributing to tensions. The immediate result of these changes was that the FLNKS effectively lost almost all its limited power in the regions, the only places where it had legal and constitutional authority, and was constitutionally reduced to an ineffective minority in the Territorial Congress. Its minor achievements had largely disappeared, though its support had not been eroded. What the moderate leader of the FLNKS and UC, Jean-Marie Tjibaou, had termed the 'green revolution' – the aim of building Kanak independence at the grassroots level while giving non-Kanaks a vision of this Kanak independence (Fraser 1987, p. 28) – had been effectively aborted, though the councils had briefly demonstrated their ability to provide responsible government. Militant goals had temporarily given way to more moderate pragmatism, but Pons only gave the experiment a brief life. Tjibaou consequently argued that a 'new strategy' must be adopted, since New Caledonia had moved into a situation like that of Algeria before independence, and consequently sought further support in Europe and elsewhere. With reference to the proposed referendum he stated: 'The people concerned by independence are the Kanak people. The French are independent. I don't think it is necessary to consult them or know whether they wish to remain French or not.' The FLNKS recognized that under Chirac their limited range of options, and their minimal prospects for gaining independence, had been effectively destroyed as the new government rejected the FLNKS as the legitimate representative of Kanaks.

The Lands Office was abolished but the land reform program continued. Though land was of emotional, symbolic, and economic importance to Kanaks it was also of crucial concern to *Caldoches*. For many settlers, the emotional ties with the land, developed sometimes over three or more generations, were strong. The FLNKS, under the guidance of Eloi Machoro, had sought to accommodate settlers' ties of affection with the land by offering long-standing and sympathetic settlers the freehold of their homestead and some nearby hectares for their lifetime, in addition to a leasehold of the remainder of the land they had farmed. Some settlers who had grown up in amicable relations with neighboring clans and understood their wish to recover the land from which they had been dispossessed did in fact enter into such arrangements. Under the Mitterrand government these arrangements were in the process of being legally ratified under the land reform laws. Under the Chirac government this was stopped and the land returned to Kanaks was mostly given in state leaseholds, not freeholds, with efforts being made to return settlers to land which they had quit under Kanak pressure. Moreover the structure of transfer had dramatically changed; in 1987 more newly-purchased land board (ADRAF) to Europeans, mainly around Bourail and La Foa. This was an extraordinary situation, given the existing pattern of land ownership, where 1,500 settlers already held more land than 35,000 Kanaks, and one that produced obvious Kanak concern and resentment. Nonetheless land issues are no longer quite so important as they have been in the past, in part because of migration out of the agricultural sector, and in part because of the range of political issues that now compose the agenda for independence.

The French government moved forward with plans to hold the referendum that Pons had promised in 1987 and the proposal was finally approved by the French Senate in May 1987, in the face of socialist opposition. The sixth annual FLNKS Congress predictably voted to boycott the referendum in order to 'destabilize the strategy of the colonial government' and subsequently embarked on a series of pre-referendum protests leading to strong repression from the French riot police against a peaceful, if technically illegal, Kanak demonstration, a ban on a march from the east coast, and press censorship. Tjibaou argued that 'to take part in another statute of autonomy is to walk on the bodies of the dead' (Fraser 1987, p. 31). Predictably the referendum, in September 1987, resulted in 57 percent of the electorate voting in favor of remaining with France. Almost all the remainder of the electorate did not vote. Once again, but for the first time in a referendum, the population had firmly voted against decolonization.

Not all the Melanesian population (who make up about 45 percent of the electorate) heeded the FLNKS call for a boycott. Perhaps 20 percent, at most, voted for France. In the northern strongholds of the FLNKS, like Belep, Pouebo, and Hienghene, site of the violent massacre of Kanaks in December 1984, very few Melanesians voted, though in Lifou and the Isle of Pines there was a high Melanesian participation rate. By contrast almost all Europeans and Polynesians voted against independence. The French Polynesian population, at least around Mount Dore and Noumea where their main concentration lies, believed that failure to vote would mean that the administration would regard them as independence supporters and that this could entail some form of reprisal. The referendum results emphatically demonstrate the extraordinary geographical divide that increasingly characterizes New Caledonia and is virtually a mirror image of the population distribution: a west coast, and especially Noumea, almost the 'white town' that graffiti proclaimed it to be, characterized by a predominantly European (and Asian and Polynesian) 'loyalist' population and an east coast and islands largely dominated by the minority Melanesia 'independentist' population. Growing polarization of the population and a high mountain range emphasize increasingly acute political divisions.

After the referendum nothing had really changed in a political sense; Kanaks remained convinced that though the struggle might be long, history was Tjibaou has maintained: 'As long as one Kanak remains alive a problem for France remains' (quoted in Connell 1987a, p. 445). Certainly the demand for independence was not meekly cast aside. In this acutely and bitterly divided territory France began to attempt what was intended to be a policy of reconstruction and reconciliation. Though Chirac declared in New Caledonia that 'it is on this empty page that I invite you to start to write your future', it is the history of New Caledonia that is so crucial to contemporary development. Many walls have slogans stressing that the present struggle is exactly that of the bloody 1878 revolt, the most violent in the island Pacific, now partly mythologized as a colonial war that sought to turn the tide of French colonization; in this history of violence and alienation was the genesis of land disputes and the birth of the nationalist movement. In the aftermath of the referendum President Mitterrand, speaking from Paris, emphasized the historic divisions, opposing 'injustice' and noting that there were substantial inequalities in income levels, access to jobs in the bureaucracy and, above all, land ownership, between Melanesians and *Caldoches*. In terms of quality of life, Melanesians have a life expectancy of 59 compared with 73 for Europeans and, in the rare cases where health statistics are recorded by ethnicity, their health status is significantly below that of Europeans, and their access to most basic needs (Garraud et al. 1986, pp. 2047–50; see also Connell 1987a, pp. 162–9) is also less adequate.

Soon after the referendum the French Minister for Overseas Territories, Bernard Pons, introduced a new statute for New Caledonia, a clear indication of Chirac's determination to remove the four regional councils, established by the socialist government, which had given some degree of power to Kanaks in three regions. Executive authority was transferred from a High Commissioner, responsible to paris, to a ten-member Executive Council, consisting of the Presidents of the four regions and six other members elected by the Territorial Congress. The power of veto would, however, remain with the High Commissioner. The boundaries of the regions were redrawn to produce new regions so that FLNKS would be likely to control only two rather than three regions, and communes were given extra responsibilities. Though undemocratic, in that much more than half the population (in Noumea and the South) had only one council, the regional councils enabled Kanaks to play some formal part in the development of the rural areas in which they lived and suggested the promise of future constitutional evolution. The demise of the regions, alongside Kanak electoral boycotts, left Kanaks powerless and with little incentive to take part in a reconstruction devised in Paris and determined in Noumea.

Other elements of the new statute also threatened Kanak interests. There would no longer be a Council of Customary Chiefs, but there was to be a new Customary Assembly of nominees who were not traditional leaders. Melanesians would no longer have any distinct civil status but would be entirely subject to French law and traditional land rights (the basis of the clan system) were to be radically transformed. Such threats to land and society were bitterly resented. Tjibaou accused France of 'cultural genocide', noting that the statute 'would mark the end of the Kanaks as people'. The proposed change in civil statute implied that Kanaks seeking independence would now become an illegal secession movement, comparable to Basque and Corsican groups, indeed the 'separatists' that conservatives called them rather than 'independentists'. Tjibaou consequently called for opposition to the statute which offered no concessions to Kanak aspirations, and the FLNKS refused any dialogue on the terms of the statute. The Seventh Congress of the FLNKS in February 1988 vowed to oppose the Pons statute and, in a press conference after the Congress, Tjibaou concluded that the future will bring 'a strong and determined mobilisation, as a function of a different balance of forces to that of 1984. The responsibility for working out the ways and means belong to the Struggle Committee in each commune' (quoted in *Kanaky Update* 1988, pp. 3–4). Once again, acute frustration and despondency had brought the FLNKS to a position where it appeared that only a violent struggle could convince France of the gravity and legitimacy of their claims and again draw the attention of the world to one of the last inconclusive struggles for independence. Only one channel of dissent and debate was left open.

In the lead-up to the May 1988 Presidential elections, Leopold Joredie, the General Secretary of the UC, warned that New Caledonia was 'sliding into an Algerian-type situation' and that 'the government will have to face the consequence of what will happen in the territory next month'. Thus warned, but without any information on what 'muscular mobilisation' might imply, the French government took precautions. It was to little avail. On 22 April, two days before the elections, a commando group of Kanaks made a dawn raid on the gendarmerie at Fayaoue on Ouvea island, killing four gendarmes and taking 27 hostages. Some were released a few days later but other police and officials who subsequently arrived were taken hostage, and held in a coral cave. Kanaks then made demands for the cancellation of the Pons statute and the regional elections, the withdrawal of the military from the island, and a new United Nations supervised referendum. The hostages were deemed 'prisoners of war' and it was claimed that they would be held for six months.

Conservatives in New Caledonia were incensed at the turn of events. Jacques Lafleur, President of the RPCR and *député* in the French parliament, who had already called for the dissolution of the FLNKS, accusing it of 'terrorist' methods and describing its members as 'subversives', repeated the demand. Pasqua and later Pons called for Mitterrand to dissolve and ban the FLNKS and Chirac spoke of 'savagery and barbarism'. Dick Ukeiwe, the most prominent Melanesian in the RPCR, a Senator and now President of the newly elected Territorial Congress, declared that the 'Kanaks who are holding the hostages should be treated as outlaws'. Justin Guillemard, founder of the Patriotic Action Committee of New Caledonia and a leader of the local Front National, claimed that he would rather die with a gun in his hand than submit to the Kanaks. Guy George, also of the Front National, pointed out that if Mitterrand won the elections there would be civil war, since settlers had well-organized 'self-defense groups to meet fire with fire'. The right therefore appealed for adherence to the law, stressed the necessity to give the new Pons statute a chance to succeed, campaigned against the election of Mitterrand, and sought to detect Libyan involvement in New Caledonia. Mitterrand refused to respond to the right-wing, before or after the first round of the Presidential elections. Jean-Marie Tjibaou, interviewed shortly after the hostages had been captured, observed:

82

My first reaction is that it is saddening to observe the results of the partisan, cynical and despicable policy of the RPCR . . . This Pons statute is the final touch in a system of refusal to take the Kanak people and its claims into consideration. And this has been ordered by the local clique which has grabbed the land, hunted the Kanaks from their own homes, taken control of the mines and of commerce. It is these people who are setting themselves up to give us a lesson and who made no such fuss over the murder of 10 people at Hienghene . . . They must bear all the consequences for what is happening, just as they bear the consequences of 1878, 1917, the deaths at Hienghene, of Eloi Machoro and the others. Attention must not be drawn away from the real problem.[3]

Once again the FLNKS sought its legitimacy in history and traced the history of nationalism back over 110 years. Violence was attributed to those who, over the decades, had provoked this 'colonial war'.

As the hostages remained in Ouvea the first round of the presidential elections went ahead, disturbed by a series of violent events in different parts of New Caledonia. Ouvea was not therefore an isolated incident but was probably the catalyst for further actions. Other local 'struggle committees' effectively forced the evacuation of Canala, barricaded many highways, and later executed Jose Lapetite, one of the defendants of the Hienghene massacre. Police were fired on, a Melanesian woman was killed by a stray bullet at Canala and pro-France Melanesians were harassed. The French government continued to refuse to yield to the demands of the militants. Just two days before the second round of elections, with Mitterrand well ahead of Chirac, crack army units stormed the cave, rescued the remaining 23 hostages and killed 19 Kanak militants, including their leader, Alphonse Dianou. Two soldiers were also killed. Once again violent events had led to the deaths of many more Kanaks than any other ethnic group and, as the evidence trickled in that three militants were killed after their capture, napalm bombing had been contemplated, and that the hostages were almost certain to have been released after the second round of the presidential elections, Kanak anger and resentment resurfaced. In the immediate aftermath of the deaths a FLNKS press release promised that 'neither deaths, tears, suffering or humiliation will stifle the cry for freedom'. Bitterness was the dominant sentiment.

The calculated political expediency of the raid failed to convince the French electorate which returned President Mitterrand with a substantial majority. The re-election of President Mitterrand under the slogan 'La France unie' brought an immediate end to the hard-line confrontationist policies of the Chirac era. Michael Rocard became Prime Minister, following Chirac's resignation, and the uncomfortable era of cohabitation was ended. Tension remained high though there was increasing dialogue between the French government and the FLNKS. The right-wing preferred their own form of comment on the new dialogue. A Front National rally, stressing their new slogan, 'we have two choices, our suitcases or our guns', burnt a number of suitcases to indicate the choice that they had made, and emphasized their own historic ties with New Caledonia. Their leader, Guy George, stated:

> In New Caledonia our roots have been irrigated with the blood of our dead. Our roots have germinated in every plain, in every house and we will not let the FLNKS pull them out. The criminals who have lit the inferno in New Caledonia are among those who wish to surrender old Europe and the world to the monstrous dictatorship of Soviet Communism, sharks who are already in our waters and waiting for their prey (Connell, 1988, p. 17).

The words were a deliberate echo of Kanak works and the sentiments were shared by many outside the Front National itself.

Early in June, Rocard achieved further diplomatic success by holding talks in Paris, attended by both the FLNKS leader, Jean-Marie Tjibaou, and the RPCR leader, Jacques Lafleur, the first time that the two had met officially since 1983. These talks eventually led to the signing of the Matignon Accord, which transformed the basis of political change in New Caledonia. The key elements of this were, first, to establish three new regions and regional councils, second, to focus economic development strategies on Melanesian rural areas and, third, to have a future referendum on independence in 1988. Finally, there would be a general amnesty for those (Kanaks, Europeans, and others) convicted of all crimes not involving bloodshed. Rocard later described the set of proposals as 'decolonization within the framework of French institutions' (Connell 1988, p. 17).

The proposal to establish a new set of regions again represents a movement to return some degree of power and authority to Kanaks in the regions. Three-quarters of new government investment would go directly to the rural regions and only a quarter to the South. With political power, and significantly increased levels of finance, it should be possible for Kanaks to move much closer to the 'green revolution' in two-thirds of the territory, establishing a version of a post-independence Kanaky political, economic, and social structure. There are a number of views of what the result of this kind of decentralization might be. First, this delegation of authority and finance should enable the FLNKS to consolidate development in prominently Melanesian areas. Much of the new French financial support is directed into training Kanak bureaucrats; without the policies of localization that have existed elsewhere in Melanesia, Kanaks are conspicuously absent from the higher echelons of the public service. 'The training and placement of Kanak cadres must surely be a pressing priority for a new government. For it is in giving the settler community the direct experience of sharing power with, and being governed by, Kanaks that independence becomes a viable option.' This was also the optimistic view of those in the FLNKS, like Tjibaou, who saw in the new regions the opportunity to demonstrate their legitimacy and effectiveness in power. A second view is that if further substantial economic resources are diverted from Paris to rural New Caledonia, an even more artificial economy would be created and many Kanaks will have achieved positions of power, status, and high income in the emergent economic system, so creating more Melanesians (like their predecessors from parts of the Isle of Pines and the Loyalty Islands) who are no longer interested in independence, especially with certainly reduced foreign aid levels. Jacques Boengkih, Director of the Kanak Agency for Development in Sydney, feared that the creation of a Kanak middle class would lead this class to identify their interests with continued French rule, since the '16 per cent of Kanaks' who were already pro-French had gained from public service employment and other sorts of patronage. Third, based on the precedents of failure and disappointments in economic and political development in other parts of Melanesia, closely monitored and catalogued by the sole daily newspaper, *Les Nouvelles Calédoniennes*, it is widely believed by conservatives, including Jacques Lafleur, that Kanaks would so misrule loosely autonomous Kanak regions that a decade under this rule would show Kanaks the benefits of the French flag. There is certainly absolutely no guarantee that this limited autonomy, in the part of the territory where Europeans are a minority and the economy is more agricultural, would encourage its wider acceptance in the other heavily populated region of Noumea and the South (Connell 1988, pp. 17–18).

The Matignon Accord was thus similar to the earlier French national policies that have sought to effectively 'recolonize' the island on a more secure basis with a new urban-industrial growth pole at Poindimie, a free trade zone, and a more modern transportation system,[4] policies that can only incorporate a dependent island territory more firmly into the metropolis. Whereas the previous FLNKS strategy, repeated at regular intervals over the past decade, was to set an independence date and then seek to construct the basis of an independent state before that deadline, the accord reversed this procedure and in so doing reduced the probability of independence. The specifics of the referendum also generated considerable debate and there was serious concern in the FLNKS that, at the very least, independence had been postponed for a further decade, and this delay was unacceptable after so much previous discussion, violence, and bloodshed, in which the victims had been predominantly Melanesian. Ten more years, a long time in politics, seemed to represent minimal commitment to the principles and practice of decolonization. The FLNKS thus sought firm guarantees that the plan could not be overturned with a change of government. The signing of the Accord brought a period of peace to New Caledonia, peace that was rudely broken in May 1988 when the two leaders of the FLNKS, Jean-Marie Tjibaou and Yeiwene Yeiwene, were murdered at Ouvea, during a custom ceremony to mark the end of one year's mourning for the victims of the Ouvea violence. The two leaders were shot by Djoubelli Wea, a local pro-independence leader who was opposed to the signing of the Accord and concerned that independence had been postponed indefinitely. The FLNKS, already fragmenting into its coalition components because of disputes over the Accord, was shattered by the blow and took almost a year to choose a new leader. The demand for independence was fragmented and weakened.

LAND, LIBERTY AND IDENTITY IN NEW CALEDONIA

The genesis of nationalism and the struggle for independence in New Caledonia were rooted in the past, primarily in the struggle for land rights. Throughout the struggle for independence land issues were at the fore, though land was of greater importance psychologically than economically. As Tjibaou has suggested

> One cannot stress too much the importance of land for any tribe . . . The Melanesian environment is not only the agricultural land or the land to which clan history is tied, it is one of the basic elements of the whole of societies. Alienation of land has not only displaced tribes, but left them wholly disintegrated. A clan that has lost its land is a clan that has lost its personality (Tjibaou 1976, p. 285).

Land claims steadily expanded, eventually to incorporate the whole of New Caledonia; because 'ancestral land is the basis of the Melanesian personality, every lineage must be re-established where it was at the time of European contact' (Doumenge 1983, p. 441) though there was never any intention that Europeans would be displaced from this land or that Melanesians physically return to their ancestral land. Ownership of land was the key.

Even where land has been lost for more than a century, Melanesian society remains 'written on the ground' (Leenhardt 1979) and mountains, streams, and rocks bind the ancestors and the living together into an inseparable whole. The link with the ancestors ensured that land was also perceived as being thoroughly involved with the blood of the dead. The sacred and secular relationship with the land is a continuous thread throughout Melanesian history. The massacre at Hienghene in December 1984 was, in many respects, the outcome of a lengthy struggle for land and, after the subsequent death of Eloi Machoro and Marcel Nonnaro, Tjibaou commented 'these are not perhaps the last whose blood will enrich the land of our ancestors'. Even were Melanesian land claims to have been considered more sympathetically by the French administration, no effective response could have met demands based on more than a century of colonial history and millennia of oral history.

The combination of blood and land was nowhere more apparent than in the 1878 revolt. Atai, 'has attained legendary status among Melanesians seeking independence, and the 1878 revolt has become their most powerful symbol' (Douglad 1980, p. 22), apparent in the title of a recent book by the Melanesian priest, Apollinaire Anova-Ataba, *D'Atai a l'indépendance* (1984). In death, more than in life, Atai has achieved fame and renown, increasingly becoming a symbol of struggle, a man now more myth than reality. In the manner of Atai's death symbolism was ensured, an inevitable symbolism in a society where ancestors were revered as the links between land and life. Although Atai and other Melanesians of that era were primarily traditionalists seeking to re-establish intact a disrupted order and recapture the past, their considerable unity in defense of Melanesian identity and territorial integrity has ensured strong connections between these 'primary resistance' movements and modern nationalism (Ranger 1968, p. 437). Contemporary Kanak nationalism has succeeded in achieving mass commitment partly by using similar methods (appealing to Melanesian consciousness, stressing land losses and so on) and partly by appealing to memories of this historic revolt against colonialism, where Melanesians were far from passive spectators in their colonization.

As in other nationalist movements and even perhaps in Atai's struggle a century earlier, there has been a simultaneous appeal to both progress and tradition, modernity, and a 'Melanesian way' combined. Through tradition, Melanesian identity can be reclaimed and, through modernity, reinvested with vitality and relevance. In this there is necessarily some biblical or even millennial element: 'The image of nationalism is that of restoration and renovation, of return from spiritual exile to the promised land. The past has not been in vain. Out of its sorrow and tragedy the nation has rediscovered itself and seeks its rebirth' (Smith 1977, p. 42). Nationalism promises a new social order, based on customary legal precepts, in which a better life might be achieved free from European or other alien administration. Tradition is the ideological arm of those who defend a threatened status quo or those who seek to revert to real or imagined past where 'tradition' prevailed alone. It is therefore a vital and valuable resource for those who support collective action and resist the encroachments of an external authority. Alienation is met with the reconstitution of the past and the restitution of identity. There is then a blending of the old tribalism, based on social structure, with a new and more anonymous nationalism based on shared culture, which partly supersedes tribalism. Resistance to alien authority has resulted in a degree of populism: an anti-urbanism with *coutume* (tradition) and agricultural work being invested with positive qualities, as resistance has led to a reaffirmation of Melanesians values. Yet, at the same time, there is no hesitation about developing a new social and political regime with many innovations and institutions consciously derived from European society. It is an illusion to pretend that Melanesian cultures could ever regain the integrity and self-reliance that they once had; 'the return to tradition is a myth' (*Les Temps-Modernes* 1985, p. 1601). Many Melanesians, say nothing of *coutume*, are distrustful of a past history of social inequalities, and anxious to progress forwards, not backwards. There can be no wholesale reversion to tradition, more a transformation through tradition: 'Our identity is in front of us' (*Les Temps-Modernes* 1985, p. 1601).

What such a transformation would entail is necessarily unclear; Melanesian societies are now different from precontact times, a result of modernization, education, Christianity, and migration. Traditional authority, the chiefly system, is still respected by the young as, more generally, is the authority of the old. Other customs, such as parental choice of marriage partners, are widely seen as outdated, though male superiority is maintained. Traditional values are more vibrant in the larger, more homogeneous Loyalty Islands and less apparent among the urban employed and school youth. Yet, at the same time, Loyalty Islanders have long had better access to urban jobs, high incomes, and some degree of modern status.

It is very much out of uneven development in New Caledonia that nationalism has emerged; uneven development, especially in the post-nickel boom years, is regional and socioeconomic and, though classes are no more than incipient in Melanesian society, social differentiation is sometimes substantial. At the same time, the fact that the vast majority of Melanesians were subordinate to Europeans, ensured that under colonialism Melanesians themselves were virtually in the process of becoming a class. The conjunction of class and emergent nationalism is basic to the struggle for independence. Such modern divisions cannot be legitimized within old cultural contexts; they therefore create an illusion that consequent tensions and uncertainties can be remedied by granting 'national' independence.

Christianity has influenced the lives of all Melanesians and the course and structure of the independence movement; Tjibaou, for example, stresses both his Melanesian identity and Christian heritage even though 'the Catholic church in Caledonia is opposed to our struggle' (*Les Temps Modernes* 1985, p. 1599). Indeed there is no perceived antithesis between Christianity and custom, as there is in parts of Vanuatu. Linguistically, Kanaky could only remain Francophone. Two major colonial institutions, language and religion, would be unchanged. Though *coutume* is a powerful ideological tool, its practice is viewed very differently from those who see any loss of modernity (wages, imported foods, clothes, television, cars, and so on) as an unacceptable withdrawal and reversion, to those who genuinely seek to build a more self-reliant and agricultural society, organized according to custom on a more communal and egalitarian basis, and those few who anticipate a genuine 'socialism in one country' solution, with withdrawal from the world capitalist system. Differences are considerable, cutting across a range of conservative and radical parties, united in the struggle against colonialism, but likely to lead, as in the Melanesian states to the North, to a system that has scarcely broken with the colonial heritage in society, economy, or politics.

In order to build a new society for an independent Kanaky it is argued by most Kanaks that it would be necessary to construct a new order where elements of tradition might co-exist with the best of the modern world. The principal indication of the different organizational structure in Kanak is in the schools and cooperatives that now exist in parts of the countryside. Kanak schools are a definite rejection of a colonial educational system, though, other than in encouraging greater respect for Melanesian custom, languages, and life, they may not have a real influence on the direction of change. The establishment of a Melanesian education system throughout New Caledonia, divided between many Melanesian language groups, and with a majority of the population knowing no Melanesian language poses massive problems, far beyond those in the Melanesian countries to the North, where education systems broadly remain as they were inherited from pre-independence days. The objective of Kanak schools to maintain and develop Kanak identity through an emphasis on Kanak language and culture presents real paradoxes, and appropriate education, through Melanesian languages, may remain a chimera (Néchéro-Joredié 1985, pp. 198–218).

Cooperatives, on the other hand, have been supported by Melanesians for more than a quarter of a century and have grown in recent years throughout New Caledonia. The independence movement has stimulated commercial activities, such as cooperatives and stores, which have led to reflections on the most appropriate form of rural

development for the Melanesian environment. Chosen rural development strategies are often orientated to maintaining traditional social
ties, especially communal ownership of land, to oppose tendencies
towards individualism. Yet, in attempting to resolve these problems,
Kanak independence parties have run into the same kinds of organizational problems that they faced earlier. Throughout Melanesia cooperatives have gone through phases of growth and decline and many
have been replaced by a more individual form of production and
entrepreneurial activity, just as land tenure, under pressure from long-
term cash cropping and population increase, has increasingly moved
from a more corporate to a more individual form. In short, throughout
the Pacific, there has been a movement towards the establishment of
both a peasantry and a proletariat, accompanied by the withering of
cooperative organization. The movement towards a more capitalistic
organization is seemingly inevitable. Both Kanak schools and cooperatives combine idealistic and real responses to the problems of colonialism and capitalism. In opposing the present structure they have an
important role to play, but in an independent Kanaky their role might
quite quickly disappear. The practice of development may well turn
idealism into irrelevance.

Land tenure presents a similar situation. Despite the fervor and
directness of land claims, concern is essentially about title rather than
about occupation and use of the land. Though some rural land would
revert to Melanesian use, as most regained land is currently doing,
and minor migration movements would restore some Melanesians to
their traditional lands, establishment of title would probably be followed by most land being hired out to its present users: those *Caldoches*
who chose to remain as tenants of Melanesian landlords, a wholly
capitalist outcome. If land is crucial to the struggle for independence,
it is less crucial to its outcome. Almost a quarter of all Melanesians
live in Noumea, and for many the future is urban. In a territory
dominated by capitalist enterprises, in a region dominated by capitalism, albeit at the periphery of the global economy, where French
institutions (missions and schools) structure a part of social life, it
would be surprising if some Melanesians had not found a substantial
and apparently permanent stake in the capitalist system, either in
commerce or the bureaucracy. Such men have achieved power, status,
privilege, and wealth that might have been denied them in the Melanesian world. A few, secure in their positions in a lavishly inflated
bureaucracy, are hostile to the independence movement that offers a
more uncertain future, and unwilling to reject the colonial system
through which their advancement has been secured.

The future economy of Kanaky, as perceived by the FLNKS, would
be virtually unchanged. Indeed it is potentially divisive within FLNKS

to discuss the structure of post-independence economic change, especially when that future remains remote. Effectively the FLNKS has called only for greater Melanesian participation in the economy without any structural transformation. There is little radical socialism here. As Tjibaou has suggested

> We Melanesians are only socialists in terms of the traditional principles of ownership and reciprocity. If you have something it must be shared. We are not opposed to institutions that make profits and benefit the whole of society . . . Currently we lack economic models in every sphere (Tjibaou 1983, p. 16).

Such views based on Tjibaou's concept of traditional life where 'prestige is in giving, giving much and giving widely: the opposite of the capitalist world'. (Tjibaou 1981, p. 87). As Edgard Pisani well recognized, 'the socialism of Monsieur Tjibaou is a communal socialism, a socialism close to the earth, a socialism based on tribes and clans. It is a kind of cooperative, an economic expression of social realities' (Pisani 1985), a socialism then that no one need fear. For the FLNKS, even with its internal divisions, socialism is about reactions and not revolution. Differences between Melanesians, and between *indépendantistes*, are as important as between themselves and Europeans. Those differences, in a small island state, and the lack of cohesive strategy for a post-independence era that remains elusive, ensure that approaches to economic change are broadly conservative.

NATIONALISM AND TRADITION

In New Caledonia, and throughout Melanesia, much of what is valued in traditional Melanesian life remains: a sense of community in association with a particular tract of land, shared beliefs and values, a rough equality of material conditions, direct and multifaceted relations between members of a community, reciprocity, and some degree of community control over the means of production. Such values, and virtues, which have promoted localized autonomy and self-reliance, are not transferable into the modern world as much of anything. This is a question of scale. Nationalist sentiment has grown through the inevitable failure of assimilation and the attendant growth to maturity of a Melanesian society with its own identity, in terms of a distant yet common ancestry, related languages (and a particular colonial lingua franca) and distinctive cultural and economic structures. In time this culminated in the demand that nation and state should coincide, the only means of giving true and full expression to nationalist sentiments. The pressures of the French colonial system have created a nation where one had never existed before. The barest fragments of this were apparent at the time of Atai's rebellion, and the legends and glorification of this century-old rebellion, allied to recent disappointments and frustrations, have forged a widespread awareness of a common political destiny. Despite contemporary differences within the FLNKS, and some Melanesian support for France, the desire to mobilize resources against French colonialism has created a unity and consciousness of New Caledonia or Kanaky significantly different from most other parts of Melanesia where centrifugal forces are stronger. Nationalism engendered the nation, returned tradition and history to Melanesians, legitimized the struggle for independence, and provided an identity in an alien world.

The demand for independence has been associated with such themes as cultural and ethnic identity, uniqueness, minority status, the desire for national freedom, and the innate right to independence: the hope of achieving the restoration of a 'degraded' community to its rightful status, dignity, and authority. The merit of *coutume* is thus defined with reference to a distant past, and invokes the notion of the past itself as more or less an inversion of the present, where what existed is now felt to be lacking. In New Caledonia this referred to a live and integrated practice, with Melanesian languages, traditional knowledge, and codes of order, and, in its most dramatic fluorescence, in the united and violent struggle against colonialism of 1878. In the struggle for independence material goals have played only a minimal part, hence policies for the development of an independent Kanaky have been given little consideration. There is no revolutionary ideology; expressions of commitment to 'socialism' are essentially the stimulation of a revolutionary spirit and fervor rather than a socialist practice that is unknown in the Pacific. Kanak socialism is a broadly-based mixture of populism and nationalism, partly based on the rejection of the contemporary structure of development, partly on an idealized vision of an imaginary past and an anticipated future, and partly on a genuine desire to combine the best of traditional and modern worlds.

The claims to unity that have characterized Kanak nationalism, and Melanesian nationalism in general, are relatively permanent features such as ethnicity (color), values, territory, language, and history. The FLNKS has used the racial dimension to heighten its appeal to all Melanesians. The stress on 'Kanak' (and indeed the restoration of the word itself from obloquy) within the FLNKS is essential to the struggle for legitimacy and the restitution of tradition, rather than the racism that its conservative critics elaborate upon. However, some supposed characteristics of unity are neither as factual nor as ubiquitous as the proponents of nationalism would prefer, hence history and tradition must be continually re-assessed and re-invented. The Great Revolt of Atai has become a 'colonial war' (Douglas 1985, p. 73), the same that is now being waged, where its local scale, the significant support of many Melanesians for the French cause, and doubts over the extent to which the objective was significantly 'political' are tacitly omitted. The violence of the past has given the nationalist struggle in New Caledonia a unique legacy of martyrdom. This legacy and the extent of land alienation have meant that although all 'the newly independent nations of Melanesia are developing post-colonial identity largely by resuscitating a pre-colonial past – on the one hand reaching back to a certain view of the past, while at the same time forging ahead with and instituting the full apparatus of the modern nation-state' (Philibert 1986, pp. 2–3), nowhere is this process more vivid than in New Caledonia, where the prospective rulers of an independent state have few means of taking over the present institutions of government. Consequently the past plays a much greater role than the indeterminate future.

Nationalism gains greatest legitimacy when ethnic boundaries do not cut across political ones (Gellner 1983, p. 3) hence it can only develop under conditions of greater homogeneity, where regional and tribal differences are welded together. In New Caledonia this fusion was less difficult than elsewhere in Melanesia, primarily because of a well-entrenched form of colonialism that engendered opposition and contributed to a common history, and partly because of more limited regional distinctiveness than in the larger Melanesian states. Nonetheless the number of language groups, and their separate identities and traditions, have hindered the emergence of identity; this is most apparent in the Loyalty Islands, excluded by geography and the absence of white settlement from participation in a common history. Throughout Melanesia a number of superficially similar characteristics are equally divisive; territory is an arbitrary function of scale, a common history is rarely shared, with friction and tension between local groups rather than cooperation and alliances, characterizing most of the history of Melanesia. Values are often individualistic and languages exceptionally localized. In other words, exactly the same characteristics that have suggested Melanesian calls for unity have proved to be the basis of regional micronationalist and secession movements. Nationalism, despite its apparent orientation to local personalities and local goals, is fundamentally a colonial and neocolonial construction and 'a mechanism of adjustment and compensation, a way of living with the reality of [uneven] historical development' (Nairn 1981, p. 334). Colonial rule emphasized ethnic nationalism by strengthening and enlarging ethnic allegiances, centralizing administration, bringing ethnic groups into wider fields of interaction and exposing them to various facets and resources of modernization. Consequently nationalism emerged in opposition to the incursion of colonial or peripheral capitalism and was an inevitable outcome of that, determined by its past rather than its present or future; 'romantic nationalism responded to the damage likely to be caused by modernism by providing a new and larger sense of belonging . . . new social ties, identity and meaning and a new sense of history from one's origins to an illustrious future' (Nipperdey 1983, p. 15). In different places, at different times, there has been a substantial 'culture of resistance' to colonialism; to enact 'custom' is at once to follow the rules of everyday life and to assert political autonomy (Keesing 1982, pp. 39–54). Indeed the various millennarian movements, or cargo cults, may be seen as responses 'to the crisis of indigenous society provoked by the disruptive effect of foreign capitalist penetration' (Barnett 1979 p. 776). As Jean Guiart has described them in New Caledonia they were the 'forerunners of Melanesian nationalism' (Guiart 1951, pp. 81–90).

Crucial to the construction of nationalism was the appeal to the legitimizing role of tradition, and the manipulation of that tradition to emphasize and establish common elements. In New Caledonia appeals to Kanak nationalism and *coutume* have not created a unity of purpose; a significant proportion of Melanesians (probably around 20 percent) do not regard the pervasive French presence as a form of domination but see it as normal, even welcome, to be French citizens in a distant ocean. Ethnic consciousness has not been transformed into national consciousness. Underlying the rhetoric of unity is a reality of distinctiveness and separation, bridged by alliances, trading ties and contemporary friendships, but separated by mutual suspicion and linguistic diversity. The pressures and attitudes that have fostered secession movements in the independent states are also apparent in New Caledonia, notably in the Loyalty Islands and especially in Lifou, where there was suspicion of islanders from the Grande Terre and firm beliefs that the other would gain a dominating role in an independent state.

The struggle for and achievement of independence in Melanesia abundantly demonstrates that, despite the concept 'nation-state', there can be no presumption that where there is a state there is necessarily a nation. Throughout Melanesia nationalism has been exceptionally weak, founded primarily on a basis of opposition to colonial (and neocolonial) political and, to a lesser extent, economic development (or nondevelopment) strategies. Attempts to engender a nation, through appeals to linguistic and ethnic unity, a common history and traditions, have been fraught with uncertainties, notably in that such attempts may be made at the level of quite different geographical regions. Indeed where geography has encouraged separateness and homogenity within an island, especially in the case of the Grande Terre, a greater degree of regional consciousness has emerged. It is the particular form of colonialism therefore that has constructed the particular variant of nationalism. It is in New Caledonia therefore, where the territory remains incorporated into France, and where an unusually rapacious colonialism (in terms of land alienation, mining, and the construction of reservations) has constructed substantial relative deprivation across ethnic boundaries, and fostered a degree of historical unity absent elsewhere, that Melanesian nationalism is most vibrant, though New Caledonia (officially 'new Caledonia and its dependencies') is scarcely more of a regional geographic unit than any other part of Melanesia. The similar history (of land alienation) in some respects ties the experience of the Grande Terre closer to that of Vanuatu than the Loyalty Islands, suggesting ultimately that the nation is an 'imagined political community' (Anderson 1983, p. 15) and that, within Melanesia, the construction of nations from within is fraught with difficulty. Similarities of analysis, in rhetoric and development plans, throughout Melanesia, point to the manner in which the independence movement in New Caledonia is essentially a Melanesian movement, largely uninfluenced by alien ideas and aspirations, that has continued to focus on land issues, cooperative and communal development strategies, the restitution of *coutume* and the necessity for a more appropriate education. Such aspirations, necessarily balanced against demands for control of the modern economy and local resources, demand the re-invention of tradition and the reconstruction of history. This is no more than has occurred over and over again in other similar contexts in Melanesia as the challenge of combining politics and tradition continues.

A COLONIAL DESTINY?

Though the political future of New Caledonia necessarily remains in doubt the prospect of independence has been deferred to the distant future and future social and economic change is almost certain to occur while New Caledonia remains an overseas territory of France. Power and authority are being restored to Kanaks; for some this will lead to a greater certainty that independence will follow economic development, but for others future economic links with metropolitan France may triumph over more nebulous political aspirations. For conservatives, most of whom are Europeans, no real political privileges have been lost, especially around Noumea, and there has been no economic decline, and no lack of subsidies, to discourage future residence in New Caledonia. In short, though the Accord has dramatically revised economic and political structures it has not disturbed key economic and political realities. It is not the first stage of a process of decolonization but rather a brilliant cosmetic change.

Independence in New Caledonia appears unlikely as long as France retains a global strategic vision in which its role as a nuclear power has strengthened its hold over French Polynesia. It gave up power unwillingly in Vanuatu (even though its presence was no more than half a condominium with the United Kingdom). This loss led France to strengthen its control in the region to ensure that this was not the thin end of the wedge for a wider decolonization. France has stepped up its military presence in New Caledonia in the past few years, and New Caledonia is seen as a critical link between other territories in the Pacific and Indian Oceans from which France might exercise a balance of power. France also has other technological, cultural, and ideological interests in the South Pacific region. The new Exclusive Economic Zones may even offer France a source of unexpected revenue in the South Pacific, which could dampen the prospects of independence even further (Aldrich and Connell 1990).

As France continues to subsidize the artificial economy at increasingly high levels, increasingly in the North and the islands, so converting New Caledonia more firmly into a 'consumer colony', more Melanesians are likely to move towards pragmatic anti-independence positions, especially as the disappointments of protracted and violent struggle mount. Conceivably young, educated, urban, and radical Kanaks may become more militant (Ward 1982, p. 68) but their numbers are still few. The FLNKS coalition may fragment into individual parties as political structures take on the extraordinary pattern of fission and fusion that occurs elsewhere in Melanesia. Tjibaou has commented that within FLNKS

> Unity is always limited by time. There is unity in our struggle for independence . . . this unity today might be different tomorrow. Each group has its own autonomy, as far as method is concerned, the way to work, to conceive independence, there isn't a unity on this yet. It is still being discussed (Quoted in O'Callaghon 1988).

Indeed in constructing the 'green revolution' there will be divisions between those Kanaks who seek to construct and participate in a modern economy and those who wish to re-establish a more traditional Melanesian order. At the very least, the future will be more difficult for the FLNKS, in support of dramatic change in New Caledonia, than for the French government, anxious for steady progress or the RPCR, determined to thwart any substantive change. Such a divided settler colony as New Caledonia will not be united through legislative action in the foreseeable future. In many respects the closest parallel to New Caledonia is Northern Ireland where different 'loyalists' also seek to maintain ties with what others see as a distant, alien, and colonial state. Many Kanaks see an affinity with Palestinians. Similarly, in New

Caledonia, just as the *Caldoches* seek to remain with France, so Kanaks also look beyond New Caledonia for assistance.

In the unlikely event of New Caledonia becoming independent there might be little substantial change. France has maintained strong economic and political ties with other former colonies and would be likely to do the same with New Caledonia despite the threat of Chirac's government to break off aid if New Caledonia chose independence. The fragility and artificiality of the economy would enable France to exercise control even with relatively low levels of aid. The demands for independence grew through the inevitable failure of assimilation, the belief that, as elsewhere in Melanesia, nation and state should coincide with the demand for social justice, status, and power in a Melanesian land, over Melanesian society, economy, political system, and destiny. 'Socialism' is no more than a rejection of colonial capitalism from which Melanesians were largely excluded. In an independent Kanaky, Melanesian rights to land would be restored and agriculture would be more important, but otherwise the economy would be unchanged; tourism and mining would continue, and further, and more diverse, foreign investment would be encouraged. Kanaky would thus be much like the independent Melanesian states to the North where, despite a similar loose socialist rhetoric, capitalist development has continued and intensified and foreign investment everywhere increased. Similarly the stress on 'Kanak' within the FLNKS is essential to the struggle for legitimacy and the restitution of tradition, rather than the racism that conservative critics imply. There is no prospect of a radical post-independence transformation, but these conjectures are inevitably hypothetical; New Caledonia is not about to become independent Kanaky. Despite widespread assumptions that the independence of New Caledonia is inevitable and that only doubts over the timing and structure of that independence remain, there is currently much greater certainty that this will not occur. There are a number of other colonial territories in the island Pacific that have reached a satisfactory negotiated dependence based on high levels of aid (and access to migration opportunities). They may be more appropriate precedents than the nearby independent states and, in this century at least, the era of decolonization appears to have drawn to a close in the South Pacific region.

4

Social Change in Vanuatu

Jean-Marc Philibert

Introduction

Covering over one hundred years of social change, even in a country as small as Vanuatu, is no easy matter. Besides the logistics of space, there is a structural problem which faces social scientists interpreting colonial history, that of the relative weight to assign to various historical factors when (1) The forces confronting one another are at first sight disproportionate in military or material terms and (2) When the written sources are produced by only one side. To top it all, popular accounts à la Moorehead (1971) of the fatal impact of Western colonialism on the South Pacific have found such widespread acceptance that it is now hard to argue for anything else. Yet, in my opinion, this view amounts to a denial of any significant historical role to Pacific Islanders, a denial all the more pernicious for being disguised under layers of sympathy for their supposedly lost cultures. Finally, there is a sense in which the South Pacific itself is so much part and parcel of Western mythology, in the form of an idea of a state of grace and the subsequent fall from it, that it is hard to know whom one is really describing, the Islanders as they see themselves, or an inverted version of ourselves and our own culture.

To refer to the process of interaction between representatives of Western societies and Islanders by the term "culture contact" is now seen by all as wrong. It is predicated on a bowdlerized view of colonial history, which has all but completely expurgated power from its description before naturalizing the process as modernization. On the other hand, the use of the muscular World System approach almost precludes the discovery of small forms of resistance or accommodation in remote islands of the Pacific, adaptations inconsequential to all except for those who live there. In both of these approaches, Islanders find themselves offstage, the subject of irresistible historical forces: their cause is lost from the start. This is why Islanders are usually portrayed in such accounts with the sympathy reserved for extinguished species.

Anthropological discretion may be in such cases the better part of scholarly valor and the wiser counsel might be to leave this sort of work to historians. And yet, this would not in itself take care of the problem: historians writing about social change must themselves use sociological models to account for what they describe. Moreover, if we want to know how some of the social actors thought and felt, what they believed they were up to when they acted as they did, we must turn to anthropological materials.

One way of getting round this problem is to look for a middle term between, on the one hand, descriptions couched in such local terms that they ignore the fact that local actors are a manifestation of larger historical forces and, on the other, worldwide economic and political mechanisms so powerful that there is no role left for indigenous groups. This middle term can be the ideologies developed by local peoples to understand the colonial situation. Peel (1973) suggests we analyze the ideas and perceptions of social reality which participants hold in given social situations; more precisely, it is a question of knowing the terms in which members of non-Western societies recount their collective experience. For Peel, indigenous ideologies constitute a cultural source of social change and they, of course, abound in the Third World where there is so much to explain and do to narrow the gap between aspiration and reality. Studying a Yoruba concept of development, *Olaju*, he discovered 'the steady matrix of ideas, existing in relation to a relatively enduring social context through not evidently a mere reflection of it, upon which subsequent responses to social change were developed' (1978, p. 159). Anthropologists can analyze the indigenous people's own conceptions of the social phenomena affecting them as a partial means of reconciling *grand theories* and the local reality. These cultural factors are the product of a unique social experience, a unique cultural body of knowledge at grips with a phenomenon of a universe nature, namely colonialism or neocolonialism. By studying indigenous symbolization of what lies ahead, one can grasp a particular cultural script of what is virtually universal history. To use a description made by Worsley in a different context, 'Such an analysis links past and present and provides a cognitive map, a set of value commitments, and a project of action' (1981, p. 248).

I propose to analyze what I consider the two poles of indigenous reaction to the coming of Europeans in Vanuatu. At one pole, I shall examine briefly the ideological context found among island groups that have been restrained in their borrowing from Europeans. This strategy will be analyzed during two historical periods: in the early twentieth century first and, second, during the period immediately preceding independence. I shall consider at greater length the strategy of adaptation pursued in a peri-urban context by villagers who chose early on to adopt European ways. The villagers' own ideology of modernization will also be studied during the same two historical periods.

It is important not to perceive these broadly contrasting strategies of adaptation as, in one case, praiseworthy resistance and, in the other, cultural sellout. The leaders of various Melanesian communities developed strategies meant to give them the upper hand against Europeans, and these strategies in turn were constantly changed and adapted when they did not produce the results expected. I do not wish to create a dichotomy, which can only be misleading, between *traditionalist* and *modern* groups. There are groups now in existence in Vanuatu that fit these opposite categories, which the majority of communities do not, but such groups are the result of a series of past decisions taken by or enforced upon various indigenous communities which were not made in the first place according to this opposition; in other words, rather than accounting for the decisions, this *traditionalist-modern* opposition follows from them.

I shall try here not to reintroduce by the back door what I have rejected by the front, that is, the commonsense distinction between traditional and modern groups in use in all island groups. It is of course a distinction that speaks clearly to all participants, be they members of rural or elite groups, European advisers or tourist authorities, to refer to a particular cultural condition. These descriptive concepts have limited explanatory value because they do not allow for synchretisms of any kind, such as the recent 'invention' of traditional culture by modernizing elites in newly independent Melanesian nations. The traditional-modern contrast – or what I prefer to call the *closed-open* stances because the terms have a different valence – will only be used to refer to the poles of the continuum of ni-Vanuatu reactions to colonialism. At these poles, we will try to find a series of images, contrasts, ideologies developed by ni-Vanuatu to explain their historical situation to themselves and to derive from it a program of action. For instance, we will investigate among groups who resisted European influences the notion of place and the centrality of this idea for Melanesian social and cultural identity. It would be wrong, however, to conclude from this that the same notion is of no account to those who accepted changes. What takes place is not giving up but a rephrasing of this notion within a system of ideological contrasts different from what is found in the previous case.

I wish to point out a final advantage of such an approach in Vanuatu. This small country is linguistically, socially and culturally fragmented to an extent rarely seen anywhere else. Some 105 different languages are spoken by a population of about 140,000. The social organization of the northern islands is dominated by graded societies absent in the rest of the group. The center has hereditary chieftainship not found anywhere else, while in the south almost everyone has a title, but power is attributed differently there. Descent is patrilineal in some cases and matrilineal in others, and probably not the central principle of social organization in most instances.

If one were foolhardy enough to try to catalogue the various cultural forms of knowledge, art, religion, medicine, found in Vanuatu, one would be confronted with a mass of the most varied data. On a few islands, reaction to Western implantation has been in the form of nativistic movements and one full fledged cargo cult has been in progress without interruption since 1942. Yet such a reaction is atypical for the archipelago as a whole. We are dealing with such a small stage in this country that the personality of the actors involved, be they planters, missionaries, district agents on the one side, and various indigenous leaders on the other, cannot be taken for granted or held constant. The history of colonization on various islands, and that of missionization in particular, varies according to the individuals who promoted and opposed it: social processes bear clearly the imprint of those once responsible for them.

Under the circumstances, one would be hard put to it to come up with generalizations that apply to all islands. Yet there is a way, a naive and superficial way perhaps, though not without some truth, in which all these social and cultural forms can also be said to belong to a Vanuatu cultural ensemble. I hope that some cultural constants will emerge even when dealing in the last part of the chapter with the post-independence period and the attempt by the political elite to develop a 'generic' culture for the whole country.

I. THE CLOSED STRATEGY

Though at first sight it seems logical to start with the reaction of those who refused European ways in order to have a baseline from which to measure the changes most Vanuatu communities made during the colonial period, it is nonetheless essential to keep the following in mind. First, that the isolationist response was adopted by a minority of groups in the archipelago. Second, that such groups were also deeply affected, albeit in a less obvious way, by Europeans: it is quite wrong to consider their present way of life what Vanuatu would have been like today if Europeans had not come. Let me explain. Although we shall never know for sure, it is safe to assume that the strategy of resistance evolved in a manner which might have been different had it not been also a reaction to what was being done by other ni-Vanuatu

101

at the time. Indeed, the opposition between open and closed groups covers a traditional opposition between coastal (*salt water*) groups and inland (*bush*) people, a political divide which could be bridged only with the greatest of difficulty and never for very long.

Coastal communities were immediately affected by colonialism – demographically, economically and politically – and had to take action more quickly than the inland groups insulated by rough terrain. Their reaction explains in many cases the opposite solution adopted by inland people: if inland communities could only obtain European goods and ways through the intermediary of coastal groups, the effect of which would be to tilt the balance of forces in favor of the latter, was the game worth the candle?[1] If, instead of the proud refusal partly foisted on them by age-old indigenous oppositions, inland people had had direct access to the cards Europeans dealt, would their cultural history of the last 100 years have been the same? Merely raising this question amounts to admitting that it is unlikely. It is safe to say that the rejection established a cultural dynamic which was no longer the precontact one, and that it became part of a more restricted cultural whole, turned much more inward than it ever was under traditional conditions.[2]

One of the difficulties of reconstructing this strategy of resistance, let alone the set of ideas which justified it, is that it cannot be done for the archipelago as a whole because the social and cultural coordinates vary so much between islands. Resistance was carried out through political institutions which differ markedly between groups.

An expeditious way of getting round this problem is to concentrate on only one island to obtain the 'traditional' cultural configuration. I have chosen the island of Tanna for this purpose, the reason being that it is, in the words of Bonnemaison (1986, p. 381), 'une ile philoso-phique', an island on which the suitability of the open and closed strategies is still being debated today. On Tanna, to an even greater degree than elsewhere in Vanuatu, either because of their own propensity or that of their ethnographers to intellectual tidiness, cultural themes common to the archipelago seem to have been taken to their logical conclusion.

I shall first give a skeletal outline of Tannese social organization and move from there to the myths justifying it and finally to the ideology of place. Next, when considering the pre-independence period and the troubles surrounding it, I shall argue, after Bonnemaison (1985a; 1985b; 1986; 1987) that it was in part the Tannese traditional notion of place which led some of them to try to stop their island from becoming part of an indigenous nation-state.

The End of the Nineteenth Century

There is on Tanna, one of the main islands of Vanuatu with a population of some 17,000 people, a complex form of social organization
not found in Northern and Central islands. According to Brunton
(1979), Tanna seems to have developed political characteristics found
throughout Vanuatu into, on the one hand, a near inflexible spatial/so-
cial grid controlling the movement of people and goods and, on the
other hand, a hierarchical system of titles so devoid of social authority
that it is unable to contain the centrifugal forces at work in the society.
As he succinctly puts it, 'The outcome is an atomistic society (Guiart
1956, pp. 10, 107) with a lot of structure' (Brunton 1979, p. 102).

Tanna comprises well over 200 *nakamal* (dancing grounds or *yim-
wayim* as they are called locally), cleared, open spaces where men
assemble at dusk to drink *kava* (*Piper methysticum*) and where dances,
rituals, and other public activities take place. The people connected to
these *nakamal* live in near-by hamlets of a few households made up
mostly, though by no means exclusively, of agnates.

Tannese have a Dravidian kinship terminology (Brunton 1979),
whereby male kinsmen of ego's generation call one another either
brothers or cross-cousins (*taniel*), the effect of which is to produce
exogamous moieties. *Nakamal* normally include people belonging to
both moieties. There is another dual division, political this time and
not corresponding with the former, which results from belonging to
one of two canoes (*niko*). Membership in these political moieties had
in the past a military purpose, although in many parts of the island
members of the opposite canoes inconveniently shared the same *nakamal*.

Hamlets associated with *nakamal* all own a stock of titles for men
and women. These titles, transmitted patrilineally, confer on men
residence rights in the hamlet as well as rights to land, access to
various rituals and political offices, and the right to use specific alliance
roads (*swatu*). Some of these titles also carry other privileges, such as
the ownership of magical stones, the right to catch or eat sea turtles,
etc. The two highest ranks are those of *yremira* and *yani niko*. The
yremira title is associated with certain ritual privileges and uncertain
secular ones. The problem is the abundance of such high titles: Guiart
(1956) states that there are over 470 *yremira* titles and some 120 *yani
niko*. All in all, Guiart counted over 1,100 titles on Tanna in 1953 when
the estimated adult male population was 1,790 (as reported in Brunton
1979, p. 103).

The hamlets and the titles they comprise are linked together by a
complex system of roads (*swatu*) through which all transactions must
circulate, whether they be messages, food, magic, rituals, help in

warfare, ceremonial exchanges and women. These are passed on from one hamlet to another through the men allowed on account of their title(s) to use the road. Skipping a link in the chain would anger those whose rights had been ignored and could well lead to fighting and the road being closed for some time. These roads, the work of deities in the past, cannot be modified or expanded.

How these various kinship, economic and political institutions actually fitted together was a puzzle that the main monograph written on Tanna by Guiart (1956) did not solve. One of Bonnemaisons contributions has been to eliminate the confusion surrounding Tannese social organization by focusing on *swatu*.[3]

The main *nakamal* are linked by roads, most of which have a name; in addition, the most important roads are accounted for by myth. The Tannese say that originally land was soft and formless. Magical stones (*kapiel*) which could move and speak landed one day on Tanna. The stones for a while ran all over the island fighing one another often to the point of exhaustion. After they completed their task, they stopped and became silent. Where they stopped space emerged and from space came out magical places. Later men emerged from the stones to find a space already organized into territories, criss-crossed by the roads left by stones. Individuals are thus linked to the territory, the magical place and the stone from which they sprung.

> The immobile stones that became places and the places in turn became magical spaces . . . each more or less powerful according to the magical power of the stones themselves. Such magical powers are varied and specialized, ranging from mastery of climatic phenomena to that of the growth of certain plants and, more generally, the powers of sudden materialization and curing sickness. Some *kapiel* are also the seat of political or psychological powers, such as the stones of intelligence, courage, and skill; of kept promises; of success in enterprises undertaken in rituals, war, or love (Bonnemaison 1984, p. 121).

At first men were not fully human; it was only later that they acquired houses, women, and fire. A myth documenting the killing of the monster *semo-semo* and the dispersal of pieces of his body all over the island accounts for the name and territory of Tannese political groups.

> Since that day, each independent geopolitical group on Tanna is assimilated within a particular canoe, or niko, consisting of one or several clans and organizing its space around a network of strongholds and magic places that constitute the territory of its identity (Bonnemaison 1984, p. 123).

To this organization is grafted the idea of first appearance. Each group is attached to a root-place where founding ancestors are said to have first appeared. By providing a chain which hierarchizes places according to the location of the primordial magical appearance of, say, food, men or women, and the subsequent diffusion of these to other places, the social groups associated with these places are as a result also hierarchized in relation to one another.

> The strength, both political and traditional, of clans and individuals mirrors the authenticity of their root-place and their greater or lesser primacy in the chain of magical appearances (Bonnemaison 1984, p. 128).

Traditional Tannese society represents social space and defines social positions through the imagery of the canoe. The position of *yremira* is that of aristocrat seated at the prow and, on land, he is the 'gate' controlling the circulation of goods along *swatu*. The *yani niko*, the helmsman or canoe master, steers in the name of the *yremira*; it falls to him to guard the territory and prepare for war; and finally, the *naotupunus*, or nurturer, is the magician responsible for garden magic.

> Each clan, lineage segment, and hamlet on Tanna is attached to one or the other of these specialized titles, and each large canoe congregates on the same territory as complete and diversified a range of them as possible (Bonnemaison 1984, p. 132).

The canoe imagery also evokes the idea of passages and safe havens extending outside the territory to allied groups, those with which sisters are exchanged and cooked food shared. Groups are involved in cycles of ritual exchange which are carefully carried out along *swatu*. One such exchange is that of the sea turtle – a secret food belonging to *yremira* – between coastal and inland groups, the effect of which is to reaffirm internal and political ties.

What do we find if we use Tannese ideas to stand for some of the spatial, social, and cultural understandings with which ni-Vanuatu tried to make sense of Western desire for land, labor, political, and religious control? We find societies which were in essence 'geographic societies', predicated on an organic unity of culture and place (earth, land, safe territory, and religious landscape rolled into one) and where humans were said to grow from the soil like plants.[4] This notion of place is also central to other parts of Vanuatu (Rodman 1986; Philibert 1988).

This is how Bonnemaison expresses this idea:

> Culture is in a way an extension of the earth, a 'law of the earth' inscribed on the territory and bound up with magical powers

sprung from the sacred ground; it can only be practised by men whose forefathers were bred out of that territory. There is thus a profound identification between blood (kinship) and earth (territory), hence the foundation of a very strong territorial ideology. (Bonnemaison 1980, p. 183, my translation) What is called 'custom' [cultural tradition] in the New Hebrides hinges on this law of the earth (Bonnemaison 1980, p. 183, my translation).

I believe that Bonnemaison makes the case that Vanuatu culture is molded by territorial ideology too forcefully: it almost implies cultural immutability. There were, after all, forms of mobility permitted outside the territory, even though these were always rigidly controlled (Bonnemaison 1979; 1984). He has reconstructed the inner logic, the official ideology of yesteryear on Tanna, a difficult task for which we are indebted to him. Care should be taken, however, not to read directly from ideology to social action, or to imagine that this internally coherent set of ideas matches the conception of all Tannese. This amounts to taking culture as given, rather than something which is challenged and remade daily, with contradictions of its own, making this reproduction problematical. This contrasts with the remarkable work of Lindstrom (1982; 1984) on Tanna who believes that inequality there rests on knowledge and its control, rather than wealth, as in other parts of Melanesia. His work focuses in particular on the mechanisms of transformation of private knowledge to public knowledge and its use as political resource.

I shall return to the issue of how to conceptualize traditional culture in the third part of this article, but for the moment I wish to turn to the period immediately preceding the 1980 Independence.

Tanna in the 1970s

So far, following Bonnemaison, I have distilled Tannese culture a great deal to unearth precolonial dominant cultural themes. However, I would not like to give the impression that it is possible to explain Tannese reaction to colonialism solely as the outcome of the cultural crucible presented above. Although the Tannese are steeped in a cultural tradition in which, on account of the close link between identity and the territory of origin, the spatial dimension is singularly stressed, it did not prevent them from developing a range of responses to colonialism during which, amazingly, few cultural bridges were ever totally burned.

Over the last hundred years, the reaction of the Tannese to colonialism followed various strategies, ranging from outright rejection to total acceptance, whether these were embraced enthusiastically or reluctantly. It led to the formation of 'pagan', Presbyterian, Roman Catholic and synchretic religious groups; to 'Westernized', 'traditional', and neotraditional cultural enclaves; to British-educated, English-speaking, anglophile groups and French-educated, French-speaking, francophile groups; to British and French-aligned political factions. Nor was this process of accommodation irreversible: surprising reversals or leaps forward took place over the years. The perenniality of the John Frum movement, a Cargo Cult in existence on Tanna since 1942, testifies not only to the flexibility, but also to the creativity of Tannese cultural reproduction. To be sure, such lines of cleavage often followed the same old fault lines, thus leading to larger groupings, such as the Presbyterian, anglophile, progressive Vanuaaku Pati. I now wish to turn to the period between 1975, time of the first national election ever held in Vanuatu, and the rebellion of 1980 which preceded independence, and to focus in particular on the political discourse of moderates[5] or 'rebels' on Tanna. I wish to make clear at the outset that I do not consider that a phenomenon as complex as the Rebellion can be understood simply by outlining the political discourses of those involved. In fact, the political stances taken had more to do with the positions of other Tannese, in other words, with pre-colonial conflicts and past attempts at political control of one group over another, than with the rhetoric of the national parties. Moreover, between 1977 and 1980, a good number of acts of provocation and intimidation took place, which antagonized the forces confronting one another to the breaking point. The opposition was also fuelled nationally as well as locally by the two Residencies.[6]

National elections were held in Vanuatu in 1975, in 1977 when internal self-government was granted, and in 1979, to form the government which would lead the country to independence. The national and local proportion of votes between the Vanuaaku Pati and the Moderates did not change substantially during these elections.[7] But party representation in Parliament did. On Tanna, the 1975 election had given only one seat to the Vanuaaku Pati, two went to UCNH candidates and the others to Kapiel and John Frum. However, the 1979 election saw the Vanuaaku Pati increase its popular support on Tanna from 46.5 per cent to 50.6 per cent, electing in the process three of the five members of parliament. Suddenly, the Tannese faction affiliated to the Vanuaaku Pati found itself in power, both locally and nationally, and in a position to impose its own view of what the independent country would be.

The arrival of two ni-Vanuatu as district agents precipitated the rebellion of June 11, 1980. Tannese had become used to having their political and judicial affairs administered by French and British administrators who, as foreigners, could be relied upon to be impartial or

at least in opposition to one another. The resulting neutrality could•
no longer be assumed when local administration was to be carried out
by ni-Vanuatu working for a government without legitimacy for many
Tannese. The rebellion of June 1979 ended for the rebels in one dead
and twelve wounded and 256 men sentenced to varying fines and jail
terms.

The reason for focusing on the moderates' political discourse is that,
paradoxically, the discourses of minorities often allow us to take the
measure of hegemonic ideas which all too often appear only natural
or commonsensical. If, as Bonnemaison believes in this case, Tannese
moderates wanting nothing less than total opposition to a political
form that has become second nature to us, that of the modern nation-
state, then we must closely examine such views.

Bonnemaison sees this conflict as dominated by the issue of cultural
identity and sovereignty, a cause lending itself particularly well to the
spatial concerns of Tannese. This is how he frames it:

> The champions of traditional culture were rejecting the modern
> State in favor of their existing culture and identity. The movement
> did not, therefore, rest on ethnicity or territory but on an idea
> (Bonnemaison 1985a, p. 231, my translation).

The arrival of national politics upset the balance that had been
reached on the island between the Protestant anglophones, and the
other, Catholic, non-Christian, neotraditional groups who were francophone. The creation of a nation-state was in the main the project of
young and educated ni-Vanuatu nationalists who wanted a strong
central government in order to create a new country in the Western
mold. Almost half of the population of Tanna feared the ascendancy
that the other half, belonging to the Vanuaaku Pati, would gain in
implementing this project.

Groups sharing the same fear had regrouped themselves in the Tan
Union which stood for a federal, decentralized administrative and
political structure, and for French and English bilingualism, as they
were sending their children to French schools. They countered the
nationalist political discourse with a cultural one (Bonnemaison 1985a,
p. 235). Tannese culture thus entered the political debate, each camp
justifying its political project in the name of a claimed fidelity to local
cultural themes. At first the notion of *kastom* was flexible enough to
be used on the national scene by the nationalists who demanded the
independence of Vanuatu in the name of a cultural identity which
they sought in a precolonial past. Needless to say, the Western-educated national leaders of the moderates were also nationalist, though
having recently found in the French Residency a powerful ally, they
were in no rush for it to leave. Neither did they want to forego

immediately the recent economic benefits of the French presence in the archipelago. In a country as dependent as Vanuatu is, and will remain for the foreseeable future, on foreign sources of income, they asked, what is political independence worth without economic independence?

Bonnemaison again:

> The traditionalists then reinvented Tradition, endowing it with a philosophical and political project which they thought capable of preserving the autonomy of their own society . . . In their use of the word Tradition, local groups were in effect claiming their right to choose freely a political organization different from the Western State model: they were claiming their right to 'nativeness' ('*autochtonie*') (Bonnemaison 1985a, p. 245, my translation).

This author presents Tannese rebellion as the radical refusal of a Western modern political form and, though he does not mention it, Iran comes to mind as another attempt to move away from Western-style modernization. It is by drawing on the Tannese notions of space and time seen in the previous section that the moderates were able, not only to counter the Vanuaaku Pati's anti-colonial, nationalist and North – South political rhetoric, but also to refuse the modern form of political power.

Social space is in traditional Tannese culture conceptualized as a route linking economically and politically independent places forming a network. There is no center, no capital to which peripheral zones are related.

> The lived space of traditional society, multi-centered in as many places and segments of roads as there are local groups and territories, thus at first appears narrower and more segmented than modern space. But on further examination, it also appears more open. The space-road of Tradition is a space without frontiers; it stretches from island to island, with no theoretical boundaries unless it is the infinity of known horizons (Bonnemaison 1985a, p. 242, my translation).

This overdeveloped notion of space is matched by an underdeveloped notion of time which appears little more than a 'reified' version of primordial time. The time of *kastrom* is situated outside history; it is the golden age described by myths of the creation of the world.

> The traditional project in its firmest ideological expression consisted, therefore, not only in denying the colonial episode, but Melanesian history as well. The project of returning to Tradition consisted of a return to the Time of Greatness (Mircea Eliade's

109

Grand Temps). . . . This dream contained a religious vision of the world and a fantastic will to stop its course and freeze it in the reconstituted space of its origins (Bonnemaison 1985a, p. 243, my translation).

Kastom clearly did not have for some of the embattled diehards of the Tan Union the meaning it had for progressive Vanuaaku Pati members. In fact, the more relevant, in moderates' hands, *kastom* became as a map of the political future of Vanuatu, the less specific the *kastom* corpus became in Vanuaaku Pati rhetoric.

The complex religious and cultural make-up of Tan Union makes it hard to know the extent to which members of each of these groups espoused the political ideology outlined above. Nonetheless, common political issues, such as the future administrative structure of Tanna, led the leaders of such groups to search for alternatives to Vanuaaku Pati political answers. In the process, relying on varying forms of knowledge about the past, real or imagined, they reinvented *kastom*, giving 'traditional knowledge' a new contour. But it was not the traditional knowledge described in the previous section so logically, almost Cartesian-like, by Bonnemaison. How could it be? This new discourse had to speak to John Frum millenarists as well as Western-educated Catholics and 'pagans', all of whom were involved in the development of an ideology that would act as a counterweight to the Vanuaaku Pati's discourse. Such *kastom* had to address the current political issues, while remaining sufficiently vague to win the adhesion of very different people.

Although Bonnemaison uses the term invention to categorize this construct, he nonetheless presents it as contained in traditional culture. I prefer instead to focus on the creative dimension of this process, as I consider problematic what Bonnemaison takes for granted, the reproduction of Tannese culture. He posits an entity, traditional Tannese culture now manifesting itself as *kastom*, to account for the behaviour of opposition groups on the island. Having previously developed a highly integrated and internally coherent model of ideas which he calls traditional Tannese culture, he is naturally led to state that it is in the name of this corpus that people rose against the Vanuaaku Pati. As stated earlier, I believe that Tannese cultural knowledge was never as tightly integrated as Bonnemaison's model of it. He only arrives at such a systematic, totalizing set of ideas because he considers them outside daily practices of Tannese, where the meanings and implications of those ideas are contested. Secondly, these bits of knowledge were not already constituted outside of the moderates' political activity: such ideas were historically put together in the course of varied people confronting the nationalist discourse.

In this description of the rebellion, which relies almost exclusively on Bonnemaison's account, our different positions on traditional culture do not significantly change the interpretation of the empirical data. However, I believe that we hold very different views on culture, reflecting no doubt the use of this concept in our different disciplines. This in turn leads to a different understanding of agency in social life. To an anthropologist, Bonnemaison appears to give his heuristic device, his model of traditional culture, an acting force and a role that it cannot have by turning it into the cause of people's resistance. For all the perspicacity of his analysis, it appears to a skeptical reader as a case of misplaced concreteness.

I shall return to the issue in the third part of this chapter. I now turn to the study of the opposite strategy of adaptation to colonialism.

II. THE OPEN STRATEGY

In Vanuatu, probably more than elsewhere in the Pacific, it is difficult to maintain a Eurocentric view of colonial history. In the 1870s and 1880s, at the start of European settler colonization, France and Great Britain were both reluctant to assume control over the archipelago in spite of the fact that French and British nationals living there were pressing hard for annexation. The two powers agreed in 1878 and again in 1883 to maintain a 'hands off' policy toward Vanuatu: neither of them wanted it so long as the other could be relied upon not to claim it as its own. Indeed, given the strategic importance of the *Entente Cordiale* between the two countries and the European scramble for Africa in the last two decades of the nineteenth century, there was little concern for Vanuatu in French and British colonial offices.

At the time of the signing of the Anglo–French Convention establishing the Condominium of the New Hebrides in 1906 by making the archipelago a region of joint influence and providing for its joint administration, the European settlement amounted to only 640 people – including missionaries and their families – and a full third of these Europeans were settled on one island alone, while the ni-Vanuatu population was around 60,000 people. The 1914 Anglo–French Protocol allowing for the setting up of a system of native administration and of native courts – and this is typical of the colonial administration of the time – was not ratified before 1922 or implemented before the late 1920s. Laxity has been the dominant administrative quality in this country, perhaps because the condominial graft between the two colonial administrations was given little chance to succeed, though, one must admit, not without some reason. The Condominium required that all decisions regarding ni-Vanuatu be taken jointly by the local French and British administrations, while each one pursued from the start different national strategic and economic aims and followed different colonial policies toward native peoples. The net effect of this bicephalous administrative system was that precious little was done by either power until World War II was over (MacClancy 1980; Philibert 1981). Education, for instance, was left entirely to the missions until the 1950s.

The year the Condominium was created, the inhabitants of the indigenous village studied here, Erakor, had already been Christian for over 30 years; they were literate in their own vernacular with many of them speaking bislama and a few, English; a good number of men had also returned from work in the sugar plantations of Queensland or had traveled in the Pacific as seamen. In 1884, villagers had ceded their land to the Presbyterian mission which acted as trustee over it in order to prevent further land spoliation. Encouraged by the same mission, villagers had already turned to the production of copra in their own groves as an alternative to work on European plantations.

And yet, Erakor's colonial history reflects closely that of the archipelago as a whole: villagers lived through the sandalwood trade of the 1840s and 50s when, in one instance, natives were slaughtered on their island; they suffered like others from the epidemics which decimated parts of Vanuatu's population; they lived through the worst excesses of the labor trade; they experienced land spoliation as well as heavy-handed tactics by European settlers on their own island; finally, they were evangelized by Polynesian teachers from 1855 onward, and by European missionaries from 1864.

It is clear that one will not understand social change in Vanuatu without first acknowledging the part played by ni-Vanuatu themselves. Given the small size of the European settlement and the comic opera quality of the colonial administration, it is hard to explain the relative case with which ni-Vanuatu adopted Christianity and education and turned to migration and cash cropping. One must acknowledge that in many cases ni-Vanuatu themselves took the initiative in establishing links with the newcomers. This presupposes that they had very early on an understanding of this new social and cultural order; that they developed their own ideas about modernity. Erakor villagers were undoubtedly among the modernization autodidacts of Vanuatu. It is thus important to try to grasp their ideology of modernization and this is what I now turn to.

Ethnographic Setting

The village of Erakor is situated on the island of Efate, some 10 km from Port Vila, the capital of Vanuatu, which has a population of about 15,000, including the peri-urban villages. Erakor, with a population of

1,000 inhabitants (1983), is a prosperous urbanized community, settled on land covering an area of 1,400 ha. The Erakor villagers have throughout their history shown great receptivity towards new ideas: they were the first on Efate to embrace Christianity and they played a large part in the evangelization of the island; they began very early to take part in the market economy through commercial copra production, and today they are active in the Port Vila food market and in the tourist industry.

The End of the Nineteenth Century

Of all Europeans who have entered the South Pacific, the missionaries are no doubt those whose influence on the indigenous inhabitants has been the most profound and lasting.[8] Christian churches have played a particularly important role in Vanuatu. It can even be argued that the Condominium Government owed its existence and survival to the Presbyterian Mission, which intrigued so long and so determinedly for the maintenance of a British presence in the archipelago. Throughout most of colonial history the mission not only carried out their religious functions but also provided the only public education and health services available. They were the main social, political, and cultural broker between the indigenous people and the British government.

The desire to be converted – leaving aside for the moment the problem of discovering what that term meant to the ni-Vanuatu of the period – originated with the people of the islands themselves. Power (political) and knowledge (religious) are closely linked in traditional Melanesia, so in attempting to assimilate the newcomers' knowledge, often at any cost, the people were seeking moral, economic, and political equality with them. Now true knowledge in Melanesia, as Lawrence (1964) tells us in his classic work, *Road Belong Cargo*, is, above all, sacred knowledge, revealed knowledge; secular and technical knowledge are of little importance and carry little prestige, as everybody possesses them. So it is not surprising that ni-Vanuatu sought in Christianity the keys to the Europeans' material and spiritual wealth. Conversion to Christianity was also an admission that traditional religion could no longer adequately account for a cultural horizon which had been considerably expanded with the arrival of the Europeans. What is more, Christianity presented itself with the prestige that attaches to the institutions of a 'conqueror'.

Christianity had another attractive feature. In Melanesia sacred knowledge was jealously guarded as a treasure, secrecy and dissimulation being intended to preserve the power connected with exclusive knowledge. Melanesians were therefore struck by the missionaries' apparent reckless generosity.[10] Accustomed as they were to keeping strict account of any exchanges because of the importance of maintaining reciprocity in a social system where the notion of equivalence is fundamental, these Melanesians could only interpret the missionaries' generosity either as a sign of childishness or else of bottomless wealth. Having opted for the second interpretation they set themselves to acquire this knowledge as rapidly as possible.

Not only did it hold out the promise of infinite riches, Christianity also offered solutions to the problems with which the people were faced: galloping mortality as a result of epidemics which traditional remedies were powerless to control; land spoliation and other forms of economic exploitation by various Europeans; lack of a government which could deal with disputes between natives and Europeans.

Ni-Vanuatu were looking for a framework, both intellectual and social, within which they could react effectively to the new order. The new religion provided first of all a new political framework. On Efate, hamlets formerly scattered throughout the island were brought together by Presbyterian missionaries into coastal villages. People were regrouped into a much larger social whole, a sort of parish, under the authority of church elders and a European missionary. This was accompanied by a widening of indigenous social solidarities. Equally important was the ideological underpinning that enabled them to think of these social realignments in terms of a universal morality of human brotherhood, in place of the traditional attitude of mistrust and hostility.

The influence of the Presbyterian Mission in Erakor was particularly strong from 1864, when the first European missionary arrived in the village, until the late 1920s which saw the beginnings of a native administration. The Presbyterian Mission introduced immense changes during those 60 or so years.[11] The cost of the people's conversion was high; they had to give up many customary practices in order to accede to the *new life*. The loss of traditional institutions was partly compensated for by the fact that they became literate. They were given a mainly religious education, but they also learnt the rudiments of arithmetic, English, geography, and music. The women were taught how to wash clothes and to sew. The mission introduced the first cash crops (maize and arrowroot) to offset the cost of publishing a number of books in the Erakor language, and it later encouraged people to establish coconut plantations. The political organization of the village was altered, power passing into the hands of adepts of the new religion. The sons of influential men were sent to school to complete their religious education and this reinforced the mission's power in the villages.

This alteration of customary social organization went hand in hand with profound ideological changes. The mission defined the new order for the villagers in the following terms:

1. The universality of religion: the European world was presented as obeying divine precepts, even if these were not always respected by all Europeans, and the religious domain as the determining principle of the white man's social system.[12] In a way, this was similar to the indigenous people's own concept of religion. In the anarchic situation which prevailed in Vanuatu in the nineteenth century, the mission laid down a set of rules, thanks to which the people could avoid getting into trouble: it was only much later that the villagers made a distinction between the mission's 'law' and the administration's 'law', in other words, between the spiritual and temporal domains.

2. *In terms of salvation*: the Christian religion was an entirely new social universe which could not tolerate *diabolical* customary practices and which had to be accepted without reservation if one wished to accede to the *new life*.

3. *In paternalistic terms*: the missionary, representative of Europeanm society, left his country and his people to come and save God's children, who were thus constrained to play the part of pupils or apprentices in the faith with a long way to go before they could attain the moral equality promised by the Europeans' religion.

4. *In modernistic terms*: the changes the missionaries introduced were not only of a spiritual order; the mission also concerned itself with the material well-being of its flock, bringing in medicines, cash crops, education and new needs such as books, clothes, manufactured tools, and so on.

5. *In messianic terms*: the new religion and the salvation it promised were a kind of experiment such as one finds in cargo cults where suddenly the traditional social categories are shaken up in the hope of finding new formulae, new equations to account for the social order; many villagers espoused the new cause body and soul by becoming teachers and often risking their lives to carry the gospel to the other ni-Vanuatu.

One can see that during this period there was a desire to experience new ideas and forms of behaviour. It was not until the Second World War that another revolution of such importance was to occur. The inexhaustible wealth of the Americans, some of whom were black like them, the generous ease with which they gave away their goods to the local people, posed fresh problems of comprehension. In the military camps, the hospitals and the 'PX', Erakor villagers found new standards by which to measure the distance they still had to cover. They discovered the American's *Klin leven* (high standard of living).[13]

By adopting Christianity the people of Erakor obtained a new tool, a new model, for thinking about themselves, for defining themselves, and for guiding their actions. However, between the *time of darkness* and the *time of light* it was not always a case of complete discontinuity.

The political organization set up in Erakor by the Presbyterian missionary in the 1880s and 1890s resembled the customary conception of the social universe. Villagers recall traditional society as a repressive society based on fear. According to them, the only sort of power acknowledged to be real was power backed up by force, and today the people of Erakor still use penal terms to describe the power of traditional chiefs. Offenses such as theft or adultery were severely punished, and the chief had power of life and death over his subjects, either directly or through the intervention of *supe* (ancestor spirits), or magical forces which the chief was in touch with through his *tabu place*.

According to MacDonald (1892) the ancestor cult was the traditional religion of the Efate people, and it explained sickness and death as being the result of intervention of *supe*, at the instigation of a *tabu man* (sorcerer). These magical powers were transposed into the supernatural power of God. The Christian God became a vengeful God who, like the *supe*, punished his children's offenses with mortal sickness. Had he not struck down those pagans who had refused to be converted by sending epidemics? Belief in frequent intervention by a wrathful God in the lives of his flock, intervention guided by the village religious authorities, can be largely explained by reference to the ancestor cult.

From the turn of the century until the setting up of native courts, the chief and his counselors, church elders, and other influential community members were responsible for maintaining law and order in the village. People say that at that time the chief's power came to him from both 'tradition' and religion. In the latter case, it was based on the ever-possible recourse to divine anger. Chief and church elders had the power to call down the divine wrath on the heads of villagers. At that time people thought God was closer to the community than today, and anything having to do with religion, from acceptance of the function of church elder to sentencing to divine punishment, inspired fear. Church elders themselves were not in full control of the sacred power they had at their disposal: the role of church elder also carried some risk because that same power might turn against anyone who did not lead an exemplary life. Divine justice was not only severe, it was remarkably swift.[14]

Because there was no division between the 'work of chief' and the 'work of church elder', in other words, between the spiritual and temporal orders, when the influential men of the village met their 'words were dangerous'. Those now over 50 remember well the tragic end of those delivered into the hands of divine justice. There was

116

thus a cozy convergence between the Presbyterian and the traditional indigenous theocracy, the two societies sharing an understanding of power as having a spiritual origin. Erakor villagers in a sense simply replaced one set of sacred figures by another.

The influence of the elders did not only come from the fear they could inspire; it also issued from their knowledge. Since everything there was to know in the universe was to be found in the Bible, only the old men possessed the sacred knowledge which comes from long familiarity with the holy books. Only they could explain the parables and place them in the context of village life. It was for them to interpret God's designs for his earthly realm, and to make the world intelligible through their knowledge. Furthermore, the Presbyterian mission acted as go-between between the village and the government authorities. In the people's mind the British and French administrations constituted an arbitrary authority dispensing justice according to its own whim, on the basis of a poorly understood set of legal rules. Observing the Ten Commandments and respecting instructions issued by the church elders was enough to keep villagers out of trouble with the government.

Punishments inflicted on wrongdoers by village authorities also had a religious character. Not only could a person sentenced to forced labor not associate freely with other villagers, but he could not take communion either. Others forfeited their right to participate in religious life, and they were forbidden to enter the church – an important place of assembly for the community – for periods of up to six months. The whole social personality of the guilty person was punished by means of this social and religious ostracism.

Supernatural sanctions were not, however, the only ones available to the elders. Prescribed marriage still existed at this period, and the old men's help was essential to obtain a wife. They were still the heads of clans in a kinship system with a far more direct incidence on the lives of villagers than the present system.

From the 1930s to the 1950s, the history of Erakor's political organization reveals the loss of the elders' supremacy over the young in legal, marital, and economic matters, while at the same time the central government was widening its sphere of influence. The religious gerontocracy set up by the mission at the start of the century has long been a thing of the past.

To sum up, we have seen that traditionally the elders controlled the social life of the village. They were the heads of clans, those who regulated the exchange of women, who knew the supernatural forces governing the world, who communicated with the ancestral spirits, since they were themselves on the way to becoming *supe*, and who, finally, instructed the young men in the men's house. This social order

117

was replaced by the mission which imposed its domination over all aspects of village life. A second gerontocracy replaced the first, this time based on a new type of sacred knowledge, that of the Bible and the Christian religion. Social control remained as repressive as before, the people just as regimented. There was a curfew in the village every evening from 9 p.m., and on Sundays it was forbidden to light a cooking fire or to go to the beach to wash oneself. Sickness and death were still attributed to supernatural forces: a wrathful God, ever ready to answer the prayers of the church elders, intervened in the everyday life of the villagers to enforce the authority of his earthly representatives and to bring about a new social order.

For some time the old men maintained their sway over the young through the exchange of women, the inheritance of coconut plantations, and the dispensing of justice. Now things have changed. God no longer manifests himself so frequently in the lives of the faithful, people say, and the young who have not lived through the time when those who disobeyed the village authorities were struck down by God, no longer fear their elders. The alliance between spiritual and temporal powers is no more.

But we are getting too far ahead. Let us return to the villagers' perception at the turn of the century.

With the certainty that all customary institutions were the work of the devil, and that thay were in opposition to the new order, traditional knowledge became superfluous. People adopted the Christian myths, which seemed more immediate, and the old order fell into oblivion. Where Christianty did not entirely supplant traditional knowledge, it absorbed and reinterpreted it. At this time, the notion of modernity was expressed in a Christian idiom, particularly through the well-known missionary metaphors of *world of light* and *world of darkness*. The world of light was the new and better world promised by Christianity and the world of darkness, that of tradition, was a world now past and superseded.

According to my dictionary, *light* means, in the figurative sense, that which enlightens, illuminates the mind, that which makes clear, explains and, in the absolute, God, the Truth, Good. I believe that for the people of Erakor the notion of light must have included all these meanings. They borrowed this image from the missionaries to indicate that the evangelical message had come to tear down the veil of obscurity, if not of obscurantism.

What precisely did these expressions, world of light and world of darkness, signify at the beginning of the twentieth century? I would like to present two texts in an attempt to circumscribe these notions. The first is a mythical account of the settlement of Efate, one of the few myths collected in Erakor; the second is the substance of a short skit performed by villagers.

The myth is part of a myth cycle depicting the adventures of a legendary hero, Roy Mata. The name Roy Mata was familiar to people in Erakor.[15] They knew the part he had played in establishing the clan system, even though in general they were far from knowing the legend in detail. The myth of origin of the men from the south as told to me in Erakor gives a singular interpretation of these events and illustrates the historical importance of Christianity in the village:

There was a great canoe which came ashore at Forari, at Manioura point, where the water has continued to this day to be *tabu*. This great canoe had come from Africa, and the people would reach an island, stay there for two or three generations, and then their children would set out again for another island. They first went to New Zealand. There they forgot everything they knew, even what they knew about medicine, and they became natives [savages], using tree bark to clothe themselves in. They were not satisfied with their life because they still remembered that they had once known God, even though they had forgotten where God was, and had taken up magic. Satan was with them [in their hearts] and the people used his power rather than God's. Their chief was always called Roy, from father to son.

So, Roy landed at Manioura point. He settled Devil's point and set up lesser chiefs on Efate. Roy moved about with his men, and there were sharks at *tabu* places to carry him wherever he wanted to go (sharks are often found at *tabu* places). Roy had three children, two boys and a girl. These people never settled in the bush but followed the coast from Efate to Nguna, to Tongoa, and finally to Epi. This is why today all the people of those islands speak the same language; not exactly the same, but their languages all belong to the same family. Roy's son, Roy Mata, created the *namatrao* (clans). At that time Christianity had not yet come to bring changes, but the *namatrao* spread rapidly. The significance of the *namatrao* is this: people belong to different villages but at the same time they are all brothers. Taro people or coconut people are the same everywhere (belonging to a clan confers on its members certain personality traits which are 'in the blood'). The *namatrao* did not stop wars, but they provided refuges in every village where a visitor could safely go.

I wish to set aside for a moment the analysis of the terms of this legend told in a Christian idiom and in which the thread of the story is reminiscent of the wanderings of the Old Testament Jews to move to the second text, a skit caricaturing three social types well-known in

119

the village, the European traders, the acculturated, *évolué*, sophisticated villager, and the *man bus* (literally, the man of the bush, the savage). The symbolism in this case is all the more revealing in that the villagers are poor in rituals or public representations, so they have little opportunity to gloss the various aspects of their social life or comment on them publicly.

On New Year's Day, the villagers gathered in the center of the village at about 3 p.m. to watch a performance. The women were presenting short scenes about aspects of daily life: the strange behaviour of certain customers at Vila market, the village council hearing a case of adultery, and a domestic scene between a drunken husband and his wife. Then came the performance that met with the greatest success. Three young men played the parts of European trader, sophisticated villager in shirt and long trousers, and a *man bus* wearing only a piece of calico around his waist.

The action begins with the *man bus* who wants to buy his first pair of trousers from the trader. The latter demands one pound sterling for the trousers. The *man bus* gives him three pounds, asking if that is enough. The storekeeper gives him his money back saying the trousers only cost one pound. The man bus replies that he only has three pounds and forces the storekeeper to accept them; the European, pretending to act solely out of the goodness of his heart, hands over the trousers and pockets the money.

The *man bus* can hardly contain himself for joy at now owning a pair of trousers, and tries them on. He gets one leg on but is unable them to find the second leg. Turning to the storekeeper he tells him his trousers are too small. The trader and the 'man of the world' show great contempt for the poor *man bus*. The latter again tries to put on his trousers, and succeeds this time in getting both legs on, but the trousers are inside out.

He seems satisfied with this result until the other two point out to him that he has put them on inside out. He takes them off, turns them the right way out and puts them back on, this time back to front. The European and the sophisticated villager again have a good laugh at his expense, without his understanding why, and tell him to try again. In the end, after all these attempts, he gets the trousers on properly.

His apprenticeship to modern living does not end here, however. The storekeeper and the sophisticate decide to teach him the basics of modern dancing. The *man bus*, who has no sense of rhythm, tries to copy them, but only succeeds in waggling his body from side to side. The 'man of the world' makes fun of him; he knows all the steps and dances gracefully with the European as his partner. The *man bus* is frightened by the second native and refuses to be held by him. He

120

dances alone, his spear in his hand, continually tripping over it and falling down. He looks completely stupid and is an object of derision to the other two. Throughout the whole scene the audience, from the youngest to the oldest was splitting its sides with laughter at the tribulations of the poor savage who knows nothing of civilization.

This farce, so greatly appreciated by the villagers, showed the *man bus*, totally ignorant of the refinements of civilized life – how to wear clothes, the value of money, modern dancing – in an unfavorable light. His ignorance and his nervousness were in sharp contrast with the self-assurance of the sophisticated villager who could manage all these things with ease.

On no other occasion did villagers display so clearly their ideology of modernity, their satisfaction at having become what they are, or their disdain for their ancestors' way of life. The performance contrasted the attitudes of the *man bus* and the sophisticate towards clothes (first sign of conversion to Christianity), money (participation in the market economy), and dancing (a reflection of the new relations between men and women). The villagers were thus giving themselves a picture of the changes they had been through and congratulating themselves on the distance they had come.

The modernization of the village, the transformation of the villagers' way of life are conceptualized in these two texts in terms of the following dichotomies:

World of light	*World of darkness*
Christianity	tevel [spirits]
God	Satan
true knowledge	false knowledge
clothes	nakedness
easy life	hard life
white man	*man bus* (bush man)
money	mats and pigs
medicine	magic
civilization	savagery

It is not only that change is formulated in terms of categories introduced in the nineteenth century by the missionaries, but most importantly that these categories are mutually exclusive and cannot tolerate any middle term. The world of darkness and the world of light are perceived as integrated sociocultural universe, entitles which are in opposition to each other. The modern world, the white man's world, forms a whole, the parts of which are indissociable from each other: Christianity, modern medicine, corrugated iron houses, consumer goods, etc. All this is in opposition to what must be rejected once one has opted for the light.

There is, however, a social context in which Christianity is not perceived as the antithesis of tradition. Christianity is considered to have completed the social transformation begun by Roy Mata when he introduced the present kinship system based on clans. People say that before the clan system was set up all the people were 'mixed up together' and that it was difficult to live in peace. With the emergence of clan organization, people were divided into groups and everybody's duties and responsibilities were clearly defined. There was a clear line of command from the principal chief to the heads of the local clans. People could 'see clearly' with this system of social organization.

Such a type of organization meant that, going beyond village ties, all the members of one clan were brothers. This idea did not put an end to wars but did create in every village a refuge for the visitor in need. A supra-village organization had emerged. Christianity has made possible a reorganization of indigenous society in such a way that loyalties are no longer limited to village or clan groups. As one informant stated,

> The teachers taugh us a good thing: they taught us all to be friends. Before, people were afraid of one another; there was no contact between them. Now everything has changed. People can go freely to other villages and even live there.

The Christian religion has made a broadening of the social fabric possible which in the traditional context would have been unthinkable.

Until recently Christianity was still the principal means of access to the white man's world. According to an informant now in his late forties,

> The difference between the whites and the blacks is that the whites have had a long civilization. They have been serving God for a long time now and this is why they have an easy life. But the blacks have not yet proved themselves. They must follow God's message and the 10 Commandments to the letter, otherwise they will never attain the white man's way of life. White people can excuse themselves from going to church now because their forbears did God's work for so long. Anyway they have other ways of learning: they have good schools and a set of laws which put the 10 Commandments into practice. The black people do not have these things. They have only just emerged from darkness. They have only God to teach them how to attain the good things in life. And then the blacks do not think like the whites who 'know the way'. The black people have come after the whites, the Chinese and Vietnamese, ever since Noah's son laughed at his father. American blacks are not so badly off, but

then their history goes back to the 1800s whereas ours began in 1900. A white boy knows more than a grownup black as he already knows the law. The blacks only know the whiteman's religion. Religion is their only form of education. And everything you need to know in order to live your life well is to be found in the Bible.

This sounds obsequious, 'uncle Tomish' even, and few would use such a discourse today. Yet I believe it does reveal a correct assessment of the impact of the mission during most of the colonial period.

In the nineteenth century, faced with demographic, economic, political, and cultural problems as a result of the arrival of various representatives of the Western world, indigenous society reacted by extending its own epistemological notions to Western culture and sought in Christianity the keys to Europeans' knowledge – that is to say, to their power. What they found was an ideological framework requiring the adoption of new relations between men and women, the formation of new social groups such as Christian villages under the aegis of a missionary, and a broadening of the ideology of mutual assistance found in the kinship system, thus leading to a universal morality based on human brotherhood (though in practice this was restricted to co-religionists). The acceptance of Christianity by ni-Vanuatu led to a restructuring of their society.

At the turn of the century, when the Presbyterian mission was in the ascendancy, the institutions were devoloping which were to put into practice the ideology contained in Presbyterian Christianity, particularly the fusion of spiritual an temporal power. The idea proved fertile, partly because it was an extension of local political notions. The missionary and his representatives, the church elders, supplanted the *taboo man* and customary supernatural forces were abandoned in favor of a single source of power, the God of the Old Testament, who came to punush his children's wrongdoing with sickness and death.

A second epistemological change took place at this time when people discovered and recognized the value of a new type of knowledge, namely Westerm secular and technical knowledge, and the villagers applied themselves to acquiring the secular education offered to them and came eventually to reproach mission schools for not offering enough of it.

Erakor in the 1970s[16]

As a result of the two colonial powers' activities in the field of education, the young are now better educated than the old, and they do not hesitate to remind them of this when the latter attempt to remonstrate with them or reprimand them. Religious life has also become less fervent than it was, since missions schools and Christianty have ceased to be the principal means of access to modernity. There is now, in any case, an idiom different from that of religion in which the people can express their aspirations.

The images of modernity that some villagers are now introducing are linked to the consumption of development goods, both at the individual and communal level. It would, in fact, be more correct to talk of extravagant or over-consumption.

What, then, is the current ideal of the *good life*? In 1973, this is how a well-off individual was defined in Erakor: someone who owns a car, a refrigator, a tape recorder, and a lawn mower; who keeps chickens and pigs in his yard, owns a new house made of corrugated iron with a place to shower close by; who has a gas stove, electricity, and a water tank; someone whose children are at secondary school or studying overseas and who, finally, consumes a lot of tinned goods, and regularly eats curry and tomato sauce with his meals. That year, nearly one-third of all households met most of these criteria.

Wealth is defined in terms of material comforts obtained outside the village, never by the ownership of goods available within the village itself, such as coconut plantations, fruit trees, or large houses made of native materials. What is even more striking, however, is the ease with which villagers accept these things without ever challenging their utility, while at the same time treating them with the carelessness they reserve for things of little practical value. Between 1973 and 1983, this tendency to 'consumerism' only increased, even though after independence the acquisition of Western goods became harder (Philibert 1984; 1988). Whereas in 1973 Erakor boasted only one house made of concrete, it had 31 by 1983. There was also 44 cars and light trucks that year, whereas ten years earlier there were only 14. Villagers have also heavily invested in the purchase of electrical and gas appliances and other durable consumer goods.

What we have here is clearly a sort of conspicuous consumption. Villagers' lavish expenditures on such purchases as cameras, record players or refrigerators reveal much more than levels of individual wealth. They indicate that the village is no longer the exclusive or dominant social field for an individual. Prodigality reveals a person's social aspirations and constitutes an element of upward mobility. This explains why the objects of conspicuous consumption are prized even after they have lost their usefulness. It is better to have a broken-down refrigerator than none at all, though it is worth less than one which works, even if it is rarely used. A car no longer in running order is not immediately scrapped because its value as a sign remains. This extravagant spending signifies membership in the modern world as they understand it. Such are the signs of the *good life* at the individual level, reflecting in this a form of social differentiation found in a colonial political system. Prestige lies in imitating one's social superiors, as likeness forms the basis of a claim to associate white status.

Considerable development also occurred during the same period at the communal level: a water supply was installed and running water was made available to most houses; the village was given an electricity supply and new roads were built; a large, modern community center and town hall linked by telephone to the capital was put up. People coming from other islands to Vila take a taxi to visit Erakor in the same way that we like to gape at wealthy neighborhoods.

While some of these developments, such as the water supply or the new roads, can easily be justified at a practical level, others can only be explained as a kind of collective conspicuous consumption of development goods, as the consumption of modernist ideas. This consumption amounts to a text which has much in common with the skit described above, even though the change from a sacred to a profane idiom had inverted the polarity of the discourse, because the 'text' still opposes the same protagonists, addresses the same audience, and carries the same underlying message.

The propensity to consume found in Erakor is partly explained by the emergence of an intermediate term between the *man bus* and the *European* models. While during the earlier phase, identity and behavior were determined by a dichotomous world view and could, therefore, be conveyed by a series of binary oppositions, there is now another (sociological) way of being, this time an intermediate position, that of the *enlightened* native. The identity of this sociological type is no longer bounded by mutually exclusive terms. In fact, the opposite is true, as the representations have become essentially indeterminate and elastic. Hence the ambiguity, the equivocal nature of the self-image of Erakor villagers who find themselves halfway between a customary cultural universe, their knowledge of which is now imperfect and their feelings towards which are, to say the least, ambivalent, and the Western culture they have not yet fully assimilated. The identity of the *évolué* is full of potentialities barely hinted at in the character of the modern native found in the skit.

It still has to be explained why these consumer goods are considered appropriate signs of the cultural identity villagers are building up for themselves. We can reject out of hand the explanation that the people of Erakor are singularly materialistic and highly preoccupied with their material comfort, since many of these goods are in fact of limited use to them and owe their existence in the village to their sign-value. And why manufactured goods? The reason has to do with the fact that it

125

is not so much the goods in themselves which count so much, as the ability to procure them and to use them for their intended purpose – an ability acknowledged by both Europeans and ni-Vanuatu. These goods reveal a new articulation of village culture with the world outside the village: the consumer goods so blatantly flaunted are literally the material proof of equality between villagers and Europeans, since they demonstrate an equal competence in the modern world, and are a mask of villagers' superiority over less progressive ni-Vanuatu.

Prior to independence, overcomsumption could be explained as the behavior of affluent people deprived of the means of asserting themselves politically. It is almost as if the villagers, considering all political activity closed to them because the only national structures were those of the administration and the missions, had directed their political action in on itself so that it became metaphorized into a symbolic discourse in the form of conspicuous consumption. Overconsumption could be considered a sublimation of political conflict, not to say a dramatization of social reality, the effect of which was to invest manufactured goods with an overload of symbolism, making them particularly suitable to define the cultural identity of the village.

This explanation is less satisfying now. Perhaps it is the entire cultural existence of Erakor people which acts as the mediating term between the *man bus* and the *European*. Such an interpretation would at any rate be in keeping with the innovative, not to say prophetic role the villagers have claimed for themselves on the national scene ever since they were the first to be converted to Christianity on their island.

I now wish for the final part of this chapter to move to the emergence of a national culture in Vanuatu.

III. NEO-TRADITION, GENERIC CULTURE, AND THE NEW REPUBLIC

I am attempting here to provide a bottom up view of the adaptation of ni-Vanuatu to colonialism and its aftermath by examining some of their ideologies. These ideologies,[17] developed as a way of bridging indigenous categories of thought and understanding of the world and Western concepts of religion, self, gender, the family, community, politics, etc, were subsequently incorporated into programs of action. My premise is that in the land of the lax Condominium of the New Hebrides, the initiative was often left to and taken by ni-Vanuatu. I have documented so far what I take to be opposite strategies of adaptation to the modern world.

I now wish to analyze a dominant ideology *cum* political discourse that has been in existence in Vanuatu since the 1970s, that of *kastom*,[18]

126

a construct that has seen yoeman service not only in this country, but also in other recently indepenent Melanesian micro-nations.

While there have been in the past attempts at rehabilitating traditional culture, such as the John Frum movement on Tanna in the 1940s, this was only achieved in the 1970s by people who, height of irony, had lost theirs a long time ago. They were the young ni-Vanuatu pastors or civil servants who had been educated overseas in secondary or techincal schools or in universities in Fiji or Papua New Guinea. There, with other Islanders, they learned about Africam decolonization and found in the rhetoric of African independence an idealized version of precolonial indigenous societies (Keesing 1982, p. 297). Like others before them, they turned to their precolonial past to find a cultural identity, a set of ways and values which owed nothing to the European colonizers, on the existence of which they based their claim to independence.

For these young militants, the founders of the Vanuaaku Pati, *kastom*, was a vague compendium of traditional practices which could not be those of any one group in particular, if they were to act as a nationalist cement. It was a romantic version of the anthropological construct of Traditional Culture in the Pacific. Given the geographical, linguistic and cultural fragmentation of the country (a fragmentation little diminished throughout the colonial period by the existence of numerous Chruches, two colonial Powers, schooling in two European languages), it is hard to imagine what else but a so-called shared traditional culture could have served as a symbol of national unity.

Vanuaaku Pati activists also became aware of a second necessity, an organizational one this time, if they were to persuade other ni-Vanuatu that the time for independence had come. They, who were Protestants or Anglicans, saw in these largely indigenized Churches the only national organizational structure they could control. As Tonkinson remarks,

> [these Churches] were long established, reasonably decentralized and well organized – *and* they embraced a clear majority of the country's population. Additionally, most were self-governing and locally led . . . By 1973 the General Assembly of the Presbyterian Church of the New Hebrides, representing more than half the country's population, was calling for the co-operation of the colonial administrations in achieving self-government without delay (Tonkinson 1982, p. 308).

Yet churches, in particular the Presbyterian Church, had been in the past fiercely opposed to traditional culture and the relation between the two was in the main still expressed through the nineteenth century metaphors of light and darkness. Even if the militants never envisaged anything more than 'kastom *within* Christianity' (Tonkinson 1982, p. 313), a re-evaluation of traditional culture still had to take place. The majority of ni-Vanuatu, who are devout Christians, had now to go through a semantic conversion which would remove the antithesis between these two categories of knowledge. This is what Lindstrom reported from Tanna:

> Although [Vanuaaku Pati] leaders come principally from the pastorates of different Christian churches, they bring a message that *kastom*, as a whole is wise; that *kastom* knowledge and Christian, or modern, knowledge need not contradict; that all are of the light (Lindstrom 1982, p. 323).

But this message, devised in urban areas, raised problems of comprehension for many rural populations. For *kastom* to work as a unifying symbol, it had to be a nebulous concept which was ideally all things to all people, a 'contentless symbol' as Keesing calls it. This is not the way the message that *kastom* must be followed was received. As Tonkinson notes,

> The masses for whom this consciousness-raising was designed tended not to interpret the message ideologically: instead, they grappled with it in terms of practicality, and much confusion resulted. People in some rural areas took the message quite liteally . . . One important reason why this adjustment proved difficult for the Christian masses was that their leaders, the promulgators of the rallying cry, had never spelled out in detail what kinds of *kastom* were worthy of revival and promotion . . . it was essential that no one became too specific about *kastom*, yet it was generally taken for granted by church and party leaders that some kinds of *kastom* would remain unresurrected, lest the world at large brand Vanuatu 'primitive' (Tonkinson 1982, pp. 310–13).

This bewilderment was not limited to rural areas. If *kastom* was now compatible with modernity, the contexts in which each set of rules applied must be spelled out. The sophisticate urbanites who had until now felt on the winning side of history suddenly found themselves on shaky ground. What will happen to people who have no *kastom* of their own left? Following the 1979 Arts Festival (Philibert 1986), an Erakor villager felt compelled to send a recorded message to Radio-Vanuatu to account for the villagers' lack of *kastom*. He said that they had sacrificed their traditional culture so that the new world could come about. Ni-Vanuatu from other islands now working and living in Port-Vila were the beneficiaries of this sacrifice and should not forget it. If those who mock peri-urban villagers for their lack of *kastom* feel so strongly about it, he added, then perhaps they should go back to their island to devote themselves to the practice of their beloved *kastom*.

Concerns about a future in the shadow of *kastom* also took more practical forms. Erakor women wondered if money would be available, what sort of schooling their children would receive, and if they would still be able to find laundry detergent.

While these difficulties had on the whole been surmounted by the end of the 70s, other problems had developed. As we saw on Tanna, the symbol of *kastom* was complex enough to be Janus-faced, to stand for difference as much as similarity. Tonkinson again:

> [*kastom*] stands as a powerful symbol of national unity but carries within itself the seeds of divisiveness at all levels of society . . . [it] is an excellent device for separating different groups and for making ethnic boundaries. Because the question of what constitutes *kastom* is open to negotiation and competing factions use it to justify or legitimate their actions, it becomes inseparable from politics (Tonkinson 1982, p. 312).

Lindstrom describes in the following terms the increasing politicization of the symbol on Tanna:

> Men of all factions on Tanna brandish their *kastom* as symbolic spear in political oratory and debate. The contest has shifted, with the collapse of the moral opposition between *kastom* and modern, from competing black and white evaluation to the field of definition and control. In that *kastom* is generally useful, its definition is a political act. Rival groups argue over the boundaries of *kastom*, the application of the lexeme to various sorts of behavior, as they do over the ownership of *kastom*, as political symbol, itself (Lindstrom 1982, p. 325).

He goes on to say:

> Depending on context, a man may one minute claim to be more modern than his rivals . . . the next minute, he claims to be more *kastom* than they. As national leaders rhetorically proclaim a melding of tradition and development, so do individuals combine the two as personal faculties. Rival political groups may claim both correct knowledge of modernity and correct knowledge of *kastom*. This is a question of control. Whose symbol is *kastom*? Or, more exactly, which intepetation of *kastom* is the real *kastom*? (Lindstrom 1982, p. 327).

Given the context of its production, justification, and appropriation, it seems to me to be facing the wrong way to search the past in order to understand *kastom*. Lindstrom's approach highlights what seems wrong to me about Bonnemaison's attempt to find the bedrock of traditional culture on Tanna and to consider it the source, to mix my metaphors, of the political infighting on the island. There is no base line from which to measure true, authentic traditional knowledge, which itself was once invented. Had *kastom* not existed today in a politically expedient form, some Tannese would have been bound to 'rediscover' an invaluable piece of knowledge so old that it had all but been forgotten.

Kastom is a form of neotradition, some of which is invented, some not, and this is equally true of the intellectualization of the past put forward by the Vanuaaku Pati and the Tan Union. In both cases, the symbol underwrites an attempt to gain control over one group's construction of reality as the natural order of things. I beleive that one is getting closer to understanding the production of *kastom* by examining carefully the link between symbolic codes and power. This is how Lindstrom puts it:

> Consensus, or public codifications of the definition and domains of *kastom* and its interface with the modern . . . exist within political organizations. Groups (and leaders) compete by attempting to establish some particular interpretation of *kastom* as orthodoxy. These rival interpretations . . . establish political distinctiveness . . . *kastom* knowledge, like other disputed bodies of knowledge, is both an idiom and a legitimation of political dispute: a language of divergent practice (1982, pp. 327–8).

In the hands of the Vanaaku Pati ideologues, *kastom* has become a civil discourse, a way of socializing unwitting ni-Vanuatu to the idea of a modern nation-state in the guise of faithfulness to a common past. In other words, this discourse is part of an indigenization of the state through the development of a generic, no-name brand culture to promote nationalistic sentiments toward the country as a whole.

As such, *kastom* is part of an hegemonic discourse put forward by the proto-bureaucratic class now in power. The model of traditional practices has been transformed in the process in that it no longer stands in opposition to Christianity and modernity. New limits have also been set on such practices: they now require the *imprimatur* of those in power before being accepted as *kastom*, and they must challenge neither the Churches nor the government. What constitutes Vanuatu social and cultural identity, the image of its past as well as that of it future, is now controlled like never before by one group of ni-Vanuatu.

This Western-educated political elite is also putting forward a definition of national leadership modelled after the enlightened chief who governs by virtue of his knowledge and wisdom. Modern political culture is presented as a continuation of traditional ways, in other words as culture *tout court*. This is how I described what is taking place:

> When political culture becomes culture itself, that is to say when the set of rules, values, and symbols at work in the political arena are presented as no different from those at work in social life in general, the by nature contentious dimension of politics is evacuated from the field. The political discourse is no longer seen as political which means opposition becomes difficult as there is no terrain on which to stand. It becomes virtually impossible when those in power also supply the vocabulary which allows people to speak about *kastom*, that is to say when they hold the seal that authenticates events as traditional. Moreover, when the capillaries of power are distributed all over the social organism, *kastom*, this *son-et-lumière* orchestration of the 'past', largely conceals the differential distribution of power and its exercise (Philibert 1986, p. 9).

It is not my intention to explain the successful invention[19] of *kastom* as something entirely stage-managed by capable party ideologues. Conspiracy theories do not work because the link between culture and power is never an unmediated one. The construct would not have found ready acceptance had it not articulated ill-defined sentiments and unformulated ideas already in existence, both in rural and urban areas. As teleologically-inclined social scientists used to say, it answered a need. Ideological conflicts of the kind seen on Tanna between the Moderate and Vanuaaku parties are central to the cultural reproduction of social groups. However, there is nothing new in this as Keesing reminds us:

> . . . long before Europeans arrived in Pacific waters, Melanesian ideologues were at work creating myths, inventing ancestral rules, making up magical spells, and devising rituals. They were cumulatively creating ideologies, which sustained male political ascendancy and resolved contradictions by depicting human rules as ancestrally ordained, secret knowledge as sacred, the status quo as eternal. We err, I think, in imagining that spurious *kastom* is radically different from genuine culture, that the ideologues and ideologies of the post-colonial present had no counterparts in the pre-colonial past (Keesing 1982, pp. 300–1).

Still, there remains an odd irony: Melanesian intellectuals have borrowed the anthropological concept of custom at the very moment when anthropologists are abandoning it. Let me explain. The demise of the notion of customs to denote shared habits and values of traditional peoples corresponds to a recent reappraisal of the concept of culture, which has signalled a departure from our old totalizing way of conceptualizing shared norms, values, ideas, habits, perceptions, etc. Anthropologists have been slow to realize they had bowdlerized culture, though in this case power was left out, not sex: the offical culture portrayed in monographs is not, as suggested, everyone's culture and, what is more, cultural knowledge does not exist in a political vacuum. Knowledge is power; as such, it is not evenly shared and what constitutes acceptable knowledge in the first place is itself partly a reflection of power relations within a society. Although different social groups may possess different definitions of reality and carry different social projects, such ideas do not all have the same persuasiveness, nor the same legitimacy.

Anthropologists now believe that cultural production always serves someone's interest. They no longer regard social systems as moral-jural orders which can be broken down into statuses and roles, into which members of society are socialized during youth and which they are bound to uphold later in life. Anthropologists now take a dynamic, conflicted view of culture: they consider cultures as the terrain on which social groups vie with one another for the recognition and acceptance of their own views of society as the natural and, indeed, the only sensible way of doing things. In other words, they are interested in the manner in which ideas held by various groups are transformed into 'customs', a process riddled with tension, contradictions, and a great deal of resistance. Social groups compete with one another for cultural space; they strive to have what they consider right and proper inscribed in a transient cultural landscape. Ideologies are at the center of cultural reproduction, a process never assured, nor completed. As we saw on Tanna, hegemonic discourses lead to counter-hegemonies as people debate the adequacy of particular symbols of *kastom* to reflect changing experiences.

Larcom, one of the most perceptive writers on *kastom* in Vanuatu, concludes her analysis by saying:

> . . . there is reason to doubt that *kastom* will easily assume the legitimating status of a rule-bound charter or precedent in Vanuatu . . . one may expect *kastom* to remain unfinished and supple . . . it is important to bear in mind that, as it seeks to empower an authentic Melanesian past, Vanuatu builds on invention itself (Larcom 1982, p. 337).

Far fewer customs appear in Melanesian ethnographies now, at a time when *kastom* is flourishing in Melanesia, than used to be the case. It is hard to tell which of the two is the more cynical: the anthropologist who no longer uses custom or the Melanesian ideologue who still uses *kastom*.

Conclusion

The problem raised at the beginning of this chapter was how to analyze social change in peripheral countries in a way that will restore to the local people the historical role denied to them by modernization theory or a world system approach; how to portray local actors as agents, even though they may not always be the determinant cause of their fate. The solution adopted here has been to focus on the indigenous ideologies through which actors orient themselves in the world and from which they derive programs of action.

This approach seems particularly applicable to Vanuatu, a country so peripheral that colonial powers did not deem it worthy of sole ownership. Between colonial neglect and administrative laxity, ni-Vanuatu found plenty of room to maneuver and we have examined the strategies marking the two poles of indigenous adaption to colonialism.

While I hope that placing local actors squarely in the middle of the colonial stage has shown the resourcefulness and sagacity of ni-Vanuatu, just as importantly, it has led me to reflect on anthropological explanations. The study of various ideologies, that of *kastom* in particular, our subjects' own notion culture, brings to the surface the unwarranted assumptions underpinning the anthropological concepts of custom and culture. Anthropology has not until now paid sufficient attention to the hierarchical context in which the production of culture takes place.

It seems to me that if Melanesian ethnographers are to contribute at all to emerging debates in the discipline, they will have to move away from the usual study of self-contained groups to the cultural production and reproduction of more complex social formations in town.

5

Social Change and the Survival of Neo-Tradition in Fiji

Vijay Naidu

Introduction

This chapter uses the articulation of modes of production approach to analyze the evolution of Fiji society from the early nineteenth century to the present period. As this is a rather long period of time (some 180 years) attention will only be focused on what are considered to be the most significant structural dimensions and most crucial changes in the different historical periods. The central argument of this chapter is that Fijian society had undergone very considerable and fundamental social changes, but because of the survival of neotraditional structures and ideology, it is not unusual for people to think that Fiji is caught up in the divergent pulls of tradition and modernization. We begin by describing the contemporary Fiji society.

Contemporary Fiji

Fiji has been in crisis since the military-led coup d'état in May 1987. From a relatively peaceful multiracial society, Fiji has been transformed into a society where militarism, racism, and religious bigotry are profoundly affecting the social order. At the last population census in Fiji, in August 1986, the population of the country was as follows:

Table 1 *Ethnic Composition of Fiji's Population*

Ethnic Fijian	329,305
Indo-Fijian	348,704
Other	37,366
Total	715,375

Source: Fiji Census Report, 1986

Three aspects of this table are noteworthy: (1) The multiracial character of Fiji society; (2) The preponderance of Ethnic Fijians and Indo-Fijians and the numerical majority of the latter over the former; (3)

134

The relatively small number of whites in Fiji (included in the 'other' category). Structurally, the bulk of the people of Fiji are workers and peasant farmers. Political and economic power is vested in a small multi-ethnic oligarchy in which Ethnic Fijian chiefs and Europeans are particularly prominent.

Since the coup, the existing stream of emigration from Fiji has increased significantly. Although members of all the ethnic categories are represented among the immigrants, Indo-Fijians have been predominant. According to the Fiji Government Bureau of Statistics, The Ethnic Fijian population is now the largest population component because of this massive outflow of Indo-Fijians.

Fiji's sugar, tourism, gold, timber, copra, ginger, and other primary products-dependent economy has suffered as a result of the recent political events. In 1987 the economy declined by 12 percent and the anticipated contraction of production for 1988 was eight percent. Redundancies and unemployment have increased at a time when inflation and devaluation have eroded the real incomes of most Fiji people. For the moment, because of the stable sugar prices and a remarkable recovery of tourism, the economy has stabilized somewhat. Although investor confidence remains low, the establishment of Free Trade Zones or Tax Free Zones with 13 tax holidays a year has attracted garment manufactures who will export their products to Australia, New Zealand, and the United States. These enterprises are unlikely to make any significant contribution to the Fiji economy because of the low wages paid, the absence of taxation and the inclination to repatriate profits (Prasad 1988).

Meanwhile Fiji has developed stronger relations with a number of Southeast Asian countries, Japan, and France. Among the Southeast Asian countries, Malaysia has been prominent in supplying consumer goods to the Auxiliary Unit – the commercial arm of the Fiji military. The Japanese have brought out a number of large tourist resorts and Fiji's national airlines, Air Pacific, and the Fiji Visitors Bureau are promoting Fiji as a tourist destination for the Japanese. France has taken advantage of military-based rule in Fiji and has given direct material and financial aid to the military of some $5 million. However, the basic structural dimensions of Fiji's economy and society have remained unchanged and the subsequent section examines the historical origins of contemporary Fiji society.

PRE-COLONIAL AND EARLY CONTACT PERIOD 1800–1874

When Abel Tasman sailed through Fiji waters in the late seventeenth century, the archipelago was inhabited by between 200,000 and 300,000 islanders. The arrival of other European explorers, whalers, sailors, sandalwooders, beachcombers, missionaries, and traders simply inten-sified and reinforced the processes of social change that were already underway in Fiji. Europeans came at a time when the chiefly modes of production had become dominant in the eastern parts of the archi-pelago. The communal mode of production continued to exist in the west and central parts of Fiji.

The communal mode of production was characterized by common ownership of land and an authority system based on the rule by elders of each clan. Access to women and elite goods (such as salt and whale teeth) were controlled by the elders. Heredity forms of authority were weakly developed. Political units were small and akin to what has been described as the Melanesian type. It is not clear if a system of big-men existed (Sahlins 1958). However, what is known is that the chiefs present did not have the almost absolute authority of eastern chiefs. The men's club house was an important arena of decision making. Very little social distance existed between the leaders and their people.

In sharp contrast to these arrangements in the communal mode of production, the chiefly mode of production exhibited a tendency towards the deification of paramount chiefs. Although apparent communal ownership in land continued to be a feature of even the chiefly mode of production, access to land and other resources and the control of labor and the fruits of labor became vested in chiefs. By the time of regular contacts with whites, and especially beachcombers, chiefs, in such polities as Bau, Bua, Cakaudrove, Lau islands, Lomaiviti islands, and Rewa had almost absolute power. The priestly class or *bete* acted to balance some of the chiefs' power.

Surplus appropriation took three forms in the eastern chiefdoms:

1. Tributes or *sevu* such as the 'first fruits' at harvest, the best produce of the land and the best catch from a fishing or hunting expedition (including particular species) came from sub-ordinate groups and tributary arenas.

2. Labor services or *lala* were rendered to the chief, and these included work in his gardens, building or repairing his house or canoes, and contributing to the preparations for feasts at moments of life crisis, and the entertainment of visitors.

3. Appropriation of various items through the chiefs' redistributive role. This role portrayed the chiefs as 'givers', thereby attenuating their obviously exploitative relationships with commoners.

The relations of exploitation that existed in pre-European Fiji between elders and their junior kinsfolk and between chiefs and com-moners, were also obscured by kinship ties and the ideology of kin-ship. Social relations based on kinship ties (whether actual or putative) governed all aspects of one's life, including status in society, marriage relations, access to land, wealth, and political influence. Generally in Fiji, descent was taken along patrilineal lines although matrilineal ties were strong, particularly in the interior and western polities. The relation of *vasu* allowed a sister's son or *vasu* to make demands on her brothers for material objects, and in the case of a chiefly *vasu* for services, including assistance in war.

Chiefly power was both constrained and reinforced in the eastern chiefdoms by the existence of a division of labor based on lineage affiliation. That is, certain lineages were allocated specific responsibilities in relation to the rest of society. Roth (1973) listed the following lineages for the chiefdom of Bau:

Turaga – chiefly lineage
Sauturage – chiefly executives
Matanivanua – heralds
Mataisau – carpenters
gonedau – fishermen
bete – priests.

These functional lineage-based units were ranked with the chiefs and priests at the apex of the hierarchy.

Trade developed between communities with differential endowments (Thompson 1940). Coastal dwellers provided sea-derived commodities in return for products of the inland areas. People on atolls and raised limestone islands which were poorly endowed specialized in producing high quality mats, wooden bowls, and canoes. In return for these, they received foodstuffs and raw materials of the volcanic and continental high islands. In Fiji interisland and intervillage trading known a *solevu* were important. Area specialization occurred, with Rewa specializing in pots; Nadroga in salt; Vatulele in native cloth (*masi*); Kadavu in mats; Colo in *yaqona*, timber, and bamboo; Serua in Kauri gum; and Lau in rope fiber, canoes, and wooden bowls (Mac-Naught 1982, p. 60). Eastern Fiji, Tonga, and Samoa were connected by trade and intermarriage among the chiefly groups. Tongans acted as middlemen, exchanging fine Samoan mats for scarlet parrot feathers, spears, timber, and canoes which came from Fiji. Samoan and Tongan craftsmen resided in the eastern islands of Fiji where they built, among other artifacts, the massive double-hulled *druas* or war canoes.

Contradictions existed in both the western and eastern social formation. In the communal modes of production, the social units were smaller and more cohesive. However, the appropriation of surplus by male elders at the expense of junior kinsmen and women formed the basis of dissension. The competition over resources between neighboring polities also led to disputes and warfare. In the eastern chiefdoms contradictions took a cyclical fashion. Chiefdoms expanded as a consequence of surplus siphoned off from the commoners and from the tributes of subjugated adjacent groups. Ambitious chiefs engaged in warfare to extend their control over neighboring regions and entered into strategic marriages to bolster their influence.

As the chiefdom's influence expanded, a punitive bureaucracy emerged to facilitate tribute extraction, organize ceremonies, and receive emissaries from neighboring polities. The residential village of the paramount chief became a major political, economic, military, religious, and ceremonial center. Such villages were fortified, and in the Rewa Delta fortification included moats. For the upkeep of the leading chief and his retinue, more and more surplus was needed. When the appearance of reciprocity between the commoners and their chiefs became seriously undermined by the extraction of tributes and services, conflicts became endemic within the chiefdom. Moreover, subjugated communities who paid tribute were also likely to rebel when the demands upon them became excessive. The structural limits of an expanding chiefdom were reached in this way:

> . . . with the realm now stretched over distant and lately subdued hinterlands, the bureaucratic costs of rule apparently rose higher than the increase in reveue so that the victorious chief merely succeeded in adding enemies abroad to a worse unrest at home. The cycles of centralization and exaction are now at their zenith (Sahlins 1972, p. 245).

The prehistory of Fiji was dominated by three major processes. (1) The movement of people, as a result of conflicts arising from competition over resources, to virtually all the inhabitable corners of the archipelago; (2) Struggles for power within social formation between powerful lineage; (3) The waxing and waning of chiefdoms as they went through the contradictions described above. Until the arrival of the white man, there was no Fiji-wide authority and it took direct colonialism to create one.

Early Contact Period

Following the early explorers, traders, whalers, beachcombers, and missionaries promoted exchanged relations (barter and trade) with the islanders in Fiji. In the first half of the nineteenth century sandalwood and beche-de-mer, or sea cucumbers, became valuable items of trade. Profits of over one thousand percent were not uncommon. By the 1850s both these commodities were depleted and coconut oil exports had become significant. The production of coconut oil was especially encouraged by missionaries.

Levuka, on the island of Ovalau, emerged as a port town. European shopowners, agents of larger overseas-based merchant houses, and shipwrights, established themselves and vied for the trade in provisioning ships and an increasing class of settlers. Besides the 'grog shops' that proliferated in Levuka, rudimentary European legal and political institutions began to emerge. With the establishment of consuls of foreign powers that followed the increased European presence, the demands for more stable legal and political arrangements intensified.

Among the indigenous polities, the erratic process of political consolidation was accelerated with European contact. The uneven distribution of tradable goods among the different polities meant that the balance of forces between them was affected. Islanders, especially eastern chiefs, 'endeavored to turn the coming of the whitemen to their best possible advantage' (Shineberg 1967, p. 216). Through the customary exchange networks, durable goods such as axes, whale teeth, and guns accrued to the chiefs and local dignitaries.

Between 1828 and 1850 well over 10,000 muskets and rifles were introduced into Fiji. According to Ward,

In the wider scene, the forces of Mbau greatly increased their fire power during the beche-de-mer period as Mbau had close relations with many coastal groups in the main production areas. This was to be an important factor enabling Thakobau to establish Mbau as the dominant political unit in the 1850s. What pork did for Tahiti, beche-de-mer did for Fiji (1972, p. 111).

This process of consolidation of political power and influence over wider areas was facilitated by whites – beachcombers, missionaries, traders, and settlers. Beachcombers provided the firepower in battles; Wesleyan missionaries contributed to the development of Bauan Fijian as a lingua franca among the islanders, and also reinforced the dominance of chiefly rule (Koskinen 1953); traders and settlers supplied arms and ammunitions in return for more land and also to have 'their' chief take a dominant role.

Generally, in Eastern Fiji, tyrannical theocracies were established in which the morals of lower middle-class Victorian England and close supervision of Catholicism were the basis of social control. A system of laws about alcohol consumption, dress, Sabbath, and the relations between the sexes were rigorously enforced. Punishment in the form of hard labor, fines, and banishment was meted out by the judges who belonged to chiefly lineages. In Western Viti Levu the missionary compound with its mission house and adjacent residences of converted indigenes became the bases of control from which the missionary sought to ease out more followers from the non-Christian neighborhood. Often open conflict existed between those inside the compound and those outside.

Capitalist Penetration

A major initial impulse to export-oriented activities were the missionaries themselves. This was recognized in 1836 by the pioneer missionary, John Williams, who declared 'wherever the missionary goes new channels are cut for the stream of commerce' (quoted in Koskinen [1953, p. 132]; see also Grathen [1963, p. 201] for similar remarks).

The earlier church collections were in kind and included coconut oil and tortoise-shell, but by the 1860s demand for money was not unusual. Islanders were pressured to work consistently instead of in their erratic fashion. Missionary insistence on the wearing of clothes, daily shaving, contributions to the church, and purchase of bibles and hymn books created profits for traders (Graves 1984, p. 122). Missionaries seized the rivalry between individuals and families within a village and at the level of villages in a district as well as interdistrict rivalry to increase church collections. Prestige was equated with the amount of contributions given to the mission.

Other whites objected to the paternalistic and proprietary attitude adopted by missionaries. As some of the latter opposed the sale of alcohol and arms, transactions in land, sexual relations between island women and white men, and the labor trade, they conflicted with traders and settlers. Further, as middlemen and as appropriators of some of the best land and users of island labor, missionaries were seen as competitors. In the 1860s in Fiji, the conflict between the missionaries and settlers intensified.

European Merchant and Settler Capital

The number of settlers continued to increase in Fiji from 30 to 40 in 1860 to 4,000 by 1870. Cotton had become a highly demanded commodity with the disruption in cotton production and trade caused by the American Civil War. Settlers in Fiji tried their hands at producing the south seas island variety of cotton. The inception of settler cultivation had serious implications for the archipelagoes. Whereas trade relations tended to reinforce traditionally powerful groups, the establishment of plantations, with their demands for land and labor, threatened their basis of existence. Most eastern chiefs sought to cooperate with the settlers.

Enormous land claims were made by settlers and companies formed to speculate on land. The Melbourne-based Polynesia Company attempted to draw leading Fijian chiefs into land deals. Although land was communal property and not a commodity for sale in the precapitalist modes of production, it is recognized that many chiefs sold land with a clear understanding of the nature of alienation (France 1969, p. 52). Such sales were made possible by local chiefs who needed guns in their bid for military and political ascendancy. Ratu Seru Cakobau of Bau, for instance, ratified land sales at 25 cents an acre. In many cases land that did not belong to a chief but to neighbouring groups was transferred to Europeans. From the proceeds of land sales, ascendant chiefs also built European-type timber houses and purchased luxury schooners (Derrick 1950).

Alienated land was the subject of speculation or development of trade stores, ports, residences, and was used to plant crops that had a market overseas. A market for land had also emerged where none had existed before. More generally, 'nearly every known tropical plant of economic value was tried out in Oceania at one time or another, but only copra, sugar, coffee, cocoa, vanilla, fruit, cotton, and rubber had any real significance' (Oliver 1961, p. 121). In the 1860s cotton cultivation expanded, as was shown by the dramatic increase in the value of cotton exports. In 1864, copra brought in £13,000, which was much higher than the £3,260 received from cotton exports. By 1867, however, copra accounted for £3,260 and cotton for £34,000 of the revenues. In percentage terms, cotton increased from 15.2 percent to 82.1 percent and copra declined from 77.5 percent to 8 percent of the total export revenue (Sutherland 1984, p. 65).

Settlers with little capital were dependent on credit from the larger companies and were eventually overshadowed by merchant houses such as the Hamburg-based Godeffroy and Son.

The Labor Trade

Settler success in establishing plantations relied entirely on a supply of cheap labor, which resident islanders were not predisposed to provide. Planter capital, therefore, resorted to imported labor. The *Fiji Times*, during the 1860s and 1870s, expressed the widely-held settler view that only black slave labor could be profitable in the tropics. Such labor was extracted from the Solomons, Vanuatu, Tokelau, islands off New Guinea, Kiribati, and Tuvalu. These islands became the labor reserves for plantations in Fiji, Samoa, Hawaii, Tahiti, Australia, and even South America (Burns 1963, p. 109).

The living and working conditions of these islanders were dreadful. Although 70 percent were between 15 and 35 years of age, their death rate was between three and six times that of whites. Both in Fiji and in Queensland the islanders died like flies in the peak years of the 1880s (Scarr 1967, p. 48). The high death rate on the plantations, the deception and violence used in recruitment, and the great many deaths aboard some 'blackbirders' led to condemnation of this form of slavery by the Aboriginal Protection Society and various anti-slavery groups. With the murders of Bishop Patterson in 1871 and of Commodore Goodenough in 1875, the outcry against islanders suspected of attacking whites escalated.

In the last two decades of the nineteenth century this so-called labor trade came under increased surveillance by the British navy at a time when many of the islands of the western Pacific were exhausted as sources of labor. A number of the blackbirding ships flew foreign flags and continued to ply this trade in island-labor well into the second decade of the twentieth century. However, the islands were never a completely reliable source of labor and, with their depletion, more adequate sources of labor had to be sought if plantation agriculture was to survive.

THE EARLY COLONIAL PERIOD 1824–1920

With the annexation of Fiji, the British Imperial state acquired yet one more colony in its vast empire. The task of establishing Fiji as a viable self-paying colony fell on the first governor, Sir Arthur Gordon, who was an aristocrat and an experienced colonial administrator. Gordon was given a loan of £100,000 to organize the colonial state apparatus. His decisions on such matters as the system of administration, land tenure, labor policy, and economic development were crucial in the evolution of colonial and postcolonial Fiji society.

Colonial Administration

Shortly after Gordon's arrival in Fiji, a measles epidemic killed 40,000 Fijians and provoked the highland people of interior Viti Levu to take up arms against settlers and Christianized Fijians. In the so-called 'little war' of 1875, the chiefs of Bau and other eastern areas joined the British to quell the rebellion by the subsequently ridiculed *Kai colos* or devil tries (Brewster 1922). This chiefly expression of 'loyalty' and their ceremonial acceptance of Gordon as high chief contrasted with white settler demands for land, labor, and Anglo-Saxon hegemony. With the Anglo-Maori wars and establishing a cheap administration in mind, Gordon instituted a system of indirect rule, long before Lord Lugard in Nigeria.

He took the existing eastern chiefly hierarchy and associated confederacies as the basis for his native authority system. He was encouraged in this policy by John Thurston, the colonial secretary, and also by the fact that there already existed a system of 'native' administration in the 1871–73 interim settler government. By co-opting the chiefs into the administration, Gordon was not only able to utilize traditional rulers but also to obtain some legitimacy for his own colonial administration. The incorporation of the chiefly order within the colonial state meant that although the chiefly structure remained ideologically traditional, in reality it was being transformed into a foreign-controlled instrument of dominance over commoner Fijians. But this point should not be taken too far since earlier governors were mindful of chiefly interests. Thus Governor Des Voeux (1880–85) maintained that:

> The chiefs represent the army and navy and practically the police of the country. The maintenance of their interests is therefore necessary even on these somewhat selfish grounds alone (*Fiji Royal Gazette* 1884, Vol 10).

Administratively, four significant changes affected the precapitalist polities. First, the selection of chiefs was no longer subject to exclusive local control and heredity principles increasingly overtook those based on service to their people (Nayacakalou 1975, p. 14). Second, the great complexity of Fijian polities was simplified and the numbers reduced to 12 (Legge 1958). An idealized model of what constituted traditional authority was derived entirely from eastern Fiji and extended all over Fiji (Roth 1973; France 1969). The relative autonomy of many territorial entities (*vanuas* in later terminology) and political conferations (shifting alliances between these entities called *matanitu*) were overlaid by the colonial administration. Acquiescent high chiefs were made provincial governors, or *Roko Tui*, local chiefs designated *Buli* headed 80 districts (*tikinas*), and below them were the *Turaga ni koro*, the village headmen. In the early twentieth century there was a gradual trend towards having *bulis* who were not necessarily local chiefs.

Parallel with these administrative ranks were councils. The colony-wide Great Council of Chiefs (*Bose Vakaturaga*) was at the apex, followed by the provincial and district councils (Roth 1973; MacNaught 1982). At the base of these institutions was a villagization process which aggregated Fijians into easily administrable village units (Thompson 1972, p. 60).

Third, during the colonial era, Bauan Fijian became the lingua franca of the emergent chiefly class (consolidated by the Great Council of Chiefs) and many other aspirant Fijians. Earlier forms of centralization, through the federation of more or less equal partners or through conquestion of neighboring areas, were replaced by centralization through colonial dictate.

Finally, the hegemonic position of eastern chiefs in the colonial society was administratively reinforced through their differential influence, as illustrated by their unequal salaries. The Bauan Cakobau was given an annual pension of £1,500 followed by his rival Ma'afu who received a salary of £600 as *Roko Tui Lau*; the *Roko Tui* Tailevu received £340 and the chiefs from the western side of the main island, *Tui Nadroga*, *Tui Ra*, and *Tui Ba* received £100 each (Samy 1977; *Fiji Blue Book* 1877). Although parochial concerns continued to influence Fijian politics, the establishment of the native administrative began to stabilize a hierarchy of vested interests tied to the colonial administration. The government comprising British officals had loyal allies in the chiefs.

Land Tenure Arrangement

Another important factor in the articulation of capitalism with the precapitalist mode of production was the colonial land policy. The Native Land Ordinance of 1880 made Fijian-owned land inalienable except to the Crown. Fijian land could be leased, subject to the Governor's approval, for 21 years. Settlers claimed 845,000 acres of Viti Levu, Vanua Levu, and the Rewa delta. A Lands Claim Commission was established to examine settler claims. Although the Governor and the Commission took the view that every inch of Fiji's land was 'traditionally owned by *mataqali* and that all land was, according to immemorial customs of Fiji, inalienable', this principle of inalienability was not applied (France 1969, pp. 117, 122). European possession of over 414,615 acres or ten percent of Fiji's land area was confirmed, an affirmation of permanent ownership which was foreshadowed by Gordon's message to the settlers:

> I confess I do not like to see a man put upon on account of his colour; but this is certain – I have no sympathy with those whose philanthropy demands that they should think little of their own race and colour. My sympathy for the coloured races is strong; but my sympathy for my own race is stronger (Stanmore 1880, p. 184).

The more settler-oriented Governor, Sir Everard in Thurn, allowed the further sale of indigenous lands, and alienated 20,184 acres in fee simple between 1905 and 1909. Strong opposition from Lord Stanmore (Sir Arthur Gordon), in the House of Lords, brought this policy change to a halt.

In response to the diverse indigenous tenurial forms that existed in Fiji, Governor Gordon had urged the Great Council of Chiefs to identify the main land-owning groups. After prolonged discussions, this body agreed that the *mataqali* was the traditional land-owning body. But the demarcation of the boundaries of land claimed by various land-owning groups involved along and disputatious affair. The Native

Lands Commission, established in 1880, incorporated the views of 'interested native chiefs' and standardized the diverse regional tenure forms for centralized administrative efficiency (Nayacakalou 1971, p. 218). In *The Charter of the Land* (1969), France points out that after some 37 years of investigating the constitution of native tenure, the colonial regime appointed G. V. Maxwell to create such a tenure (see also Samy 1977, p. 49; Clammer 1975, p. 201).

Accordingly in 1913, a model of traditional Fijian social structure was stipulated:

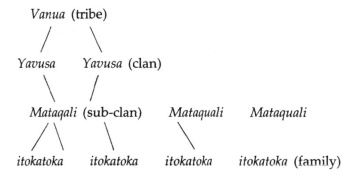

This British standardization of customary tenure was accepted by a later generation of chiefs as their 'tradition' (MacNaught 1982). Under colonial rule the formerly flexible system of land ownership was recorded and frozen (Nayacakalou 1975, p. 12).[1] Population changes, henceforth, would result in an inequitable distribution of land between *mataqalis* and within *mataqalis*.

The trend towards land becoming a commodity was halted, but by this stage land had acquired a new value as the means of production of commodities for exchange. Previously, tree crops planted in clusters of about one quarter of an acre were sufficient for a family's subsistence, but with the cultivation for trade and increasing dependence on purchased goods, 'several acres of tree crops [were] now in use' and there was 'a need for more land per head' (Crocombe 1971, p. 5).

In Fiji, by codifying what were deemed to be customary forms of tenure, colonial administrators also reinforced the titles of chiefs and the control that these titles gave over indigenous lands. Colony-wide bodies such as the *Council of Chiefs* consolidated a supra-local landed oligarchy of *ratus*. These colony-wide bodies brought chiefs together and imbued in them a sense of common identity and, on matters such as titles and lands, a definition of mutual interests.

From the point of view of political control over the majority population, the use of established hierarchies of chiefs proved extremely effective. The colonial state opposed the destruction of 'native' society and the proletarianization of indigenous inhabitants in Fiji (Legge 1958; Knapman 1983, Chapter 2).

Having resolved the land problem, the colonial state enacted legislation to enshrine the separate administration of the indigenous population. Thus in Fiji the Native Regulation of 1877 codified Fijian customs as propounded by leading Fijian chiefs and their European associates. These regulations were part of the 'native administration' which dealt with the indigenous people quite differently from other residents. The net effect of the close supervision of islanders was the 'fossilization' of archaic structures, which were defended by chiefs who had vested interests in perpetuating them (Spate Report 1959).

The detailed controls affected all aspects of islanders' daily lives including sleeping (curfew at sunset), working, playing, and church attendance. A hierarchy of local authorities was empowered to take punitive measures, including corporal punishment, 'communal' labor (on roads, paths, houses), and to collect head tax.[2] Christianity combined with pre-existing norms produced an authoritarian regime in the villages. Until the 1950s these detailed regulations – illustrative of an 'administrative society' (in Worsley's term) – reigned supreme (1967, pp. 31–44). Restrictions on movements (see Regulations made by the Fijian Affairs Board under the provision of the Fijian Affairs Ordinance, CAP83, Colony of Fiji 1948) ensured that villages were maintained as the primary mode of Ethnic Fijian residence. Ethnic Fijians were required to pay a provincial rate, an education rate (25 percent of resident teachers' salaries) and commutation fees in lieu of communal services. In 1958, 11,628 cases of nonfulfilment of these communal duties were dealt with in the courts (Mayer 1963, p. 94).

Such constraints impeded the rise of individuals or groups who were able to accumulate independently of colonial administrative structures. Besides control over movements there were restrictions on financial undertakings by islanders. With limited access to credit until the 1950s, the productive base of the peasant economy suffered. A preoccupation with the status of those who showed any incentive blocked the emergence of Ethnic Fijian entrepreneurs (Belshaw 1964; Watters 1969). Meanwhile, such specific cultural forms as respect for authority, vernacular speech, ceremonies, and communal labor were given support. Other aspects of island cultures, including travelling parties, exchange of traditional valuables, and dispute settlement (through mobilizing kinfolk, and even sorcery) were discouraged.

Not unnaturally the presence of corporate and settler capital within the colonial societies, with their demand for labor, raw materials, markets, and state support, led to growing contradictions. Thus, while colonial officials and the chiefs sought to maintain the structures of the chiefly mode of production by regulatory measures, their very actions affected the availability of labor and the output of commodities which in turn affected the profits of merchants and planters and the revenue of the state.

GENERALIZATION OF COMMODITY PRODUCTION

Export commodity production was the linchpin of colonial state policy. State revenues were largely dependent on import and export charges. Raw materials were exported in return for manufactured consumer and capital goods. This orientation of the colonial economy suited the interests of the planter class, the merchant houses, petty shopkeepers, money lenders, and the landed chiefly class. The peasantization of the islanders begun in the nineteenth century, was completed under colonial rule (Graves, 1984, p. 120)

The colonial state introduced a system of communal taxation to keep the Fijians in the villages while at the same time inculcating in them the ethics of hard work to bring 'the native within the orbit of the western economic system' (Legge 1958, pp. 10, 46). Between 1875 and 1902 produce from the 'government's garden' in the village was collected by the *Rokos* and sold by tender. Part of the proceeds were credited to the government as tax dues and the remainder was accumulated in the village funds. This tax in kind was levied according to the per capita male population in the province. Ethnic Fijians in Macuata paid 12/2d; in Lau 10/11d; in Lomaiviti 9/9d and in Colo West the average tax was 6/4d (Knapman 1983, p. 112). Copra and sugarcane made up the bulk of the produce, but maize, tobacco, *yagona*, beche-de-mer, cotton, and coconut oil also contributed (Burns 1960, p. 117).

Compulsive communal labor, based on the *lala* services by commoners to chiefs, was used to get Fijians to work in the tax gardens. Expansion of this system of production is reflected in the increases in revenue from this source for the colonial regime. In 1875 native taxation brought in £3,499 or 21.5 percent of a total revenue of £16,255, and by 1879, £19,885 or 29.3 percent of £67,771 (Legge 1958, p. 46).

According to Samy, up to 20 percent of the native tax (in kind) was made up of copra. During the depression of the late 1880s, when copra prices fell from £15 per ton in 1880 to £9 to £10 per ton, the Fijians bore the brunt of low prices because native taxes were not varied. He also points out that the government had arranged with European trade stores for the marketing of Fijian copra and many traders relied heavily on these arrangements (Samy 1977, p. 48).

In Eastern Fiji, particularly in eastern Viti Levu, the northern and southern coasts of Vanua Levu, the Lomaiviti Group, Lau and Taveuni, copra producted increased. This cash crop could easily be grafted on to the production of subsistence products without much disruption, and with little labor adequate amounts of copra could be produced to fulfill the cash needs for tax, ceremonials, and personal consumption. By 1958, of the 168,000 acres under coconuts in Fiji, 84,000 acres were owned by Ethnic Fijians, 76,000 by Europeans and mixed race people, and 3,000 acres by Chinese (Burns 1960, p. 195). Table 2 shows the acreage under various crops planted by the different ethnic categories in 1958.

Table 2 *Acreages Under Crops – 1958*

Crop	Total acreage	Fijian	Indian	European and Part-European	Chinese and others
Sugarcane	128,863	8,448	118,184	2,231	—
Coconuts	168,000	84,000	5,000	76,000	3,000
Bananas	5,000	4,600	380	20	—
Rice	31,200	400	30,150	250	400
Roots (food)	35,933	31,696	2,877	—	1,360
All other crops[a]	9,997	4,860	3,672	210	1,300
TOTALS	378,993	134,004	160,263	78,711	6,060
Approx. Farming[b] Population		28,000	23,000	600	950

Note: [a] Other crops include vegetables, fruit, cocoa, pulses, tobacco, etc.
[b] Persons aged 15 years and over; figures are approximately but based on 1956 Census.
Source: Burns Report 1959, (CP 1/9160), p. 35.

Sugarcane production was made compulsory in the tax gardens in Tailevu South, Rewa, Naitasiri, Navua, Ba, Ra, and Macuata provinces as sugar mills were established in these localities. Ethnic Fijian production of sugarcane increased to an output of 8,884 tons in 1884 or 12 percent of the total cane crushed in the colony. By 1900 Ethnic Fijian cane output was 15,447 tons or six percent of the total Fiji cane crushed (Knapman 1983, p. 115). With the adoption of taxation either in cash or in kind, Fijian cane growers quickly divested themselves of this onerous occupation (Moynagh 1981).

Another export crop produced by Ethnic Fijians from the first decade of the twentieth century was banana. It was particularly well-suited for southeastern Viti Levu but also grown in Kadavu, Cakaudrove and Nadroga. Altogether, 4,600 acres of bananas were planted by Ethnic Fijians in 1958 (Burns 1960, p. 35). It fulfilled local consumption and was sold in widespread roadside village stalls. In the late 1950s and early 1960s it was Fiji's third highest export earner. Table 3 gives the export value of the main agricultural products in 1936 and in 1958.

Table 3 *Value of Principal Agricultural Exports (Fijian £)*

Commodity	1936	1958	% of total value of exports in 1958
Sugar and molasses	1,331,710	7,806,837	63.75
Copra and products	406,393	2,429,970	19.93
Bananas	83,548	163,464	1.33
Hides and skins	3,047	17,867	0.14
Totals	1,824,698	10,428,138	85.15

Source: Burns Report 1959, (CP 1/1960), p. 36.

Despite impediments to non-village farming, there were 1500 *galala* or independent farmers in the late 1950s. They constituted less than five percent of the Ethnic Fijian ratepayers but exemplified the willingness of many islanders to break away from the 'native administration'. The *galala* were required to have an annual gross income of £100, own three acres of cultivated land and two acres of pasture per head of cattle, and pay provincial and commutation fees. Further, although exemptions were for five years, they could be rescinded in the annual review (Spate Report 1959; Mayer 1963, pp. 76–8).

By the late 1950s the policy of keeping Ethnic Fijians in villages had been discredited and attempts were made to create independent (of the village) petty commodity producers. A Land Development Authority was established to subdivide land in various parts of Fiji. This strategy was linked to the goal of strengthening the postwar British economy. Small agricultural loans were made available through the Agricultural and Industrial Loans Board established in 1952 and cooperatives were encouraged. Although ethnic Fijian acquisition of loans lagged behind those of other ethnic categories, the number and value of loans to Fijians increased from nine applications for £2,390 in 1952–3 to 123 applications for £27,290 in 1961–2.

An ambitious resettlement scheme involving predominantly Ethnic Fijian peasants was begun in Lomaivuna, in the eastern side of Viti Levu. Eight hundred hectares of land were taken by the Fiji Development Company (formed by the Land Development Authority). Half the area was for rubber, tea, and oil-palm trials, and the whole area was subdivided into 195 blocks of 4.1 hectares. Of the 190 settlers, 180 were Ethnic Fijians, mainly Lauans, the reminder comprised Rotumans and Indo-Fijians in equal numbers. This initially successful scheme produced 70 percent of the export banana crop by 1967, but because of diseases, production declined thereafter (Brookfield and Hart 1971, pp. 133, 283). Ginger cultivation, taro production for the Vatukoula gold mine, and pig farming have become the mainstay of the Lomaivuna peasants (Brookfield and Hart 1974, p. 284; Lasaqa 1984).

149

Ethnic Fijian also constituted ten percent of the peasants producing sugarcane. With state support, the Colonial Sugar Refining Company (CSR) encouraged the establishment of Ethnic Fijian cane farmers. A school exclusively to train young Ethnic Fijian men in all aspects of cane cultivation was started in Lautoka in the 1930s. A few graduates of this institution took up cane farming (Moynagh 1981).

Plantation Production

The colonial state in Fiji adopted fiscal, labor, land, infrastructural and industrial relations policies that favored plantation agriculture. Thus the areas formerly devoted to plantation production are amongst the most developed in Fiji. But licensing, taxation, other incentives, and labor policies of the colonial state created the conditions for the domination of large plantation and agency house companies. When larger planters, and particularly corporations, established themselves as the producers, millers, suppliers of equipment and credit, it was inevitable that small European settlers came to rely on them as produce suppliers, receiving credit and loans in return. The vagaries of world market prices and nature, together with the terms in which they received credit, resulted in many settlers eventually becoming heavily indebted and losing plantations to their creditors.

The history of sugarcane production is largely the history of CSR in Fiji. CSR originated in Sydney in 1855 and expanded its refining activity to Melbourne, Queensland, and New Zealand. By 1880 CSR-owned mills were producing a quarter of Australian sugar and the company was well situated to enter the nascent colonial economy in Fiji (Narsey 1979, p. 79). The colonial state's appeal and promise of sympathetic treatment to CSR, the absence of another commercial crop that might compete with sugarcane in the future, the presence of settlers already planting and milling this crop on a small scale, and the possibility of a guaranteed source of labor, convinced the company that its investment in Fiji could be profitable. Its General Manager, E. W. Knox, observed in 1879 that:

> Sugar will be produced in Fiji soon which will come into competition with ours; it is a question of whether it will not be better for us to take a share in the development of the industry and the profits that will be realized (Wollin and Roberts 1973, p. 74).

The two large landowners in Rewa, Carl Sahl and J. C. Smith, contracted with CSR to produce 500 acres of cane each for ten years to guarantee the company satisfactory supplies (Moynagh 1981, p. 24).

The depression of the 1880s and 1890s, together with the difficult conditions on the Rewa delta, led to the demise of most small planters at the hands of their creditors, which included the CSR Company. Meanwhile the company built five mills in Fiji between 1880 and 1903. The period between 1873 and 1886 saw 24 sugar mills open and 12 close with CSR's two mills producing 60 percent of all sugar in Fiji. A further eight mills closed in the 1890s and the Fiji Sugar Co. in Navua sold out to the British Colombia Sugar Refining Co.

At the turn of the century, when CSR was contemplating the construction of a mill at Lautoka, the Colonial State promised not to levy export duties on sugar in the foreseeable future. Further, supplies of mill equipment and machinery were exempt from wharfage duties. This concession saved the company £26,493 between 1901 and 1912 (Moynagh 1981, p. 25). By 1900 CSR had invested £1.1 million in the colony. Its three mills crushed sugarcane from 20,000 acres and produced 87.5 percent of sugar and spirits exported from Fiji in 1899, which accounted for 65 percent of exports (Knapman 1983, p. 39). Of the six mills in the colony by 1903, four were owned by CRS which contributed financial and managerial assistance to the British Colombia Sugar Refining Co. in Navua. Only the mill at Penang, formerly owned by the Chalmers brothers and sold to the Melbourne Trust Co. in 1896, was independent of CSR.

E. W. Knox, the CSR General Manager, was not comfortable with this virtual monopoly position in Fiji's sugar industry. He helped the British Colombia Co. to remain solvent, and he encouraged the subdivision of CSR plantations into estates that were to be leased out to European planters – a significant number of whom were former Australian CSR overseers. The latter policy spread the risks, reduced capital tied to sugar planting, lowered wage costs, and gave the CSR valuable allies when dealing with the Colonial State. However, in 1926 the Vancouver Sugar Company of Navua collapsed and CSR emerged as the sole miller and major estate owner in Fiji (see Moynagh 1981).

The sugar industry became Fiji's single most important source of export earnings. Whereas in 1875 copra constituted over 51 percent of all exports and sugar only 4.4 percent, by 1883 sugar had overtaken copra. In 1885 sugar accounted for 66 percent of exports and copra 20 percent. Sugar's lead in the Fijian economy was to persist to the post-independence period (see Table 4).

Table 4 *Sugar Exports as Percentage of Total Export for Selected Years*

Year	Percentage	Year	Percentage
1885	66%	1948	55%
1890	74%	1958	64%
1913	73%	1965	71%
1923	75%	1970	64%
1937	65%		

Source: Adapted from Sutherland 1984, p. 88.

A process of centralization and concentration of ownership and control also occurred in the copra plantations. Formerly important planters like the Henning brothers gave way to their creditors and by the mid-1930s most planters were in debt to the Sydney merchant houses, Burns Philp and Carpenters (Oliver 1961).

LABOR POLICIES

Proletarianization, or the emergence of a category of wage earners, has a checkered but long history in the Pacific. The process was mediated by the state so that a class of itinerant peasant/laborers existed because of regulative mechanisms which determined their degree of commitment to wage labor. In Fiji, according to Thurston, the tax in kind prevented the emergence of 'bands of migratory helots' (quoted in Knapman 1983, p. 130). The maintenance of precaptalist relations meant that cheap labor could be extracted from the villagers. In effect this led to the subsidization of capitalist production by the chiefly mode of production (Fitzpatrick 1980).

The colonial regime in Fiji stopped the recruitment of Ethnic Fijian labor, previously facilitated by the settler-led Cakobau government, because 'whole districts had been well nigh depopulated, and the reckless deportation of the male inhabitants had left two provinces (Ba and Ra) almost deprived of cultivators' (Gordon 1897, as cited in Young 1970, p. 201). But the move towards 'villagizing' them must not be seen as an attempt to preserve the traditional order (Legge 1958, p. 374) or an 'end to exploitation' to safeguard 'the interest of the Fijian People' (Derrick 1950, p. 124).

Ethnic Fijians who resisted colonialism were sentenced to labor in plantations. Planters continued to use Fijian contract laborers as Table 5 shows:

Table 5 *Number of Fijian Contract Laborers*

Year	Number	Year	Number
1875	604	1878	1,249
1876	1,213	1879	1,493
1877	976	1880	1,001

Source: Gillion 1962, p. 12.

In 1880 there were 580 Fijian plantation laborers and 420 crewmen in interisland shipping. Two years later, 2,300 were indentured to labor for one year. In their dispatches to the Colonial Office, Governor Desvoux and Colonial Secretary Thurston indicated that by 1884, about a quarter of adult males worked in capitalist enterprises for varying durations under contract.

The Masters and Servants Ordinance of 1890 and the Fijian Labour Ordinance of 1895 were modified by the Fijian Employment Ordinance of 1912. The ban on Fijian men absenting themselves for more than 60 days was lifted. If a Fijian had been away from his village for more than two years he could be retained for longer if his employer continued to pay his rates and taxes. From 1902 – and especially after 1912 when the *Bulis'* permission to sign labor contracts was no longer needed – many left their villages and worked in plantation and urban centers. Chiefs were opposed to this lack of regulation in the movement of Fijian commoners but were told not to retard the process of individualization.

Increasingly, a class of Ethnic Fijian proletarians was in the making.

> The financial expedient of indirect rule compelled government to support village work programmes and chiefs' rights to personal *lala*. But in the interest of emerging individualism it clearly had created a legal framework conducive to the erosion of village society (Knapman 1983, p. 132).

It has been pointed out that with the cancellation of Indian indenture contracts in 1929 there was a demand by the CSR for Fijian labor.

> Fijian indentured men lived under much the same wretched conditions as had the Indians, but for shorter periods . . . CSR paid Fijian recruiters of each man they produced in Lautoka for engagement under the Masters and Servants Ordinance . . . (MacNaught 1982, p. 102).

During the six months that the laborers were away, food production dropped, shortages were reported, and the villages underwent physical deterioration. Upon returning to their villages penniless, the ex-workers further burdened the peasantry.

In 1911, 5,421 (6.2 percent) out of 87,096 Ethnic Fijians resided outside their villages. Most of the 3,844 males involved were in wage employment or at school and most of the 1,577 women were domestic servants (Boyd 1911). Thus, from the second decade of the twentieth century some Ethnic Fijians had become permanent wage earners.

The census of 1921 recorded 6,897 Ethnic Fijians (8.2 percent of the total population) as working outside the villages. Of this total, 4,884 were males and 2,013 were females. Again the trend, although gradual, was towards increasing employment of Ethnic Fijians was wage and salary earners. The largest category of Ethnic Fijians outside their villages was laborers (261 indentured laborers and 1,882 other laborers). A further 1,548 did not state their occupations or were unemployed. It is likely that the latter categories comprised persons taking advantage of the lifting of restrictions on their mobility.

In 1927 the Secretary for Native Affairs complained that 'a supply of labor for agricultural purposes was of greater importance than the welfare of the natives' in the government's view and feared the collapse of the Fijian Administration (cited in MacNaught 1982, p. 103). Fijians also worked as policemen, nurses (women), officers in the Fijian Administration and as clerks in the colonial administration. From their inception in 1934, the goldmines deliberately used Fijian and Rotuman labor (Brookfield 1972, p. 83). In 1936, 1,002 Ethnic Fijians were employed primarily as laborers in the gold mines. By 1940 this number had grown to 2,000 (MacNaught 1982, p. 141).

The 1936 census reported 16,729 Ethnic Fijias from a total population of 97,651 as living outside their home villages (Table 4, p. 24, 1936 census). Table 14 in the Population Census of 1936 indicates the occupations of the 7,222 Ethnic Fijians listed as employed. Of this total, 16 percent were laborers, 14 percent mineworkers, 8 percent in sugarcane production, carpenters 4 percent, government officials 4 percent, and sailors 4 percent.

Knapman pointed out that the proportion of Ethnic Fijians outside villages increased from about three percent in 1921 to 17 percent in 1936, almost entirely because of a growth in wage employment. He noted that the proportion of working-age males absent from villages was 'higher than 17 percent', with provinces such as Tailevu making a disproportionate contribution to wage earners and absentees (1983, p. 137).

In 1940, Native Regulations were tightened to maintain the integrity of villages. Mayer described the predicament of the *yasa* (those who left the village) who were required to pay the 'native rates' and commutation fees, but neither enjoyed the benefit of these payments nor had any representation. Ethnic Fijians had to get permission from the *Buli* before leaving the district for more than a week. Some 10,000 Ethnic Fijians were in Suva illegally in 1959 (Mayer 1963, p. 96). Many Fijians left their districts with the permission of the local authority, but kept extending their stay by reporting on the last day of their permit, and leaving again a few days later (MacNaught 1982).

IMMIGRANT LABOR

Although the Pacific islands labor trade had an extremely bad reputation, it was allowed to continue by the colonial administration in Fiji. In 1877, the Fijian colonial state authorized £5,000 towards the recruitment of island laborers (Narayan 1984, p. 62). It is estimated that 23,000 came to Fiji (Scarr 1970, p. 226).

Melanesian laborers, mostly Solomon Islanders from Malaita Island, worked initially in Fiji's cotton plantations and subsequently labored on coconut and sugar plantations. They also labored in government public works, in bulkstores, at the docks, in slaughterhouses, in timber yards and for such merchants as A. M. Brodziak and Company (Kuva 1974, p. 10). Usually they were contracted to work for three years at £6 per annum. There were cases of ill-treatment of laborers by employers (Gillion 1962; Burns 1963) but more research is needed to enable a systematic account of their living and working conditions. It is known, however, that in the first plantations of CSR in the Rewa district in 1880, 220 out of 587 Melanesians died within the first six months of their arrival in Fiji (Narsey 1979, p. 80).

These Melanesian and Micronesian laborers preceded Asians by more than 30 years and were to continue to work alongside indentured laborers for another 30 years. It was only when the Melanesian and Micronesian laborer reserve (Graves 1984) became depleted that colonial officials in Fiji turned to Asia for alternative supplies of cheap and regular laborers.

Asian Immigrant Labor

Indian Laborers were introduced into Fiji to work in the colony's plantations. Subsequently, when labor difficulties arose, some Japanese were introduced. Several scholars have written about the Indian indentured laborers and their subsequent settlement in Fiji (Gillion 1962, 1977; Ali 1979, 1980; Lal 1983; Mayer 1961, 1963, 1973; Mishra 1979; Naidu 1980a; Narsey 1979). Of the Indian indentured labor system, it has been said:

> This was not designed as a moral improvement on slave labour, but as a system which would in fact enable the employer to enjoy the advantages of slavery with none of its disadvantages . . . this meant a greater calculability of costs was possible for the employer. At the end of the contract, he had no incalculable obligation to the worker, and the worker indeed often found that his right to return to freedom at home was theoretical rather than real. He would thus be available at first in semi-slave conditions, and then as a dependent and impoverished tenant or share-cropper. Such was the fate of Indian 'coolie' workers in Fiji, Malaya, Assam, Ceylon, Mauritius and Natal, as well as in Guyana, where they literally took the place of the slaves (Rex 1980, p. 129).

This observation, insightful at a general level, needs to be qualified by the Fijian experience. Even during the five years of the contract period, the occupational specialization and the supervisory roles of some of the indentured laborers allowed them to escape the rigors

155

of laboring in the fields as ordinary workers. On the basis of this differentiation among indentured laborers, former *sardars* (foremen), policemen, interpreters, clerks, cooks, and domestic servants were better placed to pursue opportunities available once their indentures ended.

Most laborers were young – 86.6 percent from Calcutta were in the age cohort of 19 to 30 and a little less than 70 percent were between 20 and 30 (Gillion 1962, p. 52). Females comprised less than 40 percent of the total immigrants (Gillion 1962, p. 54). They arrived in Fiji as individuals rather than in family groups. They were not riff-raff and scum off the streets of Calcutta as 'some uninformed persons in Fiji have supposed' (Gillion 1962, p. 58).

Over a period of 41 years, 87 shiploads of Indians arrived in Fiji to do their *girmit*, or indenture contract, of five years. This period was liable to be extended if the laborer was convicted for breaching labor regulations. After the initial five years, the laborer was free to return to India but he/she had to pay for his/her return passage. If, however, he/she were to work for another five years in Fiji, then he/she was entitled to a free passage to India.

I have written elsewhere about the indenture system in Fiji and its consequences for interpersonal violence (Naidu 1980a). The following is a brief outline of some of the essential features of the system.

In Fiji, the working, living and nutritional conditions of indentured laborers were difficult and the whole experience was regarded as degrading (Ali 1978; Gillion 1962; Mayer 1963; Tinker 1974; Narsey 1979). They were treated as 'sugar producing machines' (Mayer 1963, p. 18). The statutory wages remained fixed at one shilling for men and nine pence for women per day. But very quickly planters realized that more could be squeezed out of laborers by tasking. Hard and excessive tasks made it impossible for many to earn this amount. What was advertised as the statutory daily wage became the maximum earnable by the laborers. This was particularly true during the 1880s and 1890s when plantation owners sought to minimize costs to weather the depression and low sugar prices. In 1890 one third of adult males earned less than eight pence per day and 42 percent of the women earned less than five pence per day. This situation lasted for another three indentures.

> Quite incredibly, the average earnings per working day for adult males did not reach one shilling until 1908, 28 years after indenture had begun operation. Even then this was the *average* wage and so one can expect half of the laborers to be earning less than that (Narsey 1979, p. 86).

Women workers on the average did not earn the statutory 9d a day even in 1919, a year before all remaining indenture contracts were cancelled (Table 6).

Table 6 *Average Indenture Daily Wage Rates (pence)*

Year	Male	Female	Year	Male	Female
1887	7.28	4.43	1908	12.05	6.13
1890	9.64	5.39	1915	12.05	6.54
1901	11.17	6.05	1919	15.62	8.79

Source: Narsey 1979, p. 85.

The penal sanctions of the colonial state reinforced the exacting conditions in plantations. Crimes included lack of ordinary diligence, failure to complete five and half tasks a week, and insubordination. Over the indenture period, between 15 to 40 percent of the adult population was convicted and 25 percent had their contracts extended (Naidu 1980a, pp. 53–4). Table 7 provides figures for contract extensions for selected years.

Table 7 *Immigrants whose Contracts were Extended*

Immigrant of year	% whose time was extended	Average length of extension (days)
1896	37.6	37.7
1899	36.1	48.4
1902	33.6	61.9
1904	20.4	110.9

Source: Narsey 1979, p. 91.

The prison-like institution of plantation lines (barracks) and the disproportion of 100 men to 40 women (a statutory requirement not always followed) led to high incidences of violence, murder and suicide (Table 8). Poor nutrition and difficult work resulted in high death rates among the population cohort between 18 and 30 years (Table 9).

Table 8 *Suicide Rate Per Million for Period 1902–1912*

Group	Number
Indentured Immigrants	926
'Free' Immigrants	147
Province of Origin of most of Immigrants in India	63

Source: Gillion 1962, p. 79.

Table 9 *Percentage of Adult Immigrants introduced in the Following Years, who died within five years*

Year	Percentage	Year	Percentage
1909	9.06	1912	6.12
1910	7.09	1913	8.23
1911	7.09	1914	9.15

Source: Fiji Legislative Council Paper No. 39 of 1914.

Of the total number brought to Fiji, 30,806 (or 51 percent) returned to India by 1929 (Samy 1977, p. 56). Those who stayed leased land from Europeans and Fijians and began peasant production. Others became hawkers, traders, and artisans. Still others continued working for Europeans and the Public Works Department.

After 1900 another group of Indians, known as free migrants, arrived in Fiji. From 1911, 250 arrived each year till the outbreak of the First World War. At the end of indenture in 1920, the European Planters Association, Chambers of Commerce, and the missions agitated for more Indian laborers. Free migrants, largely Punijabis and Gujeratis, increased. The former took up agriculture and the latter became artisans, wholesalers, and retailers.

Whereas the colonial state effectively incorporated the chiefly class into its administrative apparatus and reproduced their economic and social bases in the villages by regulative measures, it left the emergent Indian peasantry largely to the whims of the CSR (Mayer 1963, p. 104). Between 1910 and 1920 this company had subdivided its huge estate into 600–1,000 acre plantations which were contracted out to ex-CSR overseers and other Europeans. Within a short period, however, differences emerged between the European planters and the CSR. The division between milling and farming was designed to maintain company profits. Overtime, less and less, was offered for cane produced by the planters. At the bottom of the production structure, laborers suffered the brunt of the exploitation in the cane plantations (Knapman 1988).

Once this measure no longer served the company's need, it subdivided the plantations, and leased the (on-average) 10-acre farms to Indian tenants. Other Indians leasing native lands, state-owned land, and land belonging to Europeans,[3] were subcontracted by the company to produce sugarcane (Narsey 1979). By the mid-1930s there were 4,000 tenant farmers and 4,500 subcontract growers supplying CSR. At this stage CSR also tried to utilize Fijian peasantry on the same basis.

Samy correctly argues that although ex-indentured laborers – legally freed – had been proletarianized, they depended for their livelihood on the sale of their labor-power (cf. also Howard et al. 1983, p. 266). Their limited access to land and the predominance of agriculture in the economy ensured the dependence of ex-indentured Indians and their descendants on the CSR. This dependence, reinforced by structural and institutional arrangements, secured CSR's control over production by both tenant and contract farmers.

Through its estate sectors, under the supervision of a field superintendant and his subordinates, the company

> . . . directed the time for ploughing, drilling, cultivation, drainage, etc., and . . . coordinated harvesting and transportation of cane to the mills . . . The farmers themselves had no control over the nature or extent of such 'services' (Samy 1977, p. 78, cf. Ward 1980).

The profitability of this strategy of small tenant and contract farmers, largely at the mercy of the CSR, is seen by the rapid increase in their numbers.

> In 1925, of all land under cane, tenants cultivated 10%, contractors 31% and the company estates 52%. By 1934, tenants were cultivating 53%, contractors 40% and CSR Company's estates only 6%. By 1950, however, tenants cultivated 46%, contractors 51% and the Company only 3% of the total land under cane (Samy 1977, p. 78).

The company paid farmers according to the sugar-content of their cane, thereby ensuring that it paid for the sugar in the cane rather than the cane itself. However, despite this formula, it closed its oldest mill in Nausori in 1959 because it was not profitable. The decision affected the livelihood of a large proportion of the population in the Rewa delta (see Mayer 1973 for a discussion of the Delanikoro community before and after the mills closure).

Until late 1969, CSR (up to 1962) and its subsidiary, South Pacific Sugar Mills Ltd (SPSM), completely dominated sugar production, marketing, and profit-taking (Moynagh 1981). The company deducted all its costs (salaries, field supervision, research, maintenance of rail system, and packaging) plus a further 11.5 percent of these costs as head office expenses, and three percent of the capital value of its property and machinery as depreciation. It took its share, together with these costs, before the peasants got their share. Hence the company faced no risk of loss if prices fell. In the event of droughts or hurricanes, it simply suffered lower profits. The growers, however, stood in great risk of loss (Wollin and Roberts 1973, p. 76).

The hand-to-mouth existence of the bulk of the Indo-Fijian peasantry was reflected in and reinforced by their persistant indebtedness to landlords, money lenders, and shopkeepers (Coulter 1942, p. 98; Ali 1977; Mayer 1963). Chandra (1980) lists some of the factors that compelled the peasantry to borrow. These included the legacy of low prices paid for cane and the absence of stable contracts between growers and the company. Their total dependence on income from sugarcane made them vulnerable to price fluctuations, which led to borrowing. The credit system featured 20 percent interest on loans from moneylenders and 20 percent mark-up on goods bought on credit. Lengthy intervals between sugar-payments caused purchasing on credit. An assignment system, which allowed such debts to be deducted from the cane-payment before the peasant got his money, assured shopkeepers' returns. The 'dishonesty of shopkeepers and moneylenders' perpetuated indebtedness. Social events, especially weddings, in which a family's name might be at stake, also contributed to indebtedness (Chandra 1980, pp. 22–6).

Samy (1977) has shown how both ex-indentured Indians and the European trader class were displaced by Indian (largely Gujerati) and Chinese free migrants. The Gujeratis proved able shopkeepers and craftsmen. By using relatives from India as assistants, together with a tightly knit sense of community, they rapidly displaced ex-indentured workers in many areas. 'The Gujeratis are thrifty and hard working', wrote Gillion, 'with a strong sense of loyalty' (1962, p. 135). The white traders agitated against Gujeratis and the Chinese. However, neither the Suva Chamber of Commerce, dominated by larger Australian firms, nor the colonial state felt inclined to support restrictions upon these Asians. In 1959 other Indians called for the repatriation of Gujeratis and Punjabis (Burns 1963).[4]

According to Samy, the Gujeratis and Chinese petty trading class served the interest of metropolitan capital even better than the European shopkeepers, and became an indispensable link between metropolitan capital and the colonial economy of Fiji (Samy 1977, pp. 82–3). Similar observations may be made in the other Pacific islands where Chinese and mixed-race persons largely displaced European storekeepers (Oliver 1962, pp. 190–2).

Among ex-Indian indentured laborers in Fiji the caste system disintegrated and was replaced by status based on wealth (Gillion 1962; Mayer 1973; Jayawardena 1971). But cultural differences persisted between North Indians and South Indians, between Hindus and Muslims, and between those categories and Gujeratis and Punjabis (Mayer 1973).

By the 1950s, therefore, Fiji was transformed into a colonial economy penetrated by metropolitan capital and incorporated into the international capitalist system. The country relied heavily on one or two major raw materials for exports and was dependent on a whole range of imports. These ranged from imported foodstuffs, which the island population consumed, as well as fuel and capital goods for maintaining the more complex colonial economies.

The colonial export-import economy of Fiji was a product of certain class configurations and in turn formed the basis of class alignments that influenced state societal development. The class character of the colonial state reflected alliances between colonial officials representing imperial interest, larger commercial interests (especially big corporate ones), and the indigenous chiefly class. Small settler interests were heard only when they were in accord with the interests of the corporate enterprises. But the colonial state in Fiji did not act as a mere instrument of large enterprises. If anything, the support given to the latter was part of government policy to prevent a rapid disintegration of the chiefly mode of production. Besides the partiality of some colonial officials for protecting the native race, political expediency required that the chiefs should not be alienated. Immigrant labor and corporate capital in its oligopolistic and even monopolistic forms were integral to the policy of maintaining the chiefly mode of production. Export commodities contributed to state revenues as did the use of chiefs to extract labor services and various taxes from the indigenous population. The colonial state policy of encouraging commerce while at the same time maintaining the chiefly mode of production resulted in contradictions and uneven development. The demographic character of Fiji was profoundly transformed.

COLONIAL HIERARCHIES

Class divisions in colonial society took a racial line. At the apex of colonial social formations were Europeans as senior state officials (CP 25/1946), as owners, managers and overseers of plantations and commercial enterprises, and as heads of various church missions. However, European workers identified themselves with the dominant members of their own community rather than with non-European workers, a tendency encouraged by relatively high standards of living and by common cultural traits. One authority calculated approximate costs for white and non-white labor in Fiji in 1920 as:

Whites
 Overseers salaries per year: £200–250 plus residence
 Artisans' salaries per year: £450–580
 Cost of living per year: £60–72 plus free milk and vegetables on estates
 Cost of European house: £500
 Cost of Fijian hut: £50

Non-whites

Cost of labor per year:	£40
Cost of coolie hut:	£12
Cost of Fijian hut:	£12

Source: Chapple (1921, pp. 85–87).

It is noteworthy that the cost of living for Europeans on estates was deemed to be at least £60 per annum, exclusive of free vegetables and milk. Non-white labor was calculated at only £40. In fact, in 1920, the average income of an Indian indentured laboror was £18 per year and £30 for a free laborer (Narsey 1979, p. 92).

The colonial racial division of labor induced a pattern of racial specialization that is still observable today, with Indians and Indo-Fijians involved in sugarcane cultivation, transportation and milling, and Indigenous Fijians in copra production. This occupational specialization is starkly marked by the geographical spread of these commodities. Mainly Ethnic Fijians were employed in the gold mines and as stevedores with their chiefs acting as labour recruiters and supervisors (Plange 1985). They were absent from shopkeeping because of the restrictions placed upon them as a result of the state policy of preserving the chiefly mode of production.

Three broad areas of employment between 1920 and 1945 can be indentified: state services, religious bodies, and the private sector. The major employers were the state, the Colonial Sugar Refining Co. (CSR) and the Emperor Gold Mines. In the bureaucratic machinery of the colonial state, the higher and middle order posts were filled by expatriate and local European officers (Howard 1986), and the lower posts were filled by Ethnic Fijians and Indo-Fijians. The only exception to this pattern was the appointment of the high chief, Ratu Sir Lala Sukuna as the Secretary for Fijian Affairs (*Talai ni Kovana*) in 1940. The CSR's activities covered three separate areas: fieldwork, milling and administration. After 1920, fieldwork was divided according to sectors, with sector offices being led by one or two European overseers. In the four sugar mills, a distinct racial division of labor existed. Top managerial and supervisory positions were held by Europeans, skilled and semi-skilled occupation by part-Europeans, and unskilled labor by Indians, Indo-Fijians, and Ethnic Fijians (Mayer 1963).

Racial divisions also characterized labor at the Emperor Gold Mines in Vatukoula. Senior managerial and supervisory positions were held by Europeans, below them were part-Europeans and Rotumans, and at the bottom were Ethnic Fijians. Each level experienced differential earnings, working conditions, prospects for promotions and material benefits such as housing. According to one authority

this was translated into housing benefits and social activities so that one group could not live with or use the other's facilities. But the facilities were not all of good quality across these racial communities. A general description of Vatukoula in the oral tradition holds that expatriated live at the *Delana* (top), Part-Europeans and Rotumans at the *Buca* (slopes), and Fijian in *Ieli* (hell) (Planges 1985, p. 12).

Similar observations could be made about housing at Penang, Rarawai, Lautoka, and Labasa where CSR mill staff were accommodated. Ordinary mill workers, both Indo-Fijians and Ethnic Fijians lived in barracks called lines: at Lautoka they were referred to as 'top line' and 'low line'.

Copra estates were little different. They were owned and managed by Europeans or individuals of European ancestry. The majority of the workers were colored, Ethnic Fijians (from Lau) Indo-Fijians, and Part-Europeans. The higher status positions were held by Europeans or Part-Europeans (who were extremely color conscious) while more arduous and manual jobs were allocated to Ethnic and Indo-Fijians. The occupations of cooks and houseboys were generally reserved for Indo-Fijians.

Occupational divisions based on race were reinforced by different lifestyles. The owner, overseer, or manager resided in a large estate house while his domestic workers lived in far less sumptuous dwellings nearby. In Taveuni, in Fiji, names such as Ardmore, Eton Hall and Holmburst were given to these properties. Field laborers lived in lines with little compartments for individual families. Even in the mid–1970s there were lines that dated back to the late nineteenth century with conditions little changed (Brookfield et al. 1978, p. 39; Naidu 1980b, p. 143).

Europeans had a voice in the Legislature from 1904. Nevertheless, for several decades 'a main factor in the government of Fiji was the influence wielded by a small oligarchy of local commercial interests' (J. W. T. Barton cited in Gillion 1977, p. 11). In the period extending from 1914 to 1945, the latter comprised businessmen and lawyers such as J. M. Hedstrom, H. M. Scott, H. Marks (all later knighted), R. Crompton and representatives of the CSR company. At the beginning of his governorship of Fiji in 1937, Sir Arthur Richards noted that:

It [Fiji] is a peculiar colony – *sui generis* indeed. The presence of a resident European population, their long isolation from the world and the limitation even of recent contacts to Australia and New Zealand has bred a particular insularity of its own. A few big men have obtained a stranglehold on the place – they have won their way to the top and mean to stay there. The underdog is under-paid and powerless. A few men control everything behind the scenes and even government had been run with a strong bias (Gillion 1977, p. 11).

In 1914, European representation in the Fiji Legislative Council was increased from six to seven. Local Europeans were also allowed a seat on the Executive Council in 1912. Ethnic Fijians, whose interests were officially said to be paramount, gained representation on this body only in 1944.

European economic and political dominance reinforced their attitude of racial superiority. Socially, Europeans formed a fairly exclusive caste-like community with their own clubs, schools, and sports facilities. The Planters' Associations and Chambers of Commerce were largely European interest groups. These amenities received the highest per capita funding when compared with other racial groups (Mayer 1963; Gillion 1977, p. 177; Narayan 1984, pp. 72–4). The public libraries and the Suva Sea Baths were out-of-bounds for nonwhites.[5]

Land shortages compelled Indians to live in government designated Indian coolie settlements on Ethnic Fijian land leased by the colonial state, or on crown land. Ethnic Fijian villages were made out-of-bounds for Indians and later on for all non-Fijians (Moynagh 1981, p. 75). No attempt was made to create a nonracial, multiethnic community; on the contrary, racial differences and awareness were fostered through racial legislation and practices (MacNaught 1982, pp. 112–13; Ali 1977, Chapters 2, 3; Norton 1977; Naidu 1980b). The justification for this approach of the colonial officials was that the Ethnic Fijians were to continue their village way of life and develop at their own pace, free from the unwanted influences of other races. Ethnic Fijian chiefs, as a class, concurred with this policy of separate development because their positions were entrenched.

In the urban centers of Fiji there were white and non-white areas of residence. Usually European areas were located on the hills with the best views, breeze, and drainage. Large concrete wooden bungalows with spacious compounds together with little cottages for servants (houseboys or housegirls) characterized these areas.

Asian settlements were not as salubriously located or built. Indigenous islanders whose villagers were in the urban locality were usually moved out of the town altogether. They and the poorer mixed-race people were marginalized, although islanders employed because of their chiefly status and/or missionary education were allotted residential areas to suit their higher status. In Suva, racial residential patterns are no longer so obvious though they persist.

Anti-Colonial Struggles

Most incidents of protest against the colonial order indicated the existence of a substratum of discontent that could be tapped by the leaders of such movements. MacNaught (1982) maintains that in the Fijian case, the dissolution of earlier social relations as a result of absenteeism from the villages was a leading factor. Obviously the demand for tax, money to acquire goods (now part of the needs of the peasants), and the opportunities of alternative sources of income, all contributed to this situation. The articulation of capitalist and precapitalist modes of production thus produced strains and stresses on indigenous people.

While struggles often took political forms as support grew and religious forms when repressed, the causes of discontent were usually economic (Worsley 1957). Conflicts over economic organization included struggles by workers for better wages, and by the peasants against exactions of landlords, middlemen, and the state. These disputes were invariably aggravated by state intervention, made inevitable because of the close association between the economic and political spheres in colonial societies and the articulation of capitalist and precapitalists modes of production. Penal sanctions governed labor relations and the lives of indigenous inhabitants were closely regulated. This was made possible by the basic law and order character of the colonial state machinery. The absence of suitable political forums to express and resolve matters of concern to the colonized people also contributed to the rapid deterioration in relationships between disputing parties. This entailed state intervention.

It is noteworthy that the forms of struggle varied over time, reflecting the degree to which societies had changed. Initially anticolonial activities were designed to restore pre-existing relationships. Subsequently they took forms that sought to wrest control of the productive activities of islanders from foreigners. Later still, workers formed unions to bargain on their behalf in situations that did not encourage unionism. Trade unions became the basis for the emergence of alternative leadership to the chiefs and bureaucrats.

Restorative Rebellions

There were at least ten restorative rebellions in Fiji between 1875 and 1935 designed to return to old ways or to assert indigenous control. In 1875 and 1894, eastern chiefs combined with the British to suppress rebellions by inland people in Viti Levu and Vanua Levu, respectively. In the face of colonial state repression, later movements took a more religious form. The communal and less hierarchical mode of production of inland communities in Fiji were thus subordinated to the chiefly and capitalist modes of production. The suppression of restorative rebellions merely helped to reinforce the alliance of eastern chiefs and British colonial officials and extended their control over Fiji.

Religious Movements

As colonial rule became entrenched, opposition to the colonial state and capitalist penetration took on more religious forms. This type of protest was widespread in the Pacific, and was especially prevalent in the Melanesian region (Worsley 1957). After the repression of the *Colo* (inland) peoples in Fiji, there emerged a movement called the *Wai ni Tuka* (Elixir of immortality) cult which combined an attachment to indigenous gods with a rejection of colonialism.

Participants believed that 'Fijians would rule their own land again and the white skins would be driven into the sea' (MacNaught 1983, p. 76). Demands on Fijians in the inland areas to build elaborate houses for the leaders of the movement, the nonpayment of provincial tax, the nonperformance of state-required communal tasks, and the disrespect shown to officials of the native administration (some of whom joined the movement) resulted in strong repressive measures by the colonial authorities. Navosavakadua was exiled to Rotuma (an island approximately 200 miles north of the Fiji group). In 1892 Governor Thurston deported the whole village of Drauniivi to Kadavu, because of its involvement in the movement.

Other movements such as the *Luve ni wai* (Elixer of life), originating from the *Tuka*, remained underground for long periods. In 1914 a movement with its origins in the *Tuka*, led by Osea Tamanikoro, surfaced in the North Colo region. It envisaged an end to European rule complete with the enslavement of whites. He was exiled to Oneata in the Lau group for ten years. Again, in 1918 the leader of the *Lotu Ikavitu* (religion number seven) movement preached that Navosavaka-dua had gone to kill Queen Victoria and made Saturday the sabbath for his followers. He advocated an end to taxes and *vakamisoneri* (missionary collection), but was confined to the mental asylum by colonial officials (MacNaught 1982, p. 99).

In Nadroga province in 1961, four villages disassociated themselves from the Fijian Administration and formed their own system of local government which was called the *Bula Tale* (Resurrection). This movement had politico-economic and religious aspects and rejected the taxes imposed by the administration. It was renamed 'The Blood of the Lamb' and later 'The Lamb' (Mayer 1963, p. 100; Durutalo 1986). The emphasis of this movement was communal and even communist. Its organization stressed local level autonomy (see Durutalo 1986 for an account of this movement).

ECONOMIC AND INDUSTRIAL PROTESTS

During the period from 1912 to 1920, differential access to economic rewards in colonial society resulted in a major movement of economic protest in Fiji. Aspects of the restorative and religious movement were manifested in the organization of the Viti Kabani which emerged in 1912. Unlike earlier movements, this one was designed to uplift the lives of Ethnic Fijian peasants in modern Fiji.

In the interwar period a more or less permanent category of Fijian urban workers emerged. This trend resulted in the gradual disintegration of the chiefly mode of production. In villages, amenities deteriorated and food shortages and other hardships, particularly for the elderly, began to occur. In addition, markets for produce were insecure and the terms of business were often unfavorable because of European, Chinese, and Indian middlemen.

A sense of relative deprivation prevailed among indigenous Fijians which contributed to Fiji-wide support for Apolosi Ra Nawai, a commoner from Ra in western Viti Levu, who organized the *Viti Kabani* (Fijian company). Beginning in 1913, Nawai held meetings in the Rewa area, calling upon the *itaukei* (indigenous owners of the land) to organize a company to reap the benefit of their own labor. The *Viti Kabani* rapidly grew to be a major handler of Fijian produce. It owned punts and cutters, and successfully vied with European, Chinese and Indian traders. Fijian peasants refused to sell their banana and copra produce to non-Fijian middlemen.

The company's organization parallelled the hierarchy of the Bauan-led native administration, much to the chagrin of *Roko Tui* and *Turaga ni Koro* title-holders. The company has its own police and tax collectors. Support came from all over Fiji and thousands assembled at its meetings. The *Roko Tui* Macuata complained that there seemed to be two governments in Fiji. Nawai's movement promoted solidarity amongst Fijians and began to break their parochial concerns (Geddes 1945, p. 65).

As a nascent nationalistic movement it threatened the established alliance between chiefs and larger capital. Nawai criticized the chiefs for land lease arrangements and for living in comfort as a result of forcing commoners to labor for churches, and to build boats and chiefs' houses. He declared that 'The people of Fiji are in a bad plight because of the orders of their chiefs. Very few people are in a bad plight because of their own decisions about themselves' (quoted in Scarr 1984, p. 135). Nawai's flamboyant life-style (complete with boats, a car, and a retinue of women dubbed 'doves') exposed him to criticism. Non-Fijian middlemen were concerned about this challenge to their control of the distribution of Fijian banana and copra produce and the established Fijian chiefs were annoyed at this challenge to their authority. Nawai posed a challenge to the comprador elements in the colonial economic and politico-administrative structures. Their protests resulted in his imprisonment in 1916. Further prompting by Ratu Sir Lala Sukuna, the emerging spokesman for indigenous Fijian chiefs, resulted in Nawai's exile to Rotuma in 1917. He spent the 1920s and 1930s in exile in Rotuma and was exiled to New Zealand in 1940. Colonial officials and the Council of Chiefs feared an alliance of Ethnic Fijian miners and Indo-Fijian sugarcane farmers under his leadership (Norton 1977, p. 58). He died in exile in Yacata in 1946.[6]

Nawai coined the concept *Gauna Vou* (the new era) to impart the economic and political potential of ordinary Ethnic Fijians denied by the colonial structure. The *Viti Kabani* sought to secure higher returns for the peasantry. Ethnic Fijians refused either to work for or to supply the trader and planter classes. Under the systematic repression of the colonial state, this movement increasingly took on a religious form.

The movement gained support from Fijians throught Fiji but had a particularly strong influence in Colo, Ba, Namosi and Ra where Ethnic Fijians remained relatively backward (Scarr 1983, p. 58). Similarly the *Bula Tale* movement in 1961 had an economic content as it sought to give greater control to its members over the wealth that they produced. These peasant movements attempted to re-order the colonial society in the interest of indigenous commoners.

In contrast to the widely-held view of Fijians as quiescent, the Indians and their Indo-Fijian descendants had by the 1930s acquired a reputation of being troublemakers in the colony. The Indian Problem was a prominent concern for state officials, local Europeans, and chiefs. The reason for this state of affairs was due largely to the division between workers and capital at this stage in Fiji's history coincident with the racial division between Indians and Europeans. In mid-1920, after a strike that was forcefully crushed, the then Governor of Fiji, Sir Cecil Rodwell, made the following significant observation:

> At the moment there seems to be a gulf between the government and local Indian opinion. On the one hand, the government scarcely appear to secure as early and as reliable information about the fluctuations of Indian opinion as is desirable. On the other hand, Indians appear disposed to credit the government with ridiculous and impossible intentions (cited in Gillion 1977, pp. 37–38).

Gillion believed the government was profoundly ignorant of Fiji's Indian population (1977, p. 38). The relative absence of contact between colonial state officials and Indian workers contributed to the latter's periodic public expressions of frustration.

Indian indentured laborers, the most proletarianized element in the colony's population, resorted to industrial action in the mid–1880s as they bore the brunt of the depression of 1885–92. Spontaneous reactions to over-tasking and the rapid organization of a strikers' fund in 1886 were regarded by Europeans as a positive threat to the colony. One hundred and thirty laborers from the Rewa Sugar Company's plantation in Koronivia marched to Suva 'carrying knives, axes, hoes and sticks to protest against the raising of the shovel-ploughing task from 7 to 10 chains' (Gillion 1962, p. 83).

Nevertheless, intense lobbying by employers resulted in the colonial state adoption of stringent labor regulations. The 1883 Immigration Ordinance increased penalties for work-related offenses, made gatherings of more than five immigrants illegal, and forbade the carrying of implements while making complaints (Samy 1977, p. 72). With the channels for making protests clogged, incidents of attack on overseers and sardars as well as of murder, suicide, and other forms of interpersonal violence increased (Naidu 1980a).

Indian and Indo-Fijian workers and peasants, with the active support of their womenfolk, took industrial action in 1920, 1921, 1943, 1959, and 1960.[7] The first two strikes involved workers and the second two were brought about by peasants. The earlier strikes were less affected by internal factionalism and leadership rivalries than were the later two. The causes of these strikes were economic, though on each occasion political considerations did play a part (Mayer 1963, pp. 37, 39, 115).

Dire economic conditions led to the 1920 strike. The cost of the basic foodstuffs had increased by 100 to 200 percent since the prewar period (higher food prices coupled with the European shopkeeper's habit of charging Indians and Fijians higher prices than whites for the same commodities). During the same period, wages had remained at 1/6d to 2/6d (CP 46/1920; Ali 1978, pp. 55–8). The 1920 strike followed the cancellation of indenture contracts, which was accompanied by hopes for higher wages. The wages of town laborers had remained static – two shillings a day from 1909 to 1919 when they increased to 2/6 in Suva; elsewhere a daily wage of one shilling was the norm. Moreover, the disruptions in shipping caused by the world war resulted in food shortages and profiteering by merchants, who were then largely Europeans. Even the *Fiji Times* of 24 January 1920 maintained that at the heart of the unrest was the high cost of living, particularly the price of foodstuffs. Sharps (a by-product of flour-making) from which Indian and Indo-Fijian workers made *roti*, their basic staple, had increased in price over a two week period from 18 shillings for 120 lb to 30 shillings (Mamak and Ali 1979, p. 93). The Cost of Living Commission, established during the strike, found that the cost of living for laborers had escalated by 88 percent since 1914 (CP 46/1920; Gillion 1977, p. 30).

The strike by laborers in the Public Works Department in the towns of Suva, Nausori, Navua and Levuka provoked strong repressive actions by the colonial state. A month after the strike started the Legislative Council passed an ordinance for public safety which also restricted the movement of Indians and their right to hold meetings. Australia sent a warship, and 60 troops were despatched from New Zealand with a Lewis gun. All able-bodied adult Europeans were armed. Shots were fired in confrontations between stick-and-stone weilding strikers and the armed police and one man was killed and several wounded (Gillion 1962, pp. 18–46).[8] Methodist missionaries played an active role in separating the compliant minority of Indian Christians from the strikers. The strike was broken by the use of an Ethnic Fijian special constabulary (Mayer 1963, p. 39).

Colonial state officials directed their attention to the strike leaders, particularly the Baroda-born lawyer, Manilal Doctor and his wife, who were deported from Fiji on the grounds that they were not British citizens. Of the 165 persons arrestd, only nine were acquitted and the remainder received prison sentences ranging from two weeks to five years (Gillion 1977, p. 202). The Commission of Enquiry into the strike recommended that the wage rate of 2s. 6d be retained but food rations of about 4s. 6d per week be added to alleviate the conditions of wage earners. The colonial state responded to the findings of the Commission by subsidizing the price of rice by £25,000 and by initiating a $20,000 scheme to encourage rice production. It also gave an allowance of 6d. per day for rations to its employees (CP67/1920).

In the midst of the Suva strikes of 1920, the Colonial Sugar Refining Co. granted Indian and Indo-Fijian cane farmers a bonus of 2s. 6d. per ton of cane and a one pound per acre on land properly cultivated (Mayer 1963, p. 38). This was a factor in preventing the 1920 strike from spreading to the western part of Viti Levu. However, the scheme was only for a year and when calls by government officials and workers' representatives for higher wages and better conditions failed, a major industrial dispute ensued. The strike against the CSR was extended to all European planters and the Melbourne Trust Company mill in Penang, Rakiraki. Among the objectives of the strikers were demands for better housing and furnishings, free medicine, schools, rent-free allotments to plant foodstuffs, a five day work week, a six-hour workday, and a wage rate of 12s. a day (Gillion 1977, p. 55).

These demands were vehemently resisted by the CSR and European planters who maintained that the dispute was politically motivated to obtain 'Equality with Europeans' (Chairman of Ba-Based Cane Growers Association of Fiji to the government, cited in Gillion 1977, p. 55). The massive increase in wages from 1s. 6d. to 12s. per day was unlikely to be conceded by the CSR, even though it was widely known that CSR's profits had increased enormously in the previous year. Sadhu Bashishth Muni, the strike leader, was deported to Indian a month after the industrial action took place. The strikers subsequently refused to give evidence to the Commission of Enquiry they had earlier requested. The strike was peaceful and lasted for six months. But many laborers and small growers could not go on and, having gained some minor concessions, such as cheaper provisions, they returned to work. The strike was broken when Fijians were persuaded to work at a higher wage rate in the place of Indian workers (Brewster 1922, p. 301; Mayer, 1963, p. 39).

Methodist ministers went around the Ba area persuading Ethnic Fijians to dissociate themselves from the strikers, to evict them from their villages, and to follow the instructions of the Provincial Commissioner. They were encouraged to work as scab labor. As a result, hundreds of them lived and worked as the Indians had done. Rev. Stanley Jarvis of the Methodist Church was given a gift of £100 by the CSR Company at a dinner in his honor. 'The Manager at Ba presented engraved walking sticks to Tui Ba, Tui Nadi, and a chief from Nadroga for their "loyal support in the time of stress" ' (MacNaught 1982, p. 115).

Gillion pointed out that the 1920 and 1921 strikes 'hastened the emerging accord between the Europeans and the Fijians'. Very conscious of the dangers of a general anti-European conflict if Indians and Fijians were to turn against the European community and the colonial state, they acted to preempt any such solidarity.

> It was for political reasons mainly, rather than any need of their services, that the government deployed Fijian police in the area affected by the strike. Forty Fijian policemen were sent to Ba and 250 Fijians from Bau were enrolled as special constables and posted to the strike area to protect Indians who wanted to work (Gillion 1977, p. 60).

The solidarity between Europeans and Fijian chiefs was strengthened further by events in the post-1940 period.

UNIONIZED STRUGGLES

Unlike earlier disputes in the cane fields, the post 1940 strikes were initiated by organized and recognized trade unions. The two strikes which occured in 1943, one involving cane farmers, had a tremendous impact on industrial relations as well as on ethnic relations in Fiji.

The Kisan Sangh (Farmers' Association) formed between 1937 and

1940, saw itself as a loyal opposition to CSR, working to further the sugar industry by collaboration with the company rather than by following a strategy of confrontation (Prasad 1962). It received sympathy from the Colonial Governor, Sir Harry Luke, who was described by the CSR's legal adviser in Fiji, Sir Henry Scott, as 'a dirty little Jew' (MacNaught 1982, p. 161). This support from colonial officials and the fact that 72 percent of the cane farmers in the Ba-Sigatoka belt were members, compelled the CSR to recognize the union in 1941.[9] Once recognition was attained, the Kisan Saangh's image as an organization fighting for cane farmers' interests began to wane.

During the period from 1939 to 1940 the Kisan Sangh was able to gain several concessions from the CSR. These included a written statement of proceeds from cane farms, permission for farmers to plant food crops on fallow plots, a reduced working day in mills from 12 hours to 8 hours (in effect a 33½ percent increase in wages), the recruitment of more Ethnic Fijian labor to reduce the need for compulsory peasant labor in sugar mills and on railway lines, replacement of the system of paying for cane on the percentage of pure obtainable sugar (pocs) by one based on average sugar content of all farms, and the organization of cane harvesting gag sardars by the union itself, thereby preventing the ability of Field Officers (formerly Overseers) from penalizing individual farmers. The union also represented tenants of CSR who were being offered lower lease terms, and negotiated a ten year contract agreement with the company. These gains showed that the CSR accepted the principle of collective bargaining, and represented a major achievement in industrial relations in Fiji (Moynagh 1981, pp. 158–62; Gillion 1977, pp. 168–77). The colonial state was responsive to the aspirations of the Indo-Fijian peasantry and the moderate line taken by its leaders. Senior colonial officials were not unsympathetic to the Indo-Fijians' desire to increase their share of the surplus from the sugar industry. They were critical of CSR's poor industrial relations and secrecy over its accounts, but were unable to resolve the fundamental conflict between the growers and the millers over cane proceeds. When the crunch came, they sided with the company in case it withdrew from Fiji. Further, the CSR appealed directly to the Colonial Office in London whenever colonial officials appeared unsympathetic to the company's demands.[10]

This moderate approach to the CSR by the Sangh was combined with an intense concern for rural indebtedness. Nevertheless, collaboration with the CSR provoked suspicion, especially among South Indians who were less well-off than the North Indian adherents of the Kisan Sangh. Further, the enmity of Gujerati businessmen was aroused by the union's attempt to set up a cooperative store to sell goods at cost. The union raised £10,000 though an unpopular levy of one penny a ton on members to establish the shop. Envious Indian politicians took advantage of the resentment of Gujerati businessmen and '. . . the poorer South Indian and Fijian cane farmers, at the highhandedness of Kisan Sangh Officials, sirdars and bully-boys . . .' (Gillion 1977, p. 172) to form another union. The Akhil Fiji Krishnak Maha Singh)A;; Fiji Farmer's Union) was able to outflank the Kisan Sangh by capitalizing on growers' distrust of the CSR and the Kisan Sangh's attempt to reconcile the irreconcilable interests of the peasants and the sugar company. But the Maha Sangh was backed by the unpopular storekeepers and moneylenders and, by playing on communal differences, the Maha Sangh leaders, A. D. Patel and Swami Rudrananda,[11] succeeded only in dividing the peasantry.

> Instead of being united then, farmers on the west of Viti Levu, where unions were first formed, were split between the Maha and the Kisan Sangh, both of whom wanted to side with interests fundamentally opposed to the growers. To some extent, the unions were agents of competition between CSR and Indian businessmen for a larger share of the income from cane (Moynagh 1981, p. 164).

This lack of unity and the accompanying factions and rivalries contributed to the confusion that surrounded the 1943 cane strike. As a result of considerable increase in the cost of living between 1939 and 1943, cane farmers sought higher prices for their cane. According to an official report the cost of foodstuffs had risen by 91 percent, clothing by 256 percent, and kitchen and eating utensils by 335 percent (Report on Cost of Living, 1944). For farmers, especially wealthier ones, cane harvest labor had become much more expensive. In addition, the presence of thousands of American troops circulating much more money to purchase fewer goods than normal because of the Pacific

Table 10 *Changes in the Cost of Living, 27 June 1942 to 1 April 1946 (1939 = 100)*

Date	Cost of living index
27 June 1942	128
12 June 1943	156
1 October 1943	215
1 June 1944	215
1 April 1944	200
1 April 1945	186
1 April 1946	187

Source: Adapted from Moynagh 1981, p. 170.

War, resulted in an enormous increase in inflation and the cost of labor (Table 10). The 1943 strike occurred when CSR was making secure and higher profits selling sugar to the British Ministry of Food. Between 1939 and 1943 the cost of living increased by 115 percent, but earnings from cane production increased by only 50 percent (Narsey 1979).

Before cane farmers went on strike in July 1943, a strike by workers in the Lautoka and Rarawi mills was resolved by an arbitration tribune in the workers' favor. Using this as a mode of resolving the farmer's bid for higher cane prices, the Maha Dangh called for a Court of Arbitration. This strategy was different from the Kisan Sangh's request for a Commission of Inquiry into the price of cane. The latter union split into two factions, the right wing advocating a commission of inquiry and the left wing joining the Maha Sangh in demanding an arbitration tribunal. The delaying tactics of CSR, the refusal of the Attorney General's Commission to recommend cane price increases, the colonial state's reluctant support for CSR, and the rivalry of leading Indian and Indo-Fijian leaders and associated factionalism, the call by Europeans for the deportation of the strike leaders, and the counter-productive repressive measures of the colonial state against A. D. Patel and Swami Rudrananda, prolonged the strike for five months (Moynah 1981, pp. 170–2).

The strike collapsed in January 1944 as a result of the united stand of the financially strong CSR and the colonial state and divisions among the farmers. The opposition of Ethnic Fijian chiefs and the intervention by Ratu Sir Lala Sukuna, the leading chief who threatened that 'native' lands may be not be rented to Indo-Fijian peasants in the future, hastened the end of the strike (Gillion 1977, p. 184).[12] A rush to harvest the remaining sugar crop helped only some farmers. In 1942 Fiji had produced 140,230 tons of sugar. In 1943 the colony's sugar output was 56,140 tons. The strength of the opposition to CSE may be seen by the losses sustained by the peasantry:

> The farmers lost at least £1,000,000 in income from cane . . . Because of the neglect of planting, it was several years before the quantity of cane harvested reached pre-war levels. During the dispute more than a thousand fires were reported to the police, including sixteen on the CSR railway (Gillion 1977, p. 198).[13]

After this event, to restore relationships in the sugar industry and to stabilize sugar production in the long term, a commission of inquiry was undertaken.

In May 1944, Professor C. Y. Shepherd received submissions from the cane growers' unions, the CSR, and the colonial officials. Moynagh (1981, pp. 175.81) provides a detailed account of the Shepherd Report

(1945) and its recommendations. The CSR misled Shepherd about its profits and subsequently prevented the establishment of a Scientific Investigation Committee (to encourage mixed farming) and a Sugar Board (to advise on cane prices). CSR's success in preventing mixed farming was lamented in later years when the impact of sugar monoculture and dependence on imported foodstuffs made Fiji vulnerable to external forces. The company reluctantly agreed to include, henceforth, an allowance for molasses when paying for cane purchased from farmers.

The cane farming peasantry suffered physical hardship because of the strike. They also lost out in other ways. The strike occurred in the midst of World War II. Their 'sin of commission' against the war effort aggravated a sin of omission. Most Indians refused to enlist for service overseas during the Second World War unless given equal conditions and pay to whites. This refusal met with condemnation from the chiefly class and their European partners (Mayer 1963, pp. 69 – 71). Race relations were damaged almost irreparably because of this poor Indo-Fijian war record. Non-Indo Fijians could and have used this failure whenever there has been a need to put Indo-Fijians on the defensive.[14]

In the sphere of industrial relations, the British policy underwent an attitudinal change toward trade unions. Labor unrest and riots in the West Indies coupled with a commitment to improve the lot of colonial peoples, resulted in the passage by Britain of the Colonial Development and Welfare Act (1940). This Act insisted that only colonies with suitable trade union legislation should receive aid funds. The colonial state in Fiji established the Industrial Associations, Ordinance of 1942 to facilitate trade unions by allowing for registration, control, and protection of labour organizations. Between 1942 and 1952 nineteen associations of manual workers were registered in Fiji (Hince 1971, p. 371). Unions were established in sugar-milling, goldmining, stevedoring, seafaring, public works, and other government services. In 1952, the Fiji Industrial Workers' Congress was formed as a central umbrella organization for trade unions. In 1966 it was renamed the Fiji Trade Union Congress. The development of trade unions in Fiji is dealt with in some detail in Hince (1971), Reddy (1974), and Howard (1987), but for this chapter it is important to briefly describe the most significant postwar industrial dispute in Fiji.

THE 1959 STRIKE AND THE DEMISE OF LABOUR RADICALISM

The 1959 strike had the makings of a popular anti-colonial struggle which could have become the basis of a nationalist struggle for independence. However, the reassertion of the alliance among chiefs, European capitalists, and state officials restored the status quo. The workers' challenge was met with the divisive appeal of racial solidarity between Ethnic Fijian chiefs and commoners against the alleged ambitions of Indo-Fijians.

In August 1959, the ethnically mixed Wholesale and Retail General Workers' Union (WRGWU) sought a pay increase for its members who were working for two transnational oil companies, the Shell Company and Vaccum, referring all matters to their head offices in Australia. The WRGWU had approached the Labour Department to help in its bid to increase workers' wages from £3 a month to £6 a month. When the 'polite' letter that the Labour Department asked the WRGWU to write to the companies was met with delaying tactics, a strike was called, four months after the initial claims were lodged.

The WRGWU publicly stated that it would not prevent the supply of petrol to essential services. However, its willingness to minimize disruption to public welfare dwindled rapidly when the colonial officials condemned the strike as unnecessary and sided with the oil companies. They declared that petrol supply be maintained for everybody and proceeded to provide police protection to three service stations in downtown Suva. A racial situation was created when white management staff of the oil companies and other Europeans began to drive oil tankers and use bowers (CP 10/1960).

The strike lasted for five days, from 7 to 12 December, and deteriorated into a violent confrontation between the workers and the colonial state on the second day. Police harassment of picketing workers at the petrol stations, the taxi stand, and the bus station was followed by a baton charge and a grenade attack by a group of 80 policemen on a crowd of 4,000 Ethnic Fijian and Indo-Fijian workers and sympathizers. Violence against Europeans, already evident in attacks on European motorists, escalted with the destruction of European property.

To quell the riot, the colonial state promulgated the Public Safety Regulations and called out the Fiji Military Forces. A curfew was declared in the capital and the police and army patrolled the city. Numerous arrests of workers and rampaging youths were made by the police. On day four of the strike a meeting of 3,000 people was held at Albert Park, and addressed by two paramount chiefs of Fiji – Ratu Edward Cakobai and Ratu George Cakobau. These chiefs urged the strikers and their supporters to remain calm. On the following day, WRGWU officials James Anthony and Mohammed Apisai Tora, as well as Members of the Legislative Council (MLC), B.D. Lakshman and Andrew Deoki, asked workers and sympathizers not to indulge in violence.

The Fiji Industrial Workers' Congress mediated in the dispute between the Union and the oil companies. A negotiated settlement was accepted by the WRGWU President, Ratu Meli Gonewai and other executive members. Ratu K. K. T. Mara, an Ethnic Fijian chief who was also an MLC, helped in the mediation process. The union received an award of £4 11s. 6d., as against its £6 claim) for its members.

A significant aspect of the strike and the riots was the manifestation of anti-European feeling. Both A. G. Lowe, the Commissioner inquiring into these incidents, and the Mayor of Suva. Charles Stinson, drew attention to the animosity shown by Ethnic and Indo-Fijians towards Europeans. The Mayor of Suva in his submission to the Commission of Inquiry stated: 'It did not take long for one to form a definite opinion on the pattern of destruction. To my mind it was centered near all European businesses' (CP 10/1960, p. 27). Large shops owned or managed by Europeans, such as Carpenters, Burns Philip and Morris Hedstorm, had their windows shattered. Table 11 gives estimates of damage done to the premises owned by members of different ethnic categories:

Table 11 *Assessment of Damage in the 1959 Disturbances in Suva, Fiji*

	Europeans			Indians			Chinese			Total		
	£	s	d	£	s	d	£	s	d	£	s	d
Structural damage	8,723	5	3	760	17	6	109	16	0	9,595	18	9
Moveable property destroyed or damaged	3,765	17	11	299	11	0	30	0	0	4,095	8	11
Property stolen*	610	18	0	664	0	6	373	10	0	1,648	8	6
Vehicles damaged	1,204	2	3	68	10	0	. .	. . . .		1,272	12	3
	14,304	3	5	1,792	19	0	513	6	0	16,610	8	5

Source: CP 10/1960, Appendix IX, p. 28.
* the value of property stolen from the three ethnic categories is not commented upon by the Commission. However it does seem that either Indo-Fijian shopkeepers gave inflated figures for property loss or the 'criminal element' took advantage of the prevailing situation and looted retail outlets irrespective of the ethnicity of the owners.

It is clear from the above figures that European property was the target of the rioters' wrath. The Commissioner of Inquiry into the disturbances noted that,

The evidence suggests that the feeling was probably engendered by the fact that the Europeans own the largest shops and have, at least, an appearance of wealth and that the lower paid workers felt that such large shops were indicative of considerable profits whereas many workers' wages were low. In other words what has often been described as the attitude of the 'have-nots' to the 'haves' (CP 10/1960, p. 28).

In the 1950s class division in Fijian society had become more marked as Indo-Fijian workers in the urban centers of Suva, Nausori, Ba, Nadi, and Labasa were joined by Ethnic Fijians. The racial divide between Ethnic and Indo-Fijian workers was bridged by the WRGWU in the 1959 strike as they confronted the oil companies and the colonial state.

> The strike of 1959, then, clearly marked a watershed in Fiji's history . . . Working class solidarity had for the first time transcended racial boundaires, and the ruling class was unable to play upon racial sentiment in order to deal with the strike. Moreover, that solidarity also expressed a fairly high degree of class consciousness for it was clear that capital was the main enemy (Sutherland 1984, p. 295).

This situation was not acceptable to the owners of capital, the chiefs and the officials of the colonial state. Leading chiefs were shocked by the show of solidarity among Ethnic and Indo-Fijian workers. Experience at the gold mine in Vatukoula had shown that trade unions provided an alternative basis for leadership to that of traditional chiefs over their communal groups (Hince 1971; Bain 1986; Plunge 1985). The President of the WRGWU was a *Ratu* or chief but the union's General Secretary was a mixed-race person of Indian ancestry. The latter was opposed to the terms of the strike settlement but had to accept the verdict of the rest of the executive. The presence of Indo-Fijians in the union executive and the intervention of Indo-Fijian MLCs during the strike were labelled as a 'foreign' intrusion in Ethnic Fijian affairs. The chiefs asked 'commoner' Ethnic Fijians not to follow 'foreigners' – namely Indo-Fijians – but to use their leadership to voice their grievances (West 1960, p. 46; Burns 1963; Samy 1977, pp. 104–107).

The reaction of the colonial establishment to the WRGWU resulted in the expulsion of the union's militant leadership and led to its subsequent de-registration in 1962. Ethnic Fijian chiefs and their associates actively neutralized trade union militancy and emergent working class consciousmess by promoting racially exclusive unions. A number of short-lived exclusively Ethnic Fijian unions were formed. These included Fijian Domestic and Restaurant Allied Workers Union (1960–1963), Suva and Lautoka Municipal Council (Fijian) Workers Union (1962–1969), Fijian Engineering Workers' Union (subsequently the Fijian Government Workers' Union) (1962–1969), and the South Pacific-Sugar Workers' Union (1962–1964), (Hince 1971, p. 377).

Although the Department of Labour opposed racial breakaway unions, it could not influence the decision of senior state powerholders who implicitly approved this trend by granting recognition to such unions (Hince 1971, p. 377). The Department of Labour also weakened trade unionism in Fiji by discouraging broad-based unions and by promoting industrial unions.

In the short term, however, trade union militancy in Fiji continued as did the racial breakaway unions. The Annual Report of the Department of Labour dubbed 1960 as 'the year of strikes' because of an unprecedented 14 industrial stoppages involving 4,692 workers and a loss of 12,017 work days. The representatives of capital, expecially foreign capital, responded to the 1959 strike and the emergent militancy of trade unions by organizing the Fiji Employer's Conservative Association (FECA) 'in order to present a collective and effective voice to the Government of the day' (Kuruduadua 1979, p. 6).

The collusion between the colonial state and large capital was made explicit in the 1960 cane farmers' struggle against CSR for a larger share of the proceeds from sugar exports. The 1950–59 contract between the cane growers and the CSR, together with an increase in the price of raw sugar, resulted in an initial boost in the earnings of the Indo-Fijian peasantry. Between 1950 and 1953 average real farm incomes rose 70 percent, and this substantially improved farmers' standards of living (Moynagh 1981, p. 197). subsequently, real farm incomes declined without an immediate reduction in the higher standard of living. Not suprisingly, peasant indebtedness increased. In 1959 during a bumper harvest which sharply increased farm incomes, the CSR proposed reductions in the contract price of cane, tonnage quotas for farms, so that Fiji kept within its International Sugar Agreement quota of 179,000 tons, and penalties against burnt, dirty, and stale cane.

These proposals were strongly opposed by the growers' union which formed an umbrella organization, the Federation of Cane Growers (FCG), to negotiate with CSR. State Intervention split the FCG into two factions. The Kisan Sangh and the Fijian Cane Farmers' Association (advised by J. N. Falvey, a prominent pro-establishment lawyer) accepted a compromise suggested by the governor, Sir Kenneth Maddocks. This meant that CSR would buy the 1960 crop up to the quota level at 1950–1959 contract price scales, and that the 1961 harvest be started immediately. The other unions in the FCG (led by the Maha Sangh) insisted upon acreage quotas rather than tonnage quotas for cane over and above the International Sugar Agreement quotas. When further negotiations failed, this faction went on strike.

The leaders of the FCG, A. D. Patel and S. M. Koya, were castigated by the local press, Ethnic Fijian and European MLCs, and the Great Council of Chiefs. Patel was accused of serving Gujerati merchant interests by promoting cane farmers' indebtedness through the strike. Meanwhile, the Indo-Fijian peasantry divided between those who were harvesting cane and those who were not, causing considerable ill-will and provoking violence and arson. To safeguard harvesting farmers, the colonial state called upon the army to maintain order, simply reinforcing the view among FCG members that the Government was biased towards the company. The CSR itself was unaffected by the strike action. It was in an extremely strong position with a stock of '89,000 tons of raw sugar, 52,000 tons more than we permitted under the ISA' (Moynagh 1981, p. 207).

The CSR believed that the immediate difficulty presented by the FCG, if firmly handled, could only benefit it in the long term. With the support of the dominant classes, the European and chiefly establishment as well as colonial officials, it was likely not only to win the immediate battle over the 1960 quotas but also to strengthen its hand in the 1960 to 1969 cane contract.

The arrival of Julian Amery, the Under-Secretary of State for the Colonies, in October, provided the opportunity for Patel and Koya to call off the strike. At this meeting with Amery, striking Indo-Fijian peasants greeted the Under-Secretary with posters in Hindi and English which read 'Welcome Amery', 'Save Us Amery,' and 'Do Not Forsake Us' (Sutherland 1984, p. 310). These cane growers, after six months of strike, realized that the odds against them were too great and began to harvest their cane. 'The *Fiji Times* described their leaders' decision [to harvest] as little more than a 'face-saving measure' (*Fiji Times*, 17 October 1960, cited in Sutherland 1984, p. 310). The Indo-Fijian struggle against the CSR for a greater share of the fruits of their labor met yet again with defeat.

> So it was that once again, as in 1921 and 1943, the strong position of CSR supported in effect by government and the Fijian chiefs, contrasted with the disunity of farmers. And once again, a strike by Indian cane growers was defeated (Moynagh 1981, p. 208).

To create the conditions under which CSR would continue to make what it considered to be adequate profits and to persuade Indo-Fijian farmers that the proceeds from sugar exports were being fairly shared between them and the company, a Commission of Inquiry headed by Sir Malcolm Eve investigated the sugar industry in Fiji. The recommendations of the Eve Commission became the basis of the new ten-year sugar agreement that lasted from 1960 to 1969.

Its recommendations (CP 20/1961) are critically discussed by Narsey (1979, pp. 113–115, 131–133), Moynagh (1981, pp. 208–19) and Sutherland (1984, pp. 303–4) who all stress the favorable treatment given to the CSR. The recommendations covered organizational and financial aspects of the cane industry. Organizationally, the Commission reaffirmed the need to perpetuate the smallholder farming system, but strongly condemned suggestions by the Federation of Cane Growers that the CSR be replaced by a cooperative owned by the farmers. It followed the Shepherd Report of 1944 in recommending the creation of a Sugar Board and a Sugar Advisory Council. The Board would be composed of an Independent Chairman, Independent Vice-Chairman and an Independent Accountant who would be responsible for announcing the national sugar cane harvest quota and its division among the growers. The Advisory Council was to be made up of three members of the Sugar Board, two representatives of the state, three representatives of the mill workers, and five representatives each of the millers and growers. As its name suggests, the Advisory Council merely advised the state and the millers about matters of importance to the sugar industry. It is noteworthy that whereas the Company was quite free to choose its representatives, the Eve Commission severely restricted the selection of peasant representatives. Instead of the cane growers themselves electing their representatives, cane sardars (who were usually relatively rich farmers and who coordinated farm maintenance and cane harvests) were given the mandate to elect their representatives. 'It advised that no practising barristers, solicitors and members of or candidates for the Legislative Council should be eligible for membership of the Sugar Advisory Council' (Narsey 1979, p. 132). This was designed to exclude the more militant leadership of the FCG and thus deprive the peasantry of able leaders.

On the financial side, the interim agreement between the company and the Kisan Sangh and Ethnic Fijian growers' bodies, which stipulated that only 195,000 tons of the 1960 crop be harvested at the 1950–1959 rates, was replaced by a ten-year contract that fulfilled the company's wishes against the growers' expectations.

In 1957, anticipating a close scrutiny of its financial accounts and profits during negotiations for a new ten-year contract, CSR revalued its Fiji assets from $5,895,521 to £13,030,370. This exercise had the immediate effect of making the company's return on investment appear less (Moynagh 1981, p. 203). Subsequently, the Eve Commission ignored this accounting expedient and accepted the CSR's returns were low in relation to its investment for the 1950–59 period (Narsey 1979, p. 114).[15] The Commission's new sugar formula allowed the company to defray 30 percent of net income (gross proceeds minus marketing and other expenses) for its basic sugar-making costs. Narsey is correct in pointing out that there was built-in bias in this recommendation because it permitted the company to maintain high cost levels so as to make use of this initial allocation. The farmers made a slight gain when the molasses price was increased from 22s 6s. to £2 10s. per ton.

The remainder of sugar proceeds was to be divided on the basis of 82.5 percent to the peasantry and 17.5 percent to the company. Over-all, the farmers' share was to be 57.57 percent, although they had asked for a share of 64 to 66/2/3/ percent. The company offered 51 percent. The 57.57 percent share was 'less than the average of 62.2 percent of gross proceeds which farmers had received during the 1950–1959 contract' (Moynagh 1981, p. 216). The CSR was protected from rising costs but the growers had no such bult-in safeguard in the contract. Moynagh, Narsey, and Sutherland agree that, in the guise of impartiality, the Eve Commission had produced a report which was distinctly favorable to the CSR.

Even the recommendation to establish a local subsidiary to promote more efficient management of the sugar cane industry and also to improve its public image in Fiji was put to the company's financial advantage. In 1962 CSR formed the South Pacific Sugar Mills Ltd (SPSM) to control its Fiji operations. The company sold its sugar milling activities to SPSM 'for £10 million of which £3 million was in short term debentures and £7 million in shares to CSR' (Narsey 1970, p. 116). Another £2.5 million of bonus shares were taken by CSR and £1.25 million of ordinary shares were offered to the Fiji public. The Fiji demand for these shares was rather low and less than a fifth were purchased. 'In all CSR retained 98 percent of SPSM's shares and in doing so it took out nearly £3 million in cash' (Denning 1970, p. 16).

Unionised struggles in Fiji increased the during postwar period as the emergent working class organized and pitted itself against the capitalists. The 1959 strike reflected this growing consciousness of class awareness among workers in Suva. However, the workers were defeated by the combined strength of European capitalists, state officials, and the chiefs. These dominant classes also worked together to defeat the sugarcane farmers who were led by an emergent pro-fessional class of teachers, lawyers, and moneylenders. The tendency of the members of this professional class to compete over the leader-ship of the sugarcane farmers and workers resulted in factionalism, which ultimately weakened both peasant and worker organizations.

By the late 1950s the trajectory of future development in Fiji was well established. Challenges to the colonial hierarchy were effectively neutralized and the balance of class forces favored chiefs, capitalists, and state-power holders. These dominant elements found the colonial economy and polity acceptable and worked to further the existing forms of colonial enterprise and social structure. With the colonial economy firmly tied to the metropoltian economies, the colonial state, under pressure largely from external forces, moved toward decolonization.

COLONIAL INCORPORATION IN FIJI

The decade after the end of the Indentured Labour System saw significant changes among Indian immigrants and their descendants. CRS's ten-acre allotments resulted in the creation of an Indo-Fijian peasantry that quickly became internally differentiated along religious, cultural, and particularly economic lines. A class of shopkeepers, merchants, and moneylenders was emerging, together with a professional class of teachers, lawyers, civil servants, and white collar workers. Gillion (1977) describes these emergent classes and their quest for a greater say in developments in colonial Fiji.

This quest was seen as a challenge to European dominance. At the political level, race was made the basis for political representation and participation. In 1929, the colonial officials approved representation of Indians and Indo-Fijians through the electoral process for the first time.[16] The election of three Indian MLCs was bitterly opposed by Europeans whose representation had been reduced from seven to six. Once in the Legislative Council, the Indian MLCs, led by Vishnu Deo, asked numerous questions about the inferior status of their constituents in the colony. They demanded common roll with the other ethnic categories in Fiji which was in line with a similar struggle being waged in Kenya by the Indian diaspora. This demand was rejected by European MLCs, Ethnic Fijian chiefs, and state officials. The three MLCs resigned from their positions in the Legislature.

To preserve the predominance of Europeans against growing demand for equality of treatment by an aspirant and articulate group of Indians and Indo-Fijians, Governor Murchinson Fletcher (1929–36) formulated a plan to nominate representatives to the Legislature (Ali 1980). This proposal derived from his decision to replace the elected Suva and Lautoka Town Boards by bodies formed of persons that he nominated. This policy was implemented to prevent the domination of these bodies by Indians and half-castes (see Gillion 1977, Chapter 7).

The authoritarian constitution that lasted from 1937 to 1963 incorporated each of the three major races in Fiji on a separate racial basis into the colonial legislature and executive. Its aim was to 'leave the minority [white] more firmly established' (Governor Murchison Fletcher quoted in Ali 1980, pp. 143–144). Europeans also sought to exclude Part-Europeans (who out-numbered them) from decision-making bodies. Both European and Indian/Indo-Fijian representatives accepted communal roll and income and property qualifications for local council candidates. At the national level similar restrictions applied (Meller and Anthony 1967).

At the level of local government, Ethnic Fijians were treated as a separate group. Headed by high chiefs, the Fijian Administration was flexible enough to accommodate socially mobile commoners. For non-Fijians the Advisory Councils (as of 1931) permitted some Indo-Fijian input from their centers of settlement. The latter did not enjoy the prestige of an elected body or a policy-making organ, and alternatives suggested for a multiracial local administration were unacceptable to the chiefly oligarchy (Mayer 1963, pp. 101–104). Qalo (1982) provides a comprehensive account of the evolution of local government, and stresses the fact that it reinforced ethnic divisions.

Ethnic Fijian chiefs and mission-educated persons confirmed their role as junior partners of the colonial establishment as long as their positions were maintained in the guise of 'the paramountcy of Fijian interests'.[17] This notion provided for metropolitan domination and an alliance of settler and chiefly interests.

At the colony-wide level, racial divisions were maintained through separate electoral rolls and racial representation. Europeans who comprised less than 2 percent of the population and Indians/Indo-Fijians who made up 43 percent of the population in 1936 elected three representatives each on a racial basis and the Governor appointed two from each race. The Ethnic Fijian chiefs (numbering less than 100) in the Council of Chiefs gave a panel of names from which the Governor chose five to sit in the Legislature. In 1944 an Ethnic Fijian high chief, Ratu Sukuna, was appointed to the Executive Council for the first time, joining two longstanding European unofficial members. An Indian was added to this Council in 1946.

Racial disparity and divisions were fostered by the educational system (Gillion 1977, pp. 118–122). The colonial state became involved in education only in the first decade of the twentieth century.[18] It supplemented the work of the missions in a discriminatory way. In the sphere of education, priority was given to Ethnic Fijian high chiefs and Europeans (Narayan 1984, pp. 72–75).

Racial divisions were reproduced in the organs of the colonial state. The colonial state's repressive apparatus had a predominance of Ethnic Fijian personnel. In the police force there was significant number of Indo-Fijians but the army was an Ethnic Fijian preserve.[19]

A separate Fijian Administration, headed by the Fijian Affairs Board (in effect an executive committee) of the Council of Chiefs administered Ethnic Fijians. It also controlled the Native Land Trust Board (NLTB) which was formed in 1940 to lease Ethnic Fijian lands. The NLTB replaced the piecemeal and often unsatisfactory agreements between Ethnic Fijian *mataqali* and Indian/Indo-Fijian tenants. It enabled more standardized treatment of tenants but did not eradicate the abuses associated with lease transfers and renewal of leases which required high premiums. Whereas British officials at the top of these institutions formulated policies, Indo-Fijians were completely excluded. Ethnic Fijians in the colonial state's pay filled the Fijian Affairs Board, the NLTB, and the Legislative Council.

The British attempt to regenerate their national economy after the Second World War led to a demand for greater productivity in the colonies. A series of reports were commissioned. They include the McDougall Report (1957), which critically evaluated the financial organization of the Fijian Administration; the Spate Commission (1959) on the economic backwardness of the 'Fijian race'; and the Burns Report (1960) on the population, resource and the problems of the country as a whole. These reports exposed the colonial state's weaknesses and tendency to hold back the 'progress' of the Ethnic Fijians, urged a lessening of restrictions on indigenous Fijians, and a move towards individual peasant holdings. The idea was to break up the *communal* ties and move towards *individualization*. This theme was a repeat of what had transpired in the period 1915–1940.

The reports envisaged retaining the village or district center, comprising church, school, guesthouse, parish hall, and chiefly residence as the core of Ethnic Fijian life, but anticipated a sharing of common administration with other races (Spate Report 1959, p. 9; Mayer 1963, p. 98). Burns also recommended some representation of Ethnic Fijian public opinion through enfranchising commoners. Both the Burns and Spate Commissions considered the Fijian Administration inefficient and a burden on the colony as a whole (Ali 1980, pp. 54–55).[20]

As in the prewar period, the Ethnic Fijian establishment resisted attempts to change the institutions that sustained them. Neither the chiefly establishment nor the aspirant Ethnic Fijian bureaucrats took kindly to these criticisms (Mayer 1963). A complex of factors influenced their antagonism. First, the exclusive Fijian Administration ensured their own reproduction. Labor service and deference to the chiefs had become well established. Second, an alliance of the Methodist Church (based on *vakamisoneri* or church collections), village chiefs, and officials that dominated Ethnic Fijians at the local level wished to maintain its privileges. Third, village-based peasants wished to retain communal links because of difficulties in establishing themselves as *qalala* or independent farmers. They fell back on social relations such as *kerekere* that allowed redistribution and mutual support (Lasaqa 1984).

In the early 1960s the Governor mooted self-government with the maintenance of Ethnic Fijians in the Legislative Council unanimously opposed the suggestion fearing a transfer of political power to the numerically greater Indo-Fijian population. Ratu George Cakobau demanded that sovereignty of Fiji be returned exclusively to Ethnic Fijians (Legislative Council Debates, December 1961, p. 679). Ratu K. K. T. Mara, an Oxford graduate, Lauan high chief and ex-civil servant, presented the spectre of Congo, Algeria, Cyprus, Cuba, and the partition of India and Pakistan as instances of the 'devastation and disaster of the changing winds of constitutional change' (Ali 1977, p. 45). A majority of European representatives also rejected self-government, asserting that Ethnic Fijians must have the initiative in these matters.

In 1963 the United Nations General Assembly demanded (78 votes for, none against, and 21 abstentions) that Britain 'take immediate steps to transfer all power to the people of Fiji' (Ali 1980, p. 151). Within Fiji only the Indo-Fijian representatives (with the exception of an appointed member) sought independence.

Five factors shaped race-politics between 1945 and 1965. These were: (a) The call by aspirant Indians/Indo-Fijians for equality of treatment and a rejection of communal roll in favor of common roll; (b) The Indo-Fijian refusal to fight for the British during the war; (c) The industrial actions by workers and peasants construed as having political objectives; (d) The rapidly increasingly Indian/Indo-Fijian population, perceived in some quarters as a deliberate strategy to strengthen the campaign for one man-one vote; (e) The reservation of native lands for future generations – Some of these had been cultivated by Indo-Fijians. The fact that the implementation of this policy took so long created anxiety and provoked hostility (Moynagh 1981).

In 1960, Fiji's population was 401,018, almost half (197,952) of which was constituted by Indians/Indo-Fijians. It was estimated that with their rate of births of 43.67 per 1,000 (and mortality rate of 6.42 per thousand), compared to the Ethnic Fijian rate of 37.36 per 1,000 (and death rate of 7.06 per thousand), Indo-Fijians would soon become an absolute majority. The 1971 estimate of the total population was 548,000 with the latter numbering 314,000. The Ethnic Fijian fear of being swamped by Asians together with white agitation led the Governor to appeal to 'Indian' leaders to advise their community to practice restraint (Mayer 1963, p. 84). This speech in the legislature was surprising because the colonial regime had no policy hitherto for population control.

The 1963 Constitution allowed for 12 elected members in the Legislature, four each from the three racial categories. Two members from each race were appointed by the Governor. Representatives of each race selected two from their number to sit in the Executive Council. The 1963 General Election was the very first based on universal adult suffrage. It was only by a narrow majority in mid-1960 that the Council of Chiefs conceded the franchise to the Ethnic Fijian masses (Nayacaka-lou 1975).

Although political parties such as the Fiji National Party and Fijian Western Democratic Party stressed regional issues, including the plight of Western Ethnic Fijians as second class citizens, candidates sponsored by the Fijian Peoples' Party (or the Fijian Association) and by the Federation Party were elected to represent the two major racial categories.[21] A membership system followed in which three elected members, representing each racial category, were given portfolios and a place in the Executive. This incorporated the two leading parties.

At the 1965 Constitutional Conference in London, Ethnic Fijians (together with Rotumans and other Pacific Islanders) were allotted 14 seats including two representatives appointed by the Council of Chiefs. They numbered 228,000 or 43 percent of the population. The 256,000 Indo-Fijians or 50 percent of the population were given 12 seats.[22] The General Electors, a category that included Europeans, Chinese, and Part-Europeans numbered 28,000 or 7 percent of the population, and received 10 seats or 28 percent of the 36 elected seats in the Legislative Council. In this pre-independence constitution, racial or communal voting was retained (9 'Fijian', 9 'Indian', 7 'General Electors') and a category of cross-voting was introduced. Each elector cast a vote for a candidate from each race.

In debates on the constitution Indo-Fijian representatives questioned the disproportionate share of seats given to General Electors. Ratu Mara defended this inequality in terms of Europeans being a cultural buffer between the two non-white races, the removal of which could result in conflagration (Norton 1977, p. 59; Samy 1977, p. 108). The constitution became an issue in the elections of August 1966.

Two major political parties contested the election – the Federation Party and the Alliance Party. The former was led by A. D. Patel, an Indian-born Gujerati lawyer and landlord, who emerged as a leading figure in the 1960 cane strike. This party originated from the Maha Saungh which took a more radical position during this dispute. It was populist, anti-colonial, and anti-racial by stance. Although the Indo-Fijian peasantry suffered great hardship and a crushing defeat at the hands of the CSR and the colonial state, the lawyers and teachers who led them gained considerable political mileage (Norton 1977).

The Alliance Party was inaugurated in early 1966. It incorporated on a racial basis the Fijian Association, Suva Rotuman Association, All-Fiji Muslim Political Front, Chinese Association, Indian National Congress of Fiji, General Electors Association, Fiji Minority Party, Rotuman Convention and the Tongan Organization. Its political base was the Fijian Association formed in 1956 by members of the Fijian Affairs Board and included an unofficial European adviser of the

Board. By disassociating Ethnic Fijians, it aimed to undermine the strike action contemplated by Kisan Sangh (the farmers' union which was advised by a lawyer, N. S. Chalmers). The Fijian Association was a conservative chief-led body which sought to prevent Indo-Fijian attempts at political reform (Nayacakalou 1975).

The Kisan Sangh's capitulation in the 1960 cane strike in association with the Ethnic Fijian Ba Fijian Cane Growers' Association discredited it in the eyes of the bulk of the Indo-Fijian peasants. The union's leader Ajodhya Prasad, was an ex-teacher and its support base was the richer peasantry which tried to achieve economic gains by cooperating with the CSR. Through its political wing, the Indian National Congress, it aligned itself to the Alliance Party.

The Alliance Party brought together Ethnic Fijian chiefly interests, that of Ethnic Fijian bureaucrats and peasants on the basis of issues affecting their 'race', as well as Europeans, Part-Europeans and Chinese who were the leading representatives of merchant and plantation capital in the colony. Ali has noted that this party 'had the blessing of the colonial regime' (1977, p. 67).

The Alliance won 23 seats and affiliated the two Council of Chiefs nominees and the two independents. The Federation Party took the nine 'Indian communal' seats. While 35 percent of the Indo-Fijians voted against this party, the Ethnic Fijians largely voted for the Alliance, although the total Ethnic Fijian votes for this Party (67.26 percent) did not indicate great enthusiasm (Ali 1980, p. 155). With a large majority of seats, the Alliance Party leadership's confidence increased and in late 1967 it attempted to introduce a ministerial system. The Federation Party objected to any such step on the basis of the racially biased 1965 Constitution[23] which the Opposition Leader condemned as 'undemocratic, iniquitous and unjust'. He called for 'one man one vote' or common roll.

In reply, the Minister for Social Services (Vijay R. Singh) declared that the constitution gave confidence to businessmen and led to racial integration as evidenced by the previous 50 years under communal roll. Cross-voting had led to the formation of parties on national and non-racial lines. Common roll had been unsuccessful in plural societies of former East and West African colonies; it had caused mischief in British Guiana, dividing the 'Negroes and the Indians there'. Though elected on communal rolls, politicians will rise above parochial concerns as exemplified by the amendments to the Landlord and Tenant Legislation. The Native Land Trust Board (NLTB) was the largest landlord in Fiji, yet communally elected Ethnic Fijian members consented to modifications for the progress of the nation. 'If one man one vote is the best form of democracy then perhaps we are at present four times as democratic as anybody else because we have four votes to a man' (Ali 1980, p. 10).

To the charge that he was a traitor, Singh responded by reminding the Opposition that they had not fought in the Second World War and that to cooperate with the other races – 'Europeans, Chinese and Fijian' – for the good of the country was not treacherous. He commented that the introduction of the ministerial system less than ten months after the 1966 constitution is 'plain evidence of the Alliance Government [determination] to govern the nation with honour, dignity and responsibility and on democratic principles' (Ali 1980, p. 14). At this point the Federation party members walked out of the chamber.

In 1968, a by-election was held for the nine Indian communal seats to replace the Federation Party members who had walked out. The Federation Party regained all of its seats. The Fijian Association of the Alliance Party expressed its displeasure at the results by holding meetings in different parts of the country, threatening Indo-Fijians, reaffirming links with the British crown, calling for the deportation of Federation Party leaders who were not Fiji-born, and urging dismissal of Indo-Fijian civil servants supposedly pro-Federation. As a result racial tension mounted.[24]

This election under a lop-sided constitution was crucial, as it was the framework under which Fiji's independence constitution was created. Race won over class. Although the renamed National Federation Party (NFP) (with the amalgamation of the two western Ethnic Fijian parties [see Norton 1977]), regained its nine seats, it remained a minority and failed to significantly increase its Ethnic Fijian support.

Constitutional discussions begun in Suva in August 1969 and concluded in London in April–May 1970 'proved peaceful and conciliatory' (Ali 1977, p. 73). The Fijian masses (of all races) had no say in the proceedings:

> Further, the deliberations were secret; the public was informed of the results, not the details of the exchanges. Here the aim was to thwart elements such as the press, from sabotaging the dialogue, particularly when Fijian-Indian unity brought rapid decisions (Ali 1977, p. 72).[25]

The politics of racial bargaining was enhanced by the death in October 1969 of NFP leader, A. D. Patel, who was an advocate of the equality of the races in voting rights. His successor, S. M. Koya, took a conciliatory stance. When the Alliance Party announced that it would accept dominion status for Fiji, the NFP shelved its principle of common roll.

The Alliance recognized that the UN's Fourth Committee, because of its anti-colonial and democratic predisposition, would probably support common roll or majority rule. The fact that a Labour Government in England was similarly inclined made the local ruling class anxious. Both the *Fiji Times* and the *Pacific Islands Monthly* (owned by the same Sydney based corporation) representing commercial and finance capital, condemned the United Nations and made aspersions against Indo-Fijians. The Australian merchant house, W. R. Carpenters, threatened to withdraw from Fiji (Rokotuivuna, et al. 1973; Thiele 1976).

The Alliance sought to negate external pressures by seeking immediate independence. Thus the strategy of accepting formal independence was to entrench the status quo. It is apparent that the NFP accepted independence at its face value, hoping to bargain with the Alliance to resolve the outstanding issues later. As a result, 'the bitterness and wrangling that marred the 1965 conference was kept at bay in 1970' (Ali 1977, p. 73).

The 1970 Independence Constitution had the following provisions: the state's legislature was bicameral with the House of Representatives as the lower house and the Senate as the upper house. The heritage of race or communal electorates based on three broad categories 'Fijian', 'Indian' and 'General Electors' were retained. Table 12 shows the allocation of representation by race in the Parliament.

Table 12 *Allocation of Seats by Race in the House of Representatives Senate*

House of Representatives	
Communal seats	National seats
12 Fijians	10 Fijians
12 Indians	10 Indians
3 General Electors	5 General Electors
27	25
52 Total	

Senate
8 nominees of the Council of Chiefs (all Ethnic Fijians)
7 nominees of the Prime Minister
6 nominees of the Leader of the Opposition
1 nominee of the Council of Rotuma

22 Total

Source: Fiji Constitutional Discussions, (CP 1/1970), Chapters 4 & 5.

In the lower house, Fijians and Indians had 22 seats each (12 'communal' and 10 'cross-voting') while the General Electors had 8 (3 'communal' and 5 'cross-voting'). Indo-Fijians who then comprised 41 percent of the population had 42.3 percent of the seats, Ethnic Fijians then making up 46 percent of the population had 42.3 percent as well, and the General Electors constituting 3 percent had 15.4 percent (Samy 1977, p. 109). In this manner, the last was over-represented by more than five times their proportional population size.

The Upper House or Senate was formed by appointment. The Fijian Great Council of Chiefs, which is recognized as the keeper of Ethnic Fijian traditions (the chiefly oligarchy), selected eight members, the Prime Minister nominated seven, the Leader of the Opposition appointed six, and the Council of Rotuma, one. The Senate was to safeguard the special interests of Ethnic-Fijians, including matters of Ethnic-Fijian land rights and customs (Ali 1977, p. 73; Samy 1977, p. 111; Vasil 1972).

Other races had no such entrenched clauses to protect their rights, but the political significance of the Senate went beyond being a mere protective device for specific Ethnic Fijian concerns. No significant amendments to the Constitution could be made without the approval of the Council of Chief's nominees. It was necessary to have the endorsement of three-quarters of the members of both houses to change provisions such as citizenship, the position of Governor-General, the composition of the two houses, the amendment procedure, the judiciary and the public service commission (Article 67, 1970 Constitution of Fiji).

The constitution therefore gave

> . . . iron-clad security, short of revolution to the paramountcy of
> Fijian interests . . . The triumph of Fijian political and European
> economic interests at national level, matched by the unambiguous
> commitment of Indian leaders to national peace, allowed the
> ascendent Fijian leaders to foster multiracial participation in selec-
> ted areas of national life such as higher education and civil ser-
> vice, while accepting as historically determined the sharp racial
> boundaries in community life (MacNaught 1982, p. 159).

This constitution represented the triumph of the chiefly oligarchy and their allies, the prominent members of the General Elector category. Fiji gained political independence on 10 October, 1970, after 96 years of British colonial rule. Even on a racial basis, the colonial regime had failed to reconcile paramountcy for Ethnic Fijians, over-representation of Europeans, and the promised equality for Indo-Fijians. From the viewpoint of wider national concern, it left a dependent, divided and weak nation.[26] While 'race' overwhelmed 'class' divisions in society at the level of state incorporation and in the foreign-owned press's incessant racist preoccupation (reflecting ruling class-ideological hegemony), the economic reality of gross inequalities persisted. Both race and class co-exist; the latter reality was conceded even by the

191

Pacific Islands Monthly when it declared that there was 'a lot of poverty in Fiji' (February, 1967, p. 38).

The national-level maintenance of racial divisions under colonial rule had two significant consequences:

> Careful balancing of communal interests encouraged each community to cling to its own identity, to think instinctively in racial terms, to worry incessantly about political solidarity, and perhaps to miss the main point that Fiji's divided people would never be able to loosen the grip of the Australian and New Zealand corporations and a few local Europeans over exports, imports and the internal market system (MacNaught 1982, p. 28).

The coups themselves overthrew the fairly elected Fiji Labour Party and National Federation Party coalition government and the 1970 Fiji independence constitution. The military and its primarily, though not exclusively Ethnic Fijian supporters restored the defeated leader, Ratu Mara, as head of government. His defeat in the April 1987 General Election was due to a shift in the voting preference of Ethnic Fijians and General Electors in two urbanized south eastern constituencies which previously returned the Alliance candidates to parliament. This change in voter-preference in turn seemed to have been due to a growing class-consciousness among voters who also rejected Alliance Party policies for the second time in the post-colonial period. So what were these policies?

In the first decade of independence 1970–1980, Fiji's economy grew at a rate that provided for employment and infrastructural development. Between 1970 and 1977, the Gross Domestic Product per capita increased from $324 to $1,067 but in spite of the government's objective of redistribution of wealth to reduce income disparities between rural and urban dwellers and between the rich and the poor, social inequalities increased. The tourist industry which expanded rapidly in the first five years of independence contributed not only to the relatively high economic growth rate but also to social inequality.

Fiji's tourism industry has been regarded as an important diversifying factor in the economy. Between 1966 and 1975 $14.5 million in cash grants and investment allowances were allocated by the state (Britton 1980, p. 252). This industry enjoyed a boom in the late 1960s and early 1970s. Between 1965 and 1975 accommodation units increased rapidly from below 700 to around 3,500 (Cameron 1983, p. 15). Gross receipts from tourism rose from $3.6 million in 1963 to $79.8 million in 1978 (Britton 1980, p. 252). In the late 1970s, 9,000 tourism related workplaces employed 6 percent of Fiji's labor force and accounted for 13 percent of all wages (Britton 1980, p. 251).

This high growth rate was due mainly to foreign capital investment.

Local capitalists, land owners and hoteliers either speculated on their existing properties or joined the large operators. Several major multi-million dollar tourist resorts have been built along the Queens Road, which was upgraded to facilitate tourism. The Pacific Harbour resort cost in excess of $60 million. Other investments included the massive Hyatt and Sheraton hotels.[27]

Table 13 shows the distribution of tourist receipts by ownership of enterprises clearly illustrating the point that the industry largely benefits a selected few. More than 75 percent of the tourist receipts go to foreign-owned enterprises and of the remainder, a few Europeans and Indo-Fijians take an exorbitant share.

Table 13 *Distribution of Tourist Receipts by Ownership of Enterprises in Fiji, 1977*

Ownership category	Gross turnover $F000	% of total turnover
Overseas Owned		
Airline companies	35,335	25.4
Cruise ship companies	18,679	13.4
Duty-free importers and suppliers	15,200	10.9
Fiji-based ground plant companies	32,108	23.2
Fiji Owned Tourist Plants		
European	10,467	7.6
Indo-Fijian	10,993	7.9
Fijian	719	0.5
Other	2,564	1.9
Misc. tourist expenditure	3,898	2.9
Government Revenue	8,750	6.3
	138,713	100.0

Source: Britton 1987, p. 90.

Seventy percent of Fiji's tourist income is lost as payments for imports and profit repatriation (Britton 1980, p. 250; Central Planning Office (CPO) 1975, p. 169; Ellis 1983, p. 40). Five cruise ship companies control the inflow of sea-borne tourists. Air traffic of Fiji-bound tourists is coincidently dominated by five airlines, with Air New Zealand and Quantas making 80 percent of the seating capacity. Hotel ownership is 62 percent in foreign hands with 48 percent controlled by five international hotel chains (Britton 1980, p. 253). These hotels took 65 percent of the tourist trade compared to 18 percent going to the 38 hotels owned by Fiji's Europeans, 7.5 percent to 24 Indo-Fijian hotels, and to 1 percent accommodated in the 1 percent of the hotels owned by Ethnic Fijians.

Foreign interests took 65 percent of the retail income, local Europeans 14.5 percent, Indo-Fijians 15.3 percent, and Ethnic Fijians, Sino-Fijians, and Polynesian enterprizes 4.6 percent of the total turnover.

Generally, 87.2 percent of all tourist shops owned by Indo-Fijians and Ethnic-Fijians took only 65.6 percent for their foreign shareholders (Britton 1980, p. 254). About the wider South Pacific situation one writer has asserted that

> The complex inter-relationships between travel agents, hotel and airline operators, and banks has enabled a few metropolital corporations to consolidate virtual control over the tourist industry. Southern Pacific Hotel Corporations, Club Mediteranee, Air New Zealand, Quantas, Burns Philps and W. R. Carpenters are familiar names of corporations firmly established at the top of the South Pacific tourism pyramid. Beneath them, the Pacific Islands governments, local hotels, restaurant and transport operators, waiters, waitresses, hotel employees, dance groups, handicraft producers and others scramble for the 'crumbs off the table' (Winkler 1982, p. 52).

Similar observations could be made about the lumber industry in the region. Durutalo shows that in Fiji timber extraction is largely a foreign enterprise, although in some companies there is a significant local participation (1982, p. 8). In 1980 the Fiji Pine Commission, expatriate dominated and managed, on the recommendation of Canadian consultants gave logging and milling rights to British Petroleum on Fiji's 40,000 hectares of pine or green gold (Carter 1981, p. 102). Opposition action by the landowners, who had been employed to plant the pine, included roadblocks and mass picketing. Many were charged with committing offenses and imprisoned.

The largest and most significant productive, distributive, financial, and merchandizing enterprises are owned by metropolitan capital. Some of these are subsidiaries of transnational corporations of global dimensions; others are largely Australian or New Zealand based and a few are held by individual capitalists. It is correct to assert that the extent of development/underdevelopment has been largely the function of foreign capital investment (Britton 1980; Dommen 1980).

At the start of independence in 1970, it was calculated that 94 percent of the total paid-up share capital in Fiji was held by overseas companies, which accounted for 75 percent of the total fixed capital in all industries and took 80 percent of the turnover in Fiji (Britton 1982; Howard et al. 1983, p. 270; Annear 1973, p. 43). Put in another way, 'predominantly foreign owned companies account for 99.6 percent of company turnover in mining and quarrying, 63.4 percent in distribution and 75.5 percent in hotels (Howard 1982, p. 5).

A study by Carstairs and Prasad (1981) provides updates on foreign companies in Fiji. Table 14 reveals the origins of foreign companies and the extent of foreign equity in them.

Table 14 *Country of Origin of Foreign Companies, 1979 and Degree of Foreign Ownership*

	50%	20–50%	less than 20%	Total
Australian	247	10	2	259
New Zealand	136	18	8	182
USA	61	–	–	61
UK	48	2	2	52
Japan	10	1	1	12
Join foreign participation	48	8	3	59
Other companies	98	6	15	119
Not detailed	12	–	–	12
Total	660	45	31	736

Source: Carstairs and Prasad 1981, p. 11.

Thus, of the 736 foreign-owned companies, 660 were more than 50 percent foreign-owned, 45 had between 20 percent and 50 percent foreign equity, and 31 had less than 20 percent foreign control. These companies are extremely influential in the economy because they continue to account for 54.1 percent of the total company turnover, 65.3 percent company tax assessments, 18.3 percent of the total labor force, and 19.3 percent of emoluments (Carstairs and Prasad 1981, p. 58). Carstairs and Prasad's data also show the uneven but influential role of foreign corporate interests in the various sectors of the Fiji economy (see also Sutherland 1984). The level of foreign dominance ranges from 100 percent of the company turnover in public utilities to 99.6 percent of the company turnover in mining, 73.4 percent in tourism, 63 percent in wholesale and retail trade and up to 90 percent in the financial sector. The control that foreign corporations have in the commanding heights of Fiji's economy cannot be underestimated. Of Burns Philp it has been said:

> Its power in the Fijian economy is great and its executives and directors speak with authority in the public area of Fiji's developing economy. Its profits and dividends are high and come quickly. Its wages to Fiji employees are very low (about one-fifth that paid to workers in Australia) while the goods BP sells are kept at the inflated Australian price (Rokotuvuna et al. 1973, p. 50).

Burns Philp and W. R. Carpenters are extremely influential in every aspect of the Fijian economy outside the sugar industry. From their investments in transport, hotels, plantations, insurance, manufacturing, wholesaling, and retailing, as agents for motorvehicles and heavy machinery and shipping they make significant proportions of their profits in Fiji and in the region. Winkler points out that 94 percent of

Carpenters' net profits derive from Papua New Guinea, Fiji, Tonga, Western Samoa and Vanuatu (1982, p. 31).

Likewise, more than 60 percent of Burns Philp's new profits come from the island operations, though 63 percent of the sales were made in Australia (Winkler 1982, p. 28). The two companies have an octopus-like hold on Fiji. Together they own or lease 19,000 acres of the most accessible and fertile land in Fiji and employ some 10 percent of the workforce (Rokotuivuna, et al. 1973, p. 49). In recent years, to offset the effects of economic depression, both these companies have been selling their large estates to the government.

The flow of surplus from island social formations abroad meant that possibilities of expanded reproduction were forfeited to metropolitan capitalists. Brookfield echoes the findings of Narsey (1979) and Moynagh (1981) when he maintains that CSR was able to make 'a large transfer of surplus' to the Australian controlling company. 'In all' he suggests 'about one quarter of Fiji's export earnings between 1875 and 1939 flowed out of the country along various channels' (Brookfield 1987, p. 48). More recent estimates of transfer are provided by Taylor who calculates that foreign enterprises 'transfer overseas what comparatively has been estimated to be $F28 million each year' (1987, p. 58). This, Taylor adds, is about 2 percent of Fiji's GNP. A further F$25 million in cashed traveler's checks alone were transferred overseas in 1980 (Taylor 1987, p. 58).

Low wages and low commodity prices relative to imported goods have meant a persistence of aspects of non-capitalist modes of production. The latter have also been maintained by the native regulations which helped to freeze features of the earlier order. Thus village life and its concomitant communal complexes persist, but even the poorest peasant, living on subsistence production, is still dependent on the capitalist market for his/her basic wants.

World commodity prices are beyond the control of these small states which produce insignificant amounts of primary products in any case (Balasuriya 1973). Table 15 shows the average prices for major export products in the period 1971–1981. The fluctuations in the prices of the various commodities seem to be counter-balanced by higher prices received overtime, but real price increases can only be measured in relation to the cost of imported items. Terms of trade are declining against Third World countries in general and those reliant on one or two commodities are particularly affected. This applied to the bulk of the Pacific microstates.

Table 15 *Average Prices for Major Export Products 1971–1981*

	1971	1972	1973	1974	1975	1976	1977	1978	1979	1980	1981
Copra											
a) $US[a]					207	390	395	625	572	420	345
b) tala[a]	114	73	118	378	135	157	259	266	473	332	275
Coffee											
¢US[b]					86.7	222.4	187	130.1	184.8	120.1	122.8
Cocoa											
a) ¢US[b]					62.7	140.7	138.9	180.7	145.7	95.9	92.8
b) tala[a]	445	468	874	1030	821	1359	2734	2256	2355	2004	1618
Bananas											
tala[a]	1.5	2.0	2.0	2.5	2.8	2.7	3.4	8.4	7.7	6.2	5.1
Sugar											
a) £[a]					177	142	107	94	176	310	176
b) $F[a]	97	123	126	260	380	271	289	283	273	390	320
Copper											
£[a]					579	782	666	772	1016	797	864
Nickel											
£[a]					2415	3361	2732	2264	2800	2690	2935
Gold											
$F[c]	1045	1692	2451	4007	4007	3521	4345	6086	7190	15384	12200

[a] per ton   [b] per lb.   [c] per kg.
Source: Fairbairn 1985, p. 100.

Sugar production remains the backbone of the Fijian economy. The International Sugar Agreement and the Commonwealth Sugar Agreement have been followed by the Lome Convention between the EEC and the ACP. Fiji's quota is 175,000 tons, which is a little more than a third of her total sugar output (World Bank 1986, p. 143). Fiji sells a further 201,000 tons of sugar under contract arrangements above the world market prices. Table 16 shows Fiji's preferential markets in 1985.

Table 16 *Preferential Markets for Sugar 1985*

EEC (Lome)	175,000 tonnes
USA	16,000 tonnes
Malyasia	60,000 tonnes
China	40,000 tonnes
New Zealand	60,000 tonnes
Domestic market in Fiji	25,000 tonnes
Total Guaranteed Markets	376,000 tonnes
1984 predicted production	480,000 tonnes
To Sell on World Market	104,000 tonnes (21.56% of production)

Source: Taylor 1987, p. 11.

197

The price booms of mid–1970s and 1980 boosted sugar production. 'The area of cane harvested (about 70 percent of the total in any year) rose from 46,000 to 70,000 ha between 1973 and 1982, and sugar production rose from 30,000 to almost 500,000 tonnes' (Brookfield 1987, p. 50). Between 1970 and 1980 the domestic export-earning contribution from sugar increased from 66 percent to 81 percent, averaging 75 percent in the immediate post-colonial decade (Ellis 1983).

Fiji's economy remains dependent on the sugar industry and tourism – both susceptible to world market and overseas fluctuations. Meanwhile in the 1980s the urbanization of Fiji's population has continued without proportionate increase in employment. In 1986, 277,025 people lived in urban areas, and this figure constituted 38.7 percent of the total population. Of this total, a significant proportion, numbering 107,780, were Ethnic Fijians. It is noteworthy that the bulk of Fiji's population (80 percent) reside on Viti Levu, the main island, and therefore live within a day's journey from the main urban centers.

Urbanization had produced problems such as squatter settlements, tenements, and slums. Unemployment and underemployment have increased over the post-colonial period. Currently the official figure for the unemployed is 10 percent, but this is regarded as an underestimate. In 1976 the bulk of the unemployed (90 percent) were below 25 years of age with Indo-Fijians slightly outnumbering Ethnic Fijians (5,394) (FEDM 1984, p. 128). Since there is no state support for the unemployed, a large proportion of all unemployed (81 percent) depend on family support (FEDM 1974, p. 131).

Social inequality has increased significantly with those in the top 20 percent of income earners receiving more than 15 times as much as those in the bottom 20 percent (Stavenuiter 1983, p.120). Even in Ethnic Fijian villages income differentials in the households have increased considerably, indicating that traditional mechanisms of redistribution are no longer effective. Those earning below the poverty line have also increased. It has been estimated that 17,300 households or almost 18 percent of all households in Fiji live below the poverty line (Stavenuiter 1983, p. 45).

The social inequalities in society have contributed to increasing problems of social control and law and order. This is particularly the case in urban centers where crimes against property, personal violence, and sexual offenses have increased. Young persons between 17 and 25 are heavily represented in crime statistics. Ethnic Fijian youths form as much as 80 percent of the prison population, whereas suicide, another manifestation of social pressure, is particularly high among young Indo-Fijians.

By the mid-1980s the class nature of Fijian society was fairly obvious and the formation of the Fiji Labour Party a few months after the 'wage freeze' was imposed in November 1984 seemed to reflect increasing class awareness. The decision to impose a freeze on wages and salaries was a consequence of the Alliance government's inability to meet the economic challenges of the world-wide recession. This was coupled with a sizeable debt burden ($477 million in 1984), a large trade deficit, a huge wages and salaries bill for government workers, and an economy reeling from the ravages of hurricanes. Fiji's economy stagnated in the mid–1980s. The ruling Alliance Party seemed unable to deal with the situation, and allegations of corruption in high places increased.

With the defeat of the Alliance Party, which had ruled Fiji since 1970, in the 1987 general election, there was an expectation of change. However, the military coups of May and October that year reflected the extent to which an oligarchy of Ethnic Fijian chiefs, bureaucrats, and big business had been entrenched. These chiefs had become the gatekeepers to Fiji's economic and political resources. They used the rhetoric of racial solidarity to defend 'Fijian traditions and institutions' which maintained them at the apex of Fiji society. They were also well placed to use neo-traditional bodies such as the Native Land Trust Board (NLTB), the Fijian Affairs Board (FAB) and the Ministry of Fijian Affairs to extend patronage to a hierarchy of clients right down to village headmen.

The army, an almost exclusively Ethnic Fijian organization, was mobilized against the democratically elected Labour-Party-led coalition. With the overthrow of the Coalition, the oligarchy is back in power. However, the contradictions of being a peripheral capitalist society are exemplified by the increasingly 'free market' policies followed by the interim government. Deregulation, tax-free trade zones, privatization of parastatals (state-owned enterprises) and social services simply intensify social-class differences that ultimately will take political forms.

MICRONESIA

6

The Expensive Taste for Modernity: Caroline and Marshall Islands

Francis X. Hezel, S.J.

The Carloline and Marshall Islands extend some 2,500 miles across the western Pacific and encompass about a hundred inhabited islands. The inhabitants of these two archipelagoes, the geographical center of the area known since the mid-nineteenth century as 'Micronesia', are broken up into perhaps ten cultural-linguistic groups. (Anthropologists and linguists have never agreed completely on the number, since one group often shades into another and boundaries are blurred.) While the Carolines and Marshalls are neither by traditional nor modern geopolitical standards a homogenous unit, it would be a mistake to regard them as simply a Western artifact. The three modern political entities that make up the area – the Republic of Palau, the Republic of the Marshall Islands, and the Federated States of Micronesia – have all recently adopted, or are in the process of adopting a status of Free Association with the US. Moreover, they have shared a hundred years of colonial rule under four different powers. Finally, even before the first Western incursions into the area, they exhibited enough common features to be classified in that cultural family that came to be called Micronesia.

The coral atolls of the Marshalls and the high volcanic islands of Kosrae and Pohnpei – all of which have distinct but closely related languages – had relatively stratified societies with paramount chieftain-ships over extensive sections of these groups (Alkire 1977). Truk and the coral atolls of the central Carolines, whose dialects were part of a single language continuum, had petty chiefs but lacked the strong centralized authority of the east. The high islands of Yap and Palau in the west, which differed considerably from the rest of the area and between themselves in language and culture, showed strong village authority and a ranked network of villages. Despite their relatively rich resource base, these two island groups never developed the degree of stratification displayed by the islands to the east. They may serve as a warning against any overly rigid theory of economic determinism.

All these island societies, like those in other parts of the Pacific, produced all they needed to feed, clothe and shelter themselves; otherwise they would not have survived. They did this by a mode of production and distribution that has come to be termed the 'subsistence economy'. This is often taken to mean reliance on local resources with a 'make-do' technology, but it means a great deal more than that. One feature of the island subsistence economy in the Pacific is its relatively high productivity; families can produce all they need in '3–4 male labor hours a day' (Fisk 1982). This allows plenty of time for people to maintain social relationships within the family and community, and even to develop the elaborate social rituals that some societies practiced. Available land and sea resources were under-utilized – at least by today's standards – to allow them to replenish. There were no incentives to produce a surplus except for the occasional food distributed to the community, under a chief, as an expression of solidarity. The reward for such surplus production as was required was enhanced prestige in the community. On the whole, the 'subsistence economy' could be considered as much a mindset as a mode of production. It represents a cluster of attitudes that are inimical to many of those values associated with modern development in a monetized economy.

Micronesians engaged in some inter-island trade, but this was of marginal importance since most places had the same limited resources anyway. Only for atoll-dwellers, exposed to the fury of the typhoons that periodically denuded their islands, were the traditional trading networks essential. Their ties through trade contacts with other islands in the area were an insurance of aid in time of natural disaster (Alkire 1965).

The island societies of Micronesia are usually regarded as having been static prior to Western contact. Indeed, there is good reason for this. They had limited environmental resources with which to work – copper and iron ages demand metals and the means of extracting them. The islands were relatively isolated from one another – Sea voyages of even several days brought them to other islands with similar resources and technologies. Furthermore, the societies looked on maintenance of the system as a virtue and disruption as perilous, given the need to maintain harmony in their small communities. This is not to deny the fact of occasional change, even major change at times, usually as an imposition from without. The oral history of Pohnpei, less shadowy than most, records the conquest of the island by a force from across the sea that built the mammoth stone settlement at Nan Madol. The conquerors were presumably also the architects of a similar settlement on Kosrae. Kava and other non-indigenous cultural elements may have been introduced in the same period, which is dated by archaeologists at the thirteenth century.

Europeans first visited the islands in the early sixteenth century in the wake of Magellan's voyage. The Spanish stopovers were infrequent, short, and of little lasting impact aside from the introduction of the marvel of iron tools. Even if the Spanish had not bedazzled islanders with nails and iron hoop, however, the latter would have discovered this technological wonder on their own – as they did in fact on drift voyages to the Philippines and the Marianas in the seventeenth and early eighteenth centuries. Those islanders fortunate enough to obtain pieces of iron sequestered their treasure under their mats as they slept and sought more 'with the same longing that you have for heaven,' as a priest was told by one Carolinian castaway (Hezel 1983, p. 39). The atoll-dwellers of the central Carolines began making yearly voyages to Guam to obtain more of the metal for a time, but these voyages were discontinued when a party failed to return in 1788 and were presumed to have been imprisoned or murdered by the Spanish on Guam (Kotzebue 1821 v. 2, p. 240). The desultory European visits during the late eighteenth century, after two centuries of Micronesian isolation, produced little more than had early Spanish contacts. Finding a dependable supply of iron was a problem that persisted until the early nineteenth century when regular European trade contacts were established in the area.

When Lütke and the other naval commanders surveyed the area between 1815 and 1849, they found the atoll-dwellers of the central Carolines and Mortlocks far more cosmopolitan and sophisticated than the inhabitants of the larger volcanic islands of Yap, Truk, Pohnpei, and Kosrae (Nozikov 1946; Lütke 1835 v. 2). When offered metal fishhooks, Kosraeans inserted them in their earlobes as ornaments; and others were baffled about how to use steel axes. The Mortlockese, on the other hand, scoffed at the iron bars and hoop they were offered and insisted on knives, tinder boxes, and bone-handled knives instead. The central Carolinians had tasted strawberry preserves, pâté de foie gras, and madeira; and many of them could count to ten in Spanish – a consequence of their visits by earlier sea voyages to Guam and encounters with early whaleships on their islands. The outer-islander remains comparatively cosmopolitan – in English ability and awareness of the rest of the world – even today. Lacking the rich resource base of their neighbors on the high islands, the atoll-dwellers have always been forced to travel and adapt in order to provide for their needs, in times of overpopulation as well as after typhoons or in famines.

Itinerant trading captains engaged in the three-cornered China trade paid regular visits to several of the islands in Micronesia from the 1830s on. The islanders received much more that the ironware they so coveted in exchange for the bêche-de-mer, turtle shell, and mother-of-pearl that they sold to the traders as a cargo to Canton. They also were treated to calico, denim, and serge – the favorite colour almost everywhere being turkey red – and most important of all, tobacco. Yet not even tobacco's addictive property was able to revolutionize island production; people worked as they had before, simply allocating some of the surplus for tobacco and such luxury items. The China trade introduced another commodity to a few islands such as Pohnpei, Kosrae, and Palau – the white (or sometimes black) beachcomber. The beachcomber was almost always attached to a chief and was as much his 'possession' as the axes, dry goods, and ironware that made their way into the chief's hands. In Pohnpei and Kosrae the foreign resident was regularly called on to act as the chief's intermediary in trade with ships; in Palau he acted as diplomat to plead his chief's case against rival sections and to enlist military support from British naval ships and other foreign vessels (Hezel 1978).

The China trade may have provided a regular supply of such novelties as cloth and tobacco, but otherwise its disruption of island life was minimal. Beachcombers may have done some wild carousing, even killing one another on occasion, but none of that was unknown among islanders at that time. Trading was carried on almost universally through traditional mechanisms and the goods were distributed through customary channels. In effect, it reinforced rather than changed the old political system. Even the economic consequences on the islands were not especially significant – at least not yet. Drygoods and ironware were welcome luxuries, but one could not eat them or, for that matter, even wear European clothes on a daily basis. As prestige items they were in constant demand, but the exigencies of daily life were provided, as they always had been, from the land and the sea.

The American whaleship trade that flourished on Pohnpei, Kosrae, and a few smaller adjacent islands from 1840 to 1865 brought a substantial increase in both the volume of trade and the number of beachcombers. Between 1840 and 1855 yearly trade on Pohnpei may have doubled to $8,000, and the number of Westerners living on the island grew from 30 to 150 (Hezel 1984, pp. 13–14). Islanders were introduced to other, more dubious wonders like firearms and firewater. Pohnpeians and Kosraeans avidly sought both, while Palauans generally shunned liquor. The commoners of Pohnpei and Kosrae, who were without the normal trade commodities, displayed some ingenuity by selling the favors of their wives and daughters to seamen, thus breaking the chiefly monopoly on foreign trade. The reward for this was what was then called 'the pox' – venereal disease – which was added to the host of other diseases and maladies afflicting islanders since sustained foreign contact began. All islands suffered a serious population loss, but Kosrae was by far the worst; in forty years its population declined to one-tenth of what it had been (Ritter 1978). The sudden loss of population cracked the foundation of its elaborate political system, which vanished quickly in subsequent years and was replaced by a quasi-democratic system that retained some of the external forms of the older chieftainship (Lewis 1967).

The copra trade, which began in the 1860s, was the final and most perduring step in island commerce. Until a few years ago, copra remained the main export of the Caroline and Marshall Islands; only recently has it been challenged by tourism as the main industry in the area. The large German firms in the Pacific – Godeffroy, Robertson-Hernsheim, DHPG, and the Jaluit Company – were instrumental in establishing trade stations virtually everywhere in the area. Copra production was ideally suited for island life. It utilized a resource found in abundance everywhere, it required no new skills, and it could be done without any change in local work habits. The copra trade brought Micronesians few Western goods that they had not previously seen, but if offered them in greater quantity and in regular supply, since foreign traders maintained stations on nearly every island. By the early 1880s, the peak of the early copra trade, foreign firms were exporting a total of one-quarter million dollars worth of copra a year (Hager 1886, pp. 121–123).

The cumulative effects of this and the previous stages of trade in the area produced striking changes in the externals of life in the islands. The people of the eastern half of Micronesia, who had been missionized by the Protestant American Board, now habitually wore Western dress in place of their traditional garb – although some used both, depending on the occasion. Notwithstanding missionary objectives, clay pipes and whisky were common articles of trade. Iron pots had become a standard item in nearly every household, and ordinary tools and fishing gear were now imported. Rifles and muskets had also become commonplace items; by mid-century there were an estimated 1,500 guns on Pohnpei, or one for every third person (Shineberg 1971, p. 190).

The greatest beneficiaries of this new wealth were the chiefs, of course, especially in areas like the Marshalls where the chiefs retained traditional ownership rights of the land. Marshallese chiefs, who retained for themselves a third of all income from copra made on their land, became rich almost overnight and flouted their wealth openly. Several bought small schooners and hired foreigners to captain them as they made the rounds of their island possessions to oversee their estates and collect their tribute. Some chiefs dressed in suits and top hats and bought wardrobes of silk dresses for their wives. One had an income of $8,000 a year, more than the German governor of the Marshalls was earning (Firth 1977).

A look beneath these externals – something that foreign residents barely bothered to do – would reveal that in most important respects life was lived very much as it had been in past centuries. The congeries of traditional values and customs that were intertwined with the 'subsistence economy' were very much in force everywhere except in Kosrae. And yet the seeds of a social revolution had been planted. The chiefs had always supervised the land resources and were charged with the responsibility of redistributing surplus products. With the advent of trade goods, however, the old system was being challenged. The producer had, for the first time, an attractive alternative to putting his surplus at the disposal of the community through the chief; he could try to secrete it and trade it for consumer goods. Similarly, the chief was tempted to keep surplus goods for himself – something that would have been senseless if he were not able to parlay perishable produce into storable Western items. There was a point to hoarding, Micronesians learned for perhaps the first time. Not that they actually did – for the ethos and the structures of their old way of life were still too firmly in place to allow this. Nonetheless, this realization loosened, ever so slightly at first, the land tenure system in which the chiefs retained control of the basis of production. It was too early for a revolution just yet; this would come under the buffer of colonial rule in years to come.

Spanish annexation of the Carolines in 1886, although it marked the beginning of a hundred years of colonial rule, was no major watershed in the lives of the islanders. Spanish rule was largely ineffectual in Pohnpei and Yap, the two Spanish administrative centers, and the new government had almost no impact elsewhere in the archipelago. Yapese land tenure, rooted as it was in family estates, had proven resistant to mercantile innovations; and its network of village chiefs was never challenged by the Spanish, whose major concerns were the introduction of the Catholic Faith and the carrying out of public work projects. On Pohnpei Spanish attempts to impose military rule provoked two early armed insurrections, and the Spanish, who suffered heavy losses without redress, thereafter found themselves confined to their small colony in the northern part of the island (Hanlon 1988). The Spanish threat did not even have the effect of uniting the five autonomous kingdoms on the island; they remained as suspicious of one another as ever. Meanwhile, the Trukese who were relatively untouched by nineteenth-century foreign trade, continued to fight their intersectional wars with no interference from the Spanish.

By 1886 Palau had already experienced a century of political and military intervention on the part of the British navy. Thanks to its fortunate position near the main port and clever manipulation, Koror repeatedly succeeded in enlisting foreigners to aid it with their ships and guns in its struggle against its age-old rival districts. Foreign support assured Koror of primacy from the late eighteenth century to the present. It is significant, however, that none of the Koror rulers used these extraordinary opportunities to extend his rule over the entire island group, as Kamehameha did in Hawaii and Pomare in Tahiti. In Palau, where competititon on every level had long been a way of life, the fun was in playing the game rather than leaving the table with the pot.

In the meantime, the German government established a protectorate in the Marshalls, which it administered through the Jaluit Company on a business basis. The government's main goal was to promote the copra trade and the other commercial interests of the Jaluit Company. The chiefs, who stood to gain greatly from trade expansion, worked closely with the government and the business interests it represented for the same end. But German industry in the area was even then losing ground to the Japanese traders who entered Micronesia in the last decade of the nineteenth century. By the time Germany acquired possession of the Carolines in 1899, following Spain's sale of the islands after its defeat by the U.S., its trading empire was already in decline.

Germany was slow in deciding upon a plan for its new colony. When it finally did settle on a development policy – based on the suppression of all those traditional elements that were thought to impede increased economic productivity – it found itself hard pressed to implement the policy. German reforms included the attempt to limit local feasting, which the government looked on as wasteful, and the imposition of a head tax paid in labor. On Pohnpei land, once held in ownership by paramount chiefs, was deeded to individuals who enjoyed use rights to the land. The heavy-handed way these reforms were implemented triggered a rebellion by one of the five tribes on Pohnpei. The revolt was put down decisively by the Germans some months later when several of the ringleaders were killed and the entire population of the offending tribe sent off in exile to Palau. Land deeds were finally introduced by the government and accepted by Pohnpeians, although not for fear of German military might. Pohnpeian society had been shifting towards individual land ownership for a half century prior to this, ever since one of the most influential chiefs bestowed the land he possessed by virtue of his title on his son. This radical departure from customary inheritance won gradual acceptance by many on Pohnpei, and the German edict on land law offered a convenient excuse for adopting this new practice. Pohnpeians, while still bound to their paramount chiefs by a strong code of respect, had their land freed from chiefly control (Fischer and Fischer 1957).

In Palau the German administration suppressed the last remnants of intervillage warfare and the institutionalized concubinage that was practiced in the village men's houses (Force 1960, pp. 77–84). Both warfare and the raids on other villages to obtain women for the club-houses had been prime occasions for the transfer of Palauan money and the acquisition of prestige in the society. The prohibition of these customs, together with the suppression of feasting and levying of fines by chiefs, stifled the exchange system and froze the assets of the villages. The old channels of competition were choked off and the prestige order of villages, which had always fluctuated became rigid-ified. The nativistic religion in Palau, Modekngei, originated at this time as a protest against change under the Germans.

German reforms proved unequal to the monumental task of creating a commercial agriculture in the colonies. They did not even create a desire for one. The government's real success was in the opening of two phosphate mines – one in Nauru, which was then part of the Protectorate of the Marshalls, and the other in Angaur, Palau. By the end of German rule these two mines accounted for 90 percent of the total value of exports from the islands – 9,000,000 marks yearly (Hezel 1984). The mines also offered Micronesians the first large-scale wage employment opportunities. Several hundred islanders a year were recruited to work on Anguar and Nauru; their total wages amounted to about 200,000 marks yearly, perhaps one-fifth of the total money income that reached Micronesian hands (Hezel 1984). Moreover, people from all parts of Micronesia lived together in the mining camps and were forced to fraternize. It would be claiming too much to see in this the seeds of a later political unity. Workers from different island groups could not converse with one another, but they did disseminate songs, dances and other art forms – including tattoo designs – during the idle evening hours in the camp. The marching dance, stick dances and other stylized dance forms still performed throughout the area to this day owe their dissemination to this period.

In a pattern that was repeated several times over, the islands changed hands once again during a major international conflict. At the outbreak of World War I, Japanese forces seized Germany's pos-sessions, and for the next thirty years the Japanese flag flew over the Carolines and Marshalls as well as the Northern Marianas. Initially the change in colonial status had little impact on the lives of the islanders other than requiring that they learn a new language to deal with their overlords. Japan had long since achieved a commercial conquest in the islands anyway, and the early years of Japanese naval rule were simply a holding operation until a more permanent colonial structure could be erected.

Japan's seizure of the islands was formally recognized in 1920 when the new League of Nations awarded Japan the area as a mandate; and a civilian government was set up soon afterwards. The colonial administration promptly adopted an enlightened development policy that attempted to boost island productivity, as the Germans had, while offering the local people an opportunity to advance socially and educationally. Japan's economic development during the 1920s focused mainly on its expanding sugar industry in the Marianas, which depended entirely on the Japanese nationals brought in to cultivate the land. Meanwhile, mining operations continued on Angaur with the use of Micronesian labor. The 300–400 work slots each year were filled by local chiefs, according to quotas assigned by the Japanese administration. Nanyo Boeki Kaisha, the government-subsidized firm whose pedigree could be traced back to the Spanish colonial era, ran a network of retail stores employing some 700 Micronesians and controlled the copra trade (Purcell 1967). Copra production doubled to 12,000 tons a year during the first decade of Japanese rule, a level that was maintained up to the present.

The size of Japan's colonial operations, with branch offices in each major island group, permitted it to have a deeper impact on Micronesian life than either Spain or Germany. Japan established the first public education system when schools offering up to five years of primary education were opened on most islands. The main emphasis in the curriculum was instruction in the Japanese language. Hospitals were built and health care extended to outer islanders. The population decline that the area showed during the nineteenth century was arrested; the population was stabilized, although it did not begin expanding significantly until the 1950s. The Japanese also designated island and sectional chiefs, who were paid modest salaries for performing the duties the government imposed on them.

By the early 1930s a shift in Japan's colonial policies became evident. The push towards economic development during the previous decade had been more successful than Japan had at first envisioned. With sugar as the backbone of its economy, the mandate was exporting 10 million yen worth of produce annually and was entirely self-supporting (Purcell 1967). The large government subsidy that once paid for the administrative costs of the territory had been discontinued, and all these expenses were now paid by revenue raised from export taxes. Yet the social cost for this development was considerable. By 1930 there were over 20,000 Japanese nationals living in the islands, most of them in the Marianas. Japanese and Okinawan immigration was instrumental in the success of the sugar industry; and even more important from an imperial point of view, it provided relief for Japan's internal economic problems and unemployment pressure. If sugar and phosphate could be so profitable, there must be other resources that could be developed in the mandate by Japanese immigrants who would have found no work at home.

During the 1930s the colonial government, through its commercial giants Nanyo Boeki Kaisha (NBK) and Nanyo Kohatsu Kaisha (NKK), undertook a program of intensive resource development. Marine industries developed, including fishing, katsuobushi manufacture (fish cake), pearl cultivation, and sponge collection; while agricultural emphasis was on starch production, lumbering, sisal and a variety of food crops. During these years Japanese entered the islands in even larger numbers than before; by 1935 there were 50,000 in the mandate, and by 1940 the number was 85,000 (Purcell 1967). Such extensive immigration meant that Japanese authorities were increasingly absorbed with providing social services for these newcomers. As it tried to do so, the quality of care for the islanders declined.

The most serious effect of this wave of immigration was on land use. The Japanese colonial government had from the beginning claimed title to all land held by chiefs in virtue of their traditional powers. Large tracts of land on Pohnpei, Truk and Palau accordingly passed into government ownership. A survey of landholdings in 1932 revealed that the Japanese government had title to 160,000 acres, compared with the 60,000 under the private ownership of Micronesians. During the 1930s much of this land was leased to NBK and NKK for production and industry. As immigration continued to grow and land needs became more pressing, the government began to purchase private land. If the war had not intervened, Micronesians would have found themselves land-poor on their own islands.

As it was, the local population was relatively untouched by the economic miracle that was being worked around them. Eighty percent of the total labor force was Japanese. Perhaps about 1500 Micronesians were employed with total yearly earnings of 250,000 yen. This income, together with the value of copra they produced at this time, would have yielded them a per capita income of less than $35 (in current dollars) a year (Hezel 1989). This was roughly the equivalent of the per capita income during the sluggish postwar period of the 1950s.

Nonetheless, employment opportunities for Micronesians were far more numerous than they ever had been before. In addition to the labor in the phosphate mines of Angaur, Micronesians traveled to Pohnpei to plant tapioca or worked on the interisland vessels as sailors. Others who remained on their own islands found employment as assistant teachers in the local schools, as orderlies in the hospital, or as interpreters for Japanese officials. A new Micronesian elite, consisting of those young islanders who had attained fluency in the Japanese language, was in the making. They were the individuals who rode to work on bicycles and had the income to buy zinc roofing for their houses and rice and tinned food for their meals. Many Micronesians found wage employment an attractive alternative to working the land; but some had little choice, since they had sold their landholdings to the Japanese to pay their annual head tax. Micronesians in ever increasing numbers, the Catholic bishop noted, 'were going from being owners to laborers, while the Japanese, who in their own country were workers, became owners' (De Rego 1930).

The Japanese fortification of the islands for war in the early 1940s brought an end to most salaried employment, although work demands on islanders intensified greatly. Micronesians of both sexes and all ages were dragooned into labor brigades to dig out gun emplacements in caves, construct airfields and prepare the island defenses for the war that the Japanese knew was to come. After Pearl Harbor large numbers of Micronesians were evicted from their islands and relocated in other areas to make room for the Japanese military force. Many others chose to take refuge inland after 1943 as Allied bombing raids became more frequent. Local people suffered greatly as American submarines cut off Japanese shipping and supplies became ever more scarce. The additional pressure put on local resources by the tens of thousands of Japanese civilians and troops resulted in near famine on many islands by the end of the war.

Once again the spoils of war, Micronesia passed into American hands after the surrender of Japan in August 1945. The US, like Japan thirty years earlier, occupied the islands first, and only later was ceded title to its conquest as the trustee of an international body. In 1947 the United Nations recognized the US as the administering authority over what was now known as the Trust Territory of the Pacific Islands. The US, again like Japan, put its new acquisition under military rule for a time while it grouped for a development policy for the islands.

Whatever the US did in Micronesia, it was determined to avoid the 'indiscriminate exploitation of the meager resources of the island' that it felt typified Japanese rule (Richard 1957, v. 2, p. 518). Thousands of Japanese were promptly repatriated, including many who had married Micronesians. With this hundreds of Micronesian women lost their husbands and the island communities lost the skilled labor on which past economic success had been built. US policy-makers also decided that there would be no large development projects funded by outside capital. Any economic development that occurred was to benefit the Micronesian people and be subject to their control. This decision would set the pattern for the islands during their first two decades under American rule.

The isolationist policies governing Micronesia also served another important purpose. They protected the territory from the prying eyes of unfriendly powers as the US carried out a series of nuclear tests in the northern Marshalls from 1946 on. The populations of Bikini and Enewetak were relocated and were subsequently shuffled from one spot to another. Today, forty years later, they still remain exiles from their homeland. In 1951, as US naval rule was replaced by civilian administration, the US established a military base on Kwajalein that has been maintained until the present day. The nearly island of Ebeye, to which Marshallese from Kwajalein were removed, housed the ever-growing Marshallese workforce employed on the base. The population of this tiny island grew to 8,000 as people from other islands moved in to benefit from the employment opportunities there (Johnson 1984).

During the early years of American administration, the US attempted to keep the pace of development slow enough to ensure full local participation. Copra, handicraft and trochus – and for a short time phosphate – were exported for a total value of 2 million annually. Experimentation with small-scale production of such other crops as coffee, ramie, and cacao soon proved unsuccessful. The most significant achievement of this period was the establishment of small retail stores; by 1951 there were over 300 of them, many owned by Micronesians (Hezel 1984). For almost a century islanders had been the clientele in small businesses; now for the first time they were the entrepreneurs. From the humble beginnings of their 'Mom and Pop' stores, several went on to become prosperous businessmen and eventually rose to the top of the moneyed elite that emerged during the 1960s and 1970s.

One of the surest means of upward mobility for islanders was knowledge of English. From among those few who were able to converse in English after the war, US naval officers selected health aides and teachers to work with the new administration. These individuals advanced quickly and many of them went on to become the first island doctors and elected officials. The lesson was not lost on their compatriots, who enrolled in the schools started by the navy and continued by the civilian administration during the 1950s. The most capable of those who finished at the village elementary schools were brought into the district center to attend intermediate school, with some going on to teacher training school. Afterwards they obtained jobs as teachers, which they held until new mid-level positions in the government bureaucracy were opened to Micronesians. Teaching was the low rung on the ladder of advancement in government service then as today.

Even with the administration's slow-paced approach towards development and the small US budgets during those years, the cadre of

Micronesian government employees grew quickly. By the mid–1950s there were already 3,000 Micronesians working for the government bureaucracy – twice as many as the number employed in the heyday of the Japanese production two decades earlier. During these years the total income from government employment surpassed the total value of exports, and the gap between them has widened every year since then. Eighty percent of the population continued to support itself from the land and the sea, as it had always done, but the cash economy was gradually becoming more widespread. Palauans, who were among the first to sense the importance of education, began slipping off to Guam and the US in search of better-paid jobs. By 1970 there were an estimated 1,000 Paulauans on Guam, and ten years later the figure grew to 1,500 or 2,000 (Hezel and Levin, 1989).

Although unwilling to impose economic changes in the name of development, the US administration showed no such hesitation in promoting Western political practices from the very outset. With its usual zeal, the US attempted to set up democratic forms of government on every level of island society. In 1948 municipalities – corresponding to indigenous political units – were established throughout the Trust Territory and elections were held for magistrates. At first the people chose their local chiefs for office in elections that were a mere formality. Later, as it became clear that the magistrates would be expected to deal with the American government officials, some of the chiefs chose surrogates whom they supported in the elections. In such cases, the magistrate was the equivalent of the chief's deputy or the 'secretary of state', and his responsibility was to represent the island chief in all dealings with the US administration. Those chosen to serve as magistrates tended to be younger and more acculturated men – in contrast to the older and more traditional-minded chiefs. Given the magistrates' access to American authorities and the prestige this represented, tension was bound to occur between chiefs and magistrates. In time magistrates began to carve out a leadership role for themselves that was more independent of the chiefs. The early successful integration of the two parallel systems eventually broke down and the two became coexisting rivals for political power.

The changes that early American rule had produced seemed far from dramatic. Chiefly authority, although diminished, was still honored publicly. The traditional extended family continued to be the basic unit of social organization, even if smaller residential units were becoming more common. Nonetheless, the island were poised for a leap towards modernization. The people had tasted education, wage employment, mobility within and beyond Micronesia, entrepreneurship, and democratization of their institutions. Most Micronesians, then, applauded the reversal of direction that the US took during the Kennedy Administration in the early 1960s.

Long criticized for the shoddy state of development in what the press sometimes called 'The Rust Territory', the US decided to abandon its earlier 'go slow' policy and move towards rapid modernization. Such a step would not only silence the critics but also, it was hoped, persuade the Micronesian people to opt for permanent affiliation with the US when they eventually made their choice of political status. If any rationale for such a policy shift was needed, it could be found in the development theory of the day that supported investment in social services as the key to modernization.

The Trust Territory budget was doubled in 1964 to $13 million and escalated each year thereafter, reaching $60 million in 1970 (US Department of State 1981). New village school buildings were built and high schools opened up in each district, with American contract personnel and later Peace Corps volunteers recruited to teach in the new schools. New hospitals and dispensaries were constructed and health services personnel expanded. There was little emphasis on economic development as such; the main target was administration and social services, in addition to the infrastructure projects such as roads, docks and runways. Overall, one of the most visible effects of the increase in budgets was to offer hundreds of new government jobs to Micronesians who had previously been unemployed. Powered by this rapid expansion in government employment, new retail stores, restaurants, and movie theaters opened, creating still more jobs. Between 1962 and 1965 employment doubled from 3,000 to 6,000 jobs, and in the next ten years it doubled again to 12,000. While exports remained at the level of $3 million a year, cash earnings from wage employment reached ten times that amount by the mid-1970s (Hezel 1984).

The US change in policy also resulted in an education explosion throughout the territory. In 1960 there were barely 100 Micronesians graduating from the few Trust Territory high schools each year. By the end of the decade nearly 1,000 graduates a year were being turned out (Hezel 1974). After US Federal assistance grants were extended to Micronesians in 1973, large numbers of these high school graduates poured into stateside colleges to seek the college degree that they saw as the passkey to later wage employment. At the height of the college boom, in 1979, there were over 2,000 young Micronesians enrolled in college abroad. All expected to return to find jobs awaiting them, since government was a very high growth industry during the 1960s and 1970s.

Even as the US administration was initiating the policies that would convert the islands to a wage-and-consumption economy, it took steps to push the territory towards political maturation.

In 1965 the first territorial legislature was begun when the Congress of Micronesia was created. Representatives from all parts of Micronesia gathered for the first time ever to participate in a law-making body. The congress contributed to a growing sense of nationalism and awakened a spirit of unity among Micronesians. This was enhanced when, three years later, the congress began to explore the question of the island's political future following trusteeship. After nearly a hundred years of colonial rule, the islands were preparing for self-government – not as a scattered group of chiefdoms with different cultures, but as a single political entity.

The unit proved to be short-lived, however. Hardly had Micronesian leaders begun their negotiations with the US on the political status question when the fragile political unity began to crumble. In 1971 the Northern Marianas voted to go its separate way and conduct its own negotiations for a commonwealth status. Meanwhile, the remainder of the Trust Territory continued to define a status that would give it control over its own land and laws, while providing the financial assistance from the US that it saw as necessary for its survival. As negotiations with the US dragged on, further rifts appeared between the districts. These were only deepened when the US announced its future military intersts in Palau and the Marshalls and assured these groups of financial compensation for the proposed base on Babeldaob and the existing facility on Kwajalein. By 1977 Palau and the Marshalls had both decided to form their own governments and negotiate separately for Free Association with the US. This left the remaining four districts – Yap, Truk, Pohnpei, and Kosrae – to form the loose political entity that was to be known as the Federated States of Micronesia. In 1978 all the newly formed political units were granted self-government by the US under chief executives elected by the populace rather than appointed by US authorities.

As the newly formed entities moved towards termination of the trusteeship, they faced a troublesome dilemma. How could the aspirations for full political sovereignty, which had been enkindled since the mid-1960s, be satisfied without surrendering the relative prosperity that the islands had achieved in recent years? Micronesians had turned a critical corner during the 1960s and early 1970s. They had come to rely on wage employment that depended on a large government bureaucracy and, rhetorical assertions notwithstanding, they could not easily return to a semi-subsistence economy. Indeed, the shape of their families was being altered to accommodate the reality of a regular cash income, just as their chieftainships had been changed in an earlier age to allow for modern political apparatus. Education, the most expensive of the social services provided by the government, was regarded as a necessity. People had come to expect not six or eight years of schooling, as the administration provided in the 1950s, but twelve or sixteen years. The population had acquired an expensive taste for modernity, but it was unclear just how nation-states that lacked a substantial resource base would support these tastes under their own government.

The attempt to boost production for export has not met with very much success. The real value of exports has probably decreased in the last forty years; the only signficant addition to the list is the 'invisible export' of tourism, which brought in perhaps $3 million a year by the mid-1980s (Hezel 1984). In recent years the new governments have turned to the sale of rights as a source of income. The Compact of Free Association itself is viewed by some as a sale of defense rights to the US in exchange for the yearly subsidy provied by the agreement. According to the terms of the compact, the nation-states may not allow unfriendly powers access to their land or waters. The new governments have also been leasing to Japan, Taiwan, and other maritime nations fishing rights in their economic resource zone. This brings in a total of perhaps $10 million yearly for the three freely associated states. The Republic of the Marshalls has gone further than the other island states in the sale of rights. It has recently begun chartering vessels under its flag; and there are proposals at present to sell Marshallese passports for a quarter million dollars apiece and to lease certain islands to other nations for the dumping of trash.

Free Association was formally initiated in the Federated States of Micronesia and the Republic of the Marshalls in late 1986. Only Palau remains under trusteeship status, since the conflicts between its constitution and the Compact of Free Association are still unresolved. Even under the new status, however, much has remained unchanged. The government and the subsidy that supports it continue to be the base on which the entire economy rests. There continues as well the hitherto fruitless search for an industry that will generate the income needed to support the nation states in the future. In the meantime, an increasingly educated populace that seeks wage employment is faced with a stagnant island economy. After the rapid increase in the number of jobs during the 1970s, people are disturbed by the lack of expansion in employment during the present decade. In their dissatisfaction with the lack of opportunities in their own islands, many Micronesians are emigrating to Guam and the Northern Marianas, where jobs can be found in ready supply. Over 2,000 FSM citizens have moved out in the two years since 1986, and there is every reason to believe that the outflow will continue (Hezel and McGrath, 1989).

Acculturation in the islands has been a drama played out for the past two hundred years, with the greatest changes occurring in the

218

last twenty. Micronesians have travelled a long way since they first bartered for fishhooks and iron hoops. They now enjoy a material standard of living higher than at any time in the past, but this has altered their traditional cultural forms beyond all possibility of recovery. Having committed themselves to a semi-Western lifestyle, they face the problem of ensuring that it will continue.

7

The Militarization of Guamanian Society

Larry W. Mayo

Since its capture in 1898 by United States military forces, Guam has steadily become integrated into the fabric of American society. This was formally recognized with the passage of the Organic Act by the United States Congress in 1950, which not only established Guam as a US territory, but also extended American citizenship to the island's indigenous inhabitants, the Chamorros. The actual process of social integration, however, began much earlier. Taking the lead in this integrating process was the American military, especially the US Navy. The battle for Guam (if one can call it that) never made the annals of America's major military victories. Nevertheless, the US Navy, under whose authority Guam was put after its capture, played a significant role not only in the strategic fortification of the island, but also in the 'Americanization' of Guamanian society. In the decades that followed, the navy continued to play a leading role in the social and economic development of Guam. This chapter will consider the various ways the American military has been an important institution effecting the structure of Guamanian society, and how it has directly and indirectly influenced the social lives of Chamorros.

The largest and southernmost island in the Marianas, Guam lies at 13 degrees north latitude and 144 degrees east longitude. Although Ferdinand Magellan set ashore there during his circumnavigation of the globe in 1521, Guam was not claimed for Spain until 1565. For the Spaniards, Guam was principally a port for ships sailing between Mexico and the Philippines to replenish supplies of food and water. Thus it was not until 1668 that a Spanish settlement, a Catholic mission, was established. Efforts to convert the Chamorros to Catholicism led to warfare. Combined with hostilities, newly introduced diseases and natural disasters reduced the native population from an estimated 50,000 at the time of first European contact to less than 4,000 by the end of the eighteenth century. The Chamorro's defeat in war also led to their pacification under Spanish military rule, which lasted to the end of the nineteenth century. During this period the structure of

Chamorro society was transformed from being organized by ranked matrilineal clans, and subsisting by horticulture and aquatic foraging, to one organized by a colonial administration strongly supported by the Catholic Church, based on an agrarian economy. To distinguish the new social order that emerged in Guam as a result of these and other changes, I use the term Guamanian society.

The American Capture of Guam

When two nations go to war, news of it usually travels quickly. Not so in the case of the Spanish colonial government in Guam in 1898. Due to this remote locality in the Western Pacific, the only news from Spain to reach the colonial outpost was of an advisory nature, concerning the developing conflict between the United States and Spain over another island, Cuba. And that communique mentioned that a settlement was forthcoming. Therefore, no defensive measures were taken when four American naval vessels were sighted headed toward Agana, the island's capital, the morning of 20 June 1898. The American warship USS *Charleston* steamed ahead of the other vessels and entered the port of San Luis d'Apra southwest of Agana. It opened fire on Fort Santa Cruz, believed to be the main defensive base for the harbor. Unknown to the Americans, the fort had been abandoned. The Spaniards, still ignorant of the hostile intent of the Americans, reported to their governor that the *Charleston* was saluting the port. In response, a group composed of a Spanish army officer, a physician, and Frank Portusach, the only American residing in Guam, were sent as a welcoming party. Once on board, the Spanish officer apologized to the *Charleston*'s commander for not returning the American salute with cannon fire, but he explained that the garrison had exhausted its ammunition. Upon hearing this and realizing the Spaniards could not resist, the American commander placed the Spaniards under arrest but immediately pardoned them so that they could return to their governor and deliver a message to surrender the island. The following morning (21 June) the Spanish governor and members of his staff surrendered and were taken prisoner. Later in the day, the island's garrison was disarmed: its Spanish contingent was taken prisoner, the Chamorro members released. With all prisoners on board, the American convoy left Guam for the Philippines on 22 June, leaving no formal American representation in the island. Frank Portusach was asked to represent American interests for the time being (Beers 1944, pp. 1–3).

The signing of the Treaty of Paris on 10 December 1898 ended the Spanish-American War and Spanish rule in Guam, the island being transferred to American rule. On 23 December, President William McKinley ordered that Guam be placed under the control of the US

Navy Department, and that the Secretary of the Navy take all necessary steps to establish American governmental authority (Beers 1944, p. 21).

Historical records say little about the reaction of the Chamorro people to the change of rule of their island. Beers (1944) writes that in response to the American flag being raised above Agana on 1 February 1899, a delegation of Guam's principal citizens, who welcomed the commanding officers of the American forces, appeared pleased. On the other hand, according to a Spanish priest resident on the island many Chamorros were saddened (Beers 1944, p. 13). Nonetheless, in this matter their feelings were irrelevant. The United States wanted Guam for its strategic location in the western Pacific to use as a coaling station, a naval base for defense considerations, and as a cable station for communications (Beers 1944, p. 10).

AMERICANIZATION THROUGH NAVAL ADMINISTRATION

Government

When Guam was captured its legal status was not determined by the United States Congress, nor would it be until 1950. The island was classified as a United States possession and its inhabitants were designated as citizens of Guam (Cox 1926, pp. 67–68). Moreover, the entire island was declared a naval station and was closed for security. No foreign ships were allowed entrance to the port at Apra Harbor without permission from the naval government.

Official American administration began with the arrival of the first Naval Governor of Guam, Captain Richard P. Leary, on 10 August 1899 (Beers 1944, p. 21). In addition to his appointment as head of the United States Naval Government of Guam, the governor was also commandant of the naval station. Agana remained the capital. The governor had ultimate authority and power not only in executive matters, but in legislative and judicial matters as well. By 1939 the naval government had evolved into a bureaucracy of 12 departments, each headed by a naval officer. Since none of these officers remained on Guam longer than two years, routine administrative matters fell largely into the hands of trained and experienced Chamorros who had permanent positions in the government. Despite their duties and responsibilities, Chamorros were never appointed as department heads (Thompson 1969, p. 72). The municipal political framework established by the Spaniards was retained, but with some alterations. Spanish titles to official positions at the local level were changed to American equivalents. Thus the *gobernadorcillo* ('little governor') of each village became the 'commissioner'. Formerly elected during Spanish rule, under American rule these officials became appointees of the governor (Beers 1944, p. 24). It was through the naval government that initial efforts were made to restructure Guamanian society in accordance with American institutional norms and values. Further aspects of this process that are indirectly related to government administration will be considered later under other topical headings.

For reasons not explained in the historical record, the first Guam Congress was assembled by Governor Roy C. Smith in 1917. The congress had 34 members and was composed of municipal officials and other prominent men from various villages. All were appointed by the governor, held office at this discretion, and served without pay. Congress met once a month and functioned as an advisory committee to the governor, although he was not obliged to accept its recommendations (Carano and Sanchez 1964, p. 229). Governor Smith developed a good working relationship with the congress. Subsequent governors did not. Because its powers were so limited, during the late 1920s some governors ignored the congress altogether. To remedy this situation Governor Willis W. Bradley, Jr. reorganized the congress in 1930. Described as being 'democratic-minded', Governor Bradley instituted general elections. Reorganization did not bring more power to the congress, as its function remained advisory (Carano and Sanchez 1964, p. 230; Thompson 1969, pp. 68–70).

Throughout the first three decades and more of naval government no attempt was made to draft a new legal code for Guam. Each naval governor used his executive and legislative authority to make laws by issuing general orders or by modifying Spanish laws still in force. Although midway through this period Chamorros were permitted to advise the governor in matters concerning the improvement of Guam and the welfare of its people (via the Guam Congress), they had no authority to enact laws. It was not until the abolition of the old Spanish code in 1933 that the laws and court procedures of Guam was revised. The new law code, however, was not 'new' in the sense of being formulated on the basis of the particular customs, sanctions, and needs of the people. Instead, it was merely a modification of the state code of California transferred to Guam, and was subordinate to the governor's authority (Thompson 1969, p. 86).

American administration was interrupted by the Japanese occupation of Guam during the Second World War (from December 1941 to July 1944). As they had done in 1899, the people of Guam were forced to conform to the dictates of another military regime.

The Japanese occupation forces immediately began to treat Guam as a Japanese possession and removed all traces of American influence; the island's name was changed to Omiya Jima, and the use of English and the singing of American songs were forbidden. Schools were established, through which the Japanese language and culture were promoted. Villages were assigned daily work quotas to assist the Japanese war effort, and food contributions from family farms were required to feed the occupation forces (Apple 1980, pp. 31–33).

One can only speculate as to why the Japanese moved the majority of Chamorros to internment camps in July 1944. The result was that it placed them in areas away from the main battlefields of the American counterattack (Apple 1980, pp. 45–46; Beardsley 1964, pp. 214–218).

The American campaign to recapture Guam began on the morning of 21 July 1944 and lasted until 10 August when all organized Japanese resistance ceased (Apple 1980, pp. 54, 60). A special group of military personnel attached to the invasion forces – the civil affairs unit – was assigned to organize the administration of the civilian population once the island was secured. Guam was again placed under naval control with Admiral Chester Nimitz, commander-in-chief of Pacific forces, as the military governor. The civil affairs unit immediately issued a number of administration proclamations: (1) A code defining war crimes; (2) Formation of a military court for Guam and the Marianas; (3) Guidelines for operating a labor pool; (4) Codes regarding civilian property; (5) Regulations concerning refugee centers. Unlike the situation forty-six years earlier when Spanish legal codes were maintained with the capture of Guam, all laws and regulation established by the Japanese were rescinded (Carano and Sanchez 1964, pp. 309–310).

When all Chamorros were freed from internment the civil affairs unit established refugee camps. At their peak, these centers housed 18,000 Chamorros. As soon as the Japanese were cleared from an area, Chamorros were encouraged to return to their homes and farms. The extensive war damage and an ever-increasing demand for land by the military, however, left many Chamorros with no homes or farms to return to. To remedy this situation the civil affairs unit ordered construction of new villages with temporary housing (Apple 1980, p. 63; Carano and Sanchez 1964, pp. 310–311).

In October 1944 the civil affairs unit was reorganized and became the military government of Guam. It underwent reorganization again the following year and became the United States Naval-Military Government, with essentially the same structure as the prewar naval government (Carano and Sanchez 1964, p. 311).

During and immediately following the liberation of Guam the military administration was faced with a tremendous welfare and relief problem, as food, clothing, and shelter had to be provided for thousands of people. To resolve these shortages shiploads of clothing were acquired through the American Red Cross and other stateside agencies. Food was provided by the military in refugee centers and captured Japanese food stores were distributed to the people. Not only did the military government construct temporary housing in new and old villages, it also built and funded schools and hospitals (Apple 1980, pp. 63–64; Carano and Sanchez 1964, p. 312). Even though Chamorros would become increasingly self-supporting, they would be so in a wage economy largely underwritten by the military.

The naval military government ended in May 1946 and naval civilian government, structured much as it was before the war, but with a few changes, was re-established. An admiral was appointed governor, and, as before, he maintained supreme control in all civil legislative, executive, and judicial matters. The office of administrator was established to serve as the primary adviser to the governor in management of all civil affairs. Like other governmental department heads, the civil administrator was a naval officer (Carano and Sanchez 1964, pp. 320–321; Thompson 1969, p. 74).

Resumption of naval civilian administration allowed the Guam Congress, which was dormant during the war, to be reconvened. Considering the changes that had occurred since it last met, the congress decided to hold new elections and reapportion representation. The governor stimulated new interest in congress among the Chamorros when it was announced that he would seek permission from the Navy Department to grant it limited lawmaking powers, making it more active in local government. Congressional elections were held in July 1946, and in August 1947 a proclamation issued from the Navy Department granted limited home-rule powers to the Guam Congress. The congress was to be the sole authority in changing the island's existing laws. Although the governor retained his veto power, the veto could be overridden by a two-thirds vote in each house of the congress and forwarded to the Secretary of the Navy for final action (Carano and Sanchez 1964, pp. 346–347).

As Guam began to recover from the aftermath of war there were renewed appeals for the establishment of a civilian government in which Chamorros would have a greater voice, and for the extension of United States citizenship to Chamorros. Such appeals had been made as early as 1901, but were never granted. In part due to recognition of the Chamorros's allegiance to the United States while under Japanese occupation, the US Congress passed legislation to establish a civilian government in Guam. The new civilian government, the Government of Guam (now commonly called GovGuam), was based on the American model of separation of powers. It contained an executive branch, headed by a governor (initially appointed by the President of the United States, but in 1970 becoming popularly elected); a legislature comprised of a single house, not to exceed 21 representatives (called senators); and a judiciary, composed of various courts established by the legislature.

Although granted American citizenship, Chamorros did not gain the right to vote in presidential elections. Furthermore, their new political status did not put Chamorros on an equal standing with American citizens in the mainland. Since the Organic Act was a product of the US Congress, Chamorros were governed by a document in which they had no hand in composing; they were denied the opportunity to elect their own chief executive official (the governor) until 1970; and they had no representation in the US Congress until 1972, when they were allowed to elect a non-voting delegate to the House of Representatives. Thus, even as American citizens, Chamorros remained in a politically subordinate position.

Even though civilian rule was firmly established with the Organic Act, the military maintained a controlling influence over Guam through the naval security closure. Entry to Guam was still subject to military approval. This situation changed in 1963 when the security closure was repealed to hasten efforts to rehabilitate Guam after Typhoon Karen.

Military government, both naval and civilian, was principally responsible for promoting sociocultural change in Guam until 1950. Governors of Guam exercised their executive authority to restructure old and create new political and legal institutions in the island. And because these institutions were patterned on American models, we may ask whether or not they were intended to foster American norms and values. Key values in American politics are democracy and equality, in the sense of 'one man one vote'. But these were not achievable under military government because the military is not a democratically structured institution; subordinates follow orders; they do not debate issues. And Chamorros were indeed subordinate, not only within government administration, but also in society at large. As Thompson (1969, p. 65) characterized it, government administration in Guam was a 'military regime similar to that on a battleship or in a navy yard . . . set up over the civilian native population'. The subordination of Chamorros through military government was also evident in another institutional context – education.

Education

Some initial steps were taken by Governor Leary with the intent of separating the Church from involvement in political and civil affairs. As a fundamental institution in the society for more than 200 years (since the founding of a Spanish mission in 1668), the Catholic Church continued to have a dominant influence on the people of Guam. Throughout the Spanish period priests not only provided spiritual guidance for the people but also performed an important role in their governance through close cooperation with the colonial administration. Spanish priests were allowed to remain on Guam when authority over the island was transferred to the United States. Governor Leary, however, believing their influence over the people would be a hindrance to the inculcation of American values, later had them expelled. Along with the priests, Catholic friars were also removed. Assisted by a few Chamorros, the friars had been responsible for maintaining a school in each village. With their departure, the island school system fell into disarray. In an effort to facilitate a directive from Governor Leary that every adult resident learn to write his or her name – as well as the governor's overall desire that all island residents learn to read, write and speak English – Lieutenant William Safford (the governor's aide) organized informal English classes in Agana, which met several times a week, attracting 50 pupils. In January 1900, Governor Leary decided that the naval government should assume complete control of education. By general order he prohibited religious instruction in village schools, and ordered that instruction in English be introduced as soon as suitable teachers could be obtained. Until then, the staff of native teachers in villages other than Agana continued on their own. When it was learned that native teachers were ignoring the order against religious instruction, the governor issued new orders restating the ban and directing the commissioner of each village to remove all religious paraphernalia (e.g., crucifixes and pictures of saints) from school grounds (Beers 1944, p. 31). By 1905, with more American immigration and a few local English-speaking teachers, the school system became Americanized. English and elementary subjects were emphasized and religious instruction in schools was prohibited. More teachers were trained locally and new schools were constructed (Thompson 1969, p. 217).

A major reorganization of the educational system commenced in 1922, under the guidance of professional American educators. The curriculum and general regulations pertaining to schools were revised and patterned on the California educational system. New textbooks were introduced and schools were more adequately graded (i.e., first, second, third grades, etc.). Again emphasis was laid on instruction in English. Chamorro, still the language most often spoken in the home and for social discourse among kith and kin, was forbidden in public schools and on playgrounds. In 1923, a two-year normal school with instruction in college-level subjects was established. Attendance was compulsory for all local teachers and candidates for teaching positions. By the late 1930s, the educational system included twenty-eight schools. All were coeducational and staffed with Chamorro teachers. About thirty American children were taught in a separate school in Agana run by six American teachers; thus schooling for Chamorro and

American youth was segregated. At the primary level school curriculum included courses in English, simple arithmetic, local and American history, local and world geography, and tropical hygiene. Except for instructional material on local history and tropical hygiene, course textbooks used in Guam were identical to those used in stateside public schools (Carano and Sanchez 1964, p. 413; Thompson 1969, pp. 218, 220–222).

The educational system, like the political, was based on the American model. No effort was made to formulate a school system or an educational curriculum specifically suited to the needs of the Chamorros. Schools were used to promote learning of English and, since standard American texts were used, to convey American values.

LAND AND THE MILITARY

When the United States captured Guam, land holdings of the Spanish regime, which consisted of about one-fourth of the total area of the island, were assumed by the federal government under control of the Navy Department. Additional land came under control of the naval government when it began to acquire land from Chamorros, who lost property by foreclosure due to nonpayment of newly imposed taxes, and through direct purchases. Consequently, by 1940 the naval government had increased its total land holdings from one-fourth of the total area to more than one-third of the total area. Still more land came under military control after Guam was liberated from the Japanese. During the years following the liberation, naval and army engineers began large-scale construction of naval and air force installations on vast tracts of land acquired by the federal government, which displaced many Chamorros (Stevens 1953, pp. 90–91). Consequently, in 1946 it was estimated that the combined federal and naval government land holdings comprised almost two-thirds of the island (Thompson 1969, pp. 115–116, 118).

Following the Second World War, the military acquired privately owned land through condemnation. During my own fieldwork carried out in 1980–81, an informant described how this process affected her own family. Their land was condemned by the federal government for acquisition by the military. They had no desire to sell the land but were forced to accept the amount offered by the federal government – an amount they considered to be far less than what the land was worth. In *An Island in Agony*, Tony Palomo cites Chamorros who testified before a House Public Land Subcommittee in 1949. Their testimony conveys the general sentiment of Chamorros concerning military land acquisitions. They maintained that the market value determined for privately-owned land in Guam was too low; and furthermore, that the tactics of the Navy's negotiators were intimidating (Palomo 1984, pp. 246–248). Consequently, more than a thousand Chamorro families filed suit against the federal government for just compensation for their land. In 1983, the federal government agreed to pay the prior land owners $39.5 million in an out-of-court settlement (Eichner and Wright 1983, p. 21). Some families, however, were still not satisfied and refused the offer (Panholzer 1985, p. 14).

After 1950, lands formerly controlled by the naval government were transferred to the new civilian Government of Guam. Thus the distribution of land is now divided between the federal government, which owns one-third or 40,000 acres (35 percent of Guam's total land area of 214 square miles) (Government of Guam 1980, p. 4), the Government of Guam with one-third, and the remainder in private ownership.

One consequence of so much land being owned by government (both federal and local) is that land values have increased dramatically. To convey the extent to which land values have increased, a *Pacific Daily News* (Guam's leading newspaper) editorial relates an anecdote about a local resident who in 1930 offered to sell half the land on Tumon Bay for $4000. By 1986 the same land would be worth at least $600 thousand (Murphy 1986a, p. B–2). A report in a current issue of *Pacific Magazine* focuses on the rise in land sales, which have tripled over the past two years. Because many of these transactions involve sales by Chamorros to Japanese investors, it raises concern that more land will be permanently alienated from the Chamorro people (*Pacific Magazine* 1989, p. 18).

ECONOMY AND THE MILITARY

American Marines garrisoned in Guam in 1899 improved the island's infrastructure in preparation for constructing the naval base and cable station, which required a much larger labor force. Economically it was more profitable to use local labor rather than import American workers, so the naval government took steps to attract laborers from the local agrarian economy. Entry into this new market for natives was eased when Governor Leary abolished the peonage system that had emerged during the Spanish regime. His proclamation annulled all existing contracts under the system, thereby allowing debtors to pay off their debts with wages earned in the open market. Under conditions of the naval security closure, however, commercial enterprise in Guam was severely limited. This left few options for wage-labor to be found any place other than the federal and naval governments.

Nevertheless, to attract workers the administration offered wages much higher than the maximum salary that could be earned at the end of the Spanish period. The incentive worked only too well. With newly gained wealth Chamorros acquired a taste for many imported American goods brought in originally to supply American personnel. More and more men left their farms to work for the naval government, either in government administration or at the naval station. By 1911 nearly a fourth of all able-bodie Chamorro males were employed as laborers by the naval government. Ten years later, about a third of Guam's labor force of 3,000 men were either working for the naval government or employed in trade in Agana (Thompson 1969, pp. 128–129).

Those who worked in the naval government or on the naval base were employees of the United States federal government. As their salary was based on a scale formulated in Washington, DC, the government rate of pay did not coincide with that in the local economy, which was substantially lower. In 1939 the standard government wage for unskilled labor was $1.05 per day, while the local rate ranged from 50 cents to 60 cents per day. The great disparity in wages and the difficulty of finding high-salaried work in the local market made government employment preferable. Many were disposed to wait for a government job rather than hire out to a local employer. On the other hand, government wage rates in Guam were set on a dual standard, one rate for American citizens and another for citizens of Guam. Consequently, Chamorros were paid substantially less than Americans for similar work (Thompson 1969, pp. 147–148). The number of Chamorros employed by the naval government remained at a high level up to the end of military government in Guam. In 1949, a year before Guam became a US territory, more than a third of the island's local work force of 8,700 persons was employed by the naval government (US Navy Department 1949, p. 19).

Although the dual salary scale for immigrant Americans and Chamorros was abolished in 1950 by the Organic Act, new guidelines set wages for stateside-hired civilian military employees 25 percent higher than their locally-hired counterpart (Government of Guam 1951, p. 9). Thus a dual wage standard that was adversely discriminatory toward Chamorros persisted.

Although no longer involved in the governance of Guam, the military continues to be an integral part of the island economy. By the 1980s, the military presence in Guam had grown to include: the Apra Harbor Naval Station, which contains a ship repair facility; Andersen Air Force Base, of the Strategic Air Command; a naval air station; a naval communications station; a naval magazine; and a naval regional hospital. In terms of personnel, in January 1984 there were 10,899 services men and women on active duty stationed in Guam. With their 11,827 dependents, the military population represented a little more than 20 percent of Guam's total population, estimated to be just over 113,200 (Government of Guam 1984, p. 29).

Guam's economy receives economic benefits from the military in many ways. Civilians are employed on the bases. Jobs range from skilled mechanics, electricians, and craftsmen at the naval ship repair facility, to service workers such as clerks and cashiers at base commissaries and stores. Military construction contracts are awarded to local construction firms, and often account for the larger portion of all major construction projects on the island. The military is a consumer of locally produced goods such as fresh eggs, bread, milk, and produce from local merchants. Military agencies also contract for services such as packing and shipping, garbage and waste removal, office machinery repairs, and grounds maintenance. Island retailers benefit from purchases made by military personnel and their dependents who live off-base, and from new arrivals who on occasion make use of temporary lodging while awaiting on-base housing. The Government of Guam collects direct revenues from the military through at least three channels: (1) From federal income taxes withheld from military personnel salaries reverted to Guam; (2) From funds paid to the Guam Department of Education to support military dependent children enrolled in public schools; (3) From registration of motor vehicles purchased locally or brought over by military personnel (Pugh 1971, pp. 66–67; Government of Guam 1977, p. 23).

In addition to its impact on employment and as a source of revenue, the military played an important role in modernizing Guam's infrastructure both before and after the Second World War. For example, US Navy generators provided electricity not only for the military community, but also for the civilian community. In 1973, the Guam Power Authority – an autonomous agency of the Government of Guam – entered into a power pool arrangement with the US Navy for the production and transmission of electricity. The arrangement obliges both parties to share operating expenses based on power delivered to each (Government of Guam 1980, p. 18).

Because of its primary role in the economy the military has greatly influenced the social lives of Chamorros, particularly in terms of employment. The general pattern of Chamorro employment since 1920 has been a steady increase in non-agricultural-related occupations and, especially after 1950, a substantial increase in the public sector economy, public administration in particular. Starting in 1930, the number of Chamorros employed in agriculture and fishery declined steadily (Table 1). That year 53 percent of all employed Chamorros were in agriculture and fishery, declining to 45.9 percent in 1940, 15.8 percent in 1950, and down to 5.3 percent in 1960. Meanwhile, the number of

Chamorros engaged in public administration and service increased from only 0.58 percent of the total employed in 1930, to 1.1 percent in 1940, 24.2 percent in 1950, and up to 33.2 percent of the total employed in 1960.

Table 1 *Chamorro Employment, 1930–1960*

Industry Group[a]	1930	1940	1950	1960
A	2,625	2,747	1,036	383
B	792	813	1,108	778
C	921	1,223	985	1,194
D	212	299	1,194	1,169
E[b]	—	—	47	96
F	173	211	538	901
G	29	69	1,589	2,369
Total	4,944	5,973	6,534	7,127

Source: Adapted from Mayo 1984, p. 163.
Industrial group code:
A=Agriculture and fishery; B=Construction and manufacturing; C=Transportation, communication, personal services; D=Wholesale and retail trade, business and repair services; E=Finance, insurance, real estate; F=Professional and related services; G=Public administration and service.
[b]The industry categories wholesale and retail trade, etc. (D) and Finance, insurance, real estate (E) are not distinguished in figures for 1930 and 1940.

In the 1970s Chamorros still preferred employment in the public sector. Data from a random University of Guam survey of 700 households throughout Guam carried out in the early 1970s showed that among employed Chamorros surveyed, fewer than 17 percent (actual numbers are not reported) indicated they worked in the private sector economy. Only 0.3 percent of those surveyed gave agriculture as their primary means of support, while 90 percent of those surveyed claiming work in the private sector were engaged in retailing, construction, finance, or were self-employed. Almost half (49.7 percent) of employed Chamorros indicated they worked for the Government of Guam, and 33 percent said they were federal civil service employees, either employed at a military installation or enlisted in the armed forces (Haverlandt 1975, p. 97). This sample of Chamorro employment conforms to the trend shown for the period 1930 to 1960. The military continued to be a principal employer in the island into the 1980s, employing more than 6,000 Guam residents or a little more than 19 percent of the total working labor force (Government of Guam 1984, pp. 26, 31).

A historical fact, frequently cited by people in Guam to explain the Chamorro preference for employment in the public sector, is the practice of the Naval Government of Guam in offering wages double the top salaries offered in the local private economy. This policy persisted through the postwar period. Moreover, a 1948 US Navy report to the Secretary-General of the United Nations on labor and employment conditions in Guam stated that Chamorros were granted preference over statesiders (immigrants from the US mainland) or other foreign persons for all levels and types of employment in government service (US Navy Department 1948, p. 10). The Naval Government never promoted development of a private economy in Guam since it looked upon the island primarily as a strategic resource, rather than an economic one. Therefore, government service and employment at military installations became the role models for Chamorros seeking employment. A Chamorro put it this way:

> The island has been so inundated with colonial rule, that the only role model most of these people look toward is a government model. And so people go [off] to school; they come back, and the next thing they want to do is to work for the government. . . . They've never adequately had a private sector role model that would allow talented local people to gravitate toward those jobs, but the opportunities are there.

Among the 37 Chamorros I formally surveyed, 20 indicated they were currently employed or had been employed in the public sector, serving in GovGuam, the military civil service, or the armed forces. Eleven said that one of their parents had also been employed in full-time or part-time capacity in the public sector. None said that they took up public sector employment due to direct influence from their parents.

Some of the reasons Chamorros prefer government employment include job security, good retirement and fringe benefits, and on average higher salaries than those in the private sector. Other elements contributing to the Chamorro preference for government employment are kinship, and a sense of unity and belonging. Chamorros have been involved in government service since the establishment of naval administration in Guam, and they were hired preferentially over other ethnic groups. When new positions arose, employees undoubtedly informed their relatives and helped them obtain employment. Through the close network of family relations that persists among Chamorros, more and more of them gravitated into government employment. The long term consequence of the eventual Chamorro preponderance in local government that emerged is that they developed a special relationship with the governmental institution. Chamorros describe this in several ways. One said that Chamorros 'feel more comfortable in low [level] governmental positions due to the environment'. The implication here being that since most government employees are

Chamorro, they maintain a great sense of camaraderie. Another informant told me that his mother and an uncle worked for GovGuam and that they believed working in the private sector would have made them 'outsiders'. A third informant (a statesider) said that a senator told him Chamorros refer to the Government of Guam as 'my company'. These comments illustrate two points: First, that Chamorros express a sense of belonging in government, that is, since the institution of government symbolically represents a people, employment in GovGuam gives Chamorros a means of close and secure affiliation with the institution that reflects them as a people. Second, a form of self-imposed alienation from the private sector is implied, conceding it to non-Chamorros because Chamorros generally assume that non-Chamorros controlled the private sector. Chamorros are active in business enterprises, but the largest companies and businesses in the island are Japanese or multinational corporations. It is in this sense that the assumption of a non-Chamorro-controlled private sector is made.

It is largely because of its significant role in the development, and as a mainstay, of the economy that the military presence in Guam has been and will continue to be regarded favorably by the civilian population. This observation is derived from comments made by numerous informants, as well as from sources in the local news media. An entry in my fieldnotes from November 1980 refers to a *Pacific Daily News* article on the possible home-porting of a naval support ship at Guam. It noted that all additional military ventures on the island are considered favorable because ultimately they will provide a boost for the economy. The article reported the results of a poll administered by the Guam Chamber of Commerce in conjunction with a Guam Department of Labor survey. Of the 1,000 islanders polled on the desire to have an aircraft carrier based at Guam, 74 percent were in favor. Moreover, 84 percent of those polled thought the military presence was good for Guam, and 66 percent favored building new facilities to accommodate more military personnel.

Ethnic Diversity

A direct outcome of the military's economic impact on Guam led to an increase in the ethnic diversity of the population. When the 1940 census was taken, Guam's total population was 22,290. Chamorros accounted for 90.5 percent of the total, or 20,177. The rest of the population as listed in the census included: White, 785 (3.5 percent); Filipino, 569 (2.6 percent); Japanese, 326 (1.5 percent); and Chinese 324 (1.5 percent). In 1950, Guam's total population increased to 59,498, but the proportion of Chamorros declined to 45.6 percent of the total (or 27,124). Meanwhile, the non-Chamorro portion of the population increased substantially, as a consequence of American immigration and importation of Filipino labor during the postwar fortification and reconstruction of Guam (US Bureau of the Census 1941; 1953). Attracted to the island by new economic opportunities (mainly construction jobs contracted through the military but also other job prospects in a fledgling service economy), by 1950 the number of whites (statesiders) in Guam rose to 22,920, representing 38.5 percent of the total population. Construction firms with military contracts were permitted to import Filipino laborers to supplement the small local labor force. Once the naval government opened up the Filipino labor market, private merchants wanted to take advantage of it as well. To avoid violating federal immigration laws, employers prepared the appropriate documents for each worker and included a bond to guarantee their return passage. Under this arrangement, Filipinos were brought in by the thousands to serve as skilled and nonskilled construction workers, longshoremen, service workers, book-keepers, and accountants. By 1950, their number increased to 7,258, which was 12.2 percent of the total population (US Bureau of the Census 1953). Filipino immigration to Guam remained high throughout the decade of the 1950s, reaching a peak of 18,000. This large number was a consequence of two factors: (1) A surplus of labor in the Philippines; (2) The opportunity to earn wages in Guam that were much higher than those offered in the Philippines. When Filipinos began to compete with Chamorros for private sector jobs, the federal government stemmed the flow of immigration and began to phase-out contract Filipino workers. By 1959 the number of contract Filipino workers declined to 5,000, but several thousands became naturalized American citizens and permanent residents (Stevens 1953, p. 111; Lowe 1967, p. 409).

According to a *Pacific Daily News* editorial (Murphy 1986b, p. D–4), the population of Guam by ethnic groups in 1986 was estimated to be as follows:

Chamorros	55,000	Koreans	3,000
Statesiders	30,000	Micronesians	3,000
Filipinos	25,000	Japanese	1,000
Chinese	3,000	Total	120,000

Of the 30,000 statesiders, a little more or less than three-quarters are military personnel and their dependents. Thus the military was not only indirectly responsible for immigration to Guam of diverse ethnic groups, its own personnel, representing a significant proportion of the population, adds to the ethnic diversity.

The increase in ethnic diversity has also influenced how Chamorros identify themselves. According to Laura Thompson, who conducted research in Guam from 1928 to 1939, the island's indigenous inhabitants preferred to be called Guamanian, instead of the more correct term Chamorro (1969, p. vii). Moreover, Carano and Sanchez (1964, p. 9) note that at the present-day Guam people call themselves Guamanians and generally use Chamorro with reference to their native language. Perhaps since the 1970s, however, Chamorro has replaced Guamanian as the term used with reference to the indigenous people. One explanation for this change is that as more foreigners immigrated to Guam and began calling themselves Guamanian, Guam natives began to refer to themselves as Chamorros to maintain their distinctiveness.

Community Relations

In general, relations between the military and the civilian community in Guam have always been good, even though the relationship changed from being hierarchical during the period of naval and military government in the first half of the century to relative equality under civilian government since 1950.

By 1940, Guam had developed into what Wilson (1975) describes as a 'dual community': the local Chamorro community and the American military community. Wilson says that although interaction occurred between members of both communities, they remained separated by language,[1] culture, and standard of living (1975, p. 94). Several newly introduced structural features contributed further to the separation of Chamorros and Americans. Thompson (1969, p. 56) notes that American naval officers in Guam maintained their own social group which excluded Chamorros, in contrast to Spanish colonial officials who mingled socially with the Chamorro 'high people' (descendants of the ancient Chamorro nobility who intermarried with Spaniards). Administrative policies of both the naval and federal governments institutionalized a degree of social distance between Chamorros and Americans. American and Chamorro children were segregated in the government-run school system; a lower wage scale was set for Chamorros employed by the federal government; Chamorros were banned from appointment to top positions in the naval government; in 1938 when Chamorros became eligible to enlist in the US Navy, they were restricted to the specific job classification of mess attendant; and Chamorros were denied US citizenship (Thompson 1969, p. 72, 147–149, 221).

Most of the structural barriers separating Chamorros from the American military community were removed by changes introduced with the Organic Act. And since the military is no longer the dominant administrative authority, it has become a part of the whole island community.

Even after a civilian government was established in Guam, the military security closure remained in force until 1963. The limiting effects of the closure on Guam's economy was realized after its termination. Within five years after its opening, Guam became a major tourist destination, especially for the Japanese. Following the tourist, Japanese business interests brought in capital that led to the beginning of a new industry based on tourism (Haverlendt 1975, p. 116). Although recognizing that the military had long restricted development of Guam's private economy, this did not outweigh the fact that the military had long been the mainstay of the island's economy, providing a livelihood for a good portion of the population.

Nonetheless, there still remains some tension between the civilian and military communities. To illustrate, I refer to an entry in my fieldnotes dated August 1980. A radio program aired to entertain listener comments on the subject of a proposal to open a road running through the Naval Air Station. The idea was favored by civilians because it would provide a much shorter travelling route to get from the western to the eastern side of the island. The program host prefaced the program saying that opening the road would not only save travel time, but would also save gasoline (indirectly referring to the high cost of gasoline in the island). During the call-in portion of the program, two callers, both civilians, praised the idea. A third caller opposed the idea; he was military and resided on base. The serviceman argued that it would be too much trouble opening the road to the public because: (1) It would potentially increase crime on the base; (2) It might endanger children due to the increased traffic; (3) Base security would have to be increased; (4) It would be bothersome for base residents. The host responded to these comments by saying that crime would not necessarily rise, and that a fence could be built along the roadsides to protect children. Regarding security, he suggested that at most four more guards would have to be added to the two already posted. Lastly, the host pointed out that even though base residents would have to accommodate increased traffic in their local driving habits, opening the road would be beneficial to more than 50,000 island motorists. It was the opinion of the host that the larger number of people who would benefit from opening the road outweighed the minimal inconvenience to the smaller number of base residents. This, however, did not change the attitude of the serviceman. To conclude the program, the host noted that when there is differing opinion on issues concerning the civilian majority and the military minority, the interests of the latter usually prevail.

If this is indeed the case, it may explain why the military actively promotes programs to foster good relations with the civilian community. For example, Andersen Air Force Base in 1976 began a 'sister village program', where squadrons at the base are paired with specific villages in Guam. Air Force personnel then provide help and participate in village activities such as painting schools, raising money for village projects, maintaining village memorials, and helping with village fiestas (Sommerville 1987, p. 31). Help provided by the Air Force is not only in personnel, but also in equipment. Military trucks are used to pull floats for village parades during celebration of territorial holidays. Service personnel also take part in the parades, marching in formation. Assistance provided by the military to the civilian community does not go unrecognized. A brief article from the *Pacific Daily News* reported how the 43rd Civil Engineering Squadron from Andersen AFB helped the residents of Yona, its sister village, by providing sixty-five 55-gallon barrels of water in response to a plea for help during a water shortage (*Pacific Daily News* 1987, p. 41).

For their part, village residents invite the personnel (and their families) of their 'sister squadron' to village fiestas, religious celebrations held to honor the village patron saint. Each one of Guam's 19 villages has a patron saint, so there is at least one fiesta every month of the year. It is on such occasions that members of the military community mingle socially with the civilian community, and experience the traditions of local culture. By way of reciprocation, Andersen AFB holds an annual open house – or perhaps more accurately, 'open base'. In addition to public demonstrations of military aircraft, these events provide an opportunity for each squadron to host visitors from their sister village.

We have already noted that individuals view the military presence in Guam favorably because it provides opportunities for employment. Another way of perceiving the impact of the military on the individual is the high enlistment rate among Chamorros. Although lacking detailed information on this subject, I learned from a University of Guam professor (a specialist on issues concerning Guam and Micronesia) that in proportion to their number in the population, more Chamorros enlist in the armed forces than any other American ethnic group. Interest in military service begins at an early age among Chamorros, based on their participation in Reserve Officers Training Corps (ROTC) programs. According to a 1986 *Pacific Daily News* report, enrollment in the ROTC program at the University of Guam is four or five times higher than in those at stateside universities. That students who qualify for the program receive free tuition, books, and the opportunity for scholarships are certainly reasons for the high level of enrollment. And there are Junior ROTC programs offered at all five of the island's high schools (O'Connor 1986, p. D–22). This writer can attest to the popularity of ROTC among the youth and in the society at large, based on their involvement in community events. I observed ROTC units at fiestas, in parades, at a rosary (a prayer meeting for the dead), and doing special performances – such as short-order drills, or martial arts demonstrations – at holiday celebrations. Occasions such as these exemplify the various ways the military presence is manifested in Guamanian society.

Conclusion

The militarization of Guamanian society does not mean that all its members have taken up arms in preparation for combat. Instead, it refers to the fact that as an institution the military has played, and continues to play, a principal part in the process of social change in Guam. In the preceding review we noted some of the ways the American military has directly changed the structure of Guamanian society, and changed the pattern of living for the Chamorros. And for the most part, the pattern of change was based on an American model.

Beginning with the basic institutional framework of the society, it was through the Naval Government of Guam that American-styled political, educational, and economic institutions were established. A theme common in the early political and economic institutions is that they followed an American model and they subordinated Chamorros to Americans, in terms of holding positions of authority and in salary. In part this inequality stemmed from the fact that political and economic administrations were handled within a military structure, which subsequently caused discrimination toward Chamorros because they were civilians. But some of the same structural inequities persisted under civilian administration. In this case discrimination toward Chamorros may have been racial, an inauspicious structural feature of the Americanization of Guamanian society.

The military has affected the social lives of Chamorros in many ways, but the greatest impact on them has been the influence on their participation in the economy. During the four decades of naval and military rule in Guam, government administration and civilian military employment became the principal means of earning a livelihood for Chamorros. This pattern persisted into the period of civilian administration, with a majority of Chamorros being employed in the public sector, as well as military enlistment. The latter development raises an interesting issue, that is, the high enlistment rate and the popularity of ROTC programs among Chamorros. What is the correlation between these and the major role the military has had in Guam? Are ROTC and military service merely perceived as good avenues for employment or a career? Or are there other reasons? This is a prime topic for future investigation.

Indeed, Guam has become integrated into the fabric of American society. All institutions are based on an American model. The population is ethnically diverse. With this ethnic diversity there is also linguistic diversity, but English is the principal language in the society. Where Guam strays from the typical American pattern is in the greater reliance on the public sector economy for generating capital. This was a consequence of the strategic, rather than economic, interests that guided administrative policy in Guam. The most significant economic changes in the island have occurred since the suspension of the military security closure, which allowed the development of tourism. Thus even a reduction of the military's role in Guam was a major impetus for change.

8

Elements of Social Change in the Contemporary Northern Mariana Islands

Samuel F. McPhetres

Introduction

There are so many factors having an impact on the structure of society in the Commonwealth of the Northern Marianas Islands (CNMI) that it would be impossible to include all of them in this chapter. Some of the subjects are not dealt with here because of their relative obviousness or their treatment by others: the American educational system, the Spanish introduction of the Roman Catholic Church, the Japanese sugar plantations, and others that the reader may find important. With history as prologue, I have attempted to establish the base for discussion of contemporary social change. The issues of social change described here were selected because they are neglected in most of the research on the CNMI. If what is provided here leaves the reader feeling that there is more to be had, then it will have succeeded. There is no question that what is happening in the Marianas is unique, if only for its scale in proportion to the population.

Pre-contact History

As with most small island cultures, the Northern Marianas (or Marianas in general) were without a written tradition. Knowledge of their early history therefore is based on various remnants and such oral tradition as has survived, and on archaeology.

Archaeologists agree that there was a culture in these islands different from the Chamorro in the early settlement period, perhaps as early as 200 BC (the earliest recorded settlement in Micronesia is 2000 BC in the Marshalls). The first Marianas culture was distinguished by its complex pottery, related to that found in other parts of the Pacific and Indian Ocean communities. The earliest settlers are called the pre-latte culture in the Marianas.

241

This group was submerged by the latte group, beginning about 700 AD. These are the ancestors of the present-day Chamorros. There are some indications that the survivors of the pre-latte group were enslaved by the later migrants and lived in the interior of the larger islands, while the dominant group occupied the more desirable coastal regions.

The Chamorro culture left some significant monumental remains in the latte stone columns. Apparently chiefly house foundations, these limestone pillars measured from 3 feet to 14 feet tall in some areas, such as Tinian. The fact of their sheer size, the technology needed to put capstones weighing several tons on the larger pillars, and the distances they have been found from quarries, give substance to the theory that the cultures were highly complex, well-organized and relatively prosperous.

Early European explorers from Magellan (1521) on attested to the complexity of the cultures, the generally healthy appearance of the people, and their ability to sail long distances in what was called a flying proa, an outrigger sailing canoe capable of high speeds.

It should also be noted that the peoples of the islands were organized largely by island group and did have a tradition of warring on each other. It was not a monolithic all-encompassing tribe, but a scattering of smaller units speaking the same language and sharing the same culture with trade, warfare, and social links provided by interisland canoe traffic.

Estimates of their population in the whole archipelago (including Guam) range from 50,000 to 100,000 at the time of European contact. Three hundred years later, the indigenous and mixed or part-Chamorro population is possibly back up to that figure.

European Contact

In 1521, Ferdinand Magellan spotted the islands, particularly Guam, although there is still some dispute about which one he actually landed on. His visit was made memorable by an incident which led the island chain to be called *Islas de los Ladronas*, Islands of Thieves.

Following the initial hospitality of the islanders to the sea-weary mariners, some of the host islanders made off with one of the ship's longboats. Magellan was angered and sent a party of sailors to shore to recover the boat and punish the wrongdoers. Several men were killed, houses burned, and the boat recovered. At that time there was no recognition by the Europeans of the local custom of property sharing. Unfortunately, the name he gave the islands stuck on the maps for many years: the islands of thieves.

Although Magellan never made it back to Spain, one ship of the original three did survive and returned, beginning the opening of the western Pacific to Europe.

No formal claim to ownership was made following discovery until the mid-1500s and that was not followed by any settlement until another hundred years had passed. In the meantime, there were various casual contacts, occasional temporary settlements, and the usual coterie of sailors and beachcombers who decided that staying among the hospitable and accommodating natives was preferable to going back to an uncertain continuation of the hazardous sea voyage that brought them. It is from written accounts of sailors and missionaries who passed by that anything is known about the early cultures.

During this period the Chamorro way of life was penetrated by the West. Iron, firearms, and a variety of other commodities were intro duced in the exchange of reprovisioning of European ships.

The key date in the metamorphoses of the Marianas is 1668. This is the year when a Jesuit priest, Fr. San Vitores, after a long campaign to reach the Marianas with a permanent missionary settlement, established himself on Guam with a small garrison of Tagalog troops from the Philippines, Spanish officers, and a few other priests. With the cooperation of a leading Chamorro chief on Guam, Fr. San Vitores constructed the first permanent settlement, including a small fort and a church at the site of the present cathedral in Agana, Guam.

A Chinese Confucianist named Chaco had preceded San Vitores by several years and was actively proselytizing and countering the Christian's efforts. Chaco convinced many of his Chamorro followers that Christian baptism was the application of poison to the children and he succeeded in fomenting rebellion. Keeping in mind that infant mortality was high at the time, there was enough correlation to lead many Chamorros to accept the theory.

The revolt that followed included the 'martyrdom' of San Vitores and most of the other priests. The revolt was most intense in the islands north of Guam, where the Spanish only visited periodically.

The Spanish saw the rebellion as a challenge to their civil and ecclesiastical authority. The Spanish troops and their Guamanian allies began a sweep to the northernmost islands of the chain and brought all of the survivors of any battles – men, women and children – down to Guam. The strategy was to resettle the inhabitants where they could be easily controlled. On Rota, however, a few of the people were able to hide out in remote caves. The Northern Marianas were in effect depopulated.

Figures are hard to come by in terms of how many people were actually relocated. However, by 1720, observers variously record 3,500 to 5,000 surviving 'natives' all living on Guam: this as a result of the constant warfare, depredations of imported diseases, and the results of loss of traditional lifestyles.

During this time, when the Marianas were being administered by the Spanish from colonial headquarters in Manila, Guam was becoming an important way station for the Manila bi-annual galleon trade with Acapulco, Mexico. Many surviving Chamorros, mainly women, inter-married with the Tagalog, Mexican, and Spanish garrison troops. It was the indigenous custom for the non-Chamorro spouse to join his wife in her village. In this way the culture had some continuity, since the mother had the responsibility of raising the children.

Interim Summary

It is important at this point to digest events up to about 1720. The policies of the Spanish, both secular and clerical, resulted in the near annihilation of the traditional culture. Traditional communal organiz-ation was completely revamped. As in the Philippines, Spanish priests replaced indigenous chiefs. Nothing could be done without their involvement. Daily life revolved around the Church and its calendar, rules, and regulations. Much of the language was changed by contact with the Spanish. Other than the Church's emphasis on the role of the father as family leader, altering forever much of the matrilineal and matriarchal nature of the society, indigenous leadership was dis-couraged. Traditions and customs were evaluated against Church stan-dards and either thrown out as tools of the devil or integrated as useful means of control of the natives. Traditional sexual mores had allowed women great freedom in their choice of spouse(s). Women had the power to independently divorce. Marriage in the Church eliminated that liberty, or at least made it go underground.

The nine-day novena at funerals, unique in its form to the Marianas, was an adaptation of the Chamorro belief that the soul of the departed roamed the countryside for that period playing tricks on the living until called to judgment. The Church was quick to integrate this belief, saying that having the novena every 4 hours for nine days kept the spirit in purgatory until judgment day.

The period from about 1720–1886 is when the Chamorro culture became an amalgam of traditional Chamorro, Spanish, Spanish-mes-tizo, and Filipino cultures. The resulting people should be called, following a term given by an indigenous rights activist on Guam a few years ago, neo-Chamorro, thus avoiding the constant question: What is a Chamorro? – A question that is frequently raised in terms of special rights of the indigenous people versus the universal appli-cation of American culture and law.

During this period also, the Spanish were building up Guam into a resupply area for the Manila Galleon trade. In 1810 a group of Carolinians fleeing a disastrous typhoon from the outer islands of present-day Yap State, Federated States of Micronesia, so pleased the Spanish governor of Guam that he gave them permission to settle on Saipan, still unpopulated, if they would manage the Spanish cattle herds. Today, the Carolinian population of Saipan is descended from that and later groups that settled following devastating typhoons to the south. The presence of Carolinians on Saipan would become a nagging issue when Chamorro resettled on the island.

Repopulation

In 1886, following a papal decree giving Spain sovereignty over all of Micronesia and Germany, commercial rights throughout the government on Guam began permitting Chamorros to leave for the islands to the north. Saipan and Tinian were the primary destinations, but Pagan and some of the Northern Islands were resettled. Some of these, however, had been settled earlier by Carolinians.

With resettlement came Spanish clergy and administration. Coastal lands had been taken by the Carolinians and the Chamorros began settling inland and taking up farming and other trades taught to them by the Spanish. By this time Chamorros were practically indistinguishable from Filipino peasants in their dress, customs, and lifestyles. Long contact with the Tagalog soldiers garrisoned on Guam had introduced them to Filipino foods, some of the language, and dress styles common in the larger Spanish colonies. The Carolinians, in contrast, had maintained their tradition and custom almost undiluted except for nominal concession to the Church by being baptized and observing the rituals.

In the context of cultural change, Rota, where some Chamorros eluded the Spanish resettlement on Guam, retained the most of the original Chamorro customs and traditions. Rotanese, the people of Rota, pride themselves as being the closest to the original Chamorro culture.

The German Period: 1899–1914

Since the early 1850s the Germans and Spanish were competing for copra and the trading routes through Micronesia. The papal bull of 1886 resolved the differences to the satisfaction of all concerned at the time. The Germans never really reached the Marianas during this period, but they did have a significant colonial influence in the Marshalls and Pohnpei.

In 1898 the United States and Spain went to war. After Dewey's victory in Manila Bay, the Americans claimed Puerto Rico, Guam, and

245

the Philippines. The United States determined that the rest of the Spanish colonial empire in this part of the Pacific was to be left to the former enemy.

Because of the costs of the war, Spain had to divest itself of its Pacific colonial empire. Germany bought the balance of the islands for what would now be about $4,000,000. The basic effect of this purchase was effectively to isolate Guam from the rest of the Marianas and the rest of Micronesia, leading to the intensive Americanization of the island. This isolation would have a profound effect on the rest of the Marianas in many ways.

One of the first German administrators of the Northern Marianas, George Fritz, was one of the period's real renaissance men. He was a scholar, scientific observer, and casual anthropologist. His description of the culture of the Marianas is one of the authoritative sources even today.

The German administration was one of benign commercialism. There were never many German citizens in the islands, but they did develop policies that would forever mark the development of the culture. These policies included private ownership of land (but never more than could be cultivated by the owners), surveying, and quota production.

The Germans, who controlled the rest of Micronesia, engaged in significant movements of populations. After a revolt in what was then German Samoa (now Western Samoa), the rebels were exiled to Saipan where they were settled in the village of Tanapag. (In line with this policy, many Pohnpeians who had revolted against the Germans in the Sohkaz Rebellion were exiled to Palau, where some descendants still live.) Many Chamorros were sent to work in the Angaur (Palau) phosphate mines and to Pohnpei where they worked on the construction of a Catholic Church. The Germans also took significant numbers of Chamorros to Yap to aid in the business enterprises. Since by Micronesian standards at the time, the Chamorros were better educated and more used to Europeans than the rest of the island peoples, as a result of the Spanish colonization, they were deployed as middle-level management.

At the same time, the Germans encouraged the migration of Carolinians from Guam to the Northern Marianas, a policy that was supported by the American administration of Guam, where the 'primitive' lifestyle of this group was an embarrassment. With their broader experience in island lifestyles, the Germans had no objection to the Carolinian semi-nude mode of dress and their customs. The more puritanical Americans encouraged the departure of this group to Saipan.

Taking into account this movement of peoples, the German administration was not otherwise notable for any significant impact, except for a contribution to the gene pool and some German names still prominent among some of the families.

With the onset of World War I, everything changed. Using the authority of a secret treaty with the British, the Japanese navy sailed throughout Micronesia (except for Guam) and captured the islands from the Germans. The British took all of the German colonies south of the equator.

The Japanese took all of the German nationals off the islands, repatriated the Samoans, and kept the Chamorros in Yap and Palau.

The Japanese Civilian Period: 1914–1939

Initially under Japanese naval government, the islands of Micronesia became a class C mandate of the League of Nations in 1920 (recognized by the US in 1922). At first, the principles of the League were observed with little Japanese activity in the islands other than small-scale trading. However, as time progressed, the Japanese began intensive colonization with their own people, Koreans and Okinawans. Throughout Micronesia the peak Japanese population reached about 100,000 civilians to about 40,000 Micronesians. In the Northern Marianas the ratio was about 20,000 to 4,000.

During this period the Japanese in the Northern Marianas engaged in intensive sugarcane production, fisheries, phosphate mining, and small industries. These activities required large amounts of land, and they managed to take it from the local inhabitants in a variety of ways. The Chamorros and Carolinians never were granted full Japanese citizenship. The locals were restricted to three years of obligatory elementary school and an additional two years of advanced vocational school, if they were particularly gifted. Certain occupations were restricted and intermarriage was discouraged (quite unsuccessfully it turned out). Many of the older Chamorros remember this era with nostalgia, since there was no unemployment, no hunger, the children respected their parents (or faced the Japanese police), and they had a modicum of self-government with Japanese appointed magistrates and policemen of their own race. The Spanish clergy was allowed to remain, although subject to Japanese Catholic authority. As long as they respected authority, they were left to their own customs and generally ignored as an underclass of Japanese nationals, not citizens.

Most of the land was under sugarcane cultivation in Saipan, Tinian, and Rota. Some of the Northern Islands were also under cultivation. What was not under cultivation was urbanized to the maximum degree. There was electrical power, paved roads, railroads around the islands, a water supply, hospitals, and schools. Japanese was widely spoken by the natives.

The War Years: 1939–1944

The war in the Pacific region had begun early in the 1930s with the Japanese attacks on Korea, Manchuria, China, and other areas. The Marianas, because of their location, were considered a strategic area and fortification of the islands by the Japanese military began in the late 1930s. It was during this period that the tides of fortune began to change for the Chamorros (and Carolinians). Keeping in mind that the Chamorros of the Northern Marianas and those who remained on Guam were often related, there was frequent communication between the groups. As the war effort grew, the Japanese sealed off communications and began treating the Chamorros as potential threats to their security. The Japanese introduced forced labor, land confiscation, and a variety of repressive measures, increasing in severity as the American military moved West. The Church began to be treated as an enemy and everyone was suspect.

Loyal Chamorros from the Northern Marianas were conscripted or volunteered for the Japanese armed forces to serve in various capacities, including Japanese security forces in the occupied US territory of Guam, captured the day after Pearl Harbor was attacked. But for the most part, the Chamorros and Carolinians remained relatively neutral.

As the US forces advanced in 1944, there was increasing tension between the islanders and the Japanese. Forced labor for the construction of defensive positions increased. Summary executions became common. Many of the local people began thinking they would be better off under the Americans.

In an interesting sideline, there is the apocryphal story of a Japanese radioman, of French and Japanese blood and US citizenship caught up in the war while vacationing with his mother in Japan. After being conscripted into the Japanese army, he was assigned to Saipan. He apparently fell in love with a local girl. As a radioman, he could monitor the advancing American fleet. Just before the invasion began, he is said to have gone out to his local friends and warned them to take to the hills and caves with food and supplies. He also advised them to carry religious pictures from the Catholic Church. Since the Japanese were not Christian, the Americans would be able to distinguish them from the real enemy and would spare them as non-combatants. This is, in fact, what happened and out of the approximately 4,000 Chamorros and Carolinians who were present in the Marianas during the actual fighting, only a few hundred were casualties. Out of the 20,000 Japanese troops, only a few hundred were captured alive and many Japanese civilians jumped off Suicide and Banzai Cliffs rather than risk capture and presumed torture and humiliation at the hands of the Americans.

At the end, in 1944, the Americans fought one of the bloodiest battles in the Pacific campaign for the island of Saipan, losing about 4,000 (killed and wounded) troops. But the vast majority of 'natives' survived.

The several thousand captured Japanese civilians were interned in one prison camp and the Chamorros and Carolinians in another. Liberation Day, the 4th of July, is the day the Chamorros and Carolinians were released from the detention camps. The holiday is not related to the US Independence Day except by interesting coincidence.

In a decision that was to influence the Marianas to this day, the American government determined that the native inhabitants of the islands were to be considered non-belligerents and spared, this notwithstanding the fact that they had been under Japanese government for thirty years and were considered by the Americans as Japanese Nationals. As a corollary, the US Naval Administration determined that the civilian Japanese inhabitants were to be repatriated to Japan, even though many of them had been born in the islands, many had married locals, and had families as well as businesses. Regardless of family situation, Japanese nationals were sent to Japan, sometimes never to be heard from again. In other cases, children separated from their parents spent years trying to relocate relatives in Japan.

The Naval Government: 1944–1951

During this period several relevant events took place. For one, the Chamorro inhabitants of Palau and Yap were brought back to the Marianas. Since the island of Tinian was uninhabited (the Japanese had taken all of the local people off and moved in Koreans to work the fields), they were settled on what was by now considered to be government land. In the Marianas, government land was that which the Americans had taken from the Japanese administration, and from local citizens to build bases from which to attack Japan.

During this period, the US government had to decide what to do with their newly-conquered territory. Fearful that they might be used again to launch attacks against US interests, there was great reluctance to look to releasing them to any other power. But at the same time there was an agreement among the victorious allies that enemy colonies would not be absorbed by the victors. As a compromise measure, the Japanese Mandate under the League of Nations was declared a United Nations Strategic Trusteeship, the only one of its kind in the UN system, to be altered only by action of the Security Council where the US had a permanent veto power.

From the beginning of the Trusteeship, the Marianas held a special place. The naval administration, charged with developing self government, began forming municipal governments and district legislatures with limited law-making authority. They also began looking forward to the future. In 1951 the administration of the Territory was turned over to the Department of the Interior. In 1952, the administration of the Marianas was returned to the Navy with the exception of the island of Rota, another gesture to the individuality of the Rotanese. This began a completely new era for the developmental experience of the people of the Northern Marianas.

Naval Government: 1952–1962

It is no secret that during this ten-year period the US Central Intelligence Agency (CIA) ran a guerrilla training camp on Saipan, particularly for anti-Mao nationalist Chinese and later, perhaps, for Indo-Chinese anti-communists in Vietnam, Cambodia, and Laos.

This operation entailed sealing off the Marianas from public access, something provided for in the Trusteeship Agreement. Along with Guam, all of the Marianas except Rota were governed directly by the Navy Technical Training Unit (NTTU), the administrative arm managing the civilian population while the CIA operated its training base in the Marpi area of Saipan.

Rota was separated and continued to be governed by a presidentially appointed civilian High Commissioner along with the rest of the Trust Territory. It was completely cut off from Guam in the south and Saipan and Tinian to the north. It was known as Rota District.

For the northern islands, this period was marked by a heavy military influence. With the presence of hundreds of naval personnel, trainers, and support personnel, it was necessary to provide an acceptable infrastructure. The result was that Saipan had the best schools in the Territory (taught by Navy wives), modern jobs for the local people in public works, services, and other fields. The roads were paved and improved and water systems were installed. Housing was built on Army Hill, now Capital Hill, and Navy Hill. The American influence was pervasive. Islanders were encouraged to re-open contacts with their relatives on Guam, and were even allowed to go to the better schools on Guam if they could get security clearances. Many did.

Most of the local entrepreneurs and skilled workers received their training during this period. Many of them are now in leadership positions in the Commonwealth.

Then in 1962 another event of earth-shaking importance to the development of the Marianas took place. Under the Trusteeship system, the UN Trusteeship Council made periodic visiting missions to the Trust Territories, usually every three years. This year, the Mission toured the rest of the Micronesian islands in the Territory and the

Marianas. The UN Council issued a devastating report, criticizing the US, the Administering Authority, for the lack of economic and social development. This report made its way to the desk of President Kennedy. Kennedy determined that a new direction was necessary. After years of basing the administration of the Trust Territory in American territories outside of the Trusteeship, (Hawaii and Guam), it was decided to move the High Commissioner and all of his staff into the Trust Territory. While the original site was to have been Truk, the geographic center, the capital infrastructure on Saipan was too good to pass up. Closing down the NTTU, the Trust Territory administration was moved to Saipan from Guam, to take over the facilities left by the CIA and NTTU. Saipan became the communications hub of the Territory and its political center.

The US was attempting to carry out its responsibilities for political development of the people. It had encouraged municipal councils, elected mayors, district legislatures, and participation in local government. It was in this milieu that the Saipan Municipal Council first passed a resolution for reunion with Guam in the late 1950s.

Trust Territory Headquarters: 1962–1976

When the High Commissioner and his staff moved to Saipan to direct the affairs of the whole Trust Territory, he also, as a secondary effect, realigned the directions of the Northern Marianas. Rota was reunited with the rest of the Marianas District and brought back into the mainstream of Marianas politics, through participation in the Marianas District Legislature. The Rota delegation has, since that time, been a vocal and somewhat dissonant voice in the overall direction taken by the legislature.

In 1965, by order of the US Secretary of the Interior, the Congress of Micronesia was created. This replaced the purely consultative Council of Micronesia and other bodies of appointed leaders called to confer with the High Commissioner at his pleasure. The new Congress could actually make laws for the Territory (albeit only with the approval of the High Commissioner). The Marianas delegation was an active participant and soon took the position that the Marianas were different from the rest of Micronesia. The implied purpose of the Congress was to prepare the islands of the Trust Territory for the termination of the Trusteeship and anticipated self government as a single political entity. The Congress was offered by a status of a US commonwealth for the whole Territory. This offer was refused by all delegations to the Congress, except for the Marianas. The rest of the Congress wanted something closer to independence and had settled on a status vaguely defined as free association. In the Northern Marianas, however, union with Guam and the US was mandated by early non-binding referenda (in which Guam's voters rejected union with their poorer cousins in the North). The Northern Marianas delegation began the long and arduous process of separation. While Free Association was being negotiated by the Congress and the US, the delegation pressed for separate negotiations on commonwealth. The persistence paid off; in 1972 the US agreed and separate negotiations began.

In the meantime, the Congress, representing all of Micronesia and its ethnic and linguistic multiplicity, had its residential headquarters on Saipan. Many bonds were forged, through marriage, work, and social interaction with people from other districts. Today, Saipan's population reflects the results in the Trukese, Ponapeans, Yapese, Kosraeans, and Marshallese now resident on Saipan and who have become CNMI citizens.

In addition, the presence of large numbers of American staffers from the Trust Territory headquarters, whose work brought them into frequent and close contact with the local population, led to more social interaction than in other parts of Micronesia. There were no language barriers on Saipan. English had been adopted quickly and was the language of instruction in all of the schools.

And because the schools in the Marianas, particularly in Saipan, were considered among the best, there was a feeling of superiority over the rest of the islanders in the Trust Territory. Many of the leaders of the other island groups enrolled their children in Saipan's schools.

With the Headquarters presence on Saipan, the local people had first chance at developing a job in government. They rose quickly to positions of authority over other Micronesians.

When the Commonwealth covenant was presented to the voters in 1975, it was approved by a majority of 78.8 percent, with a 92 percent voter turnout. American citizenship and social welfare programs were large inducements in attracting voter turnout. In 1976 the administration of the Northern Marianas was removed from the jurisdiction of the High Commissioner and placed under a presidentially appointed Resident Commissioner. This followed approval of the covenant by the US Congress. Between 1976 and 9 January 1978 the Marianas were administered as a separate entity, but still under the Trusteeship. On 9 January 1978 the first constitutional government of the CNMI was sworn in and real self government began as the citizens became interim US citizens (still carrying TT Passports, however).

On 3 November 1986, President Reagan proclaimed that the Trusteeship no longer applied to the CNMI (or the Republic of the Marshalls or the Federated States of Micronesia). With this, the covenant came into full force. The qualified members of the population became full US Citizens.

Social Change in the Marianas: From 1978 to the Present

Obviously, social change is going on all of the time in every society. However, it would be interesting to find any place in the world undergoing such a thorough, intensive, and rapid change as that which is going on in the Commonwealth of the Northern Marianas.

Citizenship

After hundreds of years of somebody else's forced dominance, the Marianas voluntarily and freely voted to join the American Family. This union, defined in the covenant, established the relationship and set certain limits on both sides. But it did grant US citizenship to a majority of its inhabitants. It also erased, with the stroke of a pen, the legal distinction between people from the United States and people from the Mariana Islands. Suddenly, with a few important exceptions, equality in rights became the rule rather than the exception. Previously, as Trust Territory citizens, the people were protected (rightly or wrongly) from the encroachment of outside forces. US civilians could not come to work in the Territory unless they were Trust Territory government employees. Private sector outside employees had to have entry and work permits issued by the Trust Territory government. Land was restricted to Trust Territory citizen ownership. Micronization, a term used to refer to the policies to replace US citizen employees in the government with Trust Territory citizens, meant preferred treatment. With US citizenship for the inhabitants of the Marianas, there was no longer a basis for preferential treatment. Equal rights were assured by both the covenant and the Constitution (with the exception of land ownership).

The US citizen population in the islands is slowly increasing. As more US mainlanders try to participate in local affairs, more resentment is being expressed by the locals. Where the initial euphoria at becoming American with the approval of the covenant was extreme, now that reality is shaping a different picture with former outsiders taking active roles in the administration of policies, particularly in the private sector. An increasing 'us vs. them' climate is beginning to take form. The American spirit of competititon is foreign to the culture. American aggressiveness and a perceived lack of sensitivity to Chamorro culture are seen as threats.

Application of Federal Law

The covenant describes those US laws which apply to the CNMI as being those in force on Guam on the effective date of the covenant. Legislation after that date would have to be directed to the CNMI or to all of the States and US Territories. Federal authorities have recently been applying several criminal statutes concerning white-collar crime in government. Some convictions have been attained. Most recently, however, many corruption charges have been dropped pending a judicial review of the application of certain US laws to the CNMI. It is the contention of some local authorities that applying some laws constitutes unreasonable interference with the internal affairs of the CNMI as defined in the covenant.

For those who believe that the Commonwealth is fully a part of the US the review process has been a set-back. There is, in fact, a movement to further separate the US federal and the local government through re-interpretation of the covenant. The movement for exemption from some US laws is motivated by some of the local political leaders who are having their authority challenged by the application of federal status. In the context of challenges to US law, a confusion over what US citizenship means has developed. The publicity and discussion revolving around these questions have some people saying: 'Yes, I am a US citizen, but different from the others.' Others are wondering what they are if not full US citizens.

This issue was aggravated by the refusal of federal passport authorities to grant passports to citizens of the Northern Marianas who had parents of another nationality (usually Japanese) or who were born outside of the Northern Marianas during Japanese times and repatriated by the Navy in the 1940s. These people were stateless until a court decision in August of 1988 which resolved this issue. However, a negative impression of the federal government had been created by the initial denial of passports.

Federal Programs

The CNMI receives millions of dollars in US federal grants. These outlays cover everything from food stamps to education and construction projects. These are moneys over and above the local tax revenues. US monies are subject to federal rules and regulations. The Nutritional Assistance Program, food stamps, has a widespread social impact. Approximately 4,000 people are dependent on them. Many of them receive more value in stamps then they could earn with the local minimum wage ($2.35/hr). Some reason that the food stamp program relieves pressure to seek employment, unintentionally supporting unemployment in the face of a labor shortage (See section on alien labor).

All of these federal programs have requirements that constrain or change local government. Some require reports, audits, site visits, modifications of certain behaviors, matching funds, or other efforts on the parts of the recipient agency. There is a feeling in certain parts of the Commonwealth that federal agencies are obligated to provide money and adjust requirements to local needs. Matching fund requirements for some US programs have been waived for the Marianas.

Economic Development and Alien Labor

Under the Trusteeship, foreign investment was forbidden to non-US firms until 1974. The Marianas District government budget during the Trusteeship ranged around $10 million per year for about 15,000 inhabitants.

Today, the Marianas Commonwealth budget is $170 million. The indigenous population is estimated to be about 19,000 (Chamorros and Carolinians) and recent government statements reveal a contract labor force from outside at approximately 16,000 or nearly half of the total population of the commonwealth. There are about 2,000 hotel rooms available now with another 2,000 under construction and more on the drawing boards. The visitor industry requires an expanding labor force.

In addition, because of tariff advantages in the US market, many Asian garment factories have relocated to Saipan and the Marianas. The work force is predominantly from Asia, with mainland Chinese making up a majority. At last count, there were 26 legal factories with a total workforce of about 6,000 alien employees.

The issue of alien labor is the most significant social problem in the contemporary Marianas. In just the past five years, the population of the islands has doubled. The doubling is due to the importation of labor to fill jobs for which there are no local people available, regardless of the wages being offered. Most of this workforce is housed in barracks and dormitories.

Alien workers encounter resentment and hostility from local residents because of competititon for resources. The shortage of power, water, and other amenities, including hospital services, is often blamed on the foreign workers and the factories where they work. Efforts at controlling the number of aliens have included: (1) Introducing legislation to reserve certain occupations for local citizens; (2) Deportation of pregnant alien women; (3) Deportation of alien dependents earning below $20,000/year. These efforts are mostly symbolic. The fact that the government issues licenses for construction of new hotels guarantees that the alien workforce will grow.

A clash of cultures has become apparent between locals and the imported proletariat. Complaints from alien labor are met with local declarations that Filipino and Chinese labor must abide by the conditions they find in the Commonwealth, no matter how difficult. After all, they accepted the contract. The reluctance of Chamorro dominated local government and employers to meet the complaints of alien workers may be leading to the union organization of this workforce.

The fact that the bulk of the alien population is Filipino is historically ironic. The Chamorros and Filipinos share much of the same colonial experience under the Spanish. However, the Marianas are an insular part of the US and Japan political economies. Filipinos are willing to come to Saipan to work at menial jobs for minimum wages, and this has contributed to a feeling of superiority in the Saipanese. Local people can hire live-in Filipino maids at $150.00/month, carpenters, accountants, and drivers at CNMI minimum wages. Night clubs are staffed by Filipino girls. Filipino farmers are being brought in to work the fields. New attitudes are forming about what are respectable professions for local people.

The presence of Filipino and Chinese labor, an influx of American, Japanese tourists, and hotel management, and the increasing travels of locals, problematize the Chamorro identity. Looking ahead, however, with increasing mobility, greater education abroad, the presence of large numbers of aliens in the islands and related factors, there is a serious question about the future of the ethnic group. The islands are small, area is limited, social contacts unavoidable. Inter-ethnic marriage has become common.

Women as Factors of Social Change in the CNMI

Traditionally, women in the Hispanic culture of the neo-Chamorro occupied the role of homemaker. Women were expected to be obedient, hard working, and religious. As noted in the introduction, it was the Chamorro women who preserved the culture from near annihilation in the early 1700s. This role generally prevailed up to American occupation in the mid-1940s.

However, with greater education and mobility for women, their role is changing. These changes are having a profound effect on the society as a whole.

The 1970s was a period when many of the women in the Marianas attained higher education. Women entered professions previously held only by men. Women began to take an active role in politics. From 1974 to 1975, professional women were active in the political education program related to the plebiscite on the Commonwealth covenant. Again, from 1978 to 1979 women successfully mounted a campaign and a referendum to abolish casino gambling.

More recently, the women of the CNMI have depended less on organization and more on individual achievement. Many women are now in positions of responsibility and authority in business, government, and education. There is a Bureau of Women's Affairs in the governor's office.

The emancipation of women has been made possible by the growth of wage labor in government and commerce and by education. Participation in the labor force has changed some of the role expectations of women. Women can now be single parents and live independently of their families.

Because so many Filipinos are anxious to leave their country at almost any cost and accept almost any living conditions, it is no problem to recruit maids for $150.00 per month plus room and board. Without the availability of inexpensive domestic help, the lifestyle enjoyed on Saipan and Rota would not be possible. On the other hand, large numbers of children are being raised essentially by third country nationals with no knowledge of the indigenous language and little or no interest in Chamorro customs and culture.

Alien Females

Foreign national females are working in the garment industry and in entertainment. Young Chinese and Thai women are recruited to assemble garments on Saipan to take advantage of the Commonwealth tariff-free access to US markets. These women live in dormitories supplied by the sewing factories. Although closely supervised, they are exposed to labor organizing efforts, civil rights investigations, and the US consumer culture.

Filipino women are brought to work in night clubs and bars. They are employed as waitresses, bar girl hostesses, and entertainers. Supported largely by the tourism industry, but heavily utilized by the local male population, this sector also represents an element of social change. Local girls are kept from even the most innocuous employment in these establishments by their families. The Filipino workers are provided with living quarters, often on the grounds of the place of employment. The employers exercise rigid control of the employees' working and non-working hours.

Land Ownership

Nothing could be more indicative of social change than land ownership in the Commonwealth. Initially, with a view of preserving the heritage of the indigenous cultures for future generations, a provision was written into the covenant (Section 805) which restricted land ownership to persons 'of Northern Marianas Descent' (later defined in the Constitution as a Trust Territory Citizen with family connections in the Marianas dating back to the early 1950s). This effectively eliminated non-Chamorros and non-Carolinians from land ownership. The provision goes on to state that persons of less than 25 percent Northern Marianas blood – using 1950s citizenship as the definition of 100 percent blood quantum – would no longer qualify for land ownership, thus complicating the lives of mixed families formed after 1960. Other Micronesians who acquired land under the Trust Territory government were allowed to keep it, but they could only dispose of it by title to persons of defined Northern Marianas descent. Only people of Northern Marianas descent have the right to hold title to real property in the Commonwealth.

It would appear that development would be restricted to Marianas citizens. Foreign investors might be reluctant to put large sums of money into projects where land rights were insecure. However, soon after the Commonwealth government was installed, the long-term lease was used as the vehicle of land conveyance. Foreign investors, in particular Japanese hotel interests, began negotiating for prime beachfront land on which to construct large world-class hotels and other resort facilities. Land values escalated to the point where a small beach front property in what is now called Hotel Row might go for a million dollars. Per square meter prices are ranging up to $1,000 or more.

Cultural preservation – as it is related to the retention of land – is being subordinated to the instant wealth of the long-term land lease. The leases, for up to 55 years, have an option to buy clause in the lease, the lease fees becoming the purchase price of the land if the constitution restriction on land ownership is changed. Often the full 55-year lease price is paid up front, making instant millionaires of previously relatively improverished families.

Leases to foreigners are causing difficulties in families holding joint custody of property. Some members want to sell now, some want to wait for a better price, and some want to reject any offer. Title searches are becoming big business as the investors want to be sure that the people who are leasing the land are, in fact, the owners. Often, remote family members show up later and challenge the lease or demand a share of the lease fees. Others, once they see the wealth to be made from land they already leased to someone else, try to renege on their actions, stating that they really did not understand what they were signing or that the transaction was itself illegal because of the participation of persons of non-Northern Marianas descent. Since titles have traditionally been held by whole families, it is not a simple task to acquire a piece or pieces of land large enough to support a large hotel and resort facilities without encountering multiple party interests.

The best coastal land, on the East side of Saipan, is being taken up by luxury hotels. Beach front owners are being offered exchange homes in the interior of the islands, in addition to the lease price. Others are taking their money and running to Marianas citizen communities in California, Oregon, and Washington State.

Class distinctions are beginning to form, based not on traditional family stratification or on caste, but on whether or not the family has land for lease. The haves and have-nots are taking on new character. Classes in the capitalist sense are forming. Traditional myths of family origins and rank are being effaced by relative purchasing power.

Involved in this land exchange process are several American lawyers and businessmen working as brokers between Japanese investors and local landowners. They are making big commissions and constitute a separate caste within the resident American mainlander community. However, the acquisitive activities of these middle men is not without controversy.[1]

The Cargo Cult Liveth

Another factor influencing social relations and standing involves anticipated windfall money from several sources. Most commonly cited in this category is the War Claims awards. This is where the US has agreed to compensate Micronesians for damages and losses resulting from the invasion and occupation of the islands. Title I war claims (damage during invasion combat) and Title II post-secure claims (due to American action when advancing on the Japanese), gave many local citizens windfall awards, amounting in some cases to hundreds of thousands of dollars. Title II, paid in 1977, contributed $16 million to CNMI economy. Not only were all of the cars and boats snapped up within a few hours, but family members were falling on each other in efforts to get larger shares of the booty. In 1988, another $12.3 million was paid in Title I claims, with another $10 million due in 1989. Already the heirs of the original claimants are squaring off to see who will be the prime beneficiaries. Not all of the money in 1988 and 1989 is due to the CNMI, only about one third. The remainder is divided among the rest of Micronesia.

Unrelated to war claims, but still in the windfall income category, is the so-called Temingiil suit against the Trust Territory Administration. It promises another $25 million to about 800 former employees of the Trust Territory headquarters. The suit claimed, and the trial court agreed, that from 9 January 1978, when the CNMI Constitution took force, there should have been no wage discrimination between US and Micronesian employees of Trust Territory headquarters on Saipan. A dual, even triple, wage scale had been the official policy of the Trust Territory administration. The rationale of the unequal wages was that if Micronesians were paid the same wage scale as Americans, they could not support themselves with the resources available after the Trusteeship. No American, however, would come out to work in the islands at prevailing Micronesian wages. The $25 million represents the difference between US and Micronesian salaries from the date the Northern Marianas became a US commonwealth to the date employees terminated employment with the Trust Territory. For obvious reasons, with the Trust Territory headquarters in Saipan, the bulk of wage discrimination award recipients would be Saipanese. Pending the verdict of the Appeals Court, individual claimants will get as much as $100,000. And the Trial Court ordered that as long as the money is not paid, it must earn 9 percent interest. By the time it is paid, it could double with interest.

So, along with war claims, Temingiil may create another moneyed class. Much of that money is already obligated by the putative recipients. Other members of the community are asking what they can do to get a part of the action.

The Cargo Cult of Melanesia is predicated on getting the goods of Western society without the Westerners and without having to work for it. As we can see, it can take other forms as well.

Millionaires and the Marianas

An unusual and unique tax situation in the Commonwealth has resulted in several extremely wealthy Americans setting up homes and business operations in Saipan. Under the Commonwealth government, all federal taxes collected in the islands are returned to the Commonwealth government (known as the mirror tax system). The Commonwealth in turn has its own income tax which is much lower than the IRS. When tax time arrives, the Commonwealth refunds to the taxpayer a substantial percentage of the difference between the Commonwealth tax and the US tax. For the people in question, the savings are measured in the millions of dollars. But it takes a certain type of person to relocate to the Western Pacific from the hyper world of high finance. Those who have settled on Saipan are having a pervasive influence on the society.

One such person wanted to bring his pet dog to Saipan. He didn't like the quarantine facilities available so he bought a new facility for the government before his dog arrived. He is also building a walled castle-type home for himself and his family – a structure which will set a new unachievable model for the community.

Several other wealthy individuals are more quietly taking up residence. They are not subject to the usual machinations of local politicos and power brokers. They operate on a level of international capital and have access to people and institutions which neutralize local power structures. Their immunity to local pressures is affecting the scope of native leadership.

Future Elements of Social Change

Military Elements

One of the most obvious potential elements of social change in the Marianas is the possible military base on the island of Tinian. Under the covenant, the US has leased two-thirds of the island for ten years. So far, there has been no effort to develop it militarily, other than for the occasional training exercise. However, looking at a possible loss or reduction of forces in the Philippines, growing opposition to US forces in Korea and other components to Pacific military dynamics, it is not inconceivable that in the short-term future, Tinian might become a major multi-service base. This would bring in not only US servicemen in quantity but would also support employees. It would possibly make Saipan a location for rest and recreation. The impact of the military presence in the Commonwealth would be substantial. It would also have to be measured against the desire for Japanese tourism. There are many possible points of conflict.

Alien Labor

Given the fact that foreign national labor, made up of Filipinos, Chinese, Korean, Japanese, Taiwanese, and others, is indispensable to the operation of the Commonwealth, their political and social power will only increase as capital development progresses. Their ability to support local 'friendly' politicians, provide favors, or withhold them, cannot help but become a greater threat to the leadership status of local citizens. Since, to a large degree, the availability of skilled construction and operational labor controls the flow of investment money, their power could be substantial. This power can be enhanced by union organization. Unorganized, they can be controlled with the threat of deportation.

261

Application of Federal Law

If, in the future, the federal government determines that its laws do apply and are enforceable in the CNMI, it can be assured that corruption trials will be resumed. While on the face of it this may be positive, it pits perceived local custom against impersonal American law. The Bordallo case on Guam is a classic in the genre. The late ex-governor was convicted under federal corruption statutes relating to bribery. His defense was never that he did not receive money, but that it was Chamorro custom (*Chinchule*) and was not illegal. It was, in fact, well within tradition. His conviction on most of the charges was overturned when the US 9th Circuit Appeals Court determined that if the recipient did not solicit the money or suggest in any way that he would use the powers of his office favorably for the contributor, it could not be called bribery. Also the court stated that since the law being enforced applied to state and local officials, it did not apply to territorial officials. However, the defendant in this case, Bordallo, committed suicide while chained to a statue of a Chamorro historical figure, rather than serve time on the remaining convictions. There is a grey area in the application of US law to territories.

In the Marianas, there could be social and political backlash if popular officials were actually convicted of crimes that could be termed 'local custom,' no matter how liberally this might be interpreted. This may accelerate the review of the entire US commonwealth status by the Northern Marianas.

Conclusion

Over the past 400 years, the Chamorro race, if we can use that term, has undergone severe and manifold challenges. It was almost decimated in the early 1700s. It survived the Spanish, albeit forever changed. The Germans did not make much of a dent and the Japanese brought new and immense resources in population and economic organization. Of course, there was the impact of the Pacific War. Under the Americans, the people were protected, fostered, some would say spoiled excessively, or, to the anthropological zoo theorists, underprotected. At any rate, up until 1978, someone else was always responsible for their welfare. There was always an external force that could be called upon to solve problems, correct wrongs, ostensibly protect the people, even from themselves. At the same time, under the American administration, the people were taught liberal democracy. This form of government would be considered as natural, as emanating from a natural order. From American television and movies as well as from local resident role models, they learned to treasure consumer lifestyles. They have been given the goals and aspirations of Americans, but the means of attainment are far more difficult to work with, cost far more individual effort, and erode what has been accepted as the family and clan foundation of Chamorro culture.

Today's people of the Northern Marianas are faced with challenges that are far more important than any they have faced since the Spanish first arrived. They have, up to now, survived as an identifiable group. Enough of the language has remained in use to hold most of the people together. Chamorros hold effective political power but are losing economic and social power rapidly to overwhelming forces of international capital and US law.

Perhaps as many Chamorros are living in the US mainland as in the islands. This out migration, particularly to Southern California, is fast assimilating Chamorros into American culture. In fact, most educated (high school and college) Chamorros speak English to their children at home because this will help them to get ahead. In many cases, Chamorro is spoken only in anger or chastisement of the children. This is happening in the Marianas as well. A growing number of local children are growing up with the idea that English is their first language and Chamorro is something to swear in.

The land is rapidly being alienated to outsiders through the long-term lease. Outside of a handful of local businesses, land is the only indigenous commodifiable resource. Its loss will marginalize the Chamorro and Carolinian residents.

The situation is so unstable that any definitive conclusion is presumptuous. For example, in the covenant establishing the commonwealth, the US federal government has the authority to take over immigration in the CNMI. This could effectively reduce the threat of alien labor. It may put a halt to hotel and other development as well.

It is possible that some revision in the land law could become effective, reducing the alienation of valuable land to outsiders. This would, also, as with immigration, restrict further capital intensive development.

At the present time (early 1990) there is a growing awareness of the threat to the identity of the Chamorro people. A reaction of resentment of foreigners, as described earlier, has set in. At the same time, land continues to be converted to hotel facilities and luxury homes, bringing in more foreign workers.

It is safe to say that this is a period of confusion in the community. The opportunity to strike it rich pervades the whole society, from the poker machine addicts to the land speculators. Preservation of the culture is not a very high priority in capitalist culture. Perhaps preservation is not even possible among the vicissitudes of changing fashions in commodity production in a monetized economy.

9

Kiribati: Change and Context in an Atoll World

Roger Lawrence

The aim of this chapter is not to attempt an interpretation of Kiribati society within the framework of a particular theory or set of theories of social change. Rather, the intention is to give more consideration to the details of the space and time context in which the penetration of capitalism into the island world of Kiribati took place. In this way it is hoped to develop an understanding of why particular changes occurred in particular places and why processes observed elsewhere in the underdeveloped world did not occur, or took a different form in Kiribati. The process of incorporation will be described in the context of the traditional society, the physical environment, the ecology of the principal agricultural system, and the evolution of colonial and independent government policy in an attempt to understand how Kiribati society today attained its essential character and particular place on the periphery of the modern capitalist world.

THE ISLAND CONTEXT

The Land and the Sea

The 33 atolls and reef islands that make up Kiribati today have a total land area of only 811 km² but are spread over some five million km² of ocean. They are exceedingly recent land-forms. The accumulations of coral debris which make up the islands developed only after sea level reached its post-glacial maximum around 6,000 years ago. The only plants, animals, reptiles, and insects present on the island prior to human settlement were those capable of dispersal on the currents of sea and air. The atoll environment presents one of the most limited and difficult environments available for human settlement. Land areas are small and fragmented. Elevations rarely rise more than 5 meters above sea level. Soils are non-existent in the strict sense of the word and are composed entirely of coral debris. Surface fresh water is absent. The flora endemic to the area is restricted to a very small range

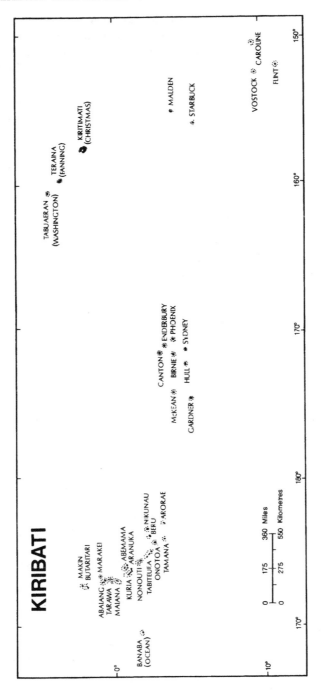

Published with permission of Roger Lawrence, Department of Geography, Victoria University of Wellington, New Zealand.

of ubiquitous strand plants. Even the lagoons are sea surrounding the islands support relatively low levels of total marine productivity compared with larger volcanic islands or continental coastal areas, because of the lack of significant nutrient run off from the hinterland. In the Southern Gilbert Island, Phoenix Islands, and Northern Line Islands, these limitations are compounded by low and unreliable rainfalls. Recurrent drought characterizes these islands. Kiribati lies outside the recognized hurricane belts, although Butaritari in the north has experienced a cyclone in historical times and other islands have storm beach ridges inland which must be the product of extreme events in the more distant past.

The Peopling of the Islands

Despite these severe limitations the atolls of the Gilbert Islands are home to some 63,000 I-Kiribati (1985) and have been for perhaps the last 2,000 to 3,000 years. The people of Kiribati are of Micronesian origin. Settlement of the island was part of the general movement of people into the Western Pacific which began about 5,000 years ago, soon after the beginning of plant and animal domestication in Southeast Asia. Both Spoehr (1957) and Osborne (1966) see similarities between the prehistoric record of the Philippines and Micronesia. Alkire (1977, p. 10) concludes that genetic, linguistic, and archaeological evidence points to Melanesia and specifically to the Fiji-Vanuatu area as having the most direct links, not only to Polynesia, but also to eastern and central Micronesia. Settlement was not a simple one-way movement of a mass of humanity. Contact between cultural groupings seems to have been continuous. The I-Kiribati language shows considerable borrowing from Polynesian languages, suggesting more recent contacts as well as their common origins in the Eastern Oceanic language grouping (Pawley 1972, p. 133). Genealogical evidence documents an influx of Samoans in the fourteenth century (Sabatier 1977, pp. 45–48). Recent archaeological excavations in Makin and Tamana led Takayama, Takasugi, and Kaiyama (1988, p. 5) to tentatively conclude that excavated artifacts from Tamana show no sharp and clear division between Polynesia and Micronesia, and that Tamana has a closer historical relationship with Polynesia, whereas Makin has closer parallels in the Marshall Islands and the Carolines to the West.

   Simmons, Graydon, Gadjusek, and Brown's (1965) work on blood group genetic variation in the people of the Caroline Islands and other part of Micronesia showed that gene patterns varied markedly between atolls and even between atolls and even between islands in the same atoll. This led them to argue that historical accident and chance time and again returned the genetic fate of these small island communities to a few new 'founders'. As a result of catastrophes such as cyclones, tidal waves, severe prolonged drought, warfare, canoe sinkings, accidental voyages, or disease, the breeding populations would periodically have been reduced to a small number of individuals whose genetic makeup in no way approached the mean or average gene pool of the population. No comparable genetic studies are available for island populations in Kiribati but there is no reason to believe that similar processes would not have operated there too.

Whatever the ultimate origins of the I-Kiribati, it is clear that a well-developed canoe and navigation technology were essential for the colonization and continued occupation of this island world. It is probable that small groups, possibly as few as one or two canoe loads, may have been involved in the initial peopling of some islands. Some voyages may have been the purposeful conscious seeking-out of new islands for settlement, others may have been accidental drifting. Intermittent contact, both within and between regions continued after settlement permitting the diffusion of new cultural traits and contributing to the genetic diversity now characteristic of the populations.

In the time following settlement the migrants successfully adapted to the very limited land areas and land resources available. They developed considerable emphasis on the exploition of the relatively richer marine resources of their environment. The small size of the islands, prolonged droughts, cyclones and other natural catastrophes resulted periodically in starvation, population decline, and at least temporary outmigration.

Thus isolation and intermittent contact characterized relationships between the islands of Kiribati in pre-contact times. No unified economic or political system encompassing all the islands emerged until the colonial period. Inter-island warfare between rival dynasties with attack, invasion, and reprisal seem to have been endemic in the central islands in the late pre-contact period. In the seventeenth century the southern islands joined forces and invaded all islands to the north except Butaritari and Makin, bringing with them the *maneaba* tradition and the *boti* system of organization but without establishing any over-arching political organization linking the islands.[1] In the period immediately prior to European contact inter-island warfare for power and land was common in the central islands. This encouraged the emergence of centralized leadership and rival dynasties there. On a more peaceful note, some trade did occur. The southern islands were centers for the manufacture of coconut and conus shell beads which were traded and used as currency in the northern islands as well as the Marshall Islands (Quiggin 1949, p. 140). Despite all these factors no enduring supra-island political or economic alliances, comparable with the *hu* and *sawei* systems in the Western Caroline Islands, emerged in Kiribati. Both these systems entailed the presentation of tribute, usually goods such as coconuts, preserved breadfruit and sea turtle, in return for rights to exploit the resources symbolically owned by the superior partner, as well as the invocation of superior magic to ward off natural catastrophes and seek assistance in the event of such disasters. Alkire (1978, p. 24) sees these systems as a response to the peculiar limitations of an island world insofar as they served to link communities on a number of small, dispersed, and vulnerable islands into a larger unit which permitted members of their ranked societies to expect aid or move between islands in times of disaster and resource shortages. For the most part the islands of Kiribati must have functioned as independent, self-sufficient man/environment systems.

TRADITIONAL SOCIETY

Livelihood

The first settlers must have eked out a precarious existence on the barren islands initially encountered. The endemic flora provided few economically important plants. Practically all of the food plants which now form the basis of the subsistence diet were introduced. The giant taro like *babai* was the only root crop cultivated. It was grown in pits excavated to the water table and enriched with compost. Through continued planting the coconut palm has become the dominant vegetation on all inhabited islands. The flesh of the nut and the toddy sap, *karewe*, were fundamental items of subsistence diet. The importance of toddy was accentuated by the fact that even in severe droughts the coconut palms continued to flower although no fruit would set. Thus toddy could still be tapped even when coconut supplies failed. *Karewe* was also reduced to a syrup (*kamaimai*) by boiling, and this could be stored for long periods. Breadfruit (*Artocarpus spp.*) were also introduced and may have been an important food source on the wetter islands. On drier islands it required careful tending and mulching to survive through droughts. The small fig *bero* (*Ficus tinctoria*) was better drought-adapted. On some islands the fruits were dried and stored as a drought standby food. However, the greatest elaboration of the plant world occurred in the pandanus (*Pandanus tectorius*). It was only in eastern Micronesia that the fibrous fruit made a significant contribution to diet. The fruit was preserved in a variety of ways for special occasions, sea voyages, and in anticipation of food shortages. The I-Kiribati recognize more than 160 varieties of fruiting pandanus. Because it does not breed true from seed, the proliferation of named varieties can only be the result of conscious selection, vegetative reproduction, and dispersal by human agents. Sir Arthur Grimble (1933–1934) describes the I-Kiribati as 'the pandanus-eating people'.

The importance of livestock in traditional society is uncertain. The pig was unknown before its introduction by Europeans. The dog may have been known but became extinct five or six generations ago (Grimble 1933–1934, pp. 28–29). Chickens were present on some islands in pre-contact times (Koch 1965, p. 71) and were totem creatures (Grimble 1933–1934, pp. 21, 28). *Baneawa* fish (*Chanos chanos*) were raised in specially stocked fish ponds.

Given the paucity of atoll land resources, it is not surprising that considerable emphasis was placed on the exploitation of marine resources. No resource went unutilized. The lagoons and ocean waters were fished from canoes using hook and line as well as trolling lures. Scoop nets and flares were also used to catch flying fish. Early whalers' reports record seeing scores of canoes engaged in fishing at one time. A great variety of techniques were used to fish the reef shallows, reef flats, and lagoon shallows. Large permanent fish traps of coral blocks were constructed on the reef flats. Nets and communal fish drives were used in the shallow waters of the lagoons. Again the constant specter of scarcity dictated the drying, salting, and storage of any surplus fish or shellfish. Even the skin and fins of sharks were dried and stored for future consumption.

Island Society

The social and political systems characteristic of pre-contact times show a variety of external influences, not all of which affected all islands equally. Underlying all is a land tenure system based on individual inheritance and ownership of land by both men and women. The northern island societies were stratified into four basic social groupings of chiefs, nobles, freemen, and slaves and these divisions survived until the colonial period. The seventeenth-century invasion of the central and northern islands by the south replaced the stratified societies with more egalitarian gerontocracies and implanted the *maneaba* tradition and *boti* system of organization as far north as the equator. In the islands from Aranuka to Marakei, the *maneaba* system was introduced without displacing the chiefly classes. The *maneaba* was a large, rectangular communal meeting house where the affairs of the community were discussed and arbitrated upon by a gerontocracy representing the constituent *boti*. It was thus the center for community affairs, the locale for formal ceremonies, feasts, and more dignified dances. Each island was divided into several *maneaba* districts, with one *maneaba* taking preeminence over others in interdistrict affairs.

269

The *boti* was the sitting place in the *maneaba* reserved for a clan and, by extension, the clan itself. *Boti* membership controlled marriage, with marriages between members of the same *boti* being prohibited. Sexual relationships between members more closely related than fourth cousins were regarded as incestuous and punishable by death.

The basic residential unit in pre-contact time was the clan hamlet or *kainga*. Rights to residence in a *kainga* could be traced through either parent with preference being shown for residence within the father's *kainga*. Thus a *kainga* would have been made up of the houses (*mwenga*) of the head of the *kainga*, his spouse, and unmarried children and possibly that of the eldest son, his spouse and children, plus the additional *mwenga* of however many sons and their families as might be accommodated. In the absence of male descendants the families of daughters and other kin might set up *mwenga* on the *kainga*. The *kainga* was thus a cluster of dwelling houses accommodating related extended families located on clan lands. It was sometimes fenced or at least surrounded by an apron of coral gravel to give warning of approaching intruders. The large consolidated village characteristic of Kiribati today emerged as the result of missionization and colonial control.

The *kainga* served as the center for cermonial activities relating to the clan. In return for rights to residence the individual was expected to provide food and other products for the *atun te kainga* (the head of the clan) and to assist other *kainga* members in the carrying out of duties and obligations connected with *maneaba* ceremonial activities. Common residence and kinship obligations also provided the basis for cooperation in deep sea and net fishing and capital-creating tasks such as house, canoe, *babai* pit and fish trap construction.

While the *boti* was the land holding unit as far as residential (*kainga*) land was concerned, bushlands (*buakonikai*) were divided into individually owned plots. Transfer of these lands was generally restricted to members within the *utu* (extended family). Land was inherited by both men and women from both parents. Sons received more than daughters; eldest sons more than younger sons. Because of inheritance from both parents and restrictions on marriage within four generations, an individual's lands must have been scattered all over the island, or even on other islands depending on family connections. Adoption was common and usually confined within the *utu*. It facilitated the redistribution of land among more distantly related kin groups, thus reducing the prospect of any individual becoming entirely landless and therefore being denied access to the means of production. Land also figured prominently as compensation for such crimes as murder, theft, breaking of betrothal, and adultery.

In some aspects of everyday life at least the *mwenga* must have functioned as independent units because the bushlands were vested in the individual rather than the clan. The individual rather than the clan would have taken basic decisions on the management of land, the planting of coconuts, pandanus, other economically important trees and shrubs, and probably even *babai*. Since these are mostly slow-maturing, long-producing crops, land management must have been concerned as much with ensuring adequately productive lands for one's offspring as with immediate consumption needs. Planting thus expressed the essence of land ownership.

In the stratified societies of Butaritari and Makin the property rights of the paramount chief and his family created a more complex pattern of land tenure. The ownership of most land was vested in commoner families. However, the high chief and aristocratic families held ill-defined residual rights over all land and *babai* pits. These rights were acknowledged by tribute.

Man/Land Relationships

Thus the traditional societies of Kiribati came to terms with the limited atoll environment. It must have been a somewhat precarious existence and few potential resources went unutilized. Equally, because continuous settlement was maintained over many centuries, it must be argued that the settlers achieved a 'successful' balance between man and nature in their new homeland. In many islands this balance must have been a fluctuating one.

On the islands in the center and north, where environmental perturbations were less severe, direct checks on fertility must have been used to maintain populations at levels below those where resources would become a limiting factor. As far as is known abortion and infanticide were practiced as a means of population control before European contact. Late marriage for men and restrictions on intercourse between husband and wife for up to 18 months during the suckling of a preceding infant may have had some effect in reducing fertility (Bedford, Macdonald and Munro 1980, pp. 207–208). The same authors consider intergroup warfare may have been a major cause of mortality among adult males in Kiribati.

However, on some islands, particularly the drought-prone islands of southern Kiribati, 'Malthusian mortality' may have been a more significant factor reflecting the fluctuating ability of the environment to sustain population levels. Starvation and death or forced emigration appear to have been a fact of life on these islands. In 1983 the whaler Navy visited Tamana and found the islanders in a starving condition. One hundred and fifty islanders were transferred to the neighboring island of Arorae. In the severe drought on Tamana from 1872–1875 the European trader on the island reported that 800 people had died of famine and disease. A missionary census of the island population in 1870 put the total population at 1,000 people. By 1878 the combined effect of famine and labor recruitment had reduced the population to 282.

Thus the traditional society came to terms with the stringencies imposed by the limited atoll environment. In the case of the drought-prone islands this must have been an exceedingly precarious balance. The islanders brought with them a maritime technology that enabled them to colonize the far-flung islands and exploit the relatively abundant resources of their surrounding seas. They brought new plants and modified the existing vegetation to increase the productive capacity of the land. In the *babai* pit they created an artificial environment to cultivate a valued crop plant. The social systems which evolved regulated access to resources and ensured that all got access to at least some land and a livelihood. On the better endowed islands conscious population regulation must have been part and parcel of successfully coming to terms with the atoll environment. Continuing contact within the island groups and between different cultures ensured that indigenous societies were not static and were constantly adjusting to new influences. Despite such contacts, at the close of the eighteenth century it is probable that individual islands functioned as independent, self-sufficient man/environment systems. Each island was a world in itself and core and periphery had little meaning.

However, the nineteenth century saw changes of unprecedented range and magnitude. Despite their isolation, tiny size, and meager resource bases, these islands became incorporated into the expanding industrialization and imperialism of the West. In the process island society was transformed. New opportunities were created, as were new wants. New flows of goods, people, and resources emerged. Independence and self-sufficiency were replaced by dependency. Incorporation transformed these island worlds into the very periphery of the modern world economic system. Many of the processes observed parallel those observed in larger, better endowed states, but often the detail reflects the playing out of these processes of change in the limitations of the atoll environment.

THE EMERGENCE OF THE MODERN ISLAND ECONOMY

At the beginning of the nineteenth century the island world of the Pacific was known in Europe and America only through the fragmentary accounts of seventeenth- and eighteenth-century Spanish, French, and British explorers. By the end of the century few parts of the region were untouched by the emerging western capitalism which sought raw materials, trade outlets, land for plantations, and labor for these, and mining. The island entities thus became incorporated, to varying degrees, into the capitalist world. The islands became the periphery of the expanding mercantile and industrial economies of Europe and America. The incursion brought other agents of change; the missionaries in search of converts to Christianity and the colonial administrators as agents of imperial control.

Despite the limited resource base of the atolls, Kiribati became part of this process. It saw the growth of the whaling industry in the early nineteenth century. As whale stocks declined through overfishing the oil-handling technology was transferred to the development of the coconut oil trade which flourished in the 1850s and 1860s. In turn this was replaced by trade in copra, the dried flesh of the coconut. Copra has remained the islands' only major export crop to this day. In the latter part of the nineteenth century 'blackbirders' and labor recruiters sought labor for plantations and mines in Peru, Tahiti, Hawaii, Samoa, Fiji, and Queensland. Closely regulated labor migration characterized the development of phosphate mining on Ocean Island and Nauru, and for some seventy years formed part and parcel of rurual life. Contract labor migration continues today with the training and recruiting of seamen to work on overseas ships. Missionaries from America, Britain, and France also arrived in the latter part of the nineteenth century. They exerted considerable influence in the field of social change, attitudes, and settlement patterns. Finally, these incursions brought colonial control. In part this was justified through concern to control the lawlessness and violence associated with the presence of traders and labor recruiters, but it also reflected the working through of power games involving the principal imperial powers in the Pacific. Britain declared protectorates over the Gilbert and Ellice Islands in 1892. In 1916 these became the Gilbert and Ellice Islands Colony and were administered as such until 1978 when the Polynesian Ellice islanders successfully petitioned to separate and became the independent state of Tuvalu. The Gilbert Islands, together with Banaba and the Line and Phoenix Island groups, became the independent Kiribati in 1979. The administrative control imposed by Britain during these 86 years exerted a considerable influence on the character of economic development of the islands. The paternalistic control of the early years tended to consolidate many of the changes introduced by the missionaries and minimize the impact of major resource developments on Ocean Island and Nauru on rural life. Changes is colonial policy after World War II resulted in increased government spending, the emergence of an urban elite, the growth of Tarawa as the major urban node in the region, and accelerated outmigration from the rural areas. Much of the funding for these developments came from aid and phosphate royalties. It was perhaps not surprising then that the granting of independence coincided very closely with the exhaustion of phosphate reserves on Ocean Island. Eighty-six years of colonial rule thus left Kiribati with no new major resource developments to replace the phosphates, and an economy where migration, remittances, and the bureaucracy determine the evolution of the economic system. Bertram and Watters (1985) designate this as the MIRAB system and claim that it characterizes the economies of many small Pacific nations.

Whalers and Whaling

The earliest sustained contact with the capitalist world resulted from the development, beginning in 1819, of the 'on-the-line' whaling grounds which stretched from the Line to the Gilbert Islands. By the 1840s the area to the southwest of Tamana and Nikunau and around Ocean Island and Nauru provided the main focus for whaling activities. The whalers had no shore bases in the islands and were away from port for long periods, sometimes even years. For these reasons, the captains were keen to procure provisions, water, firewood, replacement crew, and women. For fear of attack and the dangers of navigating reef passages, the whalers rarely landed on the islands. Instead they conducted business standing well offshore. Flotillas of up to 40 canoes travelled as far as 20 km from the islands to trade in coconuts (used mainly to feed the pigs kept on board) toddy molasses, curios, women's favors, and pigs and chickens. The pigs were introduced to the islands by the whalers. In return the islanders received tobacco, hoop iron, wire nails, knives, hatchets, and whale teeth. Firearms were also introduced on some islands.

Contact with the whalers does not appear to have produced the traumatic and far-reaching effects experienced in Tahiti, Hawaii, and New Zealand. Some violent incidents, such as when islanders led by a Spaniard attacked the Alabama off Nonouti in 1848, did occur. For the most part relations between islanders and whalers appear to have been remarkably good. The introduction of weapons may have increased the efficacy of inter-island warfare, particularly in the central and northern islands, thus aiding the emergence of secular dynasties there. The whalers' crews are known to have suffered from venereal disease and smallpox, but the impact of such diseases on island populations during this period is unknown. No whaler's logs record epidemics on any of the islands, but this is not surprising because their contact with the shore was brief and superficial. Precise evidence on mortality from epidemics in the region is lacking.

Contact with the whalers also led to the introduction of metals, as well as pigs, goats, and crops (pumpkin, squash, and watermelon). The introduction of metals led to the substitution of iron and brass for shell in adzes, knives, fishhooks, pump drills, shovels, and coconut graters. This occurred without altering the basic design of the implement. Canoes made with iron adzes still appear to have been made in the same way although construction may have taken less time and effort. There is no evidence to suggest that canoes became more numerous as the result of this. The *Rose* was met by 40 canoes when it visited Tamana in 1804 and traded what was probably the first hoop iron seen on the island for coconuts and beads. No greater numbers of canoes were reported on the water at the same time in later whalers' logs when the use of iron tools would have been more widespread. The introduction of iron tools would also have made the construction of *babai* pits enormously less arduous. However, it is impossible to assess whether these changes led to more *babai* pits being constructed or less time being devoted to *babai* cultivation. There appears to have been no response similar to that recorded by Salisbury (1962, pp. 219–221) for the Siane of the East New Guinea Highlands where, as the result of the introduction of steel axes, less time was devoted to subsistence gardening and more to social and other activities. By far the most interesting outcome of the interaction between the whalers and the islanders was the provisioning trade. This seems to have reached its peak in the 1850s and 1860s, when the whalers were most active in the area. On some islands participation in the trade may have been stimulated by the Europeans now resident on the islands. In others it was an entirely indigenous response. The islanders saw the opportunities and actively sought the trade by taking their wares (including the women) to the whalers. They produced what the whalers wanted: coconuts, *kamaimai*, chickens and pigs. *Kamaimai* was sold by the gallon or barrelful and was appreciated by the whalers for its antiscorbutic qualities. At times the whalers were able to purchase as many as 200 fowls at a time. The *Martha* purchased 455 fowls on several visits to Tamana between 17 and 25 November 1853. Jarman (1838, p. 169) records that the ship *Juno* was carrying pigs from Samoa in the vicinity of Tamana in 1838. Pigs were first recorded as trade items in 1849. They were probably never obtained in large numbers, although the *Massachusetts* recorded getting forty pigs in one visit to Tamana in 1859. The trade lasted some thirty to forty years. Overfishing resulted in the demise of the whaling industry and with it the demand for provisions.

The impact of this contact and trade is hard to assess. It demonstrated the willingness of the islanders to respond to the new opportunities created and also their interest in adopting and raising new animals for the trade. Some islanders gained wider experience and new skills from crewing on the whalers. However, major changes in social life, technology or resource use patterns did not occur. Since tobacco appears to have been the main medium of exchange down-stream impacts from the trade were limited. Hudson of the United States Exploring Expedition gives a colorful account of I-Kiribati tobacco use in 1841:

> They are truly disgusting for they eat it and swallow it, with a zest and pleasure indescribable. Their whole mind seems to be bent upon obtaining this luxury and consequently it will command their most valuable articles. (Wilkes 1845, V. p. 62)

In addition, the trade was relatively short-lived and disappeared with the overfishing of the whale stocks. Its place was taken by the coconut oil and copra trade and the main emphasis of European activity shifted from the southern islands to the wetter and more productive northern islands.

The Coconut Oil and Copra Trade

The development of new techniques in soap and candle manufacture in Europe created a demand for coconut oil. The trade in Kiribati developed initially as an adjunct to the whale oil trade. The whalers usually left a crew member ashore with a supply of trade goods and barrels to collect the oil which was picked up on the return of the whaler. With the decline of the whaling industry the coconut oil trade became established in its own right. Further technological developments led later to the switch from coconut oil to copra production. Maude with Leeson (1968, p. 242) record that the coconut oil trade was immensely popular with the I-Kiribati. The product was already manufactured by the islanders for their own culinary and toilet use from existing resources. It enabled the I-Kiribati to get coveted European goods without any real change in their traditional way of life. The same productive techniques were used. The islanders simply produced a surplus which was then bartered. New technological developments in Europe thus created a market for the islands' only major resource, the coconut palm. Coconut oil and then copra became the dominant export of the group and remained so for many decades. Copra production today remains the only major cash-earning opportunity available on the islands to many rural households.

Combined whaling–coconut oil trading soon gave way to trade centering on the resident trader. Trading stations with trade stores, storehouses, coopering, blacksmithing, ship repair and provisioning facilities were established on Makin and Butaritari. Agents were appointed on some other islands. Ships from Sydney called first at Butaritari and then spent several months trading among the islands: landing relief traders, trade goods and empty casks, and taking on time expired traders and oil in casks. In contrast to the provisioning trade for the whalers, the oil trade appears to have been dominated by resident expatriates, although on the islands with stratified societies the High Chiefs all endeavored to become principal trading agents for their islands or to levy a duty on all copra exported (Maude with Leeson 1968, pp. 258–259). No trader could afford to alienate such chiefs.

With increasing sophistication of demand, tobacco ceased to be the major item of exchange. The range of goods now available to the islanders included tobacco, pipes, firearms, cloth and sewing goods, axes, knives, and fishing equipment, as well as soap, combs, looking glasses, and cooking utensils (Maude with Leeson 1968, p. 277).

The switch from coconut oil to copra in the late 1860s resulted in the establishment of bulk handling facilities for copra in Sydney and Apia, Samoa. This reduced the need for elaborate and diversified trading stations on the islands. The same period also saw the entry of large-scale businesses from Australia, Germany, Japan, USA and New Zealand into the industry. However, independent traders continued to operate on some islands until World War II. Some islands, particular the smaller, drier, and more densely populated islands in the south, were without resident traders for long periods and were serviced directly by visiting trade ships which called every few months. With the onset of World War II all trading ceased. Private trading concerns did not re-establish after the war. Their place was taken by the government and the cooperative movement it sponsored. The government assumed responsibility for all activities associated with the purchase, handling and shipping of copra as well as the supply of goods to the newly established cooperative stores.

Thus copra emerged early as a dominant element in the rural economy. No alternative commercial crop to the coconut has been found. However, the importance of copra as a source of income in the present-day rural economy is somewhat equivocal. The Victoria University Socio-Economic Survey demonstrated that few, if any, of the households surveyed could be regarded as being dependent on commercial agriculture for a livelihood. On Tamana only two of the 16 households got more than 20 percent of their cash income from copra (Lawrence 1983, p. 142). On Abemama copra was slightly more important with 5 of the 16 households sampled getting more than 20 percent of cash income from copra (Watters with Banibati 1984, p. 106). Copra seems to have been somewhat more important as an income source on the wetter island of Butaritari, but even here only 6 of the 16 households surveyed got more than 50 percent of their cash income from copra (Sewell 1983, p. 165). Thus, although copra remains the only feasible commercial crop, crop production is, in most instances, remarkably desultory. Price has a strong influence on production. On most islands producers withdraw from the market as prices fall. It is only on some of the northern islands, like Butaritari, that price falls are accompanied by increases in production. This reflects a strong ethos of maintaining socially recognized levels of consumption of store goods. On the other islands producers appear happy to depend more heavily on subsistence production for their livelihood or rely more heavily on remittances from relatives in employment off the island (Geddes et al. 1982, p. 59).

In many ways the economic and social changes wrought by the commercialization of agriculture differ markedly from those described by Gregory and Piche (1978) in Africa or even that experienced in other areas of the Pacific (Brookfield with Hart 1971; Howlett 1973; Bayliss-Smith 1978; Bedford 1978). In these areas the commercialization of agriculture led to the introduction of new commercial crops, the displacement of traditional subsistence systems, the emergence of plantation agriculture, and increased migration to the urban centers in search of work. There are several reasons for this.

No plantation sector emerged in Kiribati as a direct result of colonial policy. When the Protectorate was declared in 1892 land sales to outsiders were forbidden and sales between indigenes restricted on some islands. Land sales occurred mainly on islands where the paramount chiefs or upper classes disposed of surplus lands. The only plantations in Kiribati occupy previously uninhabited islands, e.g., Christmas, Washington, and Fanning Islands.

In traditional society land was vested in the individual rather the kin group. Colonial policy codified and fossilized this system. This meant that no individual could preempt the resources of the wider kin group in order to engage in commercial agriculture. With tree crops in particular this meant that communal lands were unavailable for subsistence production for very lengthy periods. In Kiribati a would-be commercial agriculturist could only plant and utilize lands to which he had, or could expect to gain title to. However, in the stratified societies the high chiefs may have been able to extract tribute from subordinate classes in order to amass a surplus to engage in the copra trade. This does not appear to have denied others access to at least some resources for subsistence and trade.

In other ways colonial policy may also have reduced the impact of the commercialization of agriculture on rural life and limited the downstream effects of new developments. After World War II the colonial government assumed responsibility for copra purchasing, wholesaling, retailing, and shipping throughout the Colony. Its policies ensured that all islands had regular shipping, trade stores, and copra purchasing points. A uniform pricing policy for copra and trade goods applied, regardless of location or volume of output. This policy tended to reduce some of the disadvantages experienced by the smaller, more remote, more densely populated, and less productive islands. At the same time it removed the advantage of the more productive and accessible islands. Even the earlier centers associated with the coconut oil trade disappeared with the change from oil to copra production. In this way, no clearly recognizable centers of copra production emerged. No urban centers associated with commercial agriculture developed, and few spread effects resulting from the change from subsistence to commercial agriculture were experienced.

Perhaps the most fundamental reason for the limited impact of the commercialization of agriculture in Kiribati relates to the limitations of the atoll environment and the ecology of the coconut. In Kiribati and probably on atolls generally, we have a peculiar case where a tree crop, which can be used as both a subsistence staple and a commercial crop, makes up a substantial part of the 'natural' vegetation. Traditional crops or the natural vegetation were not displaced by the introduction of new commercial crops. Since the coconut is a slow-maturing, long-lived, and continuous-bearing tree it does not impose a cultivation cycle requiring decisions to plant, cultivate, harvest, and sell. The question of choice between production for subsistence or sale does not really arise. Instead decisions tend to focus on utilization; whether to tap the tree's flower spathes for toddy or allow it to bear nuts; whether to eat the nut or turn it into copra. A landowner is not forced to participate in the market because his crop is ready and the crop has no utility for other purposes. Nor is he forced to accept the prevailing price or lose the crop. On the drier islands nuts could still be turned into copra up to eighteen months after they had fallen to the ground. Since it takes about seven years for a coconut to reach maturity, planting strategies could only affect productivity in the longer term and were aimed more at ensuring the adequacy of resources in the longer term. The situation is reinforced by the sea's importance in the subsistence economy. Thus under the conditions prevailing on the atolls the islanders are not forced to make a choice between traditional subsistence production or production for the market; they can operate in either or both systems using the same basic resources. The lack of need for commitment to commercial agriculture is heightened by the fact that few households today derive a major proportion of their income from copra production. On many islands remittances and gifts from kin and friends in employment off the island provide a much more important source of income than commercial agriculture (Geddes et al. 1982, p. 69).

Thus there seems to be little evidence to suggest that the commercialization of agriculture in Kiribati (if that is an appropriate term for the collecting of nuts from an already widely planted tree crop) led to significant social or economic changes. The subsistence economy probably changed little. The diversion of resources to commercial production did not displace or destroy production for subsistence. New crops and land use practices were not introduced. The trade did enable the islanders to get access to the new trade goods brought by the western commercial interests which penetrated the area. However, the trade appears to have had limited wider impact. Initially much of the newly generated wealth was spent on tobacco. Even today a very large proportion of household expenditure is devoted to purchases of rice flour and tea which supplement rather than replace more traditional foods. The new technology introduced may have increased the efficiency of fishing and made such tasks as canoe building and *babai* pit construction considerably easier, but it has not produced major new patterns of resource use in the rural areas. The commercialization of the coconut simply intensified a trend initiated by the original settlers of planting more and more land in coconuts. This time the expansion may have been at the expense of other traditional tree crops such as pandanus.

The commercialization of agriculture also failed to precipitate substantial rural urban migration. Islanders were not displaced by land sales for plantations. The copra trade created few employment opportunities in the processing or service sectors. No urban centers emerged as a result of the development of commercial agriculture to become targets for migration. Migration to other centers of employment remained closely controlled as a matter of colonial policy. Isolation and poorly developed transport systems made movement doubly difficult. The combined effects of these factors led to something of a paradox. Despite their small land area and limited resource bases, the atolls have some of the highest rural population densities in the Pacific. In 1973 Tamana, the most densely populated rural island in Kiribati, had a population density of 281 persons per square km. Comparable densities in the Eastern Islands of Fiji (considering only land of agricultural potential) ranged from 37 (Koro) to 113 (Kabara) (UNESCO/UNFPA 1976).

Given the desultory nature of copra production and the rather tenuous commitment of many households to commercial agriculture and cash-earning, it is perhaps inappropriate to attempt to apply any of the stage or continuum models of economic and social change (so fashionable in the 1960s) to Kiribati. In the most widely quoted examples from the Pacific literature (e.g., Fisk 1962; 1974; 1975; Fisk and Shand 1969; and Epstein 1968) the writers tacitly assume that the process of incorporation into the market economy will be by way of increasing specialization in cash crop production for export and an increasing dependence on the market for all other goods and services required. Other possibilities for gaining access to cash go unconsidered. Epstein (1968, p. 166) justifies such an approach by claiming that:

> evidence from the Tolai, as well as from other culture areas, indicates that in the early stages of growth an indigenous population if given the choice, will prefer to sell produce rather than labor. This reaction appears to continue until urban centers in the area progress so far as to offer great attractions such as cinemas, restaurants coffee shops and beer halls not available in the rural hinterland.

This clearly does not apply to Kiribati where an established tradition of labor migration existed for at least as long as commercial agricultural production. The islanders have since the mid-nineteenth century shown a remarkable willingness to leave their homes, crew on whaling ships and work on plantations or in mining activities. In more recent years the growth of Tarawa has shown their willingness to exchange rural life for urban wage employment. The range of options is much wider than commercial agriculture alone. The options arise in part from direct investment in such activities as phoshphate mining, but also from government policy decisions on the provision of services, infrastructure, employment, and how and where aid funds are spent.

Blackbirding, Recruiting and Overseas Labor Migration

The I-Kiribati were thus no strangers to wage labor migration. In 1847 labor recruiting began in the region when Benjamin Boyd's ships recruited 65 Pacific Islanders (including 17 I-Kiribati from Tamana and five from Arorae) to work on his cattle and sheep stations in New South Wales (Maude with Leeson 1968, p. 268; Parnaby 1964, p. 6). Intermittent recruiting voyages thereafter saw I-Kiribati men and women taken to New Caledonia, Reunion (in the Indian Ocean) and Peru. There can be little doubt that this early recruiting amounted to little more than kidnapping and that many islands suffered depopulation as a result. Of the 1,200 Pacific Islanders known to have gone to Peru, fewer than 100 are known to have arrived back in the islands (Parnaby 1964, p. 14). However, as the establishment of plantations in Fiji, Samoa, Tahiti, and Hawaii generated demand for labor a more sustained and possibly more regulated trade in labor developed. Bedford et al. (1980, pp. 213–221) estimate that around 3,000 I-Kiribati adults were recruited to work in Fiji between 1864 and 1895. Between 1867 and 1885, 850 went to Tahiti. A majority of these did not return. Around 2,500 Kiribati adults worked in Samoa between 1867 and 1895. Fifteen hundred adults and three hundred children left Kiribati for

281

Hawaii. Some of these were stranded in Hawaii when repatriation came to an end in 1881. A further 1,000 went to Central America between 1890 and 1892. Only 230 of these are known to have returned. Smaller numbers went to work on the sugar plantations in Queensland.

In theory at least the recruiting agreements should have given rise to a regular circulation of islanders between overseas work and their home islands at the end of the contract period. However, the reality depended very much on the willingness of governments and planters to adhere to contract agreements. Economically difficult times, such as the financial crisis in Fiji 1873–1874, also affected the planters' ability to pay wages and meet the cost of repatriation. The Fiji ordinances of 1887 required that laborers were provided with food, clothing and shelter (comparable to that of Fijian native) and pay in kind at the rate of £3 per annum paid at the end of the contract, which was usually for five years (Parnaby 1964, pp. 31, 184).

Bedford et al. (1980, p. 217) stress that the effects of labor migration on the nature of population changes were not uniform throughout Kiribati. Some islands were depopulated, while others were less severely affected. On the drier more densely populated islands in the south, recruiting may have temporarily reduced population levels and in fact lessened the impact of the severe droughts experienced from 1871 to 1875. Increased outmigration coincided with drought periods. The missionary Pratt (1872) records that the recruiters took advantage of the islanders' plight by assuring them that there 'was plenty to eat in Fiji and no work'.

Increased mobility brought with it contact with alien diseases. The mission ship *Morning Star* and a labor vessel brought measles to the Gilbert Islands in 1891 and they spread rapidly to most islands. Walkup (quoted in Bedford et al. 1980, p. 240) claims that over 1,000 died from measles or other effects.

The social and economic effects of employment overseas are more difficult to assess. While it is clear that many I-Kiribati were keen to leave their home islands, particularly during drought years, for most it was a circular migration and the commitment to return home never diminished. Nostalgia, satisfied curiosity and disenchantment with working on strangers' plantations caused a vast majority to take advantage of the promised return passage (Bennett 1976, p. 22). Genealogies collected during fieldwork on Tamana in 1972 suggest that many recruitees never returned or died overseas. Those who did return were often left on other islands or visited newly-made friends' homes enroute. Marriages or adoptions often resulted. Friendships also gave rise to *te bo* (like kin) relationships which linked people living on different islands. Returning workers brought new planting techniques and plants, tools, clothing, household goods, firearms, and money. The London Missionary Society representative, Powell (1879) refers scathingly to 80 workers returning from Tahiti who brought with them 'Mormon delusion'. The activities of returning converts to Christianity in trying to convert others to the cause preceded the missionaries on some islands. Indigenous movements blending Christian religious ideas and indigenous practices often resulted. The *Anti n Tioba* (Jehovah) movement began on South Tabiteuea in 1855 with attempts by two Tahitian deserters to convert the island population (Geddes 1982, p. 33). The movement persisted until 1880 when Hawaiian missionaries placed on North Tabiteuea by the American Board of Commissioners for Foreign Missions (ABCFM) led an invasion force which massacred the adherents of the cult (Macdonald 1982, p. 38).

However, in total, the social and economic impacts of the labor trade were probably not great. Clearly it established the concept of wage labor as a path to economic advancement, and this has continued to be important, particularly for the drier poorer islands in the south. The wealth accumulated by returning workers in these early years was probably quickly dispersed. The missions quickly absorbed much of the ready cash and the traditional redistributive mechanism, *bubuti*, ensured that goods were dispersed through the extended family.

The disruption associated with 'blackbirding', coupled with depopulation through newly introduced diseases and starvation due to the long drought in the early 1870s may all have weakened the traditional social structure and made it easier for the missions to establish, gain control, and instigate far-reaching social changes. The problems of lawlessness arising from the presence of traders and recruiters on the islands also provided a pretext for establishing colonial control in the region.

Missionaries and Missionization

Mission activity began in 1852 with the arrival of the Protestant American Board of Commissioners for Foreign Missions (ABCFM) in Butaritari and Makin. Hiram Bingham and his wife took up residence on Abaiang in 1857 and began the evangelization of Kiribati with the assistance of Hawaiian missionaries. By 1867 their activities extended from Makin to Tabiteuea. In 1870 the London Missionary Society (LMS) extended its activities in Tuvalu to the islands of Southern Kiribati with the placing of Polynesian pastors, trained at Malua in Samoa, there. By 1895 the LMS had become an almost unchallenged power in the five southern islands of Kiribati. The activities of the Roman Catholics began somewhat unofficially about 1880 when converted laborers returning from Tahiti attempted to convert the people of

Nonouti. The first Catholic priests (Frenchmen belonging to the Sacred Heart Mission [MSC]) arrived in Nonouti in 1887. Their activities initially concentrated in the northern and central islands where progress was made at the expense of the ABCFM which gradually lost initiative and vigor. Extreme rivalry characterized relations between the LMS and MSC. In the early years the colonial administration had to act as arbiter and peacemaker between rival missions. This distinction between a predominantly Catholic north and Protestant south remains today, and while sectarian rivalry is less vociferous than in the past, it is still an undercurrent in political differences.

Whereas the whalers, traders, and labor recruiters may have contributed indirectly to social change, direct attempts to change certain characteristics of Kiribati society were the avowed aim of the LMS missionaries. After the 1870s conversion proceeded rapidly and fundamental social changes emerged. In the Protestant areas the distinction between sacred and secular tended to become blurred with time and the pastors assumed roles as leaders of local government. In contrast the Catholic priests (mostly European) arrived at a time when substantial change through contact had already occurred. They were thus confronted with an encouraging choice between competing persuasions rather than conversion. For these reasons they tended not to be more accommodating of custom; they did not oppose traditional dancing. Nor did they seek to outlaw tobacco and alcohol. Where the Protestant flock was required to support their pastor and mission activities with gifts of food, money and coconuts, the priests paid for food and building materials and food and schooling were provided free (Macdonald 1982, p. 51).

As the result of missionary activities literacy and Bible knowledge were promoted and schools established. In the south dancing and other 'heathen' practices were discouraged. Shrines, sacred relics, and totems were destroyed. Infanticide, abortion, and all sexual relationships outside monogamous marriage were discouraged. The wearing of garments of imported cloth was introduced and attempts were made to prohibit the use of tobacco and alcohol. As the church gained influence it attempted to introduce standardized codes of law and punishment. It also sought to ameliorate traditional punishments for some crimes. The traditional leaders were absorbed into a new power structure. Many *boti* leaders became deacons in the church as well as members of the village council, but their real power depended ultimately on their willingness to accept the dominant role of the pastor.

Centralized villages were built to replace the dispersed *kainga* hamlets. New house designs were introduced with the houses being uniformly spaced along a wide central road. (On non-LMS islands this change did not occur until the arrival of the colonial government). On many islands imposing limestone churches were constructed as the focal point of the village and symbolizing the dominance of the Church in island activities.

Relocation had basic repercussions for the community social structure. According to informants the pastor chose the location of the new village and exisiting landowners were not permitted to stop other unrelated people taking up residence on their land. Compensatory land exchanges were supposed to have been made. This one action broke the tie between kinship and residence and the whole significance of the *kainga* changed. It was no longer the residential estate of the descent group. Relocation and restrictions relating to the spacing and number of houses permitted on a village plot meant that the descent group could no longer function as the basic unit of economic cooperation. As a result the *kainga* as a unit of economic organization declined in importance and the *mwenga* or household emerged as the most important economic unit. The loss of the bond of common residence for the descent group must also have weakened the *maneaba* organization by substantially reducing the cohesion of its constituent groups. The situation is further confused by the fact that the land (*kainga*) names within the village were retained even though the people now living on them are no longer linked by common descent. These area divisions within the village were, and still are used to delimit work and fund-raising groups for church and island government activities. This experience of working together has encouraged the constituent households to form cooperative cash-earning groups (*mronrons*) and so the land unit, rather than the kin linkages, now determines the limits of cooperative activities in some spheres.

The church also appears to have provided a sink for much of the new-found wealth of returning laborers. Surviving data suggest that annual church collections on Tamana in the late nineteenth century rarely netted less than $200 (US dollars were then in use). In 1878 donations, book sales, and levies for the pastor's stipend totalled $684.95 (Powell 1879), an astounding sum from a total population of somewhere around 500 people.

The schooling system introduced by the mission should in theory have benefited the islanders by increasing their ability to participate in the new economic order that was emerging. However, the standard of teaching and facilities appears to have always been poor (Goward 1902, p. 2 and GEIC F34/4/15, Assistant Commissioner's Travelling Diary, 4–7 May 1960). The school syllabus was slanted heavily towards religious studies and was, in the case of the LMS, aimed at selecting a few better performers for training as pastors. Poor primary school education put many rural dwellers at a very great disadvantage in the scramble for urban wage employment that was to develop towards the end of the colonial era in Kiribati. In other respects the iron rule of the Samoan pastors no doubt prepared the people well for the authoritarian and paternalistic rule that followed the declaration of the British Protectorate over the islands in 1892.

Colonial Control and the Phosphate Era

In no sense could the Gilbert Islands have been regarded as a colonial 'plum' by any of the nineteen-century imperial powers. It lacked large areas of land, and indigenous populations were dense. No valuable raw materials were at that time evident. The islands lay in a climate zone where it was popularly believed that permanent residence by white men was impossible. It was the heavy indebtedness of Butaritari and other northern islanders to the traders, coupled with concern about drunkenness, disorder, inter-island warfare, the need to control labor recruiting, and Britain's obligations under the Anglo-German agreement of 1886 concerning spheres of influence in the Pacific, that led her to somewhat reluctantly declare a Protectorate over the group in 1892. Annexation took place in 1916 when the Protectorates of the Gilbert and Ellice Islands became the Gilbert and Ellice Island Colony.

Two themes characterized early colonial administration. The first was the overriding concern that colonies should not become a charge on imperial funds; that they should be self-sufficient or preferably make a contribution to imperial coffers. The second was a belief in indirect rule; that the administration should interfere as little as possible with customary political systems. The latter should achieve the former by reducing the need for a large administrative structure.

This penny-pinching attitude ensured that the initial impact of the new colonial administration was limited. Touring by administrators was severely limited by lack of funds and staff. The administration concerned itself principally with the collection and codification of island laws in the hope of preparing a simple local government constitution to fit all islands. 'Simple laws and regulations' were introduced for the 'better conduct of the islanders' affairs'. In many instances the most tangible changes resulting from early colonial rule were in roads, housing, etc. and in the area of local government where, in the islands lacking hereditary High Chiefs, a Chief Magistrate assisted by a council of Kaubures was appointed.[2] Modifications to this system, particularly by limiting the numbers who could serve as Kaubures, and recruiting Kaubures from the younger men further weakened the exisiting social structure in these islands by removing the role of the *unimane* (old men) in society. This paved the way for increasing control by the colonial administration.

What emerged under the 'museum policies' advocated particularly by later Resident Commissioner Sir Arthur Grimble, was control of most aspects of island life. This was exemplified in the *Regulations for the Good Order and Cleanliness of the Gilbert and Ellice Islands 1930* which sought to control, among other things, dancing and obscene gestures during dancing, adoption and the bringing up of children, the hours for fishing, the days on which an individual must work on his land, eating in sleeping houses and sleeping in eating houses, the rolling up of mosquito nets, as well as imposing a ban on interisland travel and a night-time curfew. Able-bodied men and women between the ages of 16 and 60 were required to work for up to one day each week on designated public works.

Despite efforts by younger officials to remove the more offensive regulations and reduce the degree of control, colonial policy from its inception to the outbreak of World War II tended to reinforce the authoritarian approach of the LMS mission and stultify initiative. The island governments became little more than extensions of the central administration. Even the cooperative movement, introduced in the 1920s, failed to be an important vehicle for economic and social change because overcentralized control meant that no member societies were able to take any initiative on policies of what to stock, pricing, the development of local produce sales, or other activities which might have expanded economic activity on their island. The movement did, however, ensure that all islands had trade stores and copra buying points, but it did little to stimulate economic initiative.

The early colonial administration also had little impact on other areas of potential economic and social change. Apart from attempting to force individuals to plant lands and work in their *babai* pits on specified days, no policies attempted to improve the islanders' resource bases. Taxation was imposed in part in the hope of stimulating copra production. Schooling was left in the hands of the missions. For the most part these were ill-equipped and unequal to the task. Health and other services were kept to a minimum. In 1922 a government school was established to teach English and provide the small number of candidates needed for clerical and other vacancies in the government services. In the view of the Colony's isolation and lack of potential, no system of wider education was deemed necessary or desirable. Colonial policy thus did little to promote economic change and development in the rural areas. Financial policy ensured that the bureaucracy remained small and constrained in its activities. No urban center associated with the administration emerged.

The financial constraints and operational difficulties which beset the early administration of this far-flung colony were compounded by the development of phosphate mining on Ocean Island by the Pacific Islands Company in 1900. As 'protectors' of their flock, the colonial administration assumed responsibility for labor relations and relations between the Company and the Banabans. By 1907 these activities had become sufficiently important to the economy that the administrative headquarters were transferred to Ocean Island, thus making the administration of the rest of the Protectorate even more difficult. After World War I German control of Nauru passed to Australia, and in 1920 the Australian, New Zealand, and British Governments took over the interests of the Pacific Islands Company and established the British Phosphate Commissioners to control phosphate mining activities on both Ocean Island and Nauru. The aim of the Commissioners was to secure the cheapest possible supplies of phosphate to agriculture in their countries.

Prices paid were well below prevailing world market prices and profits accrued to interests outside the colony. Royalty payments were kept to minimal levels and went into the administration's coffers rather than to the indigenous population. Commissioners were exempted from the payment of all licenses, fees, and duties (except on liquor and tobacco) in return for making good the shortfall between administration revenue from other sources and approved expenditure. In this way the Commissioners were able to exert considerable influence on the financial policies of the colony with respect to expenditure (Schutz and Tenten 1979, p. 114). Their response was to reinforce restraint. Their needs for a skilled labor force were small and expatriates were recruited. The Commission controlled all activities associated with phosphate mining; labor transfer and housing, mining, crushing, shipping, and ancillary services, even to the extent of providing the trade store from which the islanders could purchase their goods. Workers were recruited on their islands on a strict contract basis. Preference was given to the islanders from the south because of poorer conditions there. A worker was permitted to take his wife and two children only. They were repatriated at the expiry of their contracts. A very small number were rehired and trained for more skilled jobs. At the workings the islanders were housed in compounds among fellow islanders thus ensuring stronger social control and the maintenance of linkages with home. No associated urban node emerged and the spread effects of investment in mining were minimal. The fact that the commissioners had to approve and underwrite expenditure programs of the colony administration put further checks on government spending and restricted the growth of the bureaucracy.

In this way labor migration did not provide the same new-world experience that earlier recruiting did. The migrants' contacts and choices were strongly controlled. What emerged was a system of circular migration which was an integral part of rural life. A young man would go off to the phosphate workings, make new friends and experience new sights (within closely defined constraints). Above all it would enable him to earn the money to buy such capital goods as a bicycle, sewing machine, wood for a canoe, clothing boxes, and household goods: goods not usually available through an island store or too expensive to be purchased with the income that could be raised at home from copra. Older men might make further trips to accumulate more goods or save larger sums of money for school fees, etc. Remittances sent during employment did much to bolster incomes in the rural areas. On the islander's part there was never any expectation that the move would be permanent and an alternative to rural life; it was simply a part of it.

What emerged from the early years of colonial administration was a rural economy that, as well as being confronted with a very limited and intractable resource base, was smothered by overprotective and paternalistic control. The initiative shown by the islanders in responding to the opportunities created by the whaling trade and early coconut oil and copra trade were nowhere to be seen. The penetration of capitalism into agriculture did not, for the reasons discussed above, transform traditional production systems. Copra production remained on most islands a rather desultory activity and few households got a significant proportion of income from copra. Remittances emerged as a major factor in rural incomes and this tended to further dampen the need for cash-generating activities on the islands. Investment in mining thus created what Curtain (1981, pp. 189, 203) has termed 'dual dependence' and encouraged the emergence of a 'straddled economy' where the household bridges the village economy and the mining sector. This state of affairs characterizes the rural areas of Kiribati from the early twentieth century to the present. The place of Ocean Island as a major place of employment has been taken by Tarawa although the character of migration to Tarawa and the rural peoples' freedom to move and expectations of benefits from moving differ radically from the earlier circular migration. The growth of Tarawa as a major center for employment arose not from capitalist investment in production but rather from changes in government policy after World War II and increased expenditures on infrastructure and services.

Post-War Developments

World War II was a singularly important event in breaking down the isolation of Kiribati and hastening the process of change. With the withdrawal of the British the whole structure of the paternalistic administration was destroyed. On the occupied islands the rural population was first exposed to the Japanese and then the full might of American technology. The Americans, in contrast to the former colonial administration, seemed to have an unending supply of wealth, food, tobacco, and gifts. They paid laborers at much higher rates than prevailed previously. I-Kiribati men were recruited into the Labor Corps and travelled to new and different places in the Pacific. However, the most enduring changes resulted from modifications in colonial policy. In 1946 the establishment in Britain of Colonial Development and Welfare Grants meant that aid money could now be channeled to the colonies. The newly re-established colonial administration relaxed its policy of restraint on the provision of welfare and other services. Phosphate royalties and aid funds were poured into a program of infrastructure development in preparation for independence which was now inevitable because of increasing pressures for decolonization in world fora and the working out of phosphate deposits on Ocean Island. The impact of these increases in spending was very unevenly distributed.

In the rural areas the main results were the provisions of state-run primary schooling, improvements to health services, transport, and communications. Attempts to improve the productive base of the rural society came rather later in the process. Many of these were ill-conceived and failed to meet the real needs of the community. The limited atoll resource base severely limited the range of options that could be tried.

Increased spending generated increases in employment and most of the new opportunities created were in the administrative center on Tarawa. The government became the major employer. Increases in urban populations created the need for new and more sophisticated services and more employment opportunities were thus generated. The long history of neglect of education meant that skills were lacking in the local population. Expatriates were brought in and had to be housed and paid at rates and standards above those prevailing previously. Demands for imported foodstuffs were also generated. As independence approached localization policies ensured that artificially high urban wage expectations emerged among the I-Kiribati urban elite.

Hughes (1973, p. 16) estimates that in 1971 three-quarters of all cash incomes were earned on Tarawa. The government and its agencies' wages bill for the urban center topped $3 million, 15 times the sum spent on wages in the rural areas. Eighty-three percent of all government staff were employed on Tarawa. The estimate of aid requirements for the 1979 to 1982 period, presented at the Constitutional Conference in London in 1978, requested $18,219,000 for new projects. Only $460,000 of this sum (2.5 percent) was earmarked for rural development, even though the rural areas contained 65 percent of the country's population (Green, Bukhari, and Lawrence 1979, p. 103).

These developments did not go unnoticed in the rural areas. Improved transport and communications ensured this. It is not surprising then, that in the absence of tight controls on migration, massive relocation of people ensued. In 1947 Tarawa's population stood at 1,529. By 1963 it had risen to 6,101, to 14,681 in 1973 and to 17,921 in 1978. These rates of immigration are exceptionally high in comparison with any other center in the Pacific (Walsh 1982, p. 169). By 1978 no island had less than 20 percent of its dejure population resident on Tarawa; for some islands it was as high as 40 percent.

The character and context of migration to Tarawa is essentially different from that to Ocean Island and Nauru. It is not circular and reflects the conscious exchanging of one way of life for another. The expectation is that the move is a permanent one. The people on Tamana see those who move to employment in Tarawa as being under different *tibanga* or 'fate'. Even the traditional value system is reinterpreted to accommodate the change. The traditional household goal of being *onibai* (independent, self-sufficient, free from the necessity to *bubuti*), which in the past meant having sufficient land, one's own canoe, and, more recently, sewing machine and cash etc., now incorporates having paid employment on Tarawa (Lawrence 1983, pp.38–43; 1984, pp. 197–200). Thus migration is a rational response to the inequalities in living standards and opportunities prevailing between the urban center and the rural periphery. The government to date has no effective policy to control the movement. The movers see themselves as exchanging the drudgery of subsistence production and the unresponsive, unpredictable copra economy for the higher living standards, greater access to services, and the security of urban-wage employment. Access to this world is through education – particularly high school education. This explains the enormous community interest in schooling and the quality of teachers evident in many rural areas. It also explains the willingness of families to send their children to kin in Tarawa to attend primary schools in the belief that the quality of teaching there is better. Many informants during fieldwork in 1973 expressed the sentiment that the subsistence economy and cash earning opportunities available on Tamana could not provide for future wants, particularly because of increasing populations levels, and that migration to the urban center was inevitable. At that time the fact that the whole urban-wage system was, and still is, supported by an aid-dependent bureaucracy did not greatly concern them. It was simply part of the same mysterious external capitalist system that also brought the whalers, traders, and missionaries to their shores. The paternalism of the past had generated an implicit faith in the government to manage their affairs for them and ensure an expanding base to urban employment.

Bertram and Watters (1985; 1986) argue that these trends evident in Kiribati are widespread and characterize a new class of economies in the South Pacific. They coined the term MIRAB to denote such economies where the evolution of the system shows the combined effects of migration, remittances, aid, and bureaucracy. Such economies characterize countries too small to attract normal investment in productive activities, but which for a variety of historical and political reasons, are the recipients of a good deal of external aid, inputs and skilled technical personnel, agricultural training, and extension facilities, and the provision of infrastructure, particularly in urban services, water supplies, sewage, etc., and transport and communications. Thus the process of socia and economic change in the island microstates has to be viewed in terms of local adjustments to external factors rather than endogenously driven development. The 'modernization' of island societies has produced increased 'dependence' despite the countries being endowed with the trappings of 'independence'. Remittances from kin and large scale flows of unrequited aid appear capable of sustaining such systems for a considerable time into the future despite the absence of a productive base. Real exchange rates are held up by large rent inflows which further discourage the expansion of tradable-goods production. The limited size and restricted resource bases of the microstates give them a low capital absorptive capacity. Investment opportunities are few and far between. Little scope exists for utilizing aid flows for productive purposes. Aid gets spent on infrastructure, usually urban infrastructure in the 'colonial-bureaucratic town.'

Such characteristics are consistent with the processes described in the paragraphs above. With the cessation of phosphate mining in Ocean Island in 1979, the economy changed dramatically from a colonial export economy to one dependent largely on remittances and aid flows. What distinguishes the Kiribati economy today from other MIRAB economies is that Kiribati has no ready access to metropolitan centers for overseas migration. Migration to Nauru continues at reduced levels and international shipping provides the major source of overseas employment for young male migrants. Most migration is internal to urban centers and employment in the bureaucracy is largely aid funded. New policies introduced after independence in 1979 seek to redress this situation. As yet their impact on rural areas remains uncertain.

INDEPENDENCE AND THE FUTURE

In contrast to the rapid expansion of services and investment in infrastructure that characterized the twenty years leading up to independence, the main themes promoted in national development plans since independence have been self-sustaining development, containing aspirations and tapping available resources to make Kiribati more viable in the short run and to give substance to political independence.

Development of New Resources

Progress to date in generating new income sources to fill the vacuum left by the loss of phosphate earnings has been hampered by the fact that the limited atoll environment provides few opportunities. Development options that would raise rural dwellers' incomes directly are even harder to find. By default copra became the major source of locally generated export income but prices remain low, fluctuating, and unreliable. The national fishing company, Te Mautari Ltd., set up to exploit the country's tuna fish resources, showed early promise but has experienced difficulties in recent years. Because so few prospects for exports exist Kiribati can not benefit greatly from the removal of tariff restrictions under such regional agreements as SPARTECA, of which it is a member.

Other revenue-generating possibilities include fishing licenses to its exclusive economic zone, tourism, solar production of salt, the leasing of uninhabited islands for space exploration facilities, and the recommencing of phosphate mining on Banaba (Ocean Island). Apart from fishing licenses most of these remain prospects for the future. All of them may generate revenue for the government but will have less direct impact on life in rural areas.

Policies for the Rural Areas

As far as policies for the rural areas are concerned, the main thrust of the *Sixth National Development Plan 1987 91* (1988) was to improve living standards in the rural areas without reducing the viability of the subsistence sector; to encourage self-help; to improve the efficiency and coordination of local and central government inputs into all stages of rural project planning and execution; to provide for the devolution of more powers to local councils, and to increase the allocation of development resources to rural areas in the hope that these factors in combination will stem rural-urban population drift. Resettlement of rural population in the Northern Line Islands was also considered.[3] Permanent emigration overseas was not, although the government recognized that this option may have to be reconsidered if population growth trends continue unchecked. The emphasis on all matters relating to the rural sector was caution; the plan should take no action that would accelerate monetization in the rural areas. The plan states its preference for more gradual rates of change without threatening the support of the subsistence sector (p. 111). All new policies, whether local or national, were to be reviewed for their effects on growth and the distribution of population (p. 121). To date the most tangible results of this policy have been expenditure on causeway construction, water supply and sanitation projects, reef blasting, improvement to council facilities, health services, and community development projects. Local councils have received more help and guidance in preparing and programming island development plans.

Despite an avowed concern to redress the imbalance in development expenditure between the outer islands and Tarawa, Tarawa still absorbs an exceedingly large proportion of aid project expenditure, although the situation has improved somewhat from the immediately pre-independence days. Figures provided by the Finance Office show that in 1988 anticipated projects on Tarawa would absorb 31.6 percent of the A$10.3 million already promised by donors. National projects (which include most of the technical assistance) accounted for 58.2 percent, while the outer islands, including the Line and Phoenix Islands, were to receive only 10.2 percent. Because of the lumpiness of aid disbursement caused by a small number of very large projects being initiated in particular areas at particular times, these percentages vary greatly from year to year. In 1986 Tarawa's share was 89.2 percent, and in 1985, 67.3 percent. However, the share of the rural areas has continued to hover around the 10 percent level. In 1989 it is hoped to bring the outer islands' share up to 26.3 percent, with the Line and Phoenix getting 9.3 percent of the A$18.2 million promised. It should be noted that some projects located on Tarawa do have benefits for other areas, e.g., the Marine Training School, Agriculture Department research projects, and the upgrading of airport facilities at Tarawa.

Policies Affecting Tarawa

At national level the government embarked on a program of tight budgetary restraint. Its aim was first to preserve or enhance the real value of the Revenue Equalization Reserve Fund and utilize its revenue to sustain recurrent expenditure and promote development. The fund had been established in 1956 with money from the sale of assets from the Japanese occupation and from the colony's general balance. Phosphate taxes were paid into the fund in preparation for the time when phosphate incomes would cease and other income sources would be needed to fund any budget deficit. In 1988 the fund stood at around A$170 million and contributed A$8 million to the recurrent budget. Interest from the fund amounts to about one-third of all government revenues. The government also showed a keenness to move toward independence from budgetary aid support. This aid from Britain was reduced progressively after independence and was eliminated in 1986; earlier than intended and at the request of the Kiribati government. As a result expansion in government employment has ceased, staffing levels have been cut back over the last few years, and a tight rein has been kept on wages policy. The public service reached a peak in 1983 when it stood at 2,492. By 1985 it had declined to 2,299. Government expenditure on the public service between 1980 and 1984 remained fairly stable. Given an inflation rate of around 7 percent this represents a considerable decrease in real expenditure.

These policies appear to have had considerable impact on the growth of Tarawa. The departure of many Tuvaluan civil servants and their families after separation in 1978 also affected changes. In 1985 Tarawa's population stood at 21,393 and the administrative center now contained 33 percent of the country's population. While Tarawa continues to grow, the rate at which it is growing shows something of a decline. Between 1968 and 1973 Tarawa grew annually by 849. The total colony increase was only 859 so Tarawa was absorbing most of the natural increase of the whole colony. Between 1978 and 1985 Tarawa increased by 496 per year while the Kiribati population grew annually by 1,095. The decline in growth rates must be in part at least attributable to the decline in employment creation.

Since Tamana expectations in 1973 were so clearly focused on urban employment and Tarawa has remained the only major employment center, how then do these policy changes and their implications affect life in the rural areas? Preliminary impressions gained on a 12-week visit to Tamana from June to September 1989 resurveying households studied in 1971–1973 give some insights. Kin of these households resident on Tarawa were also interviewed.

New Linkages with Tarawa

First impressions of life on Tamana in 1989 compared with 1973 were how little had changed. Workers returning permanently from Ocean Island had constructed permanent-materials houses and had more disposable wealth but still lived much the same sort of life as anyone else. There were now a few trucks and more motor bikes and hand-carts than previously. Water was now drawn from wells using 'Tamana handpumps' (installed as part of the UNDP Integrated Atoll Development Project) rather than by bucket. Fifteen years had brought no new cash-earning opportunities and copra remained the major access to cash income on the island. Activity patterns remained basically unchanged. Subsistence production seems to have maintained its importance. In fact, informants claimed that the rising prices of rice and flour had led to an upsurge in *babai* cultivation and consumption. Enquiries about remittances met with polite laughs . . . 'no one sends telemos these days'. Nonetheless, most households spent more at the store than they received from copra and other traceable sources of income, suggesting that remittances and gifts still make up a significant proportion of household income. Kin in employment on Tarawa or as seamen on overseas ships are the main source of these remittances.

Expectations of employment in Tarawa have been tempered by recent changes. Most recognize that employment prospects there are limited and unlikely to expand in the future. A surprising number of younger people expressed satisfaction with rural life and no real desire to leave their home island. However, the store placed on education as a means of access to employment remains undiminished – even to the extent that parents, school staff, and the school committee devised a policy of having all form five students (the year in which selection for high school takes place) live in the school compound during the week. Electric lighting, private study, and extra tuition by the staff were aimed at improving scholastic performance and the childrens' chances of being selected for high school.

The decision to move from Tamana to Tarawa is still a constrained one depending on one member of the family getting employment and subsidized housing on Tarawa. Once attained, other members of the family are likely to make the move, for schooling, to help in the urban household and to seek work. Kin from several of the sample households had adopted this strategy and by establishing linkages on Tarawa through marriage and friendship had got access to land and built their own houses on land on the fringes of the urban center. Few had been successful in obtaining jobs although several established their own fishing ventures and others set up businesses selling cooked food to government employees at lunchtimes. Since most of the households contacted had moved since the early 1970s or more recently, it is not easy to evaluate the effect of time on linkages with kin in the rural areas. All claim to be still making regular remittances to Tamana. However, most admit that low wages and high living costs are making it difficult to maintain the real value of remittances sent. Many remit the same monetary amount as they have in the past and this must reduce effective income in the rural areas. The respondents admitted having to refuse requests from more distant kin but said they always wrote to the relatives concerned to explain their predicament.

In this way the flow of resources to the rural areas may be diminished in effectiveness but every attempt is made to maintain kin linkages to keep open the option of returning to rural life. In fact, the impact of government policies is having the effect of generating reverse flows of resources from the rural areas. Low wages and inflation are making it difficult for many households to make ends meet. Many Tamana households on Tarawa are now actively recruiting young men to come to Tarawa to fish and cut toddy for them in return for keep and a little spending money. The move supplements the urban household's food supply, ensures a surplus of income to be remitted to the rural kin, and gives the young men opportunities to experience new lifestyles. The move is not permanent. Many return after one or two years and the kin on Tamana select a new individual to go. A few respondents attempted to find work in Tarawa but none were successful or made sustained attempts to find work.

Tamana people have also attempted to increase the options available to them by sending some kin to take part in the settling of Tabuaeran and Teraina in the Northern Line Islands, while others have purchased lands on Kuria, Aranuka and other islands, thereby augmenting the resource base and providing for future generations.

Conclusion

The preceding discussion has attempted to identify the influences which shaped traditional I-Kiribati society and to trace the way in which that society was transformed by its contacts. The process of incorporation generated new flows of goods, labor, and services. It resulted in the independent, self-sufficient man/environment systems of pre-contact society becoming dependent entities on the very fringe of the capitalist world system where man/man rather than man/land relationships become increasingly important in shaping the character of society. The process of incorporation has a universality which encourages generalization. However, this process takes place in a space and time context which must be recognized in order to interpret particular societies at different points in time.

In Kiribati the first contacts with the capitalist world affected only those islands close to the main whaling grounds. Contact was intermittent. The production of pigs and chickens for exchange was short-lived because the market disappeared with the decline of whaling.

The switch to coconut oil caused the focus of economic activity to shift to the northern islands. The trade in coconut oil and copra assured Kiribati of a place in the deprived periphery of the capitalist world. However, the transformation of subsistence agricultural systems, as describd for other parts of the world, did not materialize. 'Modern' innovations were accepted into a system whose essential variables were not transformed. No new commercial crop displaced traditional food crops. Subsistence production remained viable and vital. Here it is impossible to ignore the reality of the atoll environment; its difficult physical conditions, restricted land areas, small volume of product, remoteness from markets and absence of alternative crops. The significance of the 'coconut ecology' cannot be underestimated; its importance in the evolved 'natural' vegetation, its usefulness in both subsistence and commercial production, its long-term productivity and its unresponsiveness to labor inputs in the short-term. Colonial policy also intervened to ensure that the traditional landholding systems persisted relatively unmodified. No plantation sector was established and hence proletarianization of the workforce and migration from the rural areas did not occur. Copra production on most islands has been and remains today a somewhat desultory activity. Despite the lack of alternative income sources on the islands copra production is a relatively unimportant source of income to rural households.

To try to argue that this state of incomplete incorporation represents either (a) a temporary hiatus in continuing process of change (where some new influence is needed to overcome failure of the response factor which when overcome would lead to further economic growth and social change) or (b) that it is a state of terminal development (where market participation is limited by limited resources of the inability of productive systems to achieve living standards aspired to) tends to ignore the fact that the rural economy is linked to the outside world in other ways as well as commercial agriculture. These linkages provide alternative options for achieving household goals. Individual households may consciously adopt strategies which enable them to capitalize on a number of the options available to them.

In this way a rural household may bridge the village economy, with its subsistence and commercial agricultural base and the 'modern mining/urban sector. The village cash economy thus becomes a binary one composed of two distinct elements – locally generated income from copra handicrafts etc. augmented by government subsidy payments to rural programs and the flows of remittances and capital goods generated by employment off the island. For much of the recent past these two elements have been manifestly unequal in their ability to provide income and access to capital goods. The fact that both persist through time can only be understood in terms of conscious policy decisions.

The penetration of capitalism in Kiribati also gave rise to investment in the phosphate industry. Mining phosphate generated employment opportunities and migration flows from rural areas. The fact that the system of circular migration persisted for so long and did so little to transform the essential character of rural society can be explained only in terms of policy intervention. Tight control of recruiting, living conditions, and repatriation ensured that migration was exclusively circular. No independent urban center encouraged investment in a wider range of services and further economic growth. The level of income and capital goods remitted to the rural areas was so great in comparison with what could be generated by commercial production there, that this tended to dampen aspirations and ensure that wage labor migration became an essential part of rural life.

The loss of employment with the end of phosphate mining in 1979 would have had a much more dramatic effect on rural lifestyles had Tarawa not emerged as a major employment center in the interim and if employment on overseas ships had not been promoted by the establishment of the Marine Training School. Because of these developments the rural households continue to straddle the village and 'modern' sector but now the modern sector owes its presence not to investment in production but rather to flows of aid and rent incomes which were used to develop and now sustain welfare services and the infrastructure for independence.

Again, the emergence of Tarawa and the employment it generates represent a significant break in the process of change. It still reflects linkages with an external capitalist world, but this world has also changed considerably, particularly in its attitudes toward an appropriate role for government in the economy, towards independence and the provision of aid for colonies, former colonies, and less advantaged members of the world community. The emergence and continued existence of the urban center of Tarawa can only be understood as a response to these flows. The migration from the rural areas is the response to the inequality and opportunities created and the failure of the government to institute any effective policies to control the changes. The outcome of these changes was the emergence of the MIRAB economy. In the context of this and limited options for productive investment, the structure is likely to have some permanence. The economy's viability will be determined by the aid policies of the donor countries and the success of the government in pursuing its policies of maintaining limited self-reliance through the maintenance of the subsistence economy. The rural dwellers are less sure, now, of their role in the new order of things and seem to be keeping as many options open as possible: ensuring the viability of productive resources on the island, supporting education as a means of access to urban employment, diverting surplus labor to the urban economy to supplement urban wages with subsistence production, and finally, resettlement.

POLYNESIA

10

Schooling and Transformations in Samoan Social Status in Hawai'i

Robert W. Franco

Introduction

In the traditional rural context of the Samoan way of life (*fa'a Samoa*) social status was determined by a complex consideration of both descent group (*'aiga*) membership, and personal service (*tautua*) to the *'aiga*, and its titled chiefs (*matai*). Untitled Samoan men (*taulele'a*) demonstrated their *tautua* through productive work in cultivation, carpentry, fishing, cooking, and tattooing, while Samoan women served their chiefs and families through childcare, household and village maintenance, hosting the creation of intricately woven fine mats (*'ie toga*). The skills required to effectively complete these status-enhancing forms of service were learned from other kin group members, or from peers, by observing and modeling the skill behavior.

Since World War II, the *fa'a Samoa* has been heavily influenced by rapid population growth, rural-urban migration, accelerating rates of overseas movement, and limited local economic development. Because of these changes service to one's family and chiefs can now be demonstrated through successful participation in urban-wage employment, coupled with the remittance of sizable sums of money to the home village. The skills required to successfully participate in urban employment are learned in classrooms and school systems following two different Western educational models, one from New Zealand and one from the United States.

In this chapter I intend to show that the New Zealand schooling model followed in the Western Samoan islands – Upolu, Savaii, Manono, and Apolima (now Western Samoa) – from 1920 to the present has been selective and elitist, while the American schooling model followed in the eastern Samoan islands – Tutuila, Ofu, Olosega, and Ta'u (now American Samoa) – from 1900 to the present, has been seriously disrupted by inconsistent policy formation. I further intend to show that the elitism and policy inconsistency characterizing these two educational systems has had the same negative result for Samoans in Hawaii. That is, a relatively small proportion of Western and American Samoan men and women have the skills required to successfully participate in urban-wage employment, and thereby enhance their social status within the *'aiga* by remitting money home. Finally, in the Hawaii context, I will argue that compared to Samoan men, Samoan women, because of their very low levels of educational achievement, are more severely constrained in their attempts to enhance their social status through urban-wage employment.

Methods

This chapter relies on ethnographic research conducted in Western Samoa and American Samoa in 1975 and 1979, and applied research conducted in Samoa community college as part of the University of Hawaii's Teacher Training Assistance Program, where I had daily contact with many local teachers and administrators from both American and Western Samoa. Many of their insights are reflected in this paper.

The chapter also exploits the ethnographic accounts of traditional Samoa and the literature on Samoan education. Also incorporated into this paper are educational data from the 1976 Western Samoa census, and the 1980 American Samoa and Hawaii censuses. These data were analyzed at the East-West Center Population Institute and are interpreted from my reading of Samoan ethnography, and from my ongoing research in Hawaii's Samoan community.

SERVICE AND STATUS IN SAMOA: PRE–1900

In Samoan culture, previous to 1900 and sustained Western colonial administration, social status was reckoned in relation to family titleholders known as *matai* and, further, the hierarchy of *matai* titles. These *matai*, of varying status depending on the origin and history of their titles, had *pule*, the authority to determine land, labor, and resource allocations, settle disputes, and arrange marriages. There were two distinct types of *matai*, the high chiefs or *alii* who were generally of higher status because of their perceived divineness, and the talking chiefs or *tulafale* who 'dirtied their hands' with social, political, and economic debate, and were thus extremely skillful orators.

Most untitled men (*taulele'a*) believed it was possible to achieve *matai* status through hard work and service to their various *matai* and their *'aiga*. A central Samoan proverb, *'O le ala ile pule le tautua'* ('The way to attain authority is through service'), expresses well the Samoan conceptualization of how upward social mobility is attained.

The *taulele'a* within the village were organized into a cooperative work group called the *'aumaga*, and it was within the *'aumaga* that

304

young men served their families and their chiefs in cultivation, fishing, cooking, and carpentry. Cultivation work centered on taro production in the main taro beds (ma'umaga) on family reserve lands. In addition, the taloloa, a group taro planting, was conducted by the 'aumaga, on village lands three to five miles from the residential village. Taro was the most plentiful and consistent root crop; successful yam cultivation was more difficult due to the sensitive nature of the planting materials. Because of the prevarious nature of yam planting, plentiful yam harvests demonstrated the special agricultural skills of the village 'aumaga. Tree crops, breadfruit, coconut, and banana were located in closer proximity to the residential village, and although the planting of tree crops represented major 'aumaga labor inputs, the harvesting (picking) of tree crops was much easier than root crop harvesting. In addition, the multi-annual yields from tree crops resulted in significant net energy surpluses for the 'aumaga, the 'aiga, and the matai.

Successful fishing enterprises were conducted by the 'aumaga under the supervision of the specialist fisherman (tautai). These enterprises ranged from communal net-fishing to bonito angling to major shark fishing expeditions lasting days and even weeks. Bonito and shark fishing brought major recognition to the tautai and were an opportunity for young men to demonstrate special skills.

Cooking was a daily activity for individual men, but the 'aumaga prepared the food for major events such as hosting visiting groups (malaga), chiefly installations, marriages, and funerals. These events were always accompanied by a kava ceremony where individual young men (usually tattooed and called sogaimiti) were given the honor of serving the assembled chiefs. After the kava ceremony, the food produced and cooked by the 'aumaga was displayed and served.

Finally, within the 'aumaga a young man might also work as a laborer in canoe or house construction, supervised by a specialist canoe-builder (tufuga fai va'a) or house-builder (tufuga fai fale). These carpentry specialists were organized into 'guilds' where a young man might attain a highly respected matai tufuga status.

For Samoan women matai status was possible, but apparently women rarely attained this status. Perhaps the most famous of all Samoa chiefs was Salamasina, a woman who held all four paramount titles. Salamasina's chiefly status seems to provide a charter for female matai-ship, but historically there are few references to female Samoa chiefs. Nevertheless, Samoa women, through their position in the faletamafafine ('house' or 'side' of sister's children) have always had a strong voice in the selection of matai titleholders. Further, women's productive service in making fine mats ('ie toga) often resulted in the elevation of 'aiga and matai when these mats were presented at politically strategic marriages.

The unmarried women of the village worked together in a group called the 'aualuma, the female counterpart of the 'aumaga. The 'aualuma took primary responsibility in household and village maintenance. Further, they assisted with near-village cultivation, and nearshore reef exploitation. Because their primary forms of service were concentrated in or near the village, women took a greater role in caring for, and supervizing children.

In terms of female status enhancement, the manufacture of fine mats was the quintessential women's work. The completed fine mats demonstrated a woman's work skill, her persistence in a tedious work endeavor, and her ability to work coopertively with other village women. Each village might have its own weaving house (fale lalaga) where the young girls learned the art of weaving fine mats from older women. Shore (1982, p. 105) discusses the authority structure within the weaving house:

> The weaving house possesses its own rules, authority structure, and a power to impose fines upon offenders. Authority within the house lies with two offices; matua u'u, ideally the wife of a ranking alii, is the resident weaving expert and the teacher of the younger women. Clearly here criteria of skill and experience outweigh those of specific political status in the selection of an incumbent to the office. The latu o faiva, ideally the wife of an important orator, is a ceremonial position with responsibilities for the distribution of food during common meals. The latu o faiva also organizes the ceremonies marking the completion of a set of mats by the weaving house.

Fine mats were made throughout Samoa but certain villages, families, and individual women were more renowned than others. An expert older woman might be called upon to assist in the completion or repair of a fine mat, and when so employed had a pig killed for her and was well fed. The plaiting house was occupied by expert women workers, 'who met to plait fine mats on the invitation of the high chief who kept them in food and made appropriate presents' (Buck 1930, p. 320). In addition to the cooperative work of expert women, individual women worked in producing fine mats as their time allowed.

Fine mats were worn as clothing on ceremonial occasions and marked individual status. Fine mats were, however, most valuable in the interfamily exchanges affirming and reaffirming marriage alliances. Fine mats were the most significant item in the bride's dowry (toga), which also included tapa cloth, coconut oil, and sleeping and house mats. At the marriage ceremony these gifts were presented to the talking chief of the groom's family who supplied 'oloa – food, including pigs, for the wedding feast, weapons, tools, canoes, houses, and ornaments.

An individual fine mat possessed great value depending on the time inputs of manufacture (determined by the width of the weave and the size of the mat), the age and sheen of the fine mat (older mats had a brighter sheen, making them more aesthetically pleasing), the historical events associated with its previous presentations, and the quality of red feather ornamentation. More valuable fine mats certainly played a major role in political maneuver, both within Samoa, and in relations between Samoa and Tonga. According to the Reverend John Williams in his *Journals* of 1830–1832:

> Samoans make a small kind of mat which they weave with a remarkably fine thread from a species of palm leaf. These mats are much sought after by the Tongataboons, who came from Tonga in their canoes to purchase them a distance of six or seven hundred miles as an article of dress for the Tonga chiefs. Seven large canoes had visited the Samoas just before our arrival to purchase the above articles. Four had returned again and the other three were waiting for a favorable wind to return. (Williams, 1984, p. 82)

Within Samoa, incentives for making fine mats were clearly political as well as personal. Hjarno (1979) has emphasized the role of fine mats in the hypergamous marrying strategies of lower ranking chiefs. Talking chiefs seem to have been always urging the production of fine mats, and manipulating high status marriages. Buck (1930, p. 318) states,

> They themselves (Samoans) admit that a talking chief will enter into negotiations for the marriage of his chief influenced solely by the bundles of fine mats he has seen hanging in the guest house of the girl's father.

Mead (1930, p. 77) has emphasized another feature of Samoan political maneuver associated with fine mat flows. She describes the strategies of talking chiefs and high chiefs and stresses that the talking chief 'waxes rich' in fine mats while the superior high chief becomes poor. She concludes:

> It is only the chief with a very large and skilled family, or the chief who is a master craftsman, or fisherman, who is not forever fighting a losing battle with his rapacious subordinate. The talking chief, weary of squeezing dry improverished lords, turns heartily to the nouveau riche.

The nouveau riche were the skilled specialists (*tufuga*) who could attract fine mats for their carpentry or tattooing, or the warrior chiefs who could wrestle fine mat wealth and political power away from those previously in power.

Fine mat presentation also figured prominently at funerals, and especially at the funerals of chiefs (*lagi*). Blood relations of the deceased collected fine mats, and among these a mat of great age or historical value would be chosen as the 'mat which gives distinction to the property' (Buck 1939, p. 317). The presentation of this valuable mat brought special distinction to the family; conversely, without this mat the status of the family suffered. The most valuable mat stayed within the village of the deceased chief, and was usually presented to the talking chief presiding over the *lagi* ceremony. From the remaining fine mats, the deceased chief's sister's son received the first pick. Additionally, talking chiefs, if they could demonstrate a blood relationship to the deceased chief in the *lagi* ritual, also received fine mats, and usually in accordance with their status. Funerals were crucial events where political activity leading to the naming of a new chief was initiated. The *lagi* fine mat presentations sent a clear signal to all *'aiga* members, titled and untitled, indicating who would have a strong or weak voice in selecting the new titleholder.

Fine mats were also the currency of reconciliation and peace making. When a culprit begged for pardon for a major offense, he presented himself in humiliation (*ifoga*), wrapped in a fine mat to the offended chief. Special fine mats called *'Ie o le Malo* (fine mats of victory), were presented by defeated armies to the victorious side, as tokens of complete surrender.

To conclude, in the process of making fine mats women were creating the most valuable currency in Samoan social, economic, and political maneuver. The making of *'ie toga* was thus a crucial form of service to the *'aiga* and the *matai*, and this service, combined with her closely associated roles as *faletamafafine* member, wife and mother, served to define and enhance a Samoa woman's social status.

After 1900, and the sustained colonial administrative presence of Germany and then New Zealand in Western Samoa, and the United States in American Samoa, came the introduction of government-supported schooling. By 1945, educational trends were established in both Western and American Samoa. Since World War II, Samoan culture has been heavily influenced by major demographic and economic changes, and the two school systems have attempted to respond to these changes. In the sections that follow, after a brief discussion of mission and village schooling, I will examine the development of government supported school systems in Western and American Samoa in terms of their ability to prepare young Samoan men and women for the transformations in social status they are experiencing today.

MISSION AND VILLAGE SCHOOLING: 1836–1900

The first large party of missionaries reached Samoa in 1836, and by the 1840s mission schools were established and using a curriculum consisting primarily of 'Bible reading, writing in the vernacular and simple arithmetic' (Sanchez 1955, p. 78). From 1840 to the turn of the century the Congregational (London Missionary Society) and Catholic (Marist) missions developed primary education, while in the villages 'Samoanization of the Christian Church' resulted in the creation of the *faifeau* (minister's) schools (Baldauf 1981, p. 232). While the mission schools were established 'on the assumption that change is desirable,' the village *faifeau* schools served an important function in Samoan culture by 'providing the channel whereby Christian truths and principles accepted by the Samoans as their own were transmitted to the young' (Sanchez 1955, p. 120). The *faifeau* schools provided many Samoan children with a conservative, Christian, Samoan linguistic and cultural element in their education (Baldauf 1981, p. 232).

In 1845, the London Missionary Society (LMS) founded the theological seminary at Malua on Upolu, and education at Malua created an entirely new route to enhanced social status. According to Davidson:

> Combining general education with theological training, it produced teachers and (after the right to ordination had been conceded) pastors who possessed both a training in Christian doctrine and pastoral duties and a social standing partly derived from it. Many of these men gave life-long service to the church, not only in Samoa but also in other parts of the Pacific in which the L.M.S. was working. In the villages they came to possess an influence that the (chiefly) *fono* did not choose to ignore. (1967, p. 37)

Over the course of the second half of the nineteenth century, as the *faifeau* and mission school system evolved, the pastors took increasing pride in their own superior education, and, as Davidson points out:

> . . . many of them made sure that their children, in turn, should share their advantages. From pastor's families came not only future pastors and school teachers, but government clerks and secretaries, and the wives of many important chiefs. The church added an aristocracy of education to the Samoan social structure. (1967, p. 37)

309

SCHOOLING IN WESTERN SAMOA

From 1900 to 1914, the German administration in Western Samoa concentrated its efforts on establishing political control and expanding plantation agriculture. Little seems to have been done to develop a government-supported educational system, and the mission and village scchools maintained their primary role in Samoan education.

From 1920 to 1944, New Zealand administrated Western Samoa as a League of Nations Mandated Territory. During this period New Zealand was committed to a system of 'dual control and development of education with the mission,' and was content to give the mission 'major responsibility for educational expansion' (Barrington 1973, pp. 256–257). In 1945, an official education commission, headed by Dr C. E. Beedy, New Zealand Director-General of Education, described the Western Samoan educational system as 'beyond words muddled and messy,' and drew attention to

> an extreme shortage of schools, severe overcrowding in classrooms, teacher's lack of education, and an almost complete absence of facilities for postprimary and technical education. (Barrington 1973, p. 256)

In December 1945, the New Zealand Cabinet placed Western Samoa under the United Nations International Trusteeship System, and a Trusteeship Council Mission was sent to Western Samoa in 1947. According to Barrington (1973, p. 256), the Mission declared that a 'drastic attack' would have to be made on the existing school system if Western Samoa was to achieve self-government within a reasonable time. The Mission praised the fact that few non-self-governing territories had achieved such a high level of literacy in the vernacular. But it also drew attention bluntly to the fact that the Territory was 'almost completely without postprimary schools'. The Mission also commented on insistent Samoan demands for more secondary schools to permit their most able primary school leavers to qualify for employment in government departments and the private sector. It agreed with the findings of the 1945 New Zealand education commission that compulsory education would have to be delayed until teacher training could be expanded and improved. Finally, it recommended that greater cooperation should be established between the Christian missions and the administration in the preparation of textbooks, setting of examinations, and teacher training, which, it said, 'lay at the very heart of the problem of Samoan education' (Barrington 1973, p. 257).

From 1947 into the 1950s the school system was comprised of separate expatriate and Samoan schools. Within the Samoan schools a limited number of students from chiefly families received English language instruction, and some of these students were able to take advantage of overseas educational opportunities (Kessing 1956, p. 217). In the mid–1950s the educational goals for the Territory were to maintain cultural integrity, while attempting to direct Western Samoa toward eventual self-government. In essence this meant that children from chiefly families would receive the education they needed to maintain political power when independence arrived.

By the late 1950s, 'acceleration schools' were developed to educate the brightest people. An increasing number of government schools, including Samoa College, were opening, and a number of scholarships were given for the best students, male and female, to study in New Zealand. As Western Samoa approached independence, it was becoming increasingly clear however that not just the best students were longing for opportunities in New Zealand. According to Dr. Beebe, in his 1954 *Survey of Education in Western Samoa:*

> I have the strong impression that the need for education is greatest among the young men of the village . . . It is the young men who first feel the pull of Aipa, and beyond it, of New Zealand. The reason is not just the bright lights . . . but more deeply the desire for personal freedom which the Samoan Communal System denies.

Before major emigration flows to New Zealand developed it was anticipated that an educated younger elite would return from their overseas experience to direct the movement to independence and change the 'Samoan Communal System'. However, when the highly educated Western Samoan returned home he was soon offered a chiefly title, and through his coalescence he became, more often than not, an advocate of traditional Samoan customs. For example, one 'top elite leader', who had just returned from years of education in the United States remarked on the *fa'a Samoa:*

> There is a good distribution of responsibilities: the hardest work is for the young men in the prime of life; the leadership is for the mature who should no longer do hard physical work; there is only gentle work for the women; and for all, work is often made into pleasure by working together. And for the children, and the old people there is complete security. (Keesing and Keesing 1956, p. 220)

As Western Samoa moved toward independence, well-educated chiefs remained strong proponents of the *fa'a Samoa*, while less-educated Samoan men were leaving the 'hardest work' of the village. Along with these Samoan men, less-educated Samoan women were

311

leaving the 'gentle work' of the village for the 'bright lights' and 'personal freedom' of Apia and beyond.

Since independence in 1962, the educational system in Western Samoa has continued to be highly selective. Further, high levels of educational attainment, coupled with government service, have become an important form of service to the family, and thus a major route to authority as a chief. However, while the new government offered new employment and career opportunities for some, the vast majority of Samoans were not provided with the educational skills to participate in government or other forms of employment in Apia and beyond in 1963 and 1964, and the American UNESCO adviser, Dr Chester Williams, was critical of the New Zealand approach to educational development in Western Samoa. In September, 1963, he stated:

> The present state of education in relation to population growth, and the growing needs of the country for trained manpower and for development generally – economic, political, social – can only be described as critical . . . only 12 percent of the school age population was receiving education to the 5th and 6th grade (Form 1 and 2), and only 3 percent was receiving secondary education. (*Pacific Islands Monthly* 1963, p. 18)

In August, 1964, Williams' criticism was even more severe:

> New Zealand is paying out for education and it seems New Zealand wants to exercise control over it . . . The *matai* who pass Standard Four are the really important people, not the few who pass School Certificate . . . We have to get away from the philosophy of education for the elite to education for the masses . . . This place is both a paradise and a hell. The people have more leisure than almost anyplace in the world. But what do they do with it? They are often mere vegetables. Until this enormous waste of human potential is utilized – until the creative ability of the people is given expression, education is sterile . . . New Zealand aid will have to be increased and provided for far longer than is apparently expected if education is to be substantially improved. (*Pacific Islands Monthly* 1964, p. 45)

Data from Western Samoan government documents also pointed to educational problems during the 1960s. In 1967 only one child in every two who passed Standard Four was able to enter an intermediate school, and even fewer gained access to secondary school. In 1970, only 26.9 percent of all Samoans from 15 to 19 years of age were enrolled in school (Western Samoa's Second Development Plan, 1975,

p. 62). The 1973 Western Samoan Department of Education Annual Report stated:

> In theory, every child of school age is to attend the primary school in his own village, but many children remain outside schools because their parents had not been converted to the concept that formal schooling, at least at primary level, is a 'human right', or because there is not enough money to register their births, pay school fees, contribute to demands from school committees, or pay for labour at home or in the family plantation. (p. 1)

Through the mid-1970s, the Western Samoan educational system maintained a highly elitist profile. According to the 1976 Western Samoan census, less than 40 percent of Western Samoans 20 to 24 years of age, less than 25 percent of Western Samoans 30 to 34 years of age, and just 11 percent of Western Samoans 40 to 44 years of age, had attained a Form 3 level (approximately grade 7–8) education (Western Samoa Census 1976, Table 27.1, p. 161). While these data support an elitist interpretation, they also point to recent changes and increased access to higher education for a larger proportion of younger Samoans.

In an effort to increase educational opportunities for a larger number of Western Samoan youth, a major structural change within the educational system occurred in 1975. The number of school placed in Forms 3 through 6 was doubled and this enabled over 70 percent of Western Samoan children to progress to secondary school.

The 1981 Western Samoan Census noted 'substantial progress made in raising enrollment at the primary and secondary levels since 1971' (Western Samoa's Fifth Development Plan 1984, p. 100). In 1971, 7,507 individuals were attending higher secondary education, while in 1981, 13,571 persons were achieving these higher educational levels. 'The actual proportion of the 15–19 age group attending school for 1971, 1976, and 1981 is 43.8, 53.9, and 64.9 percent respectively'. Today, although education is niether 'free nor compulsory, most children attend school up to Form 2 level at the insistence of the village authorities' (Western Samoa's Fifth Development Plan 1984, pp. 100–101).

In sum, the Western Samoan educational system has historically been selective and elitist, while in recent years concerted attempts have been made at the national and village level to expand schooling opportunities to a greater proportion of the nation's youth. The social status outcomes of this educational history will be examined after a discussion of the history of schooling in American Samoa.

SCHOOLING IN AMERICAN SAMOA

Since 1900 pastor's schools and government schools have provided a unique educational experience for American Samoans. The pastor's schools attract Samoan children from the ages of four through middle adolescence, and these schools have continued to 'produce faithful Christians who could read the Samoan version of the Bible, sing hymns, and do some basic arithmetic, geography, and history' (Thomas 1981, p. 42).

While the pastor's schools have generally maintained their scope and direction, government schooling has been adversely affected by the lack of well-directed and consistent policy. Inconsistent policy has resulted in 'seventy-five years of starts and stops, new policies tending to follow each governor, and not infrequently completely contradictory objectives being sought simultaneously' (Beauchamp 1975, p. 24).

From 1900 to 1932, the primary purpose of the government schools was 'Americanization,' that is, to provide English language instruction, and some formal academic book learning. In 1932, a team of American educators, commissioned to assess the American Samoan public schools, concluded that

> It is recognized that much in Samoa ways and life is good in itself and admirably adapted to the people of these islands, but that American Samoa is undergoing change, specifically through the influence of Western civilization. In view of this changing condition, which is likely to go much further as time goes on, therefore, the objective of education is to conserve the best of Samoan culture and at the same time give acquaintance with the intellectual tools and social concepts and institutions of the west, to the end that Samoans may retain respect for their native heritage and skill in their traditional arts and crafts, and at the same time may learn to meet on modern world. (Wist 1933, p. 256).

In 1933, the United States Naval government in American Samoa took its first steps toward a bicultural educational policy by expanding its educational objectives to include the conservation of local customs, crafts, and culture. In the 1930s and 1940s, the bicultural objective 'was interpreted to mean that the social studies should include units of study drawing on Samoan life' (Thomas 1981, p. 42).

While bicultural objectives were working their way into the American Samoan curriculum, bilingual objectives were encountering more serious barriers to implementation. By the mid-1950s it was concluded that

> . . . the practice of translating into the Samoan language is costly of time, and of far more serious consequence a distinct handicap

to the learning of English. Therefore, no other language than English shall be used in the public schools. (Sanchez 1955, p. 141).

The requirement to use only English was an unreasonable goal because students spoke 'only Samoan outside of the classroom, and the teachers themselves had a very insecure grip on English usage' (Thomas 1981, p. 42). Still, by 1959, American Samoa's schools had 'drifted back to a stateside academic content and approach' (Baldauf 1981, p. 232).

In addition to an uneven application of bicultural and bilingual approaches, schooling in American Samoa suffered from 'a lack of funding and from continual staff turnover' (Baldauf 1981, p. 232). By 1960, the school system was characterized as a collection of

> . . . irrelevant academic practices, inexactly copied from lack-lustre traditionalist models on the mainland and grafted onto a listless U.S. paternalism which conveniently mistook large-scale attendance and dutiful departments for universal learning. In the little ill-equipped village schools, children group-memorized much but individually learned little that mattered. And their Samoan 'teachers' had come through the same crippling school experience themselves . . . Instructional content was largely derived from text- and reference-books which were topically out-moded and culturally outmoded and culturally inappropriate. Teaching methods were simplistic and rigid variants of the typical classroom mannerisms of most American schools. (Hall 1969, p. 18).

In 1961, Governor H. Rex Lee arrived in Tutuila totally unprepared for the educational problems he would confront. Less than half of the children under the age of 18 were attending school, and most of the teachers responsible for teaching courses in English were not fluent in it (Baldauf 1981, p. 234). Lee decided on 'an explosive upgrading' of the schooling system and he brought in some statesides experts to recommend how this upgrading might be achieved (Kaiser 1965, p. 58). These experts recommended increased funding to improve existing programs, or the replacement of Samoan teachers and staff with American educators. Lee recognized the need for increased funding, but felt strongly that replacing Samoan teachers with American teachers would destroy the morale of local teachers and be culturally disruptive. According to Baldauf (1981, p. 234):

> Ideally, he wanted to create a system which would both improve the standard of education for every child, using only a limited

number of teachers from the U.S., while at the same time maintaining and training the Samoan teachers in the classroom.

Between 1964 and 1968, Lee received large sums of money from the US Congress and consolidated the 'existing 50 ramshackle elementary schools into 24 modern ones' (Thomas 1981, p. 43). In addition, three high schools were constructed and numerous teachers and administrators were brought in from the US mainland. Finally, an expensive educational television system was developed to beam lessons into all elementary schools, and into selected secondary school classrooms. Thomas (1981, p. 43) discusses the educational objectives of the television education experiment:

> The new TV-oriented system was intended to equip the great majority of Samoan children to 'live and work and make a better life for themselves in Samoa'. In the process the schools were to 'increase competency in use of the English language to the point where language is no barrier to educational development'. Samoan youths who wanted to go beyond high school would gain 'a sound educational base that will enable them to enter American colleges and universities competitively on a level equal with other American students'. (Bronson 1968, pp. 7–8)

A complete assessment of the educational television experiment is provided by Schramm, Nelson, and Betham (1981). However, it should be emphasized here that a total restructuring away from 'grades' to 'levels' and 'cycles,' a continued high rate of contract staff turnover, and the non-enhancement of Samoan teacher roles, worked against the system meeting its ambitious goals.

In 1969, a new Governor, John M. Haydon, appointed Milton de Mello as Director of Education. Mello was not committed to television education, and after a complete review of the educational system, the gradual dismantling of the television system was initiated. In 1971–1972, the amount of television was reduced in some elementary school subjects, and nearly totally eliminated in the high schools. By 1974–1975, the television language arts curriculum was still being used at the elementary level, but classroom teachers were given much more latitude in their use of the televised curriculum. By 1979.

> . . . a school service that had once produced 6,000 programs a year, was producing only one series of programs, 40 minutes of oral English. By contrast, an out of service (entertainment TV) that had once offered only 28 hours a week on a single channel, was filling three channels everyday from late afternoon until late evening. The balance of education and entertainment had been reversed. (Schramm et al 1981, p. 150).

By 1980, television, which had entered Samoa villages as an instrument of elementary education, was 'an attractive pitchman for a broader cultural invasion' (Baldauf 1981, p. 242). Eventually, even the pastors had to succumb to popular pressure for Sunday evening television entertainment.

In 1973, Chief Nikolao Pula, who would become the first Samoan Director of Education in American Samoa, summed up the American educational direction for the territory:

Our educational system, contrary to the systems of Western Samoa, and other countries has been subjected to so many abrupt changes, so many beginnings all over again, that we have not been able to make as much progress as countries with less experience and less money. Perhaps, the most important thing of all would be continuity. (Schramm et al 1981, p. 21).

By 1974, thorough bicultural educational goals were restored to the forefront of American Samoan schooling policy. In *Think Children*, the American Samoa Department of Education (1974, p. 5) restated their objectives:

Inherent in this commitment is the obligation to preserve the cultural heritage and to foster the economic well-being of American Samoa, while at the same time, to prepare each individual for a personally satisfying and useful life wherever he chooses to live.

In American Samoa, between the mid-1960s and the mid-1970s, educational objectives changed from helping Samoan youth 'make a better life for themselves in Samoa', to preparing each individual for a 'personally satisfying and socially useful life wherever he chooses to live'. In this decade, as in Western Samoa, American Samoan educational officials were becoming increasingly aware that not only the 'educational elite' were interested in the bright lights and personal freedom that lie in Pago Pago and beyond. Thousands of American Samoans were opting for urban life in Hawaii, and California. By 1974 it was clear that 'almost 40 percent of Samoan high school graduates will migrate to Hawaii, or the mainland, in search of jobs or further education' (Nelson 1970, p. 16).

In 1974, Dr Mere Bethan became the second Samoan Director of Education, and held that position until 1985. She brought the school system through the difficult period after the 'Bold Experiment' of television education. Further, she emphasized that classes be taught in the Samoan language through the elementary grades. Dr Betham also implemented an extensive teacher training program with the University of Hawaii College of Education, which provided courses through the American Samoan Community College. The training enabled teachers to stay on-island, and in their classrooms, while completing courses toward a State of Hawaii Teacher Certification.

In 1985, with the election of a new Governor, she was replaced by Dr Pita Sunia, who again changed instruction policy, English again became the primary medium of instruction in the elementary schools. In November, 1988, a new Governor was elected, and as of this writing, whether or not the appointment of a new Director will again implement new English instruction policy remains to be seen.

From 1900 to 1988, changing educational policy with respect to 'Americanization', bicultural and bilingual education, local versus mainland curriculum, pedagogical models, and teacher training have severely hampered the educational abilities of a generation of Samoans who now find themselves struggling in Honolulu and other urban labor markets. Without educational skills a large number of Samoan men find themselves in constant circular movement between Samoa, Hawaii, and the US mainland in search of viable employment. Without educational skills a majority of Samoan women find themselves unable to participate in the Honolulu labor market. These women do, however, make an economic contribution to their families by maintaining public housing in the central Honolulu area. The social status outcomes of Western Samoa and American Samoa schooling policy will now be examined in the broader context of urban migration and labor force participation.

SCHOOLING, MIGRATION, AND TRANSFORMATION IN SOCIAL STATUS

Large-scale movements of American Samoans to Hawaii and the US mainland, and of Western Samoans to New Zealand, began in the early 1950s. From 1951 to the mid-1970s, the primary flow of Western Samoans was to New Zealand. However, since the slowing of the New Zealand economy in 1975 and concomitant tightening of immigration control, the major Western Samoan movement has been into American Samoa, and then on to Hawaii and the US mainland. Samoan men have gone to the cities of Apia, Auckland, Pago Pago, Honolulu, and Los Angeles, and served their families and chiefs through wage labor and with remittances to Samoa. Essentially, they went to these cities instead of to plantations and the sea to garner economic resources. Their contemporary social status is determined by their success in educational and employment endeavors. In the process of international migration Samoan women have seen their social status decline because in urban contexts they have discontinued the manufacture of fine mats, and they do not have the educational background to enhance

318

their personal status through gainful employment and cash remitting to Samoa.

In 1980, for the first time, Samoans were treated as a distinct ethnic group in the US census, making it possible to compare the educational and employment status of Samoan men and women in American Samoa and Hawaii. In the 1980 American Samoan census, data were available for Samoans by place of birth, and comparisons between Western Samoa-born (assumed to be Western Samoans), and American Samoa-born (assumed to be American Samoans) men and women, in terms of education and employment, are illuminating. In the following discussion of the 1980 census data for the populations of American Samoa and Hawaii, the education data are for all persons 25 years old and over, and the employment data are for all persons 16 years old and over.

In American Samoa in 1980, approximately 27.1 percent of Western Samoan women, and 30.5 percent of Western Samoan men had completed high school, compared to 38.3 percent of American Samoan women, and 52.4 percent of American Samoan men. These data support the more selective profile of Western Samoan schooling and also point to a rather strong bias toward higher male educational attainment in the American Samoan school system. This male bias in education will also be reflected in the Hawaii data.

In American Samoa in 1980, only one-third of Western Samoan women (33.6 percent), and American Samoan women (36.9 percent) were active in the labor force; this compares to over one-half of Western Samoan men (51.7 percent) and American Samoan men (55.8 percent). For Western and American Samoan women it appears that a low educational level constrains their ability to be active in the labor force. For Western Samoan men, low educational attainment does not appear to be as serious a constraint to labor force participation. For American Samoan men, educational attainment appears to improve their access to the local labor force.

In the 1980 Hawaii census, educational and employment data are available for Samoans but are not broken down by place of birth. Only 44.3 percent of Samoan women had completed high school, and only 37.7 percent were in the labor force. This labor force participation rate was fully 20 percentage points below that of any ethnic female population in Hawaii. For Samoan men, 58.4 percent had completed high school, and 68.1 percent were in the Hawaii labor force. The Samoan men thus had much higher educational attainment levels, and their labor force participation rate was fully 30 percentage points higher than that of Samoan females. The Samoan male labor force participation rate compared favorably with the all-Hawaii labor force participation rate (78.4 percent). However, when we look at the income generated from

Samoan employment, the data do not present such a favorable picture. In 1979, the Samoan per capita income of $2,729 was the lowest of any Asian or Pacific Islander group in Hawaii, and only 35 percent of the state average. Of the 2,481 Samoan families enumerated in the census, 931 (37.5 percent) were living below the poverty level as adjusted for family size. This percentage is nearly five times the 7.8 percent for all Hawaii families. So although Samoan men are relatively well-represented in the Hawaii labor force, their skill levels are low, and their ability to generate income to remit home is limited. Samoan women are seriously underrepresented in the Hawaii labor force, and their primary social and economic roles revolve around childcare (Samoan females have the highest fertility of any ethnic female population in Hawaii), the maintenance of public housing units (approximately 30 percent of Hawaii's Samoan population lives in public housing), and hosting visiting *malaga* groups coming to and through Hawaii from Samoan and the US mainland. These roles are similar to their traditional roles, except for one major transformation – they no longer make the fine mats.

Conclusion

In contemporary Samoan culture, educational and employment success is perceived as more important for Samoan men because it enables them to continue their participation in the Samoan status system. Educational and employment success is a major route to status enhancement for Samoan men because it can result in the generation of income to be remitted to one's family and chiefs. Cash replaces the products of traditional men's work as a resource directed to one's chiefs, and thus cash-remitting is a major form of *tautua* (service) in contemporary Samoa.

For Samoan women, educational and employment success is not perceived to be as important to their contemporary social status because they have been able to maintain many of their traditional roles without it, and in fact, have seen the elaboration of these roles with the population growth and international migration of the past 40 years. In the past 40 years, Samoan women have significantly reduced their production of *'ie toga* (fine mats), particularly in urban areas like Honolulu, and are thereby losing their primary route to status enhancement. The transformation and decline in the social status of Samoan women in Hawaii have been significant, and the Samoan and Hawaiian schooling systems have done little to help Samoan women adapt to new status-enhancing opportunities in Honolulu.

For Samoan men the transformation in social status has been less dramatic, but the elitist model of schooling in Western Samoa, and inconsistent schooling policy in American Samoa, have resulted in low educational and employment achievement relative to other ethnic male populations in Hawaii. Thus, although Samoan men attain higher educational levels, and are able to find viable employment more easily than Samoan women, their ability to enhance their social status via income earning and remitting has been limited by their experience of schooling in Samoa.

11

State Formation, Development, and Social Change in Tonga

Christine Ward Gailey

Tonga was one of the three indigenous Pacific island kingdoms that emerged in the nineteeth century; it is the only one that has persisted as a constitutional monarchy to the present day. Tongan society has changed dramatically in the past 150 years, but because there are formal resemblances to social relations before European contact, most scholars have chosen to emphasize supposed continuities, rather than ruptures. There is a tendency in these works to see progress through time, amelioration of conditions for commoners, increasing freedom, and so on. But we should be wary of associating progress with the state, if only because there is such a pervasive adherence to this model of civlization among colonial administrators, foreign advisers and development offciers, missionaries, the indigenous elite, and others whose agendas are supported by such a belief.

This chapter will attempt, therefore, to analyze the social, political, and economic changes in the Tongan Islands without assuming an inherent progress in what Darcy Ribeiro (1968) calls the 'civilization process'.[1] The view presented here of Tongan social and political dynamics before and after contact focuses on changes in the content – and sometimes the form – of social strata, land tenure, gender roles, and the relations of production and societal reproduction (not simply demographic replacement).[2] It contrasts with a national agenda among the elite, dating from the political centralization struggles in the 1880s and efforts to stave off British annexation, that has emphasized formal continuity with certain prior forms. Implied in the approach taken here is that in state formation, custom becomes a 'contested terrain': which practices come to be seen as traditional depend in large measure on shifting power relations among social constituencies. Most accounts of stratified societies privilege the views of the powerful, since 'history is written by the conqueror' (Diamond 1974). But one can discern voices of opposition in the contingencies of historical events, in the unsaid, in what could not be imposed, in what impositions remained assertions rather than practices, and in the persistence of customary actions contrary to legal codes (Gailey and Patterson 1987, p. 3).[3]

DEEP HISTORY

In the period most have called Tongan 'prehistory' – a view that presumes history began with European contact[4] – we rely on the interested and, therefore, sometimes conflicting accounts in chiefly genealogies, in archaeology, as well as in legends and accounts from Tonga and neighboring island groups. Convincing arguments have been made for the reliability of oral histories and genealogies in situating a flow of events, a relative chronology (Latukefu 1968; Ve'ehala and Posesi Fanua 1977).

Beyond this, a cautionary note is in order: Later chiefly factions, especially those supplanting earlier title holders, had a stake in portraying their predecessors in a negative light. The faction that in the nineteenth century became the nobility and royalty – and their missionary advisers – had particular interest in portraying the 'traditional' chiefs who opposed them as arbitrary and brutal, but also – to indicate their own power – as powerful rulers. Correlatively, nonchiefly people were assumed to be incapable of self-government, and depicted in most accounts as invisible or passive. So one must evaluate what may appear as pre-contact tyranny, or regular and orderly succession, or an anarchy of chiefly rivalries with an awareness of which groups were painting the picture.

A Tributary Empire

The islands appear to have been settled no later than 1200 BC, but little is known about the early inhabitants and their social relations; the emphasis on research into migration patterns has retarded social analysis. Population was concentrated along the lagoon shoreline (Groube 1971, p. 291). There is no archaeological evidence indicating significant status difference until the end of the first millennium AD (Kirch 1984, pp. 219–220).

The first chiefly genealogies date from about the mid-eleventh century; the first male sacred paramount chiefs are named, as are their mothers and some of their sisters and daughters (Gifford 1923, p. 222). Whether these women also were titled in unclear: We know the title of Tu'i Tonga Fefine, the sister's parallel title of sacred chief, existed by the seventeenth century; whether this title as also early or 'read back' for political reasons into earlier times is not known.

What we do know about the chiefly stratum in the tributary period is that high-ranking women were socially important; on some level female chiefs embodied the rank and continuity of their kindred. The first of the major tombs' *langi*, is said to have been built around 1200, for the eleventh Tu'i Tonga's daughter. Soon after the paramount chief ordered the construction of the Ha'amonga, the monumental stone structure still standing on Tongatapu (Bain 1954, p.73). His sister, Fatafehi, bore the name of the sacred paramount's kin group; we do not know if this name was also a title, although it may well have been, since it appears in later genealogies.

In the twelfth and thirteenth centuries, however, we can safely say that the society was stratified, perhaps even unstably class stratified, to an extent greater than it was at the time of European contact in the 1600s. The evidence comes from legends about the sacred paramount chiefs at that time and from corroborating accounts from Samoa, Rotuma, and Fiji. It is clear from these accounts that Tongan chiefs extracted tribute from these and other islands from 1100 until approximately 1400 (see, e.g., Krämer 1902; Ve'ehala and Posesi Fanua 1977). The only major genealogies are those of the sacred paramount kindred. It is clear that by this time were two internally ranked strata, chiefly and nonchiefly, with considerable – and temporarily successful – efforts within the chiefly stratum to consolidate authority in one kindred and to crystallize a ruling class.

Whether these strata were classes is open to question. All accounts and the material remains seem to indicate that if they were not – if instead they were estates or orders (Rousseau 1978; Gaily 1987a, pp. 54–56) – they were certainly more class-like then than the same strata were by the seventeenth century.

Ranking within the orders was intricate (Urbanowitz 1979); there were a ranked series of chiefly titles, possibly parallel of ones for men and women (Gailey 1987a). But the principles of ranking were internally inconsistent (Gailey 1980): older outranked younger, ties through male ancestors were superior to ties through females, but sisters outranked brothers. Rank, thus, created an ambiguous structure of authority and, for the chiefly order, provided a pool of contenders for most titles. Beyond the titled chiefs there were the 'houses' or *ha'a* and other kin of chiefly rank who, over time, would become increasingly distanced form the higher echelons, unless they renewed ties through marriage. Regardless of some chiefs' assertions to the contrary, the chiefly order had ties through kinship to non-chiefly groups. Moreover, although patrilineal tendencies can be seen in title succession, the chiefly 'houses' functioned as bilateral kindred.

Some authors have characterized this period in Tongan history as a theocracy. It is tempting to read the absence of chiefly-coordinated religious institutions beyond the first fruits – ceremonial at contact as an indication of an overthrow of the sacred chiefly way in everyday affairs. But the theorization of theocracies has emphasized the hegemony of religious beliefs, centered in one group, as coordination of production, distribution, custom, and so on.

The concept of theocracy seems too state-like to suit even the dynamics during the tributary empire period. The concept of *tapu* (sacred, forbidden, reserved) certainly was invoked by chiefs to reserve goods or prohibit activities and, therefore, served as a legitimation of chiefly prerogatives. But *tapu* also was backed by the threat of force, a necessity that challenges any notion of uniformly shared commitment to the religious order. In addition chiefly coordination of production appears to have paralleled local subsistence practices in most cases; when particular chiefs overrode these for any length of time, it usually sparked rebellion.

It seems from the upheavals in the wake of the empire's collapse that tribute was collected from conquered peoples abroad precisely because it could *not* be extracted systematically at home in sufficient quantity to support a non-producing, ruling class. This implies that the kinship links between chiefly and non-chiefly strata were too strong to permit such violations of reciprocity as are inherent in tribute collection. During the reign of certain Tu'i Tongas, especially the eleventh, one can discern systematic extraction from producers at home. But even here, there is an immediacy to his acts – approbation or punishment is either by his hand or in his presence – which bespeaks a relatively uninstitutionalized form of domination (Gailey 1987a).

Indirect evidence for the efforts to consolidate ruleship – that is, to divorce a small group from the claims of collateral and lower ranking kin – can be seen at this time. The eleventh Tu'i Tonga, whose paramouncy probably represents the height of Tongan hegemony in the region, is said to have raped (i.e., claimed reproductive rights to) his virgin sister. There also are references to a son of a Tu'i Tonga who marries a Tu'i Tonga Fefine, named Fatafehi Lapaha (Collocot and Havea 1922, p. 76). But the negative characterization of these events (or potentials) in the oral histories and legends indicates that such attempts to consolidate high rank within a small group met with opposition among collateral kin and other chiefly groups.

Since rank underpinned claims to goods and services of lower ranking kin and rank derived from the mother, the marriages of high-ranking women illustrate the ebbs and flows of the highest chiefly ranks' relative power. Since paternal aunts arranged marriages, female chiefs played key roles in instituting and subverting attempts to accentuate chiefly rank through strategic marriage (Gailey 1987a, pp. 72–77).

So long as there was an empire, the highest ranking women posed little threat to their brothers and nephews. By the time of the seventeenth, eighteenth and nineteenth Tu'i Tongas several high-ranking female chiefs married into Samoa (Gifford 1929, p. 54; Krämer 1902, I, p. 468). These women would represent Tongan hegemony in the marital districts, while removing them from potential contention in their natal groups. That this is not idle speculation can be seen in what happened after Tonga's tributary empire fell apart.

Aftermath: Chiefly Empires

The regional sway exercised by Tongan chiefs had been broken by the fifteenth century. None of the highest-ranking female chiefs married abroad for a time, and tensions over succession soon surfaced at home. The female chiefs seem to have brought retinues of 'foreigners' with them. While males were preferred for title-holding and while nationality was inherited through the father, rank was inherited through the mother and sisters outranked brothers. In addition, foreigners and higher-ranking relatives were free from the *tapu* on touching a chief's person. Thus sisters – especially paternal aunts – and the sons of sisters could and did call into question any pretense to orderly patrilineal succession (see, e.g. Hocrat 1915).

Assassinations of three successive Tu'i Tongas ushered in the split of scared and adminstrative functions. Around 1450 a new paramount chiefly title was created in a 'younger brother' branch of the sacred paramount line, from the Ha'a Moheofo ('House of Moheofo'). The *hau*, the Tu'i Ha'a Takalaua, was to discharge administrative duties while the technically higher-ranking Tu'i Tonga (sacred paramount) was to continue to receive first fruits, but take no part in administrative affairs. The two lines were supposed to intermarry, but did so only rarely. This indicates the capacity of other, lower-ranking and collateral chiefly groups to make demands for intermarriage with the higher ranks, i.e., that the power to consolidate rank through close-in marriage at the top remained only an assertion, subverted in practice.

A similar set of contradictions, discussed by a range of anthropologists as tensions within the 'conical clans' (Kirchhoff 1955; Sahlins 1958) – or 'pyramidal ramages' (R. Firth 1968) or 'status lineages' (Goldman 1970; cf. Gailey 1987a, pp. 48–62) – resurfaced at the end of the sixteenth or beginning of the seventeenth century. A new female chiefly title was created, the Tamaha. This chief was the daughter of the female sacred paramount, the Tu'i Tonga Fefine; as the daugher of the sister of the male scared paramount, she was the highest-ranking person in the society. The new title seems to have been an effort to shore up the waning power of the sacred chiefly 'house'. The volatile political question became – who would this trancendently high-ranking woman marry?

A generation later, around 1610, two major shifts occurred. On the one hand, the younger-brother branch of the administrative paramount eclipsed the senior branch and created another new title, Tu'i Kanokupolu, of the 'house' or Ha'a Moheofo. On the other hand, a contingent of chiefly people arrived from Fiji to bolster the position of the female sacred chief (Gailey 1987, p. 71). She thenceforth married into the Ha'a Fale Fisi ('House of Fiji'); but for the next two generations her daughter married the new administrative paramount. This degree of rank and power consolidation could not be sustained however, and eventually the Tamaha came to be unmarriageable (Gailey 1987a, pp. 77–79). The male and female sacred chiefs continued as stewards to their customary districts, to received the bulk of the presentations at the annual first fruits ceremony or 'inasi, and to receive donations in their own right. The Moheofo, titled sister of the new administrative paramount, married the sacred chief, and her brother married among a range of high-ranking kindred.

Although this arrangement attempted to ensure stability within the chiefly estate, stability remained ephemeral. Internecine chiefly rivalries continued, accentuated in the late eighteenth century by succession wars waged by the paramount chiefs from Vava'u, the northernmost island group.

SOCIAL DYNAMICS AT CONTACT: STRATIFICATION, GENDER, AND PRODUCTION

Tongan society at contact was stratified into two orders (Rousseau 1978): chiefly and non-chiefly.[5] Nevertheless, the property relations and the social connections that organized work, the distrubution of products, consumption patterns, and the replication of the society remained embedded in the multistranded expectations and security of being kin. The society, thus, can be characterized as having a communal mode of production, replete with the contradictions found in kin-stratified examples of this way of organizing production and reproduction (Gailey and Patterson 1988).

Tongan society, then, was in crisis. There were deep-seated tensions surrounding the relations between and within strata which militated for and against class relations and political centralization. While chiefs claimed rights to distribute land and requisition goods and services, in practice local kin communities retained use-rights to necessary goods, resources, and labor. One disputed area was whether what people gave to the chiefs was gift or tribute. The difference embodies basic questions of power: Who has control over what is produced? Can a donation be withdrawn or refused without threat of life or security? At contact there were no unambiguous answers to those questions.

Rank and Production

Labor claims were exercised through kinship relations, which included expectations of largesse and generosity in return for any services or donations of goods or foods. The line between donation and tribute was murky, but direct producers expressed their provision of services and goods as donations, even when some chiefs considered the presentations as tribute. Ramifying throughout the society, many claims were exercised through the *fahu*, the rights of a sister and her children (and descendants) to the goods and services of her brother, his wife or wives, and their children and descendants. Among the chiefly houses the *fahu* was invoked in numerous succession bids.

Gradations of rank within the orders depended on the rank of one's mother, seniority of birth, gender (sisters outranking brothers), depth of genealogy, and so on. For chiefly people over time the effects of these factors could shade into non-chiefliness, although this would be considered as an insult if stated openly.

The non-chiefly estate also was internally ranked. In addition to the *tu'as*, who were predominately farmers and fisherfolk, there were children of chiefs and non-chiefs, whose rank was ambiguous, and *matapules*. With the exception of the oldest group, said to be from Tonga, the several groups of *matapules* were supposed to have been of foreign origin. They served as attendents, advisers, and agents of chiefs and as (part-time) artisans. Politically they represented various attempts through time to form a wedge between high-ranking chiefly people and their collateral kin and to subvert *fahu* claims (Gailey 1985a).

Why chiefly and non-chiefly strata were not classes can be answered by examining the gender division of labor. The highest-ranking titled male chiefs (*hau*) demonstrated their position by their disengagement from men's work activities. The volatility of title-holding, however, made it unpredictable if a particular chief would be non-productive for his lifetime; lower-ranking male chiefs did till the soil or asserted claims to male relatives' labor by virtue of superior kinship, not power.

But even the highest-ranking female chiefs marked their exalted status by remaining engaged in making socially necessary goods. Women made wealth objects which were used at all levels of the society as markers of personal status and life transitions. To acquire a title, for instance, one had to acquire and exchange wealth objects (in the control of higher-ranking female relatives) of suitable value, the value of the item being derived from the rank of the marker(s) (Gailey 1987b).

The chiefly order, then, did not live off the labor and products of others on a consistent basis. Since no portion of the chiefly estate was removed permanently from direct production, we can state that class relations had not crystallized, despite almost constant efforts to estrange parts of the chiefly estate from kinship obligations through the use of *matapules* in place of *fahu* relatives, the claims of tribute rather than gift, and assertions of partilineal succession.

Value and the Gender Division of Labor

Men did most of the agricultural production, deep-sea fishing, the forming of houses, the cooking, and other tasks called *Ngaue* or 'work.' Women, for their part, did lagoon fishing, collecting reef edibles, the plaiting of mats of various sorts, the weaving of baskets, and the manufacture of varieties of plain and decorated bark cloth (generically *tapa; ngatu* is the decorated variety). Women often worked in groups, sometimes under the direction of a female chief. Women also did most of the infant care, although this task too, was not privatized. Whatever women made was considered *koloa* or 'valuables,' intriniscally of higher rank than men's work. Value still resided primarily in the rank of the maker, rather than the item per se (Gailey 1987b). Since wealth objects marked all life transitions, women were simultaneously productive – making socially necessary objects – and pivotal in the distribution and exchange of the goods which reproduced the society as stratified, but kin-organized.

For classes to emerge, little had to change about men's work, but one of two facets of women's work had to change. Either value of items had to be dissociated from the maker's rank, i.e., labor had to become alienated, allowing the highest-ranking women to be removed from direct production, or labor claims had to be divorced from status validation through the presentation and exchange of women's wealth. Both of these happened in the nineteenth century.

Gender Relations

There was no systematic gender hierarchy in the Islands prior to European contact, although tendencies toward it can be discerned. Stratification influenced sexual behavior, expressed in the sexual free-dom of unmarried, non-chiefly women and the expectation of virginity for young chiefly women. The greatest tensions can be seen between estates: *Tu'a* women were sometimes subject to unwanted sexual advances by male chiefs, and children of *tu'a* women by male chiefs had higher status than would be accorded their mothers – calling into question the inheritance of rank through the mother. Married women of all ranks were to be respected, although even they could be raped in times of war. In addition, older and higher-ranking relatives exercised some control over the sexuality of unmarried persons, as shown by first marriages being arranged by paternal aunts.

Marriage relations are said to have been amicable, domestic violence rare. The wifely role, however, was not authoritative for women. Unlike husbands, wives were supposed to be chaste. Divorce was fairly easy to effect for both spouses, especially through adultery, but technically divorce was up to the husband.

At the same time, wives were not economically dependent upon husbands. Regardless of marital status women had access to men's goods and work through their authoritative kin roles, sister and father's sister. In the event of divorce, they could rely on the brother's support.

Local Communities

Researches disagree about the nature of land tenure. Non-chiefly kindred held hereditary use-rights to lands. Chiefs claimed the right to dispossess persons, but the only recorded dislocations occurred in times of war, when non-chiefly cultivators might leave temporarily. It appears that only titled chiefs could be dispossessed of stewardship; but even in defeat, there were available arable areas for exiles – none were permanently uprooted from the lands worked by the lower-ranking kin of the 'house.' Whether these groups were lineages or not has sparked controversy; while higher-ranking chiefly kinship some-times approached partilineality, it was continuously undercut by the claims of collateral kin. Local kin groups have been called patrilineal segments by some (e.g., Maude 1971), but if one considers that people preferred to marry locally and that they also preferred to live near their mother's brother(s) to take advantage of the *fahu*, it seems likely that most of the population lived as kindred, as they do today (see Korn 1974).

Use-rights to land followed the lines of the gender division of labor: Men did most of the agricultural work, and adult/married men were allotted parcels by the *'ulumotu'a*, older man who represented the kin group to the local chief. Alongside this periodic reassignment based on houshold need was the chiefly distribution system: Paramount chiefs designated district chiefs; generally, even if the area had been recently conquered, the district chief was selected from among the chiefly ranks associated with that area.

Settlements at contact was dispersed in hamlets, with the district chiefly compounds acting as social centers. With the intensified war-fare at the end of the eighteenth century, outlying households began to cluster about the fortified chiefly compounds. But villates per se did not exist until the second half of the nineteenth century, results of the combined (and related) pressures of missionary conversion schemes and the emerging state taxation system.

When Europeans began visiting the Islands with some regularity in the later eighteenth century, they reported a series of shifting alliances among a number of chiefly factions (Marner 1827; Vason, 1810). Skirmishes, raids, and occasional full-scale attacks had fostered the growth of fortified compounds in important chiefly districts throughout the Islands, but the wars were inconclusive and there was no political centralization (cf. Cummins 1977b).

EUROPEAN CONTACT: EXPLORERS AND TRADE

The first European contact was through Schouten and Lemaire in 1616, but contact was sporadic until James Cook's voyages 150 years later. In this early contact period visiting Europeans were far from having an upper hand: Tongans frequently pillaged or scuttled visiting ships for their ironwork and other valuables. But they also traded with the visitors, who confused the spheres of value in Tongan culture and gave what Tongans would consider valuables (cloth and iron) in return for items of lesser value, such as food, or services – particularly sex – which had no place in interisland trade (cf. Cook 1777; 1784). The Europeans often considered unmarried, non-chiefly Tongan women's sexual bartering as prostitution, but professional sale of sexual services was unknown in the Islands (Gailey 1980).

By the beginning of the 1800s a few Europeans resided in the Islands; they were refugees from ships the Tongans scuttled (like Will Mariner in 1804), deserting sailors, escaped prisoners, or missionaries. Often these men were impressed into chiefly retinues as quasi-*matapules*, advisers in matters of warfare, relations with visiting Europeans, or the new episodic European trade.[6] The first missionaries arrived in the late 1700s, but were soon caught up unsuccessfully in a war of succession; the survivors left the Islands in a passing merchantman (Wilson 1799/1968).

Mercantile activities of Europeans challenged the nature of exchange. Before contact items were produced for use, whether or not they were traded in Fiji, Samoa, and elsewhere. The trade was mediated through relations of kinship or quasi-kin connections; the items received in return were destined for satisfying consumption ends. Pre-contact trade, therefore, was of use-values, not of commodities. The latter describes conditions where items were made primarily as a means to generate profits, although they may also have a use value. In this sense, then, even when at times chiefs extracted goods as tribute and traded those goods with other chiefs, the goods made or acquired were channeled into consumption.

European merchants sought not only foodstuffs, water, and curios, but also goods to be marketed in Asia and Europe. Their purpose in trading differed from Tongan practice: Tongans sought exotic goods to reinforce or enhance status, while Europeans sought profit and reinvestment. The former goal was socially closed-ended; the latter was indefinitely expandable and placed pressure on existing social relations. So, while Tongans retained control over what was to be produced, the European trade encouraged the intesification of production ad its redirection from other types of the European one. The premise of Tongan production relations was implicitly called into question.

In addition, the European trade challenged the spheres of exchange, and the relative value of men's and women's products. As I have discussed elsewhere, the shift in European demand from coconut oil to copra – that is, from a valuable/women's product to a men's good – for the same return demeaned the value of women's labor (Gailey 1987b). In addition, the preferred return – textiles, iron implements, and weapons – also skewed social relations. Chiefly people dominated the European trade, as they had other international exchanges previously. But receiving weapons gave chiefs an unprecedented means to enforce tribute extraction. Technically they could insist upon intensifying coconut oil and later, copra production, beyond the demands of superior kinship. But socially this was limited by the continuing legitimation through demonstrations of chiefliness – the generosity and insurance of local prosperity inherent in claims of superior kinship.

Imported weaponry, then, intensified the existing warfare, although chiefs still fought over succession – the social recognition of superior kinship – and, thus, the rights to the chiefly portions of local production. Fatalities mounted and fortification grew, as did the number of captives used for production until ransomed by kin.[7] In the first decades of the 1800s, then, there was a growing contradiction in warfare between claiming goods and services through superior kinship versus conquest. While compelling in the shortrun, conquest outside the establishment of kinship had no cultural legitimacy. There was no resolution of this dilemma until the establishment of the Wesleyan Methodist mission in the 1820s.

STATE FORMATION: CHRISTIANITY, CONQUEST AND 'CIVILIZATION'

For the chiefs they attracted, the Wesleyan Methodist missionaries provided two vital resources; weapons (Gifford, 1929, p. 192; Wright and Fry 1936, p. 251) and an ideology with supernatural legitimacy that justified conquest and extraction apart from the expectations of kinship. The Wesleyan Methodists learned from the failure of the London Missionary Society group that they should ally themselves unambiguously with a powerful chiefly faction (Wilson 1799/1968). They sought mass conversions by effecting a top-down strategy. They were committed to fostering the development of their image of Christian civilization: a centralized monarchy with a landed nobility and yeoman farmers. Acting as advisers to strategically placed chiefs, they also used other European residents, at first sailors and later merchants, when it suited their interests.

The chiefly faction allied with the mission was headed by Taufa'ahau, a paramount chief from Ha'apai who would become the first king, George I. He used both customary and introduced means to consolidate power, but he vacillated about converting until the utility of the new religion was clear to him (Wright and Fry 1936, p. 251; Gifford 1929, p. 192). A group of Tongatapu chiefs installed him as Tui Kanokupolu on the promise that he renounce Christiantiy. But he was baptized in 1831, bringing his faction with him into the Wesleyan Methodist fold and alienating the chiefs who had invested him. The missionary alliance assured him of material and tactical support (weapons through trade advice) in his ongoing effort to politically centralize the Islands (Erskine 1967, p. 127; cf. Hervier 1902, p. 131).

Using a combination of customary claims and force, by 1833 George's faction controlled the sacred paramount's symbolic authority by taking his primary wife (Collocott 1923, p. 184). For this act he was expelled from the church until he married the woman in a Christian ceremony a year later (Latukefu 1977, p. 129).

The 'heathen' or traditional chiefs vigorously opposed the centralization and conversion efforts. The conquest of the Islands was a long and bloody struggle. Missionaries fomented disputes between factions and encouraged violent reprisals against the pagans, including a massacre of three hundred people at Hihifo in 1837 (Cummings 1977a, pp. 33–35). A visiting American captain tried unsuccessfully to effect a truce in 1840, but in light of the 'bad faith' of the Christian faction, he came away with the sense that George and the mission would either convert the heathen by force of exterminate them (Wilkes 1852, pp. 2, 9–13). A resounding victory by the 'traditionals' at Pea brought a lull to the fighting (Morrell 1960, p. 52). In 1842, however, French Catholic priests arrived to catechize the heathens, in all likelihood being invited by some of the chiefs to stop the accusation of heathenism and still to preserve their autonomy (Morrell 1960, p. 310). The implicit threat of French intervention united the English residents behind George (Hervier 1902, p. 135). In 1845, George was invested as king, despite intense opposition on Tongatapu.

The 'holy war' continued, the Methodists characterizing Catholics and traditionals alike as 'Devils' threatening the 'hereditary and titular power of king George (West 1846/1865, p. 54). In 1847 George requested British intervention in the event of French annexations (Erskine 1853/1967, p. 133). The Marist priests, in the meantime, converted a number of the traditional chiefs clustered around the sacred paramount and his wife (Hervier 1902, p. 136). In 1852 George began an all-out campaign to defeat the remaining opposition, culminating in the siege, defeat, arrest, and exile or forced conversion of the recalcitrant factions.

Class Formation

Following the wars of conquest, George I claimed all land and distributed estates to loyal followers and amenable members of the defeated (and converted) factions. This new landed gentry was composed primarily of chiefly people, but some loyal *matapules* were made hereditary estate holders as well. A nobility was created, which included some but not all of the chiefly estate. The principles which restricted membership in the nobility and title to estates differed from customary practice, emphasizing succession through patrilineal primogeniture rather than maternal rank.

The relative power of king and estate-holders vacillated in the second half of the century. At first estate-holders were granted hereditary possession, except in cases of treason. The constitution also provided for a parliament comprised of equal numbers of representatives of the people as of the nobility, and several royal appointees.

The state represented a tense coalition of royalty and a landed nobility that did not include all the traditionally high-ranking kindred. Crosscutting the ranked orders was a new class structure, which gave the royal retinue and landed gentry unprecedented economic and political power. As the alliance between church, state, and foreign merchants grew, the nobles became more and more threatened. From 1875 until British intervention in 1900 the state came to operate financially almost uniquely on behalf of the royalty. As in other cases where states cease to present the image of neutrality, and cease operating on behalf of the preservation of class relations in general, a crisis emerged.

Overt opposition resurfaced in the late nineteenth century. The wars of conquest set the precedent for political disputes taking the form of religious sectarianism (Korn 1978). In 1885 another religious war broke out. A sizable group of nobles objected to the king's attempt to tie the Tongan mission more closely to the state (and, thus, increase the king's revenues) by forming the Free Church of Tonga (Rutherford 1977; cf. Latukefu 1974; Baker and Baker 1941). The opponents of this move, including some lower-level missionaries of the Wesleyan

Methodist mission, received support from the Australian and New Zealand branches, which benefited financially from the lucrative Tongan one (Korn 1978; Morrell 1946). The rift was mended after reforms were undertaken and religious freedom was declared.

But violence erupted again in the 1890s, when irregularities in the use of state revenues by the king and the state-linked church sparked an attempt to shoot the king's preferred missionary adviser, Rev. Shirley Baker (Rutherford 1971). The assassination attempt failed and harsh reprisals were taken against those responsible and their kindred. This was followed by repression of those who refused to join the state church (Rutherford 1971).

When George I died in 1893, his grandson became king (George Tubou II). He sought to break the power of estate-holders by claiming the right to reallocate estates and dispossessing a number of the landed gentry (Fusitu'a and Rutherford 1977). Further, he attempted to increase state revenues by reserving virtually all copra production for the state and remaining nobles.

This coincided with the call for taxes to be paid in cash, and the siphoning of these revenues into the king's private account, leaving too little money to compensate bureaucrats or to continue with road-building and other civil projects (Im Thurn 1905, p. 12). On top of this, the king alienated Tongan merchants by ordering goods through Shirley Baker, who was under investigation in New Zealand for his financial dealings while head of the Tongan mission (Fusitu's and Rutherfords 1977, pp. 177–178).

The deciding crisis was precipitated when the king refused to marry 'Ofa, a woman associated with the noble and chiefly opposition, and then dispossessed her grandfather of his estate (Thomson 1904, p. 164). The woman and her faction converted to the Church of England and called upon the British High Commission. For their part, British officials were deeply concerned about the influence of Baker on the king, especially since Baker's links to German merchant firms had caused the king to favor closer ties to the German state. In other words, the church-state-German merchant ties threatened the previously unspoken, if ambivalent, alliance between Methodist missionaries and the British state interests.

In 1900 the British intervened, insisting on financial and constitutional reforms, a resident corps of British advisers, and Protectorate status (Im Thurn 1905, p. 9). The reforms, such as the separation of state from royal revenues and the restoration of dispossessed nobles, stabilized the relationship of king and nobles. Tonga was never officially a colony, but British advisers oversaw legal-judicial, financial and diplomatic affairs until 1954, when negotiations began for the end of Protectorate status. The Islands returned to full independence in 1971, but foreign policy since then has remained steadfastly neutral.[8]

Stimulating Commodity Production

Commodity production on a mass scale was intimately tied to the creation of the state apparatus. George I and his missionary advisers imposed a law code in 1853, calling for an annual tithe in coconut oil to be donated to the mission (Morrell 1960, p. 316). Although it could not be collected systematically until after 1862, in 1853 he also imposed a tax/rent on all adult men, payable in coconut oil (Sterndale 1874, p. 2). The king then had the oil sold through Europeans; revenues were kept or disbursed by him.

The shift from in-kind to cash taxes in 1880 demonstrates the growing alliance among mission, state and merchant houses. After 1860 missionaries insisted that annual donations be paid in cash. This drew men into cash cropping and credit arrangements with those merchant houses financially allied with the mission (Gordon, quoted in Morrell 1960, p. 319). The state legal-judicial apparatus was enlisted to enforce payment of debts to the firms.

In 1878 Rev. Baker, then head of the Tongan mission, made an agreement with the German merchant house, Godeffroy and Son, to convince the king to impose cash taxes. In return Baker would receive a commission to the copra people would have to sell (Hanslip, in Maudslay 1878/1930, pp. 240–241). The king acquiesced for his own reasons and from 1880 on, taxes had to be paid in cash. Stiff fines in cash and movable property were imposed for tax evasion or default (Im Thurn 1905, p. 15).

The late nineteenth century found many Tongans facing an unprecedented landlessness, despite disease-related depopulation (Comming 1882, I, p. 34). Increasing numbers of nobles were dispossessing tenants and leasing ostensibly fallow lands in the name of the state to foreign investors; the state then received a tenth of the rent. On British advice, the king addressed the revenue shortfall caused by a shrinking tax by increasing tax/rents by 30 percent and imposing a head-tax on all men – whether or not they held an allotment (Tukuaho and Thomson 1891, pp. 61–66). Under the Protectorate some lands were restored, although the state retained the right to grant long-term leases to foreigners (Government of Tonga 1907; 1967).

Unfree Labor

There was only a negligible market for labor in Tonga until the twentieth century, primarily because the state, mission, and merchant firms used political means of extracting labor: penal servitude, debt bondage, and tenancy-related labor service. The law codes specified hard labor as the punishment for a range of sumptuary, sexual, and other offenses, and for reneging on cash fines, debts, or tax payments. Most accounts focus on the Constitution of 1862's provision freeing all Tongans from 'enforced labor and any form of slavery' (Im Thurn 1905, p. 5). But few mention the pervasive forms of legal extraction that were instituted – and the collusion among church, state, and merchants these reforms implied.

In the second half of the nineteenth century the criminalization of a range of customary activities, coupled with the enforcement of new legal forms – credit, debt, contracts, torts – created a vast pool of male and female convicts. Convicted men worked on state, royal, or sometimes mission lands, and on civil projects. Some female offenders were set to work making valuables for state officials and nobles (Erskine 1853/1967, p. 136); others were used in road building and other civil construction projects (Fonua and Fonua 1981, p. 18).

Debt bondage was eventually banned in 1891 (Tukuaho and Thomson 1891, p. 41). Until then the state intervened on behalf of foreign merchants and other creditors. Debtors or their relatives were forced to work for creditors until the debt was discharged (Maudslay 1878/1930, p. 227).

Labor service for landlords has continued, despite the 1862 ban on forced labor. Even today commoner tenants take it upon themselves to provide 'first fruits' and various services beyond rent. But it is not quite the same as pre-contract relations between chiefs and tu'as, where kin connections make labor claims subject to expectations of generosity and assurances of prosperity. Instead, there is an implicit threat of dispossession or extortion – leasing to other parties or refusing to officially register allotments. On the tenants' part, labor service in reinforced by the self-respect involved in adhering to customary expectations of rank. But while this legitimates services it also emphasizes superior kinship as a basis for donations, rather than class power. Labor service then reproduces domination, but in using customary reasons, the legitimacy of class relations is continuously called into question.

Changing Family and Gender Roles

Missionaries found the economic autonomy of Tongan women and their authoritative kin roles incontradiction to the patriarchal order they associated with civilization. Missionaries used a combination of legal compulsion and church-sponsored repression – including physical violence – to promote the new gender roles, especially among the less powerful sectors of the population. Together, church and state actions undercut women's roles as sisters and paternal aunts, emphasized their less authoritative wifely role, upheld the nuclear family household as the Christian and civilized model, promoted the sexual availability of wives to husbands after the birth of children (which increased the birthrate dramatically), instituted a new gender division of labor in the household, imposed new sexual mores, and differentially punished women and men for sexual infractions (St. Julia 1857).

The results have been the creation of a double standard for sexuality and pervasive condemnation of premarital sexuality for women. While today one can see patriarchal tendencies in the kin communities, they are contradicted by other, historically deeper practices, such as the *fahu*. Wives have become more dependent upon husbands than in the past. But as mothers and sisters non-chiefly women retain crucial control over the life crisis exchanges (especially at funerals) which act to reproduce diffuse kin networks and thus ensure local continuity and the sharing of economic and social burdens.

The civil sphere, however, is characterized by patriarchal relations. Stratification also influences the degree of patriarchy: It is much more marked in the higher echelons than among the farming and fishing people. Nobelwomen and the women of the landlord class enjoy social power through their class position. Because they are no longer involved in making and controlling goods necessary for validating status, their role within their own strata has become primarily reproductive in the biological sense. The restriction of women's legal rights, which in practice affects the elite more immediately than local people, means they have far less social authority than chiefly women did in the past (cf. Marcus 1980).

The legal fate of the *fahu* illustrates a coincidence of interests between missionaries and the emerging ruling class. The missionaries viewed sisters' material rights and labor claims as undercutting the primacy of the nuclear family as a unit of accumulation, and as fostering wifely insubordination and encouraging divorce. The nascent ruling class saw *fahu* claims as threatening orderly succession, i.e., patrilineal primogeniture – by legitimating collateral claims. The coincidence in interest of church and state meant that exercising *fahu* claims was legally banned in the first and subsequent Constitutions. Despite the ban, among the producing classes sisters and their children have continued to exercise their *fahu* rights. Today this extends to claiming portions of cash remittances from overseas brothers and maternal uncles (Gailey, in press).

Missionaries shared some interests in common with the state in promoting patriarchal relations. But the unevenness of patriarchy today in local communities attests to resitance on the part of local communities, who find the recreation of diffuse authority and bilateral ties to kin of enduring utility and value.

Transforming Women's Wealth

The effects of commodity production and exchange, coupled with legal restrictions on the manufacture of use of indigenous products, undermined the value of women's labor and products (Gailey 1987b). In addition church and state involvement in fostering 'civilization' and commerce sometimes worked explicitly in concert to skew men's and women's relative access to cash.

A ban on indigenous cloth seriously undercut the importance of women's production of wealth. In efforts to appear 'civilized' in European eyes and, hopefully, to avoid colonization, George I encouraged European dress (Scarr 1967, p. 92). The Constitution of 1875, drafted by missionary advisers (probably Rev. Shirley Baker), forbade the making or wearing of *tapa*, imposing hefty cash finds or hard labor on royal or state lands for infractions. Missionaries, it should be noted, were widely believed to have financial interests in the textile trade (Cumming 1882, I, p. 32; Rutherford 1971).

As a result, people had to purchase the textile and, through marketing copra, men had far greater access to cash than women. So women lost control over a subsistence good and became dependent upon men for the imported substitute. But civil disobedience was pervasive: in 1877 almost 90 percent of the people were fined or subject to forced labor for breaking the sumptuary laws (Morrell 1960, p. 318). The king disclaimed responsibility for the law, but did nothing to rescind it until popular protest and opposition by the representatives of the people in Parliament pressured for its repeal (Cumming 1882, I, pp. 32, 33n).

By then, however, European dress was prestigious and remained mandatory for church and school attendance. Except for the types used on ceremonial occasions, indigenous cloth became associated with relative poverty. Not until the reign of Queen Salote would indigenous cloth be reintroduced for wear by government officials and on state occasions.

Today, women continue to make cloth, mats, and baskets, but not on the same scale. Baskets are usually functional or are made for the tourist trade. Some mats are still used as flooring and bedding; varieties of fine mats are considered as heirlooms and reserved for ceremonial use. *Ngatu* and other forms of *tapa* are still made cooperatively, but today they are used almost exclusively for ceremonies and kin exchanges (see Kooijiman 1972). In other words, the non-commodity sphere of valuables persists and is still considered as vital to Tongan culture, but it no longer comprises all things made by women or all valuables.

In addition there has been some commodification of these goods. A limited national market exists among the elite and professional class commoners for old and new fine mats and *tapa*. There is also a growing international market among foreign collectors, on the one hand, and oversea Tongans who cannot otherwise obtain the requisite goods for the transition ceremonies, on the other. The need to sell one's labor power abroad, then, has also spurred commodification of wealth objects at home.

CAPITALIST DEVELOPMENT: MARKETS AND MIGRATION

In the twentieth century capitalist development in Tonga has had decidedly uneven effects. Subsistence production remains a significant sphere and the 'archaic', tribute-based land tenure situation can be held responsible for the virtual absence of chronic malnutrition in the Islands. If land were alienable, that is, if private property in land were made legal, a rapid consolidation of holdings in very few noble and wealthy commoner hands would ensue. As in other cases throughout the world, many farmers would lose access entirely or find their rents inflated; widespread malnutrition would develop alongside the rapid economic stratification.

But despite this basic food security relative to many other parts of the Third World, cash needs have grown exponentially in the past thirty years. Kindred and families deploy a plethora of income-generating strategies, often a combination within one extended family of several of the following: cash cropping, craft production, petty marketing, commercial fishing, wage labor, and small businesses.

Cash Needs

Until the 1950s it was rare to encounter food for sale, and local communities resisted the introduction of national market for foodstuffs because it violated notions of subsistence rights (see, e.g., Ruhen 1966, p. 49; Beaglehole and Beaglehole 1941). Marketplaces have grown significantly since then; in the major towns today, however, farmers' markets face competition from supermarkets and general stores owned by foreign and Tongan merchants.

Townspeople still rely on their own relatives' subsistence plots for most of their vegetables and fruits. Many keep pigs or chickens in the yard, but they must purchase other staples, such as locally baked white bread. Certain imported foods have become staples or prestige foods needed for a range of special occasions or entertaining: tinned corned beef, mutton flaps, powdered milk, sugar, tea, and jam. Other imported items have become common drains on family income – tobacco and beer.

Other cash needs include tuition and school uniforms and other supplies. Elementary education is compulsory; government primary schools are tuition-free, but secondary schools are not; parochial schools at all levels are tuition-fed. The location of secondary schools has meant that families in smaller communities who want to send a child through high school must find ways to board him or her with relatives in a town. This reality, coupled with the fact that the bulk of the available wage labor is in Nuku'alofa, the capital, has led to an accelerating influx of migrants to the major towns.

Commercial cropping

The cash cropping of copra and bananas plummeted in the worldwide depression of the 1930s, when most Tongans returned to subsistence production. This pattern continued in World War II, which found Allied troops stationed in the Islands. Later, a banana blight reduced production to almost nothing; recovery has been slow. Development projects have encouraged farmers in Vava'u to diversify: They now export vanilla and pineapples as well as copra and bananas, but the latter crops remain the dominant ones throughout the kingdom. As with other cash crops, there are boom and bust cycles in the world price. The government regulates price and quality through marketing boards (Baker 1977).

Access to Land

Access to land has become increasingly difficult in the past twenty years. Most sources attribute land scarcity to population growth (Government of Tonga 1973). But while the population continues to increase, growing landlessness is not due simply to population pressure. Part of the problem lies in the government or the king granting long-term leases to religious sects in the name of religious freedom. The Mormon church, for instance, has expanded its Tongan mission considerably in the past decade, building compounds virtually every two kilometers in more densely populated areas.

But a greater part of the problem can be attributed to the activities of some landed gentry. Certain nobles or their managing agents have become involved in cash cropping or mechanized export-oriented agriculture, and they are reluctant to keep leasing small plots to tenants. More money can be made by withholding lands or leasing larger parcels on a long-term basis to Tongan or foreign investors. Some nobles have refused to register tenants' plots or have illegally increased rents to levels where several families must chip in to share one plot. An intricate and extralegal system of subletting had developed, partly because of the need for more cash for rents and partly because tenants' cash needs often require temporary overseas migration for work.

International Labor Migration

Many of the income-generating strategies, however, require credit. Farmers need loans to expand commercial production, which, given the absence of collateral in a tenancy situation, are very difficult to obtain from banks. Entrepreneurial activities depend on acquiring capital goods: a motorboat, a truck, a taxi, the stock for a neighborhood store (*falekoloa*), building a guest house for the tourist trade. But similar problems face those seeking loans to purchase capital equipment, to construct a house or make home improvements (flush toilets, indoor kitchens, etc.). At present virtually all the money for these outlays comes from remittances by overseas relatives.

As in most Third World countries, the government is the largest employer. A government job is difficult to get, and is unavailable to those without secondary education or whose patronage ties are weak. The private sector is small and jobs scarce. Overseas migration, then, is a burgeoning phenomenon for today's Tongans and promises to become a fact of life in the next generation. Almost all families have close relatives working or seeking work abroad, mostly in New Zealand, Australia, or the United States. At this point most migrants return to the Islands, but permanent outmigration is growing.

In order to migrate, people must activate extended family connections, in terms of accumulating the air fare and choosing where to go and with whom to stay. These ties continue to be strengthened through the activities of women – conducting exchanges of valuables during life crisis ceremonies, storing and disbursing remittances – and by both women and men in loans of money and exchange labor. While churches and the state continue to emphasize the desirability of nuclear family households, economic needs reinforce ties to kindred and make household composition flexible (Gailey, in press).

Migrants who cannot obtain work permits (often dependent upon church membership or patronage) can usually get visitors' visas; they often overstay these visas until reported. Jobs open to Tongan migrants tend to be in urban areas (excepting agricultural labor); many workers are paid 'off the books' or as day laborers. In the survey of returned migrants I conducted in 1986, women obtained work as domestics, seamstresses, or assembly plant workers; men as gardeners, construction workers, farm workers, or stevedores. Beyond culture shock, living conditions for migrants include intense overcrowding in substandard housing in dangerous neighborhoods. In US cities, like Los Angeles, teenage boys are inserted into a setting of intense interethnic hostility (many poor populations competing for the same housing or jobs, with little hope for the permanent residents of upward mobility), where youth gangs are the norm and violent confrontations endemic. This articulates in complex ways with cultural expectations that unmarried young men are apt to be troublesome and can be violent toward one another until they mature.

Most migrants make considerable efforts to send remittances home to relatives[9] or to save for specific goals: an outboard motor, a taxi, a truck, construction materials. The social costs as well as benefits of migration are well understood by everyone involved. For instance, under the pressure of an average absence (for married men) of three to four years, marriage ties can attenuate. In risky situations, kin may either recall absent husbands periodically – especially for funerals – or send wives or other relatives to join them a year or so later. But wives who migrate to be with husbands face greater isolation than at home and, given the stress of migration and poorly paid, insecure livelihoods, are more vulnerable to battering.[10]

Developing appropriate responses to the migration problem has been hindered by the way that migration has been depicted in most reports. It is seen as involving mostly men, while a survey of over 250 urban Tongans showed that almost equal numbers of women migrate.[11] A range of popular articles treat migration as an adventure or a rite of passage for younger men, as inspired by the lure of acquiring luxury goods, like VCRs. None of the people interviewd, however, expressed such sentiments, stressing instead the equipment and cash needs outlined above.

State Responses

Tonga has not incurred the disastrous levels of debt that other countries have. Most official sources of revenue for development (versus remittances) have come from United Nations agencies and bilateral assistance, primarily from New Zealand and Australia. Major expenditures for development have focused on commercial agriculture (new cash crops and international marketing), commercial fishing, and recently tourism.

These ventures have favored men's over women's income-generating activities. Women's craft production was a particular concern of the late Queen Salote, who encouraged the establishment of a government-sponsored craft cooperative. Today, this and a sprinkling of private craft shops and cooperative galleries cater to the tourist trade.

343

Women sometimes manage guest houses, although they tend to use familial rather than official funding sources to begin operations.

The state remains dominated by the nobility and royalty, although under the popular Queen Salote the government began to provide social services, especially education and health facilities. Another reform has been to open the civil bureaucracy to commoners. In addition the government has not rushed to privatize land – mostly because the landlord class (and most nobles) oppose it; the result has been that subsistence agriculture remains viable and local people enjoy a relatively secure standard of living.

PROBLEMS AND PROSPECTS

Tongans confront a range of social problems today. One concerns continuing class formation. What will happen with the growing numbers of well-to-do commoners who currently have little influence on Parliament or state policy? This rising middle class dominates the mercantile and professional economic sectors, as well as the civil bureaucracy. What will happen with the growing number of chronically unemployed men and women? There are numbers of literate commoners – women and men – who have attenuated ties to the land and no hopes for any kind of stable employment unless they go abroad. In addition, a growing number of well-educated commoners see little chance for employment commensurate with their skills in the Island or abroad. At present social discontent from landlessness, under- and unemployment is siphoned off in part by temporary outmigration.

But not all social discontent has been displaced. At present there is an ad hoc, cross-class alliance of commoners – linking professionals and many bureaucrats with farmer, fisherfolk, and wage workers – around issues of government corruption and political representation. A lawsuit brought in 1986 against Parliament for gross mismanagement of public funds, which was taken under advisement by the High Court but not publicly heard, received widespread support by commoners, who read the underground newspaper accounts of the accusations avidly. More dramatically, about 2,000 people joined in a march in the capital in support of the suit (Gailey 1987c).

Alongside this kind of estate- or order-based alliance is regionalism. The various island groups and areas were incorporated into the state through different means and have been treated differently ever since; there has been uneven development. These experiences form the basis for emerging political stances. In 1986, for instance, a national sales tax was imposed without informing the people of how the revenues would be used. While most people acquiesced, a town in Tongatapu refused to collect it, and delegates from Vava'u told government representatives that the tax would not be collected in Vava'u (Gailey 1987c).

To visitors, Tonga appears to be a peaceable kingdom with a 'traditional culture' and a population well-off by Third World standards. While true in some respects, the imagery is also somewhat deceiving. 'Tradition' is neither timeless nor homogenous, as the struggles of the past centuries show. Relative prosperity is due partly to land tenure situation that, while exploitative, still affords most people some access to subsistence goods. It is also due in part to the provision of low-cost health services by the government, and an educational system that fosters widespread basic literacy. But the relative standard of living depends upon those who are absent, on the tremendous personal sacrifices people make for kin by working abroad, often for years at a time.

In sum, Tongan society has been dynamic since its origins. Struggles for and against the reproduction of kinship relations as the organizing idiom of life can be seen in its deep history. Class and state formation were catalyzed but not initiated by agents of colonialism – missionaries, merchants, and state officials. The crystallization of classes and a kingdom in the nineteenth century had profound effects on the structure of Tongan society – on property and use-rights, on community, on work and other activities, on gender relations, on kinship, on beliefs, and on the practices of everyday life.

The impact of capitalist development in the past century cannot be disentangled from class relations and the changing nature of the state. The process is ongoing – classes are still forming, the state is in flux. To weigh the possible outcomes, one must acknowledge the simultaneous roles of political and economic structures – both national and international – in concert with action by all social groups, under circumstances which are not entirely in anyone's control.

Acknowledgments

I would like to thank Karen Watson-Gegeo and David George for their support, encouragement, and sense of the importance of the issues. Tom Patterson provided me with the intellectual stimulation and critical commentary that, as always, have clarified my ideas and refined the prose. The field research could not have been done without the excellent assitance of Soane Uata and Laki Kafovalu Tonga, whose lives and relations embody many of the contradictions of Tonga society today, and whose intelligence, compassion, and humor bode well for its future.

12

French Polynesia: A Nuclear Dependency

Ben Finney

Sailing out of the center of the flag of French Polynesia is a great canoe, crewed by five human figures outlined in the style of ancient petroglyphs. The canoe symbolizes the proud voyaging heritage of the Polynesians, and the figures represent the five island groups that make up French Polynesia. Despite the Polynesian symbolism, however, this would-be nation is a nineteenth-century colonial creation of France, and its people are French citizens whose current obligation is to provide, whether they like it or not, a testing ground for developing the nuclear weapons that their metropolian masters consider essential for the security and status of France as a world power. In addition to exposing the islanders to the dangers of radioactivity, this uninvited burden has utterly transformed their lives. No longer farmers and fishermen, most French Polynesians are now concentrated in a single urban agglomeration in and around the port town of Pape'ete on Tahiti, where, primarily as wage earners, they are largely dependent upon French military expenditures and various metropolitan subsidies for their living.

This is, however, but the third in a series of radical social transformations for the islanders over the last two centuries. Before the bomb and proletarianization, there was the imposition of colonial rule by France and linking of the islanders with the world economic system as peasant producers of copra and other export crops and commodities. Before that there was the initial shock of contact with Europeans, their weapons, religious ideologies, and diseases that destroyed so much of Polynesian culture and so many of the islanders.

The Polynesian Past

The 120 or so islands that make up French Polynesia are scattered over more than four million square kilometres of the Southeastern Pacific in four main archipelagoes: the Marquesas Islands, the Society Islands, the Tuamotu Islands, and the Austral Islands. (The Tuamotu

346

archipelago is administratively divided into the Tuamotu Islands proper and the Gambier Islands thus making the five island groups represents on the flag.) To make their geographic spread comprehensible to European readers, French writers often superimpose the map of French Polynesia over that of Western Europe. With Tahiti placed over Paris, the Society chain reaches across the Channel to England, the Tuamotu chain stretches from the North Sea to Yugoslavia, while in the north the Marquesas reach into Scandinavia, and in the south the Australs extend from Toulouse into the Mediterranean as far as Sardinia. These chains are volcanic in origin, formed when magma periodically punched through the Pacific Plate as it slowly moved northeast. The Tuamotus are the oldest; except at the southeastern tip, all are coral atolls built upon the flanks of submerged volcanoes. The other, younger, chains are composed solely or primarily of volcanoes, which, though eroded, still protrude above sea level.

All the islands together have a surface area of only around 4,000 km^2. At the beginning of intense European contact in the late eighteenth century about half the islands were inhabited, with a total population of, arguably, upwards of 150,000. Tahiti dominates the islands, accounting for around one-fourth the total land area, and about the same proportion of pre contact population. Nonetheless, before the French colonial unification of the islands, Tahiti was in no sense the political or cultural center of this multiarchipelago region, for all these islands were never unified in the past. Although the settlers of the various archipelagoes shared a common Polynesian heritage, at the time of European contact each island group was culturally distinct. Indeed, within archipelagoes there were cultural and linguistic differences, and nowhere was there enduring political unity of a higher order than of a local chiefdom.

Polynesian social organization was descent-based in ideology if not always in fact, and emphasized rank according to descent from founding ancestors, and ultimately the gods. The ruling chiefs were ideally the highest ranking men, i.e., those who could trace their descent most directly to mortal and spiritual founders, although on some islands prowess in warfare could be translated into political power, and on others high-ranking women played important roles. These chiefs held sway over groups of anywhere from a hundred or so inhabitants of an atoll in the Tuamotus to several thousand people living in a section of a high volcanic island such as Tahiti or in one of the deep valleys characteristic of the islands in the Marquesas. Social stratification within chiefdoms varied roughly according to the size of the group and, ultimately, the island. For example, while in small atoll communities the chief acted like (and often was) a senior kinsman of the population, in the large chiefdoms of Tahiti the ruling chiefs and their close relatives formed a class distinct from the mass of low-born commoners. Although ruling chiefs and their chiefdoms might join together to confront a common enemy, these confederacies were short-lived. There was no state organization or central insititutions above the chiefdom level, and warfare between individual chiefdoms was common.[1]

When Polynesian voyagers settled these hitherto uninhabited islands, they brought with them a portable agricultural system that featured taro, breadfruit, bananas, and other food plants, as well as the domestic pig, dog, and chicken. Employing a technology based on tools made of stone, bone, wood and vegetable fibers, the Polynesians developed a viable economy based on agriculture and fishing that involved a good measure of surplus production and social elaboration, particularly in the fertile, well-watered Tahiti and some other high islands. Trade between islands, however, was neither long-range nor intensive. Where islands with contrasting ecologies were within easy sailing range of one another, such as Tahiti and the atolls at the northwestern end of the Tuamotu chain, some trade in high and low island specialities developed. But, at the time of European contact at least, there was no obvious evidence of long-range trade across the length and breadth of this region, or connecting it with islands or lands beyond.

In fact, these islands were among the most isolated regions in the world. Although descended from daring voyagers, by the time of European contact the people of this region had largely given up long-distance voyaging, with the notable exception of the Marquesans, a few of whom were still periodically taking to the sea to search for new lands. Nor was there any evidence of regular contact from outside of the greater Polynesian region, although the cultivation of the sweet potato – a plant of South American origin – indicates some contact with South America, either by raft-voyagers from there, or by adventurous Polynesian canoe sailors.

The Coming of Europeans

In 1521, during that first recorded crossing of the Pacific Ocean, Magellan managed to sail clear through the islands that later were to become French Polynesia, sjghting only two small and apparently uninhabited islands on the northern edge of the Tuamotus. Although the Marquesas were visited by a Spanish expedition in 1595, and during the next few decades Spanish and Dutch ships briefly stopped at a few Tuamontu atolls, intensive European contact in this region did not begin until over two and a half centuries after Magellan's transit.

Tahiti was then the focus of European attention, and no small degree of admiration, for verdant Tahiti and the gregarious Tahitians charmed the first explorers to touch on that island. Captain Cook's flattering descriptions, the glowing accounts of Bougainville and his staff, and the more fanciful portraits by writers such as Diderot who never ventured into the Pacific, were all partner to the creation of a romantic view of Tahiti as a land of plenty and the Tahitians as handsome, happy people with a minimum of onerous laws, moral rules, and cumbersome institutions to bar their way to a natural and enjoyable life of love and fulfillment. But this idyllic view cannot be solely attributed to the European search for natural man, for the Tahitians were an uncommonly attractive people who so obviously enjoyed a bounteous environment. What is more, after discovering how ineffective their spears and slingshot were against the cannons and muskets of the first European ship to visit, the Tahitians adopted a strategy of hospitality to deal with these powerful strangers. It was not that they wanted to kill with kindness. Rather, they lavished hospitality to tame these intruders and to gain access to the objects of iron and other wondrous commodities carried by these ships from outside their universe.

This strategy attracted ships' captains as well as philosophers, and Tahiti soon became a favorite anchorage for ships with depleted stores and sea-weary men. There a peaceful welcome awaited, and all things – water, fruits, vegetables, hogs and fowl, as well as comely women – needed to resupply a ship and refresh its crew were available. At the turn of the century, Tahitians added exporting to overseas markets to their nascent trading sector. England's struggling penal colony on the east coast of Australia needed meat, and for a time Tahiti was the prime supplier of salted pork for convict and jailer tables. During the first decades of the nineteenth century a modest export trade developed in coconut oil and arrowroot flour produced on Tahiti and nearby islands, pearls and pearl shell from the lagoons of the Tuamotu Atolls, and sandalwood from the forests of the Austral Islands. A handful of Tahitian chiefs briefly tried their hand at organizing exports, and some even purchased ships to carry island goods to Australia and other markets, but this was a passing episode and foreign traders came to dominate commerce with the outside world.

Tahiti and the other islands of the Society Group attracted missionaries as well as seamen and traders for the Tahitians were doubly attractive: not only were they by reputation friendly and hospitable, but sailors' account revealed that these 'licentious and idolatrous' islanders were ideal subjects for saving. The London Missionary Society, a noncomformist Protestant Group, sent the first boatload of missionaries to Tahiti in 1797. Like most European vessels then visiting Tahiti, the missionary ship anchored at Matavai Bay, on Tahiti's northwest coast, a circumstance that gave the ruling chief there an immense advantage over his chiefly rivals. These ships, brought, among other things, guns and gunpowder which this chief (who came to be known as Pomare I), and later his son and successor, Pomare II, used to conquer (with the aid of beachcomber military technicians) the other chiefdoms on the island and to establish a kingdom over Tahiti and adjacent Mo'orea in the second decade of the nineteenth century. While the missionaries at first opposed the arms trade and military tactics of the Pomares, they eventually found it expedient to link their fortunes with their militarily successful hosts who obligingly converted to Christianity and encouraged the wholesale conversion of their subjects.

A theocratic state evolved from this alliance of opportunistic Pomares and missionary intruders. Pomare II and his successors, plus allied chiefs, were the ostensible rulers, but the missionaries, acting as teachers, lawmakers, and advisors shaped the structures of government and made or influenced critical decisions. The Leeward Islanders were christianized after the conversion of a leading chief, and Pomare extended his suzerainty over them. Similarly, the people of the westernmost Tuamotus, which had erlier been taken over by Pomare I, accepted Protestant Christianity, as did the Austral Islanders. The more bellicose and politically fractionated Marquesans resisted mass conversion, however, despite earnest efforts by zealous Protestant missionaries. Later, the Marquesas and the eastern Tuamotus were to become centers of Catholic missionary efforts.[2]

As the islanders were being introduced to iron tools and other trade goods, and to two varieties of Christianity, they began to fall victim in great numbers to influenza, dysentery, measles, smallpox, and other epidemic diseases (including venereal diseases that greatly reduced fertility) imported by the Europeans for which these isolated islanders had no natural immunity. Tahiti, as the main port of call for European ships, initially bore the brunt of this biological onslaught. Even taking a conservative estimate of 35,000 people living on Tahiti at the time of initial contact, the decline was dreadful. The missionary census of 1829, a half century after the first European ship, showed only 8,674 Tahitians left (McArthur 1967, p. 261). The less visited islands in the other archipelagoes were initially spared such disasters, though as European penetration increased in the nineteenth century, island after island came to pay the demographic penalty of contact with the outside world and its dangerous microbes.

As foreign traders and whalers began to penetrate eastern Pacific waters in greater numbers, and rivalry between Protestant and Catholic missionaries grew, France began to covet the islands as bases to be part of a worldwide empire. The Marquesas in particular attracted the French. A French navy report pointed out that the Russians were developing Alaska and Kamchatka, that Americans were pushing over-land to the Pacific and were gaining influence in Hawai'i, and that the British were well established in Australia and New Zealand. A secure French outpost was needed, the report argued, to service the growing French whaling fleet and serve as a base for French commerce in the Pacific. The Marquesas, with their many natural harbors and central location in the eastern Pacific, seemed ideal, particularly as French Catholic missionaries were already established there (Dening 1980, p. 213).

So, in 1842 French naval forces, working through friendly chiefs, seized the Marquesas, raised the French flag, and started building barracks for the occupying soldiers. But the blackhulled warships of the French – which appeared so ominous to an American sailor named Melville whose whaling ship just happened to bring him into these waters at this time – did not remain long in this archipelago. They soon sailed for Tahiti where the French declared a protectorate over this island and its dependencies, including the western Tuamotus and two of the Austral Islands. In 1880, France annexed outright Tahiti and its dependencies and pensioned off the surviving members of the Pomare dynasty. As part of a diplomatic bargain with rival Britain, the Leeward Societies escaped French rule until 1888, by which time German overtures to the ruling chiefs there had so alarmed the French and British that the latter acquiesced in a French takeover. The annex-ation in 1900 of the remaining Austral Islands, again supposedly to forestall German penetration, completed the French acquisition of the multiarchipelago colony they pompously named *Les Établissements Français de l'Océanie*.

Although the French takeover may have been welcomed by some islanders – or rather by some of their chiefs as the French were skilled at exploiting chiefly rivalries – many of these new French subjects objected to the seizure of their islands. Armed opposition broke out here and there. For example, Iotete, the chief who first welcomed the French to the Marquesas, soon became disillusioned and fought them, though to no avail (Dening 1980, p. 212). It was on Tahiti and the Leeward islands, however, that the French encountered the most opposition. From 1844 to 1848 Tahitian warriors operating with guer-rilla tactics from bases in the mountainous interior of their island fought the French to a standstill till superior French arms and dissen-sion in their own ranks finally defeated them. In the late 1880s Lee-ward Island patriots attempted to forestall the French takeover of their islands, but only succeeded in delaying it briefly (Newbury 1980, pp. 179–234).

A Polynesian Peasantry

The islands did not live up to their imagined strategic potential, at least not early in the colonial era. Almost as soon as the Marquesas were taken over they became a backwater, and the colonial center of gravity moved naturally to Tahiti, the largest, most central and hospitable of all these newly-acquired French possessions. There, in the` port town of Pape'ete, the French governors, at first naval officers but later civilians, made their headquarters, as did a modest colonial bureaucracy. But there was no need of a major French commercial base in Pape'ete or elsewhere in these islands, for a French Pacific whaling fleet never materializied, and there was not that much French trade in or through the region. Furthermore, the islands proved to be of no real military value, although the Marquesas had one brief moment of naval glory when, during the Crimean War, ships of the French and British navies made their rendezvous there before embarking for their ill-fated attempt to invade Siberia (Dening 1980, p. 222).

After the French annexed Tahiti and its dependencies, they ruled their Polynesian subjects directly. In place of the suppressed tribal chiefdoms they developed administrative districts that generally corresponded with traditional political boundaries, be they of chiefdoms encompassing an entire atoll or other small island, or tribal sections of the larger islands. To help them administer the districts, the French set up local councils whose members they first appointed and later allowed to be elected. Heading the council was its president, known as a tĀvana, which is the Tahitian way of pronouncing the English word 'governor.' Early in the colonial era a number of traditional chiefs friendly to the French cause served as the tĀvana of their district. As the French consolidated their rule chiefly names drop out of the lists of district leaders. Although there was a pretense of local self-government, the tĀvana and their councils were really agents of direct French rule. While they had some authority to keep the peace, settle land disputes, and organize local public works construction, ultimately their duty was to carry out the orders of the French governor and his administrators (Finney 1973, p. 19).

In local district affairs, the local Protestant minister, an islander trained at missionary headquarters in Pape'ete, often rivaled or exceeded the tĀvana in power and prestige. The Protestant church was typically the most imposing structure in a district, and the numerous church services and Bible discussion meetings held on Sunday and at various times throughout the week had become a focus for community life in which social prominence was gained through pious participation. In the predominantly Catholic communities in the Marquesas and eastern Tuamotus, as well as in and around Pape'ete where Catholics had gained a foodhold after the French takeover, the local parish church and services were similarly important, although the lack of Polynesian priests limited local input and identification.

However, by the time the missionaries had set up churches all around the islands, and the French had gained complete control of the islands and fully established their colonial administration, there were not that many Polynesians left to minister to and rule. Imported epidemic diseases, and the social disruption of first contact then subjugation by foreigners, had taken their toll. The 1902 census revealed that the total population of the islands (including several thousand French bureaucrats, soliders, and other non-Polynesians) had dipped below 30,000 roughly a fifth of pre-contact numbers. Fortunately, by then the terrible population decline had been arrested. Numbers of Polynesians were slowly beginning to climb, except in the Marquesas where the turnaround did not come until the 1920s – and then not until after the Polynesian population, estimated to have once been somewhere between 30,000 and 80,000, dropped to a low point of 2,094 (McArthur 1967, pp. 314–344; Rollin 1929, p. xiii).

These remaining Polynesians found themselves cut off from much of their traditional past. To be sure, the people remained Marquesans, Tuamotuans, Tahitians (as all the Society Islanders are known) or Austral Islanders, continued to speak their ancestral tongues (all distinct, but closely related East Polynesian languages), still grew taro and other crops, and fished the sea in more or less the old ways. But their elaborate social structure, formal religious system, and martial tradition had been swept away. The great chiefs were no more; at best their descendants were powerless pensioners of the French. The old religion of mighty Polynesian gods and massive stone temples had been swept away, replaced by Jehovah and rival Calvinist and Catholic ways of worshiping him. Nor were there any warriors left, for their French rulers held an absolute monopoly of power. The survivors may still have been Polynesians, but they were culturally impoverished – and dispirited. Henry Adams, writing about the rural Tahitians he saw in 1891, caught this mood:

> To me the atmosphere is more than tinged by a South Sea melancholy, a little sense of hopelessness and premature decay. The natives are not the gay, big, animal creatures of Samoa who sang and danced because their whole natures were overstocked with life; they are still, silent, rather sad in expression, like the Hawaiians, and they are fearfully few in number . . . Except in the remoter places, the poor natives are all more or less diseased. They are allowed all the rum they want, and they drink wildly. They are forbidden to dance or to keep any of their old warlike

habits. They have no amusements, and they have *gens d'armes* . . . I never saw a people that seemed so hopelessly bored as the Tahitians . . . . If they have amusements or pleasures, they conceal them . . . . They do not even move with spirit. (Adams 1930, pp. 466–467, 476; quoted in Levy 1973, p. 32)

Fortunately, for these depressed islanders, the territory never developed a vigorous plantation economy, which, by flooding the islands with European planters and Asian laborers, would have completely submerged the few Polynesians left. Had the islands been as large and arable as those in Hawai'i and Fiji, had they, again like those neighbors, been made part of the more dynamic American or British colonial systems, the story might have been radically different. But, France was far away, and could get tropical products more easily from its Caribbean, Indian Ocean, and African possessions. Besides, the islands were really too small or mountainous to support large plantations, except Tahiti, and even there the amount of arable land was limited to a narrow coastal shelf and the floors of a few of the wider valleys. During the American Civil War a Scots adventurer tried to develop a cotton planation in one of those valleys, using imported Chinese labor, but failed through mismanagement and the precipitous drop in cotton prices when American production resumed after the war. Later in the century modest-sized coconut plantations for the production of copra, which was then coming into heavy demand in Europe and America as a source of vegetable oil, were laid out here and here throughout the islands. But, among other things, difficulties of obtaining large tracts of land, and of inducing the Tahitians to labor or obtaining workers elsewhere, limited these initiatives.

The Polynesians therefore survived the European takeover of their islands with much of their land intact. In fact, because of depopulation the people were relatively land rich relative to pre-contact times. They could grow all of their vegetable food they needed and still have land to spare for cultivating a variety of crops in order to buy cloth, tools, tobacco, and other exotic goods they wanted. A number of crops were briefly or regionally important to local, small-scale growers. Tahiti exported shiploads of oranges to California and Australia before a local blight and competition from California growers ruined the trade. Even after the failure of Tahiti's one major cotton plantation, Tahitians and other islanders who planted small plots of cotton kept selling small amounts of this crop for several decades. Coffee and vanilla were introduced, the latter sometimes yielding spectacular windfall profits when, because of crop failure elsewhere, the world price of vanilla suddenly jumped. In the Tuamotus the thriving trade in pearls and pearl shell won from the atoll lagoons developed early in the nineteenth century continued as a source of income for skilled divers.

But by far the most widespread and important cash crop in French Polynesia was provided by the coconut palm. Copra, the dried meat of the coconut, became the mainstay of the territorial economy, and the chief source of cash for generations of islanders (Newbury 1980, pp. 160, 238–239). Copra proved to be an ideal crop for the then land-rich islanders. The coconut palm grew well in lowland areas, especially on the coral atolls of the Tuamotus, and required little care after planting. Aside from clearing the underbrush between the trees, all the cultivator had to do was to periodically gather the fallen nuts, split them, and remove the meat, which was then dried in the sun, bagged, and sold to passing traders. These tasks allowed plenty of time for subsistence farming and fishing, so the islanders could easily participate in the world economy, at the same time providing for some if not all of their basic food needs.

The bulk of the French Polynesians came to be tied into the world economy through their sales of copra and other cash crops and products, and consumer purchases financed by these sales, while continuing to produce much of their food and to provide for other needs through their own efforts. Neither fully commercial farmers nor wholly self-reliant farmers and fishermen, they became peasants with one foot in the world economy and the other still planted in the subsistence sector.

Peasant production for the world economy required a commercial infrastructure. The port of Pape'ete developed as the entrepôt for virtually all island trade. Foreign traders and trading houses – French, British, German and American – made their headquarters there; from there schooners would sail to the outer islands to purchase cargoes of copra and other prducts which were brought back to Pape'ete and then trans-shipped to America, Australia, or Europe. Chinese traders, many former laborers whose contracts had expired, or their descendants, became an indispensable part of this commercial system. Chinese stores located on virtually every island and in every district within the larger islands, and linked to Chinese merchant houses in Pape'ete, provided the peasant cultivators with easy access to imported goods, and were ready buyers of their products (Coppenrath 1967; Moevch 1963).

In the late nineteenth and early twentieth centuries a new Polynesian middle class began to play a role in this commercial system. Biological as well as cultural mixing was and still is common in the islands, resulting in a steadily increasing porportion of islanders with some European ancestry. In the Tahitian language those regarded as true Polynesians are known as Mā'ohi, in contrast to part-Europeans, who are called *Ta'ata 'āfa Popa'ā* (literally, "persons part-white"), or *Demis* to use the shorthand French term. However, although the terminology seems biological, the classification actually mixes race, culture, and class. Those islanders who came to be known as *Demis* were not the only ones with European ancestry, for European genes spread widely throughout the island population. *Demis* were so classed because of their facility in French, and their European self-identification and lifestyle. Part of that lifestyle often involved employment by the administration or in the commercial sector, running a business in Pape'ete, or trading in the outer islands. In class terms, the *Demis* formed the nascent Europeanized bourgeoisie of Polynesian society, while the *Mā'ohi* and their outer island equivalents formed a South Seas variety of peasantry (Finney 1973, p. 22)

Prelude

While this colonial system unified hitherto disparate societies and brought peace and some structure to the shattered lives of those islanders who survived the impact of the Europeans and their deadly diseases, the stability achieved was transitory. Even if the French had not decided to make the territory a site for nuclear testing, the rural, peasant way of life that emerged in the late nineteenth century had a limited future. For one thing, as the numbers of Polynesians increased during the first part of this century, pressure on the land began to increase, making it more and more difficult for each new generation to so easily enjoy the modest degree of subsistence and cash crop affluence of their parents and grandparents in the 1800 and early 1900s. By 1956, for example, the Polynesian population (including *Mā'ohi* and *Demi*) of the islands topped 60,000, more than twice as many as at the beginning of the century, and was beginning to grow even faster (Blanchet 1987, p. 538).

Deteriorating terms of trade in the 1920s and 1930s further eroded rural life. From the 1890s to World War I, the islanders received good prices for their copra and other crops relative to the cost of the imported goods they purchased. Following the war, however, retail prices began rising while returns from exports fell (Newbury 1980, p. 282). Then, the Depression struck, and the bottom fell out of the copra market. Even those islanders who could fall back on the land and sea for their subsistence felt cheated by a world system that no longer would pay good prices for their products so that they could obtain the goods they had become accustomed to purchasing. Following World War II, copra prices rose some, but never again were to reach the levels of the early part of the century. In fact, by midcentury that period had assumed the aura of a golden age. Older Tahitans would reminisce about how much more lumber and how many more sheets of corrugated roofing they were able to buy with the proceeds from a single bag of copra then as compared to the 1950s.

As cash cropping was losing its economic allure, many rural Tahitans and outer islanders were also becoming increasingly bored with rural life. While life in the outer districts of Tahiti, and on the other islands, was probably not always as depressing as Henry Adams had described, the prospect of getting a job in Pape'ete to enjoy the amusements and excitement of this sole urban center in the territory began to attract young men who did not relish following the quiet, church-dominated ways of rural communities. Many young woman also longed to escape from the male domination promoted both by church teachings and control of cash crop production and earnings by their fathers, brothers, or husbands (Lockwood 1988). In addition, some outer islanders, ambitious for their children, were drawn to Pape'ete because of the better schools there. By 1956 the resultant movement to Pape'ete and immediately surrounding districts helped boost the population of this area to about 26,000 people, or over one-third the total for all of French Polynesia (Blanchet 1987, p. 540). However, job opportunities were limited there, and many other islanders seeking work were forced to sign up for the phosphate mines on the nearby island of Makatea, or even to go as far afield as the nickel mines in New Caledonia.

Above and beyond discontent over economic conditions, the islanders were becoming more and more politically frustrated at French direct rule continued with only minor reforms, leaving them with virtually no role in running their own affairs. While many were apparently resigned to this system, a growing number of islanders began to press for some role in government – for a voice in formulating policy and for jobs in the bureaucracy. Their ambitions were not welcomed by the French administration, even when post World War II colonial reforms allowed indigenous participation in government. Those who had fought for France during the two World Wars, and felt they had thereby earned the right to participate in their own government, were particularly bitter. When the opportunity came in the late 1940s to stand for election as French Polynesia's deputy in the National Assembly, and a few years later to contest seats in a newly formed Territorial Assembly, a Tahitian named Pouvana'a a O'opa, a decorated veteran of World War I trench warfare, emerged as the charismatic leader of the dispossessed Islanders. Despite opposition from both the French and many Demis businessmen, Pouvana'a was elected as deputy, and his party, the Rassemblement Démocratique des Populations Polynésiens (RDPT) – the first true political party in the territory – won control of the Territorial Assembly.

Pouvana'a and his party were, however, greatly disappointed by their inability to influence French colonial policy or even institute local reforms. Their frustration came to a head when in 1958 De Gaulle was called back into power and, among other things, decided to continue the decolonization process inaugurated by the previous Socialist government. In a referendum held throughout the then beleaguered French empire, the colonies were given the choice between voting *oui* to remain in the French community and to evolve politically within the French fold while receiving generous economic aid and technical assistance, or voting *non* to gain immediate independence and be cut off from French assistance. Pouvana'a, exasperated by French opposition to Polynesian self-rule, campaigned for a *non* vote. Judging from Pouvana'a's previous electoral victories and his virtually universal appeal among the Polynesian masses (but not among the *Demi* class), those voting *non* should have been in the majority. However, a split in the RDPT, and the denial to Pouvana'a by the French of access to the government radio – the only way in which he could communicate his views to followers throughout the scattered islands – led to 64 percent of the voters choosing the *oui* option.

If France's sub-Saharan African colonies were to have been the guide, political evolution, including the granting of independence within the French community, should have continued in French Polynesia. But De Gaulle evidently had other plans for the territory that his officials claimed was an integral part of France. Following the vote, Pouvana'a and his followers were besieged in his Pape'ete home by a crowd made up largely of French settlers and his *Demis* opponents. Instead of dispersing the mob, the police arrested Pouvana'a, charging him with being behind a plot to burn the town down. Although the evidence for this charge presented in court was weak (and widely thought to be manufactured), Pouvana'a was convicted and sentenced to eight years imprisonment in France to be followed by a fifteen year banishment from his homeland (Tagupa 1976, pp. 1–12; Thompson and Adloff 1971, pp. 26–47).

Why De Gaulle wanted Pouvana'a so completely out of the way did not become immediately obvious. At first, when France began to build an international airfield on Tahiti, which until then had only infrequent seaplane service, it looked like the islands were to be developed for tourism. In 1962, when officials from the French Atomic Energy Commission flew in from Paris and took great pains to announce to the people that the American atom bomb tests to be conducted that year on an atoll to the north of French Polynesia would bring absolutely no harm to the islanders, and when a French government anthropologist was expelled for disputing their claim, it became obvious that De Gaulle had in mind a nuclear future for the islands, which therefore must remain French. The realities of African decolonization had forced the French to search elsewhere for a testing ground for their atomic weapons. Domestic politics ruled out testing within France itself. Their shrinking colonial empire left few options, one of which was French Polynesia, an obscure territory which appeared to have plenty of open space and only a relative handful of powerless inhabitants. In January, 1963, after a period of rumors and counter-rumors, De Gaulle announced that the testing program would be moved to French Polynesia, a decision that has shaped the course of events there and the nature of French Polynesian society ever since (Danielsson 1986a, pp. 43–57).

The Bomb and the Boom

Moruroa, an atoll at the southeastern edge of the Tuamotu chain some 700 miles southeast of Tahiti, was chosen as the site for the French nuclear tests. If the French had wanted to conduct just a few tests over a short period of time, they could have done so by sending their ships directly to Moruroa, blowing up their bombs there, and then sailing back to France. But the rudimentary French devices needed more than a few tests to perfect them, and, besides, the French were not out to build just a token nuclear striking force. To be taken seriously as a world power, the French felt they needed a variety of advanced atomic weapons for delivery by aircraft, artillery shell, and missile, which, following the model of the other atomic powers, dictated a permanent and continuous testing program. So, not only were permanent facilities to be built on Moruroa, but a staging base for French ships and aircaft was to be constructed on the neighboring atoll of Hao. Tahiti was chosen as the headquarters of the Centre d'Expérimentation du Pacifique (CEP), which involved greatly expanding the port of Pape'ete, as well as the erection in and around Pape'ete of laboratories, warehouses, and workshops, as well as offices, barracks, and other facilities for testing personnel, military and civilian.

This massive construction program needed workers aplenty, and thousands of outer islanders from around the territory responded to the call, abandoning taro patches and coconut stands. In addition, many Tahitians and others already working on Tahiti, and hundreds of French Polynesians working in New Caledonia, left their jobs for positions in the testing program. By 1967 some 5,400 French Polynesians were working directly for the program, and many thousands of others found jobs in civilian construction and in other employment stimulated by the resting program and the influx to French Polynesia of technicians, army personnel, and others connected with the testing program (Blanchet 1984, pp. 33, 35–36). So many people flooded into

Tahiti during this period that by 1967 well over half the territory's population lived in the greater Pape'ete urban area (Blanchet 1987, p. 540). With such a displacement from rural to urban, food production for local consumption plummeted, and agricultural production for export dropped precipitously. In 1965 coffee exports ceased, and by 1967 vanilla production was down by 77 percent since 1962. More significantly, the export of copra, the territory's major cash crop, dropped by 41 percent over the same period (Blanchet 1987, p. 552). The testing program rapidly accelerated the decline of the peasant way of life based on mixed subsistence and cash cropping, turning most French Polynesians into salaried workers living largely on imported foods, and concentrating a majority of them in one urban area.

While the initial boom caused by the abrupt implantation of the CEP did not last, there has been no turning back to the old way of life. Once the main facilities had been constructed, CEP spending and the consequent demand for workers began falling, and dropped sharply when tests were temporarily suspended in 1969 (because of economic problems in France following the civil disturbances of the year earlier) and again in 1972. While employment in the program, revived somewhat after each suspension, when international protests forced the switch from testing in the atmosphere to testing underground in 1975, CEP spending and employment were permanently reduced. Some workers responded to these downturns by returning to their outer island homes, but most elected to remain in the employed sector, and if they were not already on Tahiti they moved there in search of a job. The government was therefore forced to inaugurate civilian construction projects to take up the employment slack. While public works programs were not able to completely smooth out all fluctuations, they have helped to maintain employment levels and have therefore served to further concentrate population on Tahiti. According to the 1983 census, the last one conducted in the islands, 80 percent of the work force of French Polynesia was composed of salaried workers, most of whom worked in construction, in commerce, and for the territory's major employer, the government, and lived on Tahiti, the inhabitants of which accounted for almost 70 percent of the total population of the territory and made up one urban-suburban agglomeration (Recensement 1983, p. 14; Shineberg 1986, pp. 157–158).

Affluence and Inequality

Life in French Polynesia over the last three decades has been utterly transformed. When I first visited the territory in 1956, Pape'ete was a small, sleepy tropical port town with few cars, and a long, midday siesta. A seaplane that came in from New Zealand every two weeks, a steamer from France that came every six weeks or so, and occasional freighters were the only ways in or out of the territory. Although the outlying districts of Tahiti and the outer islands were linked by truck and copra schooner with Pape'ete and through it the outside world, life moved slowly in these rural island settings. Farming, fishing, copra cutting or other cash crop activities were the focus of economic life, while frequent churchgoing, and feasts and ceremonies marking births, marriages, and deaths were the main social preoccupations.

Now, visitors and island residents alike have their choice of frequent international flights to North America, Australia, Europe, and Asia beyond, as well as to other island nations and territories in the Pacific. The port of Pape'ete is often crowded with naval warships and is visited by a continual stream of freighters bringing foodstuffs, building materials, automobiles, and other imports (but not taking away much in the way of exports), and the town area around it is filled with new buildings and shops, and congested with far too many cars and trucks for the narrow streets to handle. The formerly rural districts that extend along the coast from each side of Pape'ete proper are densely populated suburbs for commuters. Even at the extreme eastern end of the island, as far as it is possible to go from Pape'ete, many people regularly commute to their jobs in or around the port town. In fact, life on Tahiti now seems distressingly modern, complete with traffic problems, installment purchasing of cars and appliances, and too many hours spent watching television.

Outer island communities are about the same size, or slightly larger than they were thirty years ago, which means that most of the natural increase in their populations has flowed to Tahiti. In these rural communities, as in the affluent suburbs of Tahiti, old-style houses thatched with coconut or pandanus leaves are infrequently seen. Money earned working for the CEP or granted by the French after massive hurricane damage during the climactic disruptions of the major El Niño year of 1983, has been used to build modern houses made of concrete and imported lumber and roofed with corrugated iron. And, while there are pockets of rural agricultural enterprise in the outer islands – such as in the Australs where the islanders profitably grow potatoes for the Tahiti market (Joralemon 1983) – the ideal situation for many has been to have a good paying government job while still enjoying the peace and quiet of rural life as well as easy access to the fruits of the land and sea.

In terms of raw statistics, as well as surface appearance, the people of French Polynesia are well off. In 1985 the territory's gross national product per capita amounted to $7,840, exceeding that of New Zealand, and falling just 18 percent below that of metropolitan France (World Bank 1987). Income is far from evenly distributed, however.

The metropolitan French, the local *Demis,* and the Chinese, who together make up somewhere around one-third, or slightly more,[3] of the territory's population of about 195,000 are by far the wealthiest. It is they who have the best-paying government jobs, working directly for French agencies and institutions or for the territorial government, and who absolutely dominate commerce. The income of those recruited from France is generously supplemented by overseas posting allowances; all people earning high incomes benefit greatly from the total lack of income tax in French Polynesia. The remaining two-thirds of the people, basically the *Mā'ohi* Polynesians, are mostly unskilled or semiskilled laborers who work for the government in construction and in various service functions. Although exact income figures broken down by race and class are lacking, it is no exaggeration to say that these Polynesian workers earn considerably less than high-ranking civil servants and people in the business sector. Furthermore, since stiff import duties are levied on consumer goods to help make up for the lack of an income tax, these lower income earners are forced to pay a larger share of their income in taxes on imported foodstuffs and other necessities than do the well-paid civil servants and businessmen.

This Polynesian working class is far from homogeneous. In general, those with family roots and land on Tahiti are better off than the outer islanders who have emigrated to Tahiti, as are those with more education. When, for example, both husband and wife have jobs, working-class Polynesians can enjoy a fairly high material standard of living – particularly if they have family land in Tahiti on which to build a house (or, better yet, if they have inherited a house on Tahiti). People in this category commonly live in a house solidly build of modern materials, own such consumer durables as a television set with video tape recorder, various household appliances, and an automobile, or sometimes two, and daily consume a wide range of imported foods, as well as wine and beer. Even trips to Hawai'i, California, or France are not uncommon. Several times in Hawai'i I have been visited by Tahitian friends, or their now grown children, whom I had gotten to know while doing fieldwork there in the early 1960s. At that time they, or their parents, were just getting by doing unskilled or semiskilled labor, and perhaps a little farming and fishing. Now, thanks largely to the prosperity brought by the bomb and French public expenditures, they have achieved a standard of living approaching that of workers in industrial countries, even to the point of being able to vacation overseas.

By no means, however, are all Polynesians working in Tahiti so well off. At the opposite end of the spectrum from the regularly employed and landed Tahitians are the immigrants from the outer islands who, lacking in skills and connections, go from one temporary mimimum wage job to another, and are liable to be laid off with every economic downturn, and who, lacking family land or housing on Tahiti, are forced to live in slums (Finney 1973, pp. 135–139; Ringon 1971). When workers suddenly started flooding into Tahiti in the 1960s no housing was provided for them. Although some were able to move in with relatives, most rented little plots of land in informal subdivisions located on the outskirts of Pape'ete or in one of the surrounding districts, or simply staked out unused land in one of the valleys, on which they built flimsy houses out of cheap lumber, composition board, and corrugated iron. If they were lucky, they might be able to pipe water to their plots; otherwise they had to fetch water from the nearest spigot, or from a stream if they lived deep in a valley. There were no sewage facilities; out-houses over a pit in the ground were the only expedient for disposing of human wastes, with predictable results in air and water pollution. Within a few years, whole sections of Pape'ete and surrounding districts were covered by such proliferating slums, or *bidonvilles* to use the French term, which contained about 20,000 to 25,000 people, or around 20 percent of the *Mā'ohi* population.

Although some of these outer island immigrants have been able to recreate an urban echo of Polynesian family life by clustering together in groups of households made up of related persons, or at least people from the same village or island, by and large these slums are settings for anomic misery unprecedented in island experience.[4] Traditional Polynesian values of sharing food and of rendering aid so well adapted to rural living are difficult to translate into meaningful actions in a way of life built upon wages and consumer purchases (Finney 1973, pp. 78–141). Not surprisingly, social problems abound in this lumpen proletariat, although they are by no means limited to people living in the slums. Alcoholism (beer from one of two local breweries is the beverage of choice), drugs (usually home-grown cannabis, or mushrooms used to brew a kind of tea), thievery, spouse abuse, child neglect, and juvenile deliquency are common.[5]

Concerning the latter, Tahitian social workers complain that urbanization and proletarianization have shattered Polynesian family structure, leading

> . . . to juvenile social maladjustment, which is characterized by educational negligence, often accompanied by alcoholism and violence and the absence of the educational role of the father who sets a good example and perpetuates traditional values. The proletarian father hardly shares anything with his son, contrary to the traditional life-style in which he introduced his son to various techniques and activities of daily life. Thus the child has no example on which to structure his personality and consolidate

his moral conscience. (Evelyne and Vala Neuffer, quoted in Crocombe and Hereniro 1985, p. 109)

While such complaints echo similar observations made about slum life in various countries, industrial and Third World alike, Polynesian family structure may be particularly vulnerable to such disorganization. In the Polynesian conception, the period between childhood and settling down, roughly the teenage years extending perhaps into the early twenties, was considered to be a time for both work and play. A *taure'are'a*, as a youth is known in Tahitian, was expected to labor mightily, if episodically, for the household, helping his father in farming, fishing, and household chores. Yet, he was also expected to have a good time, to spend his evenings singing, dancing, and pursuing sexual adventures. Shorn of an economic role by urbanization, and usually educationally unqualified to get a good job, it is not surprising that *taure'are'a* of the slums focus on the play side of their traditional role (cf. Finney 1973, pp. 85–88). This readily translates into uncontrollable youth gangs, which are charged with all manner of crimes, including gleefully burning down part of Pape'ete during the riots of October 1988 (Réalitiés 1987; Dunn 1987).

Protest and Acquiescence

The Polynesians did not invite the French to make their islands a proving grounds for developing atomic weapons for France's *force de frappe*. Nor, judging from both public pronouncements and privately expressed fears, have any significant number ever really been in favor of having the bomb in their midst. But, the Polynesians have been divided over the issue, and even those most resolutely opposed to testing have been powerless to stop it. Despite numerous petitions from religious and political groups, the French have never put the issue to the test of the ballot box. They claim that testing is a matter of national defense, and therefore beyond the competence of any local group of French citizens, even if they are of Polynesian descent living on islands half a world away from Paris. Their response to protest has been to attempt to gain local acquiescence by applying political pressure and manipulation, by developing a concerted program to assimilate the Polynesians to French culture, and simply by trying to buy the people off with prosperity.

De Gaulle and his successors have, until recently at least, been able to manage local politics in French Polynesia so as not to jeopardize the testing program. Broadly, during the era of the bomb most French Polynesians have been of one of two political persuasions: Autonomist or Gaullist. Generally, more *Mā'ohi* than *Demis*, and more Protestants than Catholics, have supported the Autonomist cause, thereby giving it a majority. But given the lightning speed in which political allegiances and principles can shift in French Polynesian class, religion, and politics have not always been so simply linked. The Autonomists are the inheritors of the populist politics of Pouvana'a and the old RDPT party, while the Gaullists generally espouse a pro-French, technocratic line. The French have tried to build up the Gaullists through packing the electorate by allowing French military and civilian personnel to vote in local elections and gerrymandering safe seats out of widely separated Catholic communities. At the same time they have tried to intimidate and outwit the Autonomists by such moves as banning the RDPT when it came out against the bomb and then baiting the successor parties to acquiescence with promises for early release of Pouvana'a and other false pledges.

In large part, their strategy has worked. Although the Autonomists were in power during most of the 1960s and 1970s, they modified their stand on independence to a demand for internal autonomy, and did not continually press to ban the bomb. Their biggest victory, the granting of a degree of internal autonomy in 1976, was essentially hollow. Although it provided a ministerial form of government, the French remained firmly in control of the military, foreign affairs, and the purse strings; the French High Commissioner (a cosmetic change of name from 'Governor'), not the Polynesian Premier (now called the President of French Polynesia), remained the real power in the territory. When the Gaullists finally came to power in the early 1980s, it was more out of Polynesian ennui with the Autonomists, and an initially stunning strategy of their leader, the *Demi* politician Gaston Flosse, of making autonomy within the French nation their central goal, than from French manipulations. Flosse's declaration for autonomy did not at all mean rejection of France, however. In fact, for a time Flosse served both as Premier of French Polynesia, and as a junior minister on Chirac's cabinet with the charge of winning French Polynesia's Pacific neighbors over to the legitimacy of France's nuclear testing program and continuation of tight political control of both French Polynesia and New Caledonia.[6]

Up until the early 1960s, 120 years after the initial takeover of the islands, French efforts to have the islanders speak their language and appreciate French civilization had not been very successful. Although the *Demis* spoke fluent if somewhat accented French, and usually knew something of French geography and history, few *Mā'ohi* spoke French well, or knew and cared much about France. Local knowledge of French and France dropped off rapidly the farther one traveled from the center of French influence, Pape'ete. In remote outer island communities one might find only a few people who could speak any

French or knew anything about the nation that had colonized them. With the coming of the testing program, however, a concerted drive to teach French and acculturate the islanders to French civilization was hurriedly put into place. Language instruction was intensified, and the school system was greatly improved and expanded. A system of kindergarten schools was set up to take in three-year olds, teach them French, and prepare them for primary school. At the other end of the educational pipeline, promising graduates were sent to French universities and technical institutes, and just recently a university has been set up on Tahiti. To reach adult and child alike, television was introduced to Tahiti and the Leeward islands within range of the transmitter mounted atop one of Tahiti's peaks. More recently, a video cassette network has been established in the outer islands. Although the programming has become more diverse than in the first years of broadcasting (when the Tahitian language was banned and newscasts and commentary featured blatant propaganda), because of its universal popularity television has proved to be a powerful tool for teaching the mass of islanders to understand French and to grasp something of the nature of French civilization (Danielsson 1983, p. 209).

These formal efforts, plus exposure to thousands of French immigrants living in their midst, as well as the opportunities to travel to France brought on by prosperity, have been remarkably successful in acculturating the islanders. French is now widely spoken by children, even to the point where it has become the first language of children in households where the parents first or only language is Tahitian. In fact, so widespread is the use of French, that the old distinction between Polynesian-speaking *Mā'ohi* and French-speaking *Demis* is becoming more and more blurred. Although economic class distinctions have by no means been erased, with so many islanders learning French and becoming more acculturated to French ways, the category of *Demi* may be beginning to fade away as a significant social label.

Prosperity, the flooding of the islands with money and consumer goods, has been the central factor in the French effort to get the islanders to acquiesce to the bomb. The *Demis*, with their good paying government jobs and untaxed income from business and land speculation, have of course gained the most. Nonetheless, despite whatever disillusionment there may be among the mass of working-class *Mā'ohi* with this modern, wage-earning way of life, and the slum-living conditions and periods of unemployment, today only a minority of them would voluntarily opt to return to the old farming and fishing way of life; indeed, many fear the prospect of ever being forced to do so.

French officials have not been shy about pointing out that the testing program is the root of their new way of life. In 1979, for example,

Paul Cousseran, France's High Commissioner to French Polynesia, declared that:

> One can be intellectually for the CEP, or one can be intellectually against it. But the fact is that this country lives off it. Three thousand, two hundred families do so quite directly, not counting Polynesian military personnel. Above all, thousands of families live off it indirectly. (Finney 1979, p. 22)

Cousseran then went on, as French officials have been doing over and over again during the decades of testing, to emphasize that banning the bomb and granting independence to French Polynesia would mean economic suicide for the islanders.

> I have said before and I repeat: Independence is not the problem faced by this country. On the contrary its problem is its dependence. Polynesia's problem is that it does not produce the money necessary to pay for what it consumes. So someone must always be found to pay in its stead.

That 'someone' is, of course, the French taxpayer who foots the bill – now exceeding $750,000,000, or over $5,000 for every French Polynesian – for France's atomic testing program. The French taxpayer also supplies the metropolitian subsidies that finance the territory's schools, roads, hospitals and other facilities, as well as the expensive public work projects that have been instituted to take up the slack with the decline in employment after the initial boom in building up testing facilities (Henningham 1989).

Yet, all has not been smooth sailing for the French. For many years the islanders have been more or less willing to go along with the French as long as 'Mama France' was footing the bill for the new consumer society. Widespread disillusion has now set in with the problems the new lifestyle has brought, particularly among those who were basically against nuclear testing but tried to make the best of a bad bargain. For example, Francis Sanford, a Tahitian politican who fought long and hard for autonomy, pessimistically states: 'We thought that the Centre d'Expérimentation du Pacifique was going to lead to enormous wealth. It was an illusion. It brought money but destroyed our lifestyle. We have the politics and attitudes of the Arab states, but we have absolutely no resources. One day everything is going to disintegrate' (Rollat 1987, p. 33). At the other end of the political spectrum, a number of those who had been most supportive of French policies have been discredited. The pro-French machinations of Gaston Flosse were finally too much for the Polynesians; he and his party lost power in the late 1980s. As of this writing (1989) a coalition government is in control, led by a young *Demi* technocrat with a doctorate in economics, Alexandre Leontiff, the grandson of a Tsarist general who fled to Tahiti in the 1920s. The coalition government includes a number of other young and educated politicians who are aligned neither with the old Gaullists not the discredited Autonomists, and who openly challenge the status quo.

However discomforting this rise of a new type of politician may be to the French, it is probably the inevitable outcome of the policy of trying to make Polynesians more French. For Polynesians to become conversant in a world language, and to travel and study outside of their island world, has also meant exposure to political and economic ideas much more radical than homegrown ones. In the last decade or so there have sprung up a number of political parties led by young islanders who, disgusted by the ready compliance with French wishes of the local Gaullist politicians and impatient with the slow progress of the Autonomists, have started to vociferously advocate the abolition of testing and immediate independence. At first, these parties and their young, educated leaders were derided as being hopelessly radical and completely out of touch with the Polynesians electorate. However, in the 1986 elections, two of parties – 'Iā Mana i te Nunā'ā (Power to the People) and Tāvini Huira'atira Nō Porinesia (Polynesian Liberation Front) – garnered over 16 percent of the vote (Rollat 1978, p. 33). What is more, the Secretary General of the Power to the People party, a young Tahitian educated in France as a marine biologist, and converted there to both socialism and the ecology movement, is now the Minister of Health in the current government. In late 1988, to the chagrin of French officials, he managed to hold an international conference on 'Peace and Development' in Pape'ete. For the first time in French Polynesia, the health risks of nuclear testing were publicly aired and France's nuclear policies were condemned in a forum organized by the Polynesians (Robie 1988).

Despite the seemingly peaceful and happy-go-lucky stereotype of the average islander, more violent protests over French rule and the social upheaval it has fostered have also taken place. In the first incident, in 1972, a group of young Tahitians (including a part-American who had served with the US Marines in Vietnam) stole a supply of munitions from the French army to start a rebellion. Though caught in two weeks, they immediately escaped from jail, were recaptured, and then promptly instigated Tahiti's first prison riot. Although this affair had its comic opera aspects, the next incident, in 1977, was more chilling. A group who styled themselves as 'The Blood of the Ancestors' (Te Toto Tupuna) blew up part of the Post Office (after French security frustrated their initial plan to strike at naval ships), and then assassinated a French man. Again the rebels were easily rounded up, along with their alleged ringleader, a young advocate of independence improbably named Charlie Ching who was a cousin of the old leader Pouvana'a (Finney 1979, pp. 25–26). Their conviction and stiff prison sentences may have dampened whatever enthusiasm other Polynesian patriots might have had for such anti-French violence, but it has not stopped disruptive strikes and more spontaneous actions. In 1983, hotel workers went out on strike and used intimidation to close the hotels, thus damaging the one sector of the economy other than French military and governmental expenditures that was bringing money into the territory. Then, during the dockers' strike in 1987 an attempt by the administration to employ force to open the docks resulted in a wholesale riot, which culminated with the burning of part of the business section of Pape'ete by disaffected youths (Réalités 1987; Dunn 1987).

Confetti of Empire

From a global perspective, the little dots that represent the islands of French Polynesia look like mere 'confetti of empire', tiny, insignificant outposts that for some obscure historical accident belong to France.[7] To French strategists, however, the islands loom large – as a place to perfect their atomic armory, and as crucial links in a world-spanning chain of island outposts of a nation with global naval ambitions (Firth 1987, pp. 109–119; Firth, 1989). This grand conception has flourished under the Socialist government of Mitterand as well as those of De Gaulle and his successors, and has survived international protests over the spread over the Southern Hemisphere of radioactive pollution from the tests. It has made the Polynesian inhabitants of these islands hostage to an alien ideological conflict and dream of national glory.

Yet, in a sense the French Polynesians have been pampered hostages, well-paid civil servants, businessmen and workers who have benefited from the boom brought on by the bomb. Quite apart from the as yet unknown impact of nuclear pollution on the populace, this boom has transformed island society. The majority of islanders have become wage laborers and the Europeanized *Demi* class has become even more privileged and affluent. This third major social transformation of island society since the coming of Europeans has vaulted the islanders into the lower ranks of industrial societies – but only in terms of cash income and consumption. For there is virtually no production and islanders are now subject to inflation, unemployment, and other social ills largely unknown to their parents or grandparents.

Such an abrupt disruption of island society has bewildered the islanders. 'We are a fallen race,' states a Tahitian physician, who goes on to declare that things are 'getting worse', and that the Polynesians should strive 'to rediscover the values of former times' (Crocombe and

Hereniko 1985, p. 108) Such sentiments have led to efforts at cultural revival, for example, to the formation of an 'Academie Tahitienne' (*Te Fare Vana'a*) for the preservation of the Tahitian language. There has been an upsurge in outrigger canoe racing, and even the construction of a sailing canoe to follow the old migration trail to New Zealand, and a revival of tattooing with traditional designs. Though such efforts may serve to give the Polynesians a cultural sea anchor, to help them weather the rough, uncharted seas into which they have been cast, the search to rediscover the past has not included any significant attempt to go back to a life of farming and fishing.

There would appear to be no turning back the economic or social clock for the islanders, save by means of a worldwide collapse that, by cutting off the now life-sustaining imports, would force them back to taro, coconuts, and fish. But, few islanders would voluntarily return to the old wars. Among other things, the Polynesians, inheritors of a great voyaging tradition, would not want to give up access to world travel brought by their current prosperity. To be sure, some island thinkers, combining ecological reasoning with a glorification of the old culture, advocate that the people return to the old ways, but so far they are in a distinct minority.

Most Polynesians do not want a cessation of French expenditures, just the removal of nuclear testing and so many metropolitan personnel and emigres. During the 1950s when Tahiti was most difficult to reach, and the French administration seemed to have no interest in developing tourism there, it used to be said around Pape'ete that the French would prefer that rich American tourists stay in California, watch a movie about the sights and joys of Tahiti, and then send to Tahiti the money that a trip there would cost. Now, it could be said that the Polynesian subjects of France would prefer that the French go back to France and take their bombs with them, but still keep sending the francs needed to keep the Polynesian economy afloat.

Something approaching that could happen. The international outcry in the 1970s among Southern Hemisphere nations over the French atmospheric tests, and the more recent embarrassment over the Greenpeace Affair when French secret agents sank a ship (killing a crewman) in the Auckland Harbor to keep it from sailing to Moruroa to protest the underground tests there, did not stop the testing program. However, geology might. Underground tests have made the atoll of Moruroa into a Swiss cheese of radioactive caverns. The neighboring atoll of Fangataufa, where a few tests have also been conducted, is too small to sustain many more explosions. Furthermore, the cyclones that devastated these atolls in the early 1980s, and a tsunami caused when a faculty nuclear test blew part of Moruroa into the sea, have underlined the unsuitability of tiny, low-lying atolls for nuclear testing. It

370

seems unlikely, however, that the French will ever follow American and Soviet precedent by using their home country for underground tests. Thus, the hope of many islanders is that the long-rumored shift of nuclear testing to Kerguelin, a rocky island under French control far from land in the high latitudes of the Indian Ocean, might take place within a few years.

If so, the Polynesians could get their wish for a cessation of testing, and a consequent reduction in the French presence in the islands, with continued metropolitan financial support. Other small island territories under French control such as Saint-Pierre-et-Miquelon off Canada, Martinique and Guadaloupe in the Caribbean, and Reunion and Mayotte in the Indian Ocean, enjoy a fair degree of prosperity based on metropolitan expenditures and support without having to host a nuclear testing program (Shineberg 1986, p. 153). These provide ample precedent for the French to continue some level of subsidy for their Polynesian possesions. But French subsidies would not be enough to maintain living standards at their current level. Some major, non-military sources of income and employment would also have to be developed. Although many islanders fear their homeland becoming another Hawai'i, a major expansion of tourism might turn out to be the price that would have to be paid for the maintenance of their new lifestyle. While an economy based on French subsidies, tourism, and some farming and fishing may be abhorrent to those Polynesians who yearn for a return to self-reliance and cultural independence, it would not be without precedent in the Pacific, and might just be the best deal the Polynesians can get, for the near future at least.

13

Where is Social Change in Hawai'i?
The Reyn's Aloha Shirt
Albert B. Robillard

I propose to do three things in this chapter about social change in Hawai'i.[1] First, I want to formulate a position that locates social change. Then I will examine some contemporary discourses of social change in Hawai'i, a book on political economy of Hawai'i and the ubiquitous sign of Hawai'i, the aloha shirt. I will focus on one kind of aloha shirt, the Reyn's aloha shirt. This shirt is the badge of management in Hawai'i. Third, I will begin to develop the position that social change is itself a specific form of discourse, one closely associated with individualism and capitalism. I will contrast the discourse of social change in individualism and in advanced capitalism with some of the structures of identity of the immigrant Filipino community in Hawai'i.

LOCATING CHANGE IN DISCOURSE

By locate I do not mean to determine when or where an event took place, such as when Kamehameha attacked Maui and Oahu or when Captain Cook landed on the Kona coast of Hawai'i. Neither will I be concerned with the creation of the tourist industry or the rapid infusion of Japanese investment. I do not mean, either, to indicate a process of documenting, in a sense of realism, the penetration of commodified production in the Hawaiian subsistence economy (e.g. sandalwood for the China market). What I want to do with the idea of locating is to describe the discourse which embodies social change and its subjects. I do not propose to be comprehensive, describing all of the possible discourses of social change or stasis in Hawai'i. By talking about discourses of social change I mean to indicate that all people do not occupy the same discourse field or even an equivalent field of combinations. There are many discourses of social change or the absence of change in Hawai'i. These discourses are divided by class, ethnicity, nationality, occupation, education and many more everyday structures. These discourses, the combination of subjects and objects, produce the variable structures, worlds, in which social change is a describable phenomenon: whether in fact the world is changing or not, whether it is in a timeless stasis or in rapid transformation.

All of this has been very abstract. Before I ask the reader's further indulgence with abstraction and what may be a novel approach to social change, a few illustrations are in order. Let us consider the many ethnicities in Hawai'i after contact with the West, 1778. The *haoles* (Caucasians), the Chinese, the Japanese, Portuguese, Puerto Ricans, Filipinos, Samoans, Tongans, and Micronesians came to Hawai'i and collectively outnumbered the native Hawaiians. For starters, I want us to consider if all of these groups, not even considering the subcultures and occupational differentiations among them, shared the same objective world? If we answer in the affirmative, as most people will, this response involves an intellectual conceit, one that most people are unaware of but which, nevertheless, drives the 'imaginary' (Lacan 1968; 1977) of a constant world, one that the various cultures interpret differentially. The intellectual conceit, of course, is the assumption of an objective world, one that is somehow known outside of culture and is preinterpreted. I will argue just the opposite, that all of these groups and many more emergent subdivisions brought their own discourses of social structures and world and these discourses located their subjects in non-equivalent experiences. Although there is a large measure of overlap between these discourses, to the degree that they produce different structures, if not ontologies, they are incommensurate.[2] The rural, Catholic Filipino plantation workers, mostly Ilocano and Visayan, brought a very different discourse – which located agency, power, authority, obligation, work, money, to mention a few parameters – than the discourse of the urban Japanese who had come to Hawai'i earlier (Caces 1985; Beechert 1985). The contrast is even more marked with the Protestant missionaries, who had an individualistic idea of work responsibility, and salvation. While ethnicity is a colonial category (Said 1979; 1985; Kuper 1974), a distinction of difference (Babha 1983), if we are serious about alternative concrete realities, then we cannot accept an equivalence between the multiple discourses that have and continue to people Hawai'i. The Filipino sense, a Hispanic sense, of hierarchy – the differential rights of who can shape production and behavior, of who serves and who sponsors – is very different from the *haole* idea of an egalitarian responsibility, in respect to where social stability and change originate (Okamura 1982).

Yet, as I suspect the reader will say, we have social science disciplines of social change. If the people within the separate discourses of Hawai'i population groups, ethnicities, are incommensurate, do not share homologous worlds, then the social sciences in their surveys of these groups can detect objective social change and stability. This position trades upon the intellectual conceit mentioned above: that the social sciences are somehow exempt from culture, the combinatory of discourse, and have privileged access to an objective world, one that they can examine for true universal social change, comparing one world of ethnic experience against another and with an assumed constant world of functional equivalences. This is an epistemologically untenable position.

This is not the place to go into an extended exegesis of the logical contradictions of positivistic social sciences. In the introduction to this book, I have referred the reader to the methodological discussions of the social sciences by Habermas (1988; 1981) and Shapiro (1988; 1981). A few limited observations will have to suffice here. They will return us to a consideration of the role of discourse in structuration.[3] First, the social sciences stand in a unique relationship to the topics of investigation. The very language, discourse, of description and analysis is at the same time the articulation of the subject matter. This is to say, that categories of accountable, referenceable, perception, the culture of combination formative of the intersubjective world of the investigator, are simultaneously formative of the intelligibility of the topic. This observation echoes the methodological work of Alfred Schutz (1967; 1962–1966), the phenomenologist, and those working in the tradition of ethnomethodology (Garfinkel 1967; Cicourel 1964; Pollner 1987).

Second, the description, analysis, planning, evaluation, teaching, publishing, membership in professional Pacific area studies associations, consulting for governments, working for international aid organizations, obtaining government and international organization contracts and grants, employment and private enterprise, all produce a network of overlapping and interconnected reproductive relations; these are reciprocal or complimentary social institutions wherein the mutual production and exchange of objective truth is the practice of reproducing these social institutions of work. The discourses which establish the intelligibility of the topics and analysts are at the same time the structuration of the mutually reinforcing social institutions within which this work is produced and reproduced. Institutions like the university, social science disciplines, government, professional associations, market economics, international finance organizations (e.g. World Bank, Asian Development Bank, South Pacific Forum, USAID) are produced as a more or less integrated set of social institutions by the referential systems which combine individuals with careers and with the mechanisms of reproduction in the cash economy.

So, as in the case of Cantonese who immigrated to Hawai'i (Glick 1980), carrying a particular discourse of family capital development, a system of reference and validity very different from the redistributive logic of Pacific Islanders and others, the discourses of the social sciences produce and locate the individual within a discrete set of social institutions. These institutions, their origin, function and direction of development, the topicality of interior subjects, are the experience of social scientists and their public. They are not necessarily the institutions of groups with alternative constitutive discourses, like other professions and so-called ethnic groups (see Macpherson on Samoan medicine in the conclusion of this volume).

To summarize what we have said thus far about developing a theoretical position which locates social change and its members within a discourse, a referential logic, I want to enter the following: Social change will be addressed in this paper as coterminal with those accounting practices (Heritage 1984) – the discourse combinatory of subjects and predicates – which produce its membership, the social institutions of mutual reference. The social sciences will not have any privileged position among the many actual and possible[4] discourses of social change in Hawai'i. Social change will be located within the institutionalized accounts of those who articulate it, however wide the scope or ephemeral. We will not pretend to or use the image of social change as a detectable, independent, objective, and universal phenomenon.

TWO KINDS OF OBJECTS: A BOOK AND THE ALOHA SHIRT

The second thing I want to do in this paper is to describe some specific discourses of social change. There is a book, *Land and Power in Hawai'i*, by George Cooper and Gavan Daws (1985),[5] which is an analysis of how elected and appointed politicians and government bureaucrats worked with the real estate industry and the financial community to make huge profits. The profits accrued to the personal positions of all of the above-named parties, including those in government. The news in the book was how people in government used their positions to personally profit and push for land development in Hawai'i. Daws is the author of the widely read history of Hawai'i, *A Shoal of Time* (1968).

It should be no surprise that in Hawai'i, which became a state in 1959 (Rapson 1980; Fuchs 1961; Kent 1983) in the era of monopoly capitalism and a state administered economy (Habermas 1973), that people in government collaborated with private land development interests. By the time Hawai'i became a state, liberal capitalism with the separation between economy and politics had long since passed over into strong state administrations of economic life. In advanced capitalism, as in the absolute states before the bourgeois revolution, the state and those who control its administrative apparatus for private motives are joined. The stated is ultilized to socialize the costs of production, consumption, and the creation of privately appropriated surplus.

What is a surprise is that after an initial period of notoriety the book slipped into social oblivion. Some politicians and others complained about the book being inaccurate or making false interpretations. But the book soon faded from the newspapers and public debate, both in political and academic discourse. In the ensuing elections the material in the book had no place in the public rhetoric.

How can we explain the absence of impact in *Land and Power in Hawai'i*? Why didn't this remarkably detailed book become part of public discourse, perhaps rearranging the hegemony of the Democratic party in Hawai'i and rearranging public office holders? Was the reading public too small? Surely the number of people who read the book, digesting its many financial details, was relatively small. The book certainly was not read uniformly across social classes. However, even among the literati the discourse of the book did not become the means of reproduction of this social location, the cognoscenti. The book, its discourse, the many ironies it displayed of a social democratic party, the Democratic Party in Hawai'i, utilizing the state for the creation and private appropriation of surplus would not become the reproductive structuration of any but a few, mainly graduate students in the social sciences.

Where does this put the object domain of this book, as well as the book as representation? There are explanations of why the book didn't become a part of public discourse. They are explanations focused on the mystifying role of the ideology of individualism, how people in our society cannot perceive the behavior of the state in advanced capitalism because of the dissemination of the discourse of bourgeois liberalism. These explanations aside, the representation world of *Land and Power in Hawai'i* is not a representation or change in the structuration of any activity: it is not the reproduction of teaching, public politics, the news, publication, the moral order of the church, family life, the law and jurisprudence, and other social institutions.

I must add a parenthetical note here to help the reader who may be unfamiliar with this use of language. I am talking of the book *Land and Power in Hawai'i* as a calling up of a world, as in the German Romantic poets (e.g. Rilke) or in the work of Heidegger. The book, for those who read it, summons a world of objects, structuring experience. The discourse of the book produces and reproduces this structure. The structure includes the reader's presence.

If, as I contend, *Land and Power in Hawai'i* is not a significant structuration of life in Hawai'i, if this critical approach to the political economy is not the symbolic medium of will formation, then what are some of the contemporary discourses of structuration? I have deliberately formulated this question in the context of political economy because I will address these discourses in terms of a political economy of signs. This approach is taken from some of the work of Jean Baudrillard (1981; 1975; 1985; 1983). The issue is not only the identification of contemporary discourses but also to analyze where these discourses come from, their social place, and how and to what degree they structure the political economy of postmodern Hawai'i.

I will begin to discuss the political economy of signs in contemporary Hawai'i with a seemingly common item, the aloha shirt and aloha wear in general. The aloha shirt, a colorful floral printed pattern, is similar to the calypso shirt of the Caribbean. The aloha shirt was first introduced[6] in 1936 by a Chinese tailor, Ellery J. Chun, in Honolulu and quickly taken up by the hotels and steamship companies to promote tourism in Hawai'i (Steele 1984). The shirts were sold in the hotels and aboard ship as souvenirs, a practice continued today with aloha attire specifically produced and marketed to the tourist. The tourist aloha wear often features matching his and hers shirts and muumuus. The patterns are often of large tropical flowers, the hibiscus or plumeria, and also have landmark pictures,[7] like Diamond Head, Aloha Tower, the Dole Pineapple, and the word 'Hawai'i' printed everywhere. The material for the tourist aloha wear is printed and imported from Asia and cut and sewn locally.

The aloha attire marketed to the tourist today is differentiated somewhat from the aloha clothing produced and marketed for the local residents. The aloha shirt for local residents is hierarchically differentiated by class and occupation. While not everyone is aware of the meaning of the different types of shirt (differences in label, material, color, pattern, and cut) people within social groups are said[8] to consciously choose and wear a particular kind of shirt. For example, the Honolulu downtown or Bishop Street community, bankers, lawyers, accountants, stockbrokers, corporate executives, advertising agency and public relations people, government agency managers, and those who aspire or pretend to the same social strata, wear a Reyn Spooner aloha shirt, a 100 percent cotton, 'reverse pattern,'[9] pastel-colored, button down shirt costing from $40-$60 and more. These shirts are worn during business hours with the tails tucked in. They are worn with belted expensive tropical wool pants, usually of a dark gray or light brown color.[10] In this setting the shirt is always worn with leather shoes, wing-tips, or loafers. However, during the weekend and on holidays they are worn with shorts and then they are worn with the tails out.

Reyn's is the kind of store that specializes in broad cloth Oxford button down shirts, khaki pants, brass buckled web belts, and Bass Weejuns, a sort of Brooks Brothers or J. Press with Polynesian prints.

The East Coast preppy look is sold at two Reyn's outlets, one in Ala Moana Center, the state's largest shopping mall, and the Kahala Mall, located in the state's most exclusive residential section. Liberty House and J. C. Penney sell the Reyn's aloha shirt. However, the latest prints are only available in the Reyn's stores. The shirts are also sold at Duty Free Shoppers, in the Honolulu Airport, and at other airports and resorts in the Pacific (Guam, Hong Kong, Palau) and at mens' stores in Carmel, California.

While the Reyn's shirt is sold outside of Hawai'i, it is in Hawai'i where it has the most clear-cut social symbolic function, a signification of social position. The Reyn's aloha shirt is taken to mean that the wearer occupies or properly aspires to a particular position in relation to the command of capital and power. The interpretation of the shirt is for those who know; those who do not know do not count. Those who find the shirt stuffy or out of date, compared to the more widely worn and perceivedly more fashionable shirts, are looked upon by the practitioners of the Reyn's code as people who do not know how to produce or maintain the position that the shirt, as a sign, produces. It would not be going too far to say that people who do not practice the Reyn's code are treated by those who do as without the skills, without the means to produce and maintain the social position that the Reyn's fashion sign effects.[11] The shirt is the embodiment of, a signification of the knowledge, sensitivity to local hierarchy, command over resources, intention and aspirations which is, in part, the reflexive, practical achievement of being a member of the Hawai'i bourgeoisie and managerial classes.

It should be emphasized that the Reyn's aloha shirt is a local social order. By that I mean that the shirt is a sign of local social hierarchy, as I mentioned above. The shirt produces the recognition of awareness of and therein membership in a place. Just as Hawai'i pidgin English use produces membership, social solidarity in people living in Hawai'i, people who grew up in Hawai'i (Boggs 1985), the Reyn's aloha shirt *is* the belonging to a social group. The Reyn's aloha shirt is, among those who possess the requisite knowledge, a formulation of, an orientation to place, a geographically situated locale.

The Reyn's aloha shirt is not to be taken as membership in the *kama'aina* families, the mythical old monied families who are, ideally, descended from the early Protestant missionaries.[12] The shirts signify membership in something much broader than the missionary and early settler business family community. The shirts are worn, among others, by the professional management class, people who administer companies like Pacific Resources Incorporated, the local petroleum producer, the Bank of Hawai'i, First Hawaiian Bank, Bishop Estate, C. Brewer, Hawaiian Electric, Hawaiian Telephone, Alexander and Baldwin, Theo Davies, Sheraton Hotels, Hilton Hotels and an increasing number of Japanese nationals who are in Hawai'i to manage Japanese investments. These professional managers are more likely to come from the US mainland or other nations than to come from Hawai'i. The Reyn's aloha shirt is a signification of being in Hawai'i, being in a particular social position, but not necessarily having grown up in Hawai'i. The people who manage these large companies have been, for over twenty years or so, increasingly from elsewhere and the companies have been and are being purchased by overseas investors.

The Reyn's aloha shirt is always changing, like the subtle design changes in the Mercedes Benz. New patterns are always coming out. A feature of the code of the Reyn's aloha shirt is to wear a new pattern or print. There are standard or classic print patterns for the conservative end of the market.[13] However, the classic prints do not have the same social function as the subtly changing patterns. The patterns are now unique for an individual shirt; new prints and print types or styles are produced in mass.

Attached to the wearing of the latest print is the way the shirt is laundered. The more the shirt appears to have been professionally laundered the better. Bumpy collars and frontispieces are to be avoided.

The recency of the print and the freshness of the laundering move the wearer forward in the temporal order of the practice of the code and the signification of acquisition power. The contemporariness of the shirt places the wearer in a social space, a location of being a current member of the class that wearing the shirt signifies. Many people wear old Reyn's aloha shirts and wear them without the requisite laundering. They can be purchased at Goodwill and at the Aloha and Kamehameha Swap Meets. They are easily identified as Others, people not belonging to the same class as produced by wearing new and freshly laundered Reyn's aloha shirts. The temporal positioning that fashion signifies is a neglected topic in the social sciences.[14]

The Reyn's aloha shirt, taken as a variant of the general class of aloha wear, is produced, advertised, and marketed in such a way that the wearer of a copy, by buying it and signifying his knowledge of the code, replicates the phased character of the code. By that I mean that an essential feature of the code is the iterative character of continual combination and recombination of elements of the shirt. To be placed squarely in the professional management class or to be taken by members of the class to be a serious aspirant, one has to be a practitioner of the recombination of the elements of the shirt (new prints and fresh laundering). One also has to wear the proper belt and trousers, usually accompanied by loafers.

The ever-changing nature of the Reyn's aloha shirt, the acquisition and display of new prints, pushes the shirt beyond a mere covering and presents the transparency of the code within the shirt. The shirt is a vehicle and is valued for its being the medium for what Baudrillard calls simulation. Simulation is his characterization post-modern production in capitalism. The Reyn's aloha shirt is an example of simulation in that it displays not only the status of the management class but also the infinite combinatory of production. This infinite combinatory, the means of production, in post-modern capitalism has become both the means of production and the object of consumption. The means of production have far outstripped the physical needs of consumption and now manufacture both the needs (largely through advertising and media manipulation of fashion) and the products to fulfill them. The ends of production, of human activity, have been transposed from having extrinsic use value to the continual manipulation of the productive-consumptive combinatory. Production has become a simulation because it is no longer the product that is the object but the way in which objects express, simulate, the productive-consumptive combinatory and the relative social place and time occupied in the mastery of the combinatory.

Production as simulation is not new. Trade in luxury items has existed for the aristocracies of the world since the beginning of stratified societies. This exchange had little impact on the mass of the world's population (Stavrianos 1981). Luxury trade for the aristocratic market generated the simulation of fashion, especially in the royal courts. What Baudrillard is describing is the current era where even the masses find their identity by how far forward they are on the curve of the productive-consumptive combinatory: where the needs, necessities and wants of the ordinary person are as manufactured as the products to fill them. In this context, the Reyn's aloha shirt is a manufactured need for a segment of the Hawai'i population and at the same time a clear expression of the recognition of the manufactured genesis of need.

The Reyn's aloha shirt is a simulation of the power of production in that the wearer is wearing a constantly new print and freshly laundered shirt. Moreover, this code of interpretation and production is circulated in two senses. First, the shirt is advertised in the press, emphasized in the slick pages of *Honolulu* and *Aloha* magazines, publications aimed at the local bourgeoisie or at least the local managerial class. (The managerial class does not necessarily have possession of capital, as does the bourgeoisie.) Even the two local newspapers, the *Honolulu Advertiser* and the *Star Bulletin*, can be said to be targeted to the middle classes as the production run of the papers are less than a hundred thousand each for a population of over one million. The newspapers are an abstract projection of a common community, the imagined society of the readership (Anderson 1983). Both papers carry large display ads of the shirt. The local TV stations have advertisements for the shirt. Moreover, the TV news anchormen and reporters wear the shirt. The shirt is also advertised in the throw-away handout pamphlets for tourists, like *Oahu This Week*.

Second, the shirt is broadcast by the people and places in which it is worn. It is not only that these people and places, their social status and character, are reflexively produced, but within this there is a detailed discursive knowledge, cause for evaluation and prescription in conversation. People working in the places where this shirt signifies the simulation order of production and one's relative position in the order are made to know about the properties of the shirt.[15] That is, among male executives there is direct conversation about the social merits of the shirt, along with the required accessories.[16]

The aloha wear scheme is further institutionalized by such things as Aloha Friday and Aloha Week. Aloha Friday, every Friday, is when everyone is expected to wear aloha attire. Aloha Week, in September, is a celebration of the aloha spirit, a feeling of care and belonging to Hawai'i. Aloha Week was initiated in 1948 to promote tourism and Hawai'i manufactured clothing. There are many other holidays, Kamehamcha Day, Prince Kuhio Day and May Day (Lei Day), that are celebrated by wearing aloha attire. Even funeral notices in the newspaper request that aloha attire be worn.

Then there is the topicality of how 'informal' Hawai'i is in the expectations for work and entertainment clothing. The institutionlization of aloha shirts in topicality of talk, mainly among managers from the mainland, belies the formal character of aloha attire, where and with what it may be worn. An invitation to a cocktail party stipulating aloha attire does not mean shorts and beach sandals.

The aloha shirt, a commodity first resisted by the local Hawai'i population, has become an internationally recognized sign of the Pacific and Hawai'i. It is worn by almost everyone from Hawai'i in the North to New Zealand in the South and from French Polynesia in the East to Palau in the West. Moreover, this shirt has become a US mainland staple. From the beginning of manufacture, the aloha shirt has also been designed, printed and sewn in California and New York. After the War, the shirt was increasingly produced in Japan and then Singapore, Taiwan, Hong Kong, and Thailand.

What kind of object is the aloha shirt? How is it different from the object of a critical political economy of how people in government in Hawai'i took advantage of their positions to make tremendous profits from land development? I have argued that the book, *Land and Power in Hawai'i*, is an irrelevant discourse, noticed by few. Although the language of the book can be said to be true, in a truth correspondence method of reasoning, the book has a narrow impact in Hawai'i. Contrastively, the aloha shirt – a tiered or stratified social phenomenon – has a much larger ambit. The aloha shirt is a phenomenon made possible in the unique social space of late mass market capitalism. It is a medium, among others, of the community of simultaneously mass and continuously modified production and marketing. The aloha shirt was initially hand manufactured and custom made, but it quickly was taken up by the tourist industry and a local garment industry. In fact the aloha wear industry became a major exporter, along with sugar and other agricultural products. The aloha wear industry has certainly evolved into a major economic sector in Hawai'i. The aloha wear style contrasts with *Land and Power in Hawai'i*, as an object, in that the circuit occupied is much more general and iterative.

The aloha shirt and aloha wear projected by newspapers, magazines, local TV newscasters, the state recognition of Aloha week and other holidays where aloha attire is expected, tourist industry work dress codes, clothing and department store displays, private industry Aloha Friday formal and informal dress expectations, the wearing of aloha attire by private and public sector management, the promotion of aloha attire in TV movies and series filmed in Hawai'i (*Hawai'i Five-O*, *Magnum PI*, *Jake and the Fat Man*, and *Island Son*) and motion pictures filmed in Hawai'i, and the entire tourist industry promotional advertising, all these, surround the inhabitant of Hawai'i with the symbols and signs of production membership in a place. While the aloha attire began as a custom design and produced business, now in advanced capitalism it has become an omniprevalent sign, an expectation, of belonging to a place, Hawai'i. The aloha wear is a practice of solidarity, a sign of value, an index of social integration and legitimacy of two kinds of membership.

The first kind of membership is signified by the tropical pattern of the aloha wear. A copy of a shirt or a muumuu is an indication of participation in a consumer lifestyle: having been to Hawai'i for a vacation; aspiring to such a vacation or participating in a vacation lifestyle somewhere else, on the mainland US; living a tropical lifestyle in Hawai'i that is somehow different from living elsewhere. The aloha style has become the same as the safari wear, sold at Banana Republic stores, affected by Americans traveling in Southeast Asia and the Pacific Islands. While everyone else is wearing Western metropolitan street clothes, and smiling at the passing ersatz leopard band hat bedecked American, the American is wearing a sign of his difference. The copy of the khaki military uniform signifies an adventure; a charged adventure, meaning that it represents, however phantasmagorical, some script of previous military or government sponsored exploration. The people wearing safari suits are on a mission, albeit magical. Aloha attire is the same, except here the indexed mission is vacation leisure, the power to buy it, and freedom from the uniformity of mainland business clothing.

There is an element of Hollywood in aloha wear. The Hawai'i of Dorothy Lamour movies, distributed around the world, with their more than disguised exotic sexuality are replicated in the tropical print shirts. These movies and others like them are a replication of a print representation of the tropical Pacific. This literature, from the early explorers to Herman Melville (1964), to Robert Louis Stevenson (1948), to James Michener (1947; 1959) to the Fodor guide books (1968) have revealed in the differences of Hawai'i and the Pacific from European culture. To some extent the Hawai'i of the aloha shirt is a reproduction of the European perception of and attraction to Pacific exotica, if not the idea of an island Garden of Eden, and the historical mass distribution of this representation.

The second kind of membership is signified by the ephemerealness of the object. The aloha styles, pattern and print, are always in flux. It is not only the wearing of aloha attire, it is important to wear new, current, aloha wear. The producers and marketers of aloha wear are continually introducing and soliciting the consumption of new combinations. The wearer of a new copy is indicating his place in the social class structure, his relationship to the power to buy new objects which are renewed not because they wear out but because they are replaced by what is merely new. The Reyn's cotton aloha shirt, with its double measure of newness, a new print and fresh laundering, is the apex of this social class statement. The wearers of polyester aloha shirts, printed with Japanese patterns, often of a very large scale, or polyester-/cotton drip dry shirts, frequently printed with plumeria on a black background or bird of paradise on a red background, are positioned in a lesser social class. Of course, the positioning is by the practitioners of the knowledge embodied in the Reyn's aloha shirt.

There are two points to make about aloha wear. The first is that the contemporary aloha wear phenomenon is very different from making and wearing an item in ancient Hawaiian society. It is different from making and wearing a feather cape or a ritual helmet. In this context these items are manufactured and worn by age graded ascribed social standing, an inherited social caste with its privileges and obligations. It is also different from the era when aloha shirts were custom made or manufactured in low production runs. Even though the shirt had become a commodity,[17] in the 1920s and 30s the circuit of the manufacture and projection of consumption of the shirt had not become so centralized as to be able to effect a continual flux in fashion. The reign of ephemerealness or fashion had not set in (Barthes 1983). Aloha wear in advanced capitalism is a projected and consumed sign of social position and place. It is a transparent medium of the infinite recombination of the means of production and directed consumption.

The Reyn's aloha shirt is the most transparent of all the aloha wear styles. It is the most ephemeral, perpetually changing print and always appearing freshly laundered. The shirt also displays the cycle of the continual renewal of capital, the generation of money and investment in the acquisition of the latest model. This ephemerealness bears at once the production system and the wearer's relation to it. This relation, signed by the shirt, is one of cultural and material monopoly. Maybe we should talk only of self-perceived cultural hegemony. In advanced capitalism, post-modern capitalism, the production system is increasingly composed of fluctuating signification of distinction or difference in tastes. This form of production is contrasted with the idealization of production for use value.[18] It is the control of the code of signification, in its production and the universal projection, that positions the wearer of the Reyn's aloha shirt, in the context of the practitioners of the code. This position is not universal, like transcendent class categories, but is apparent only to those who practice the code. The fact that it is apparent only to them is the limited distribution of the transparency of the code of post-modern production. Obviously, there are many alternative contemporaneous codes.

The second point is that while the aloha wear style is an organizing factor of life in Hawai'i, it does not collect everyone or even most of the people all of the time. There are many people who are oblivious to the aloha wear prescription, let alone the Reyn's aloha shirt. Among these are the recently arrived immigrants from the Philippines, China, Taiwan, Vietnam, Thailand, and other groups not hooked into the mass media and work place projection of the aloha wear prescription. Often these people affect styles of clothing manufactured in the home or other Southeast Asian nation: ersatz designer label knitted shortsleeved shirts. There is a substantial functional illiteracy in Hawai'i, from 37 percent on the Big Island to half of that on Oahu. The functional illiteracy is concentrated in the immigrant population. To the extent that aloha wear is circulated by the print media, given generality, these people are unaffected. Moreover, even the people who wear Reyn's aloha shirts do not wear them all of the time. They also wear suits. They dress up; they dress down. The aloha shirt, or any other form of discourse, is simply a possibility of behavior, a institutionalization of 'probabilities of conduct honored and dishonored' (O'Neill 1983, p. 698).

There are many styles of clothing in Hawai'i. The aloha wear style is only one of them. The purpose of describing the aloha wear phenomenon, particularly in the form of the Reyn's aloha shirt, has been to adumbrate a form of social change: from a simple commodity to an increasingly centralized and universalized form of integrated production and consumption, from a symbol of being in Hawaii to a sign of the boundless equivalence of signs, including their temporality. In the Reyn's aloha shirt we have reached a point of institutionalization of the transparency of the sign form of production to itself.

The book, *Land and Power in Hawai'i*, and the aloha wear phenomenon are two different kinds of objects. Of course, the fact that they can be compared as objects, inspected as members of the same class, objects, is a modern phenomenon. That they can be compared is an achievement of the primacy of a political economy of signs, a virtual social structure of exchanged abstractions, as in this chapter. In any event, it is possible to distinguish between these two objects. The book is tied to an external referentiality, the times, places, names, amounts of money and of the relationship between real estate development and state and county government in Hawai'i. The book has a much more restricted domain of abstraction than the aloha shirt, a structure composed of components with no referentiality to any constraining external object. The aloha shirt is homologous with a social structure composed of the universal distribution of signs, a general social space projected by the abstract equivalence of signs, the space sustained by the continual combination of elements of need and products. The book is produced in post-modern capitalism but it is tied down to a limited form of distribution. The shirt, particularly the Reyn's aloha shirt, is a simulation exemplar of the transparent medium of signs, a form of production having no referentiality except to the infinite combinatory of signs.

A VARIEGATED SYSTEM OF CHANGE

The argument thus far has been twofold: First, that the social status of objects and whether they change or remain constant is located, made intelligible, in terms of a particular discourse, a specific combinatory system of differences. Second that there are different types of discourse existing contemporaneously. The book *Land and Power in Hawai'i*, an analysis of political economy of development in Hawai'i, was introduced. A description of social change, it is both an analysis and an object. Of interest was its status as an object: How this book, manufactured in post-modern capitalism, because of its ties to an external referentiality, does not produce the same social space as the aloha shirt. The aloha shirt is the same as the mass social space opened up and sustained by the general circulation of abstractly equivalent sign functions, a thereby transparent system, the essence of a system. The aloha shirt is an invented item, a merely commercial commodity, and is in continual flux; the iteration of change makes visible the equivalence and extended space of which the aloha shirt is part.

The book and the aloha shirt are two different types of social organization and change. Yet they exist in the same society. They may exist within the same individual. There are other and more incommensurate social orders within the Hawai'i setting. I will describe one of them. From the presentation of material on hierarchical relationships of obligation in the immigrant Filipino community in Hawai'i, I will develop an argument that social change is itself a production of systematic ideology of individualism, where the individual is the reductionist value of creation, a creation which involves constant change in pursuit of posited human (individual) needs satisfaction. I will argue in the conclusion that the most extreme form of individualism ideology is in the post-modern code of the Reyn's shirt.

HIERARCHICALISM AND AN ABSENCE OF CHANGE

While we like to think panoptically about society, like a unified entity, and tend to think of post-modern capitalism reticulating the entire socia structure, there are many discourses which predate and exist along side post-modern capitalism. These are, for example, the discourses of obligation and caste among Micronesians and Samoans, of whom there are an increasing number in Hawai'i, and among significant portions of he Filipino community. The Filipino population is approaching fifteen percent, in a population mix where the largest two groups, Caucasians and Japanese, compose thirty-two and twenty-seven percent.

From Northern Luzon and the Visayan Islands in the Philippines have come and continue to come a large number of immigrants. Initially recruited as plantation workers in sugar and pineapple, the post-war migration has been increasingly based on family chain petitioning (Caces 1985; Beechert 1985, p. 237), the type of immigration permitted for the purpose of family reunification. The majority of Filipinos in Hawai'i are Ilocano, coming from the Western side of the Cordillera, the mountain chain running down the middle of Luzon. For many of these people, as well as other Filipinos, the income they generate in Hawai'i is redistributed to relatives in the Philippines. The income is redistributed often to the point of sending all money above living expenses home. This is a form of unilateral gift bestowal; the power and hierarchy of family and patron structure, to extended clan and family household employees, is produced by this unilateral redistribution of income from the producers to the thereby clearly dependent. In this system of obligation, repetitive redistribution takes precedence over saving or investing surpluses. Those who have dollars are perceived as having the duty to give to those who do not; people who do not have income in terms of cash can and do unabashedly ask for money or consumer items. The people who give the money and items are constrained by a collective notion of identity and obligation, wherein failure to redistribute income would result in shame and loss of family role.

The income redistribution is hierarchical. Those having higher status are expected to support those below: parents are expected to support dependent children, whether they are minors or not, the oldest children are expected to support their parents and siblings, people in the United States are expected to support adult brothers and sisters and other family members, including distant relatives. Income redistribution is frequently spoken about as an obligation. In fact, people cannot go home without boxes upon boxes of gifts for relatives and friends. To be without gifts is to suffer shame.

The point to this excursis on immigrant Filipino income redistribution and hierarchy is to contrast it, this integration of subjects into social hierarchy and obligation, with the individualism of contemporary capitalism. Of course, individualism is not new, having its theoretical formulation in Christianity, the Enlightenment and in the history of economic liberalism. I do not want to dwell on the history of individualism (Lukes 1973) or its opposite, collectivism in restorationist thought, but the contrast is introduced to begin having a way to talk about social change being located in the discourse of autonomous individuals. The very object of social change, social change itself, should be considered a production of a system occupied by subjects who independently control their own behavior, seeking out the most efficient means to achieve autonomously selected ends. In a system of discourse of independent, rational, critical subjects the order among individuals, society is under continual inspection and is open to revision. Social change is endogenous to such a system of discourse. The ends and means of behavior are open to individual choice. Whereas among the Filipino community described above, there is less of a sense of change in a system composed of hierarchical integration and obligation. The world of obligation is not open to continual criticism and revision among significant portions of the Filipino community.

The discourse of hierarchy and obligation in the Filipino community contrasts markedly with the liberal pluralism thought to penetrate contemporary society. Yet this system of collectivism exists alongside the discourse of autonomous individuals, the ideal of advanced capitalism. The tension between these two discourses is found in comments of disapproval of Filipinos who have lost respect for elders, who do not bring gifts home, who do not send money home, and who do not recognize their obligations. These people are told they have become American and modern. The thrust of what I am proposing is that while immigrant Filipinos in Hawai'i are employed in the capitalist system – as is the case at home in the Philippines – for them the world is not composed of autonomous individuals and there is no equivalent to the iterative recombination of structure in individualism. Family and extended clan obligations are not the product of individual free choice.

The implication of this position is that social change is a phenomenon of a particular form of discourse, the discourse of individualism, including social science, and social change is not a universal phenomenon, located in equal dimensions in every group in Hawai'i. We have to begin thinking of social change as a phenomenon located within a specific discourse. Even though subjects may participate as means in one system of discourse, as Filipinos do in their employment, these subjects may not be part of the same system of representation. It is perhaps ironic that many of the cutters and seamstresses in the local production of the Reyn's aloha shirt are Filipino. They are producing the representation of an inherently unstable system, a code of infinite recombination. However, they are not members of the self-transparent system of signs, a class, that the shirt designates.

CONCLUSION

We have encountered three topics, a book, the aloha shirt, and the sense of enduring obligation and hierarchy among Filipinos in Hawai'i. The book, *Land and Power in Hawai'i*, is a fascinating account of how officials in government and developers collaborated in the transformation of the state from an agricultural to an urban-based tourist and military economy. The book names names and identifies the amounts received by government officials and private financial interests. But the book had little impact. Although digested in one of the local newspapers, the book caused no more than a ripple of public interest. As an account of social change in Hawai'i, the volume and its object world, its referent, are treated as yesterday's news. The representation of social change, what the book summons before us, in the movement and growth of capital has a limited orbit. This is not to say that the book does not report the truth. This being a sociological analysis, the move here is intended to locate any version of Hawai'i, including truth, in the social institutional structure that makes it an objective fact. The book and its referential world are produced – as an object in the world of practical activity – in the discourse of Hawai'i historians, political scientists, and a handful of social science graduate students. Collectively this is not a large group of people.

The point is not to castigate the book. The point is to locate representations and their objects in the practical institutional interaction of people who are thereby members of the structures constructed by those practical activities. This may be a strange position for the reader. We are wedded to a boundless sense of cultural realism – like knowing social change actually happens – and any proposition to locate 'actual' events in the social discourse of practical activity will sound absurd. This is what I am recommending: that we give up any realism and proceed to a thorough sociological analysis, made possible only by eliminating a priori versions of the world.

Next we encountered the aloha shirt, most particularly the Reyn's aloha shirt. The shirt was analyzed as a system of signs, a set of constantly changing combinations of print, material, laundering, and trousers and shoes. The Reyn's shirt constitutes a code among a segment of local Hawai'i wearers. The code signifies membership or membership aspirations to the professional upper middle class, the corporate managers, lawyers, engineers, architects, downtown accountants, bankers, and government administrators. The code signifies not only membership but also, through the introduction and display of new shirt combinations, a mode of production. This mode, post-modern production, is detached from extrinsic use value, producing both needs and the objects to meet them. The shirt signifies post-modern production and ones' membership and relative position by the transparency of fashion, the continual recombination of elements of the Reyn's aloha shirt.

The Reyn's aloha shirt is but an example of post-modern production and the relatively small middle class or knowledge worker group that belongs to this discourse (Ehrenreich 1989). The Reyn's shirt is the sign of the upper middle management class in Hawai'i. The fact that the purchase of the shirt by management, mostly from the US mainland, signifies belonging to Hawai'i is an indication of the 'produced', simulated, character of identity in post-modern production. The produced character of Hawai'i as a whole, as a tourist destination, extends to the substitution of Tahitian and Samoan dances for Hawaiian hula at tourist luaus, the substitution of wood carvings of tiki images crafted in the Philippines for locally produced goods, the substitution of Filipino, Japanese, Chinese, and *hapa haole*[19] women on travel posters and advertising media for Hawaiian females, the promotion of Asian cuisines and the almost total neglect of Hawaiian food restaurants and the substitution of Filipino, Japanese, Chinese and *hapa haole* young men and women, dressed in day-glow exotic lavalavas, pretending to be Hawaiian aiport greeters, for native Hawaiian greeters. At the Polynesian Culture Center, operated by Brigham Young University-Hawai'i, the supposedly authentic ethnic Marquesan and Tahitian villages are at times staffed by Samoans.[20] Pacific islanders are used interchange-ably. The ethnic villages are sanitized versions of the real island locales, without the smells of farm animals, turned earth, crops, fish, sewage, and the ever present flies. The student staff of the villages are dressed in conformity to Mormon modesty. There are no bare breasted women or naked children, not even a nursing mother and child. Even the sails on the Waikiki tourist dinner cruise boats are only for show. The sails simulate Polynesian navigation, on huge double-hulled catamarans, or the era of Western clipper ships. The dinner cruise boats are diesel powered, down to the spotlights high-lighting the fake sails. The managed identity of tourist Hawai'i has little explicit referentiality to an external reality. For those who produce tourist Hawai'i, the production is a simulation, a construction having primary referentiality to the system of capital and labor combinations which display and maintains this Hawai'i. Historical reality is inciden-tal to considerations of attracting a mass market and capital costs and return of servicing the market. This attraction replicates the most widely distributed versions of Hawai'i, the exoticism of Hollywood and European literature.

The Reyn's aloha shirt is part of post-modern man and post-modern production. For those who belong to this system, the structure is plainly transparent, including the dynamic of continual change of the combination of elements. There is a form of hyper-individualism and social change in postmodernism. This is much more than Enlighten-ment or Christian individualism, although its roots lie in the cel-ebration of individual rationality and salvation. The hyper character of the present epoch of individualism is in the calculus of the actor, the human being, as essentially a changing and instrumentally changeable ensemble. The individual makes himself. At least major portions of the world, if not all of it, are constructed and transformable by the reproductive work of humans. All is a simulation of human labor. There is no higher order. In this anthropocentric world, of potentially infinite combinations, individual ends and social structure are posited as perpetual change mechanisms. Social change is inherent, at least in potential, in every aspect of the world in the consciousness of transparent or simulation production. The Reyn's aloha shirt, for those for whom wearing it is a simulation of the productive order and, even, ontological origins, is a true representation of the world as change. The shirt represents an instrumentality and agency.

This is not to say that the practitioners of the Reyn's code are in any way avant garde. Frank DeLima, a Honolulu comedian, has a routine about how dull the people are who wear the Reyn's shirt. Wayne Harada, the entertainment editor and critic of the *Honolulu Advertiser*, writes of the DeLima's Reyn's wearing character, a local

Japanese on the way up the corporate ladder, generically named Glen Miyashiro, that he always wears his shirt tucked in his trousers, reflecting his conformity and absence of flair (1989). The intended aura of the shirt is almost ascetically conservative. The message of the shirt is that the wearer recognizes and is a master of a production order, like J. Press clothed 'masters of the universe' of New York finance capital in the Tom Wolfe novel, *The Bonfires of the Vanities* (1987). The Manhattan equivalent of the Reyn's shirt code can be seen in the J. Press advertisements in the *New Yorker Magazine, The Wall Street Journal* and *The New York Times*. J. Press is a New York men's clothing store.

The non-trivial character of the Reyn's aloha shirt is apparent when we compare it with *Land and Power in Hawai'i*. The shirt as an indicator of a production system, if not a cosmology of hyper-individualism and concomitant social change, that is the rationality of the land development reported in *Land and Power in Hawai'i*. The alienability of land, of residence, of human labor, of occupation, of lifestyles, of personal affiliations and obligations, of finance capital and beliefs are made possible and accelerated by this anthropocentric system. There are no ultimate constraints in a world of human agency. This world is thoroughly secularized, Hawai'i appears as a resource for unlimited human ends. While *Land and Power in Hawai'i* represents the movement and growth of capital in land development to a limited readership, the code represented by the Reyn's aloha shirt is the institutional structure of social change. The code is isomorphic with the living presence of social change as an inherent characteristic of existence.

The produced nature of the shirt, Hawai'i in the shirt, membership in Hawai'i by the wearer and as an indicator of a production system make the code of the shirt a prior to the book, *Land and Power in Hawai'i*. To even read the book as a true account one has to assume the anthropocentric possibility of hyper-individualism and resulting social change. This assumption is the truth correspondence value for reading or being attracted to the book. The sence of the book rests on the code. The book is a phenomenon of the code.

That the code and social change are a limited discourse, organizing some but not all, was displayed in the example of the Filipino enduring sense of hierarchy and obligation. The immigrant Filipino community in Hawai'i is largely unaware of the Reyn's shirt code. Personal relationships, particularly in the family, are not so open to individual decision. The world is not an anthropocentric place for the Filipino community. The ends provided by Philippine Catholicism, including norms of interpersonal affinity, are not projected as cognitively interpreted by every man, as in the more secularized Protestant vision. Man is subordinated to a received truth he has no role in constructing. The subordination to extra-man reality is further elaborated by beliefs in spirits who inhabit the ground, coming out of the ground after sundown, elves who live in anthills, the out of body wandering of the souls of children when they cry at night, ghosts, predestination of the time of death and curing of illness by magic, laying on of hands, psychic surgery and the deeds of angels, saints and Jesus Christ. Then there is the sense of *bahala na*, the perception that an individual is powerless to alter fate. An *ordinario* individual, a common *tao*, in distinction from the elite, does not have the means to alter his life. This list[21] is a sample of what the secularized world calls an enchanted world. This is not a world constructed by man, freely choosing and changing ends and means. This world has a stability, at least in the terminus of action, not seen in the 'productivist' world of the Reyn's aloha shirt code.

Social change, what may seem to be a universal category and object, is really embedded in a specific discourse membership. The status of *Land and Power in Hawai'i* is dependent on membership in a particular discourse. The Reyn's aloha shirt simulates the transparent discourse of post-modern capitalism, the most extreme form of anthropocentricism. This is a location of social change in Hawai'i, perhaps the most extreme social change. But this is clearly not the only discourse in Hawai'i.

14

Language and Social Change in the Pacific Islands

Donald M. Topping

Communication and change are both essential for all social organizations, human as well as non-human. All socially organized creatures, be they ants, birds, chimpanzees, or humans, must communicate in order to maintain their social organization. Continuous adaptive change is a universal feature of evolution on earth. However, problems are encountered when the communication system becomes a cause of social change, as often happens when a dominant culture imposes its language on another. The results are very similar to those of any introduced technology, particularly when the dominant language is a written one.

Changes in language and social structures of the people of the Pacific Islands have been ongoing since the time of first settlement. As new waves of adventurers set forth in canoes to discover and settle new lands, new social structures evolved, adapting to the varied conditions of small island ecosystems. New vocabulary and concepts were developed to suit the new environments. As a rule, this process continued slowly, almost imperceptibly, over succeeding generations, achieving a degree of stability and harmony among homogeneous groups. The pace of the adaptive change was abruptly quickened with the arrival of men in ships from the other side of the world who brought with them microbes, concepts, technologies, and languages that were to have a lasting, and in many cases fatal impact (Cf. Moorehead, 1966; Stannard, 1989).

Of all the innovations introduced to the Pacific Islands since the time of initial contact, perhaps none has had greater or more lasting social impact than introduced languages, particularly during the period since the end of World War II. This chapter describes some of the dimensions of that impact as seen through the eyes of a linguist who has been an observer of social change in the Pacific Islands for more than thirty years.

'The Pacific Islands' throughout this chapter includes all of the islands of the Pacific area, including New Zealand and Hawaii, but excluding Australia, Indonesia, and the Philippines. Many of the generalizations apply to most Pacific Island communities, all of which share a similar colonial history. Three of the island communities – Hawaii, New Zealand and Guam – have progressed furthest along the path of linguistic and social change, and may be harbingers of the future that awaits the rest of the island peoples as they drift into the maelstrom of westernization.

LINGUISTIC OVERVIEW OF THE PACIFIC ISLANDS

The Austronesian language family is the largest in the world in tems of number of languages (more than 800), but in terms of numbers of speakers – if we exclude the large populations of Indonesia and the Philippines – it is relatively small. The languages of the Pacific Islands constitute nearly one-fifth of the world's languages, and yet are spoken by only one one-thousandth of the world's population. If we take the population figure for the Pacific Islands to be approximately five million, this averages out to about 6,000 speakers for each language. However, the ratio doesn't work out that evenly. There are, for example, more than 250,000 native speakers of Samoan (including overseas Samoans), but only two native speakers of Korighole (in the Solomon Islands) at the last official count.

Generally speaking, each island group in Polynesia has a relatively large population that speaks a single language (e.g. Tonga, the Samoas). while comparable regions in Melanesia contain numerous languages spoken by relatively small numbers of people. For example, Vanuatu (according to Tryon, 1971) has a population of 112,000 people who speak approximately 110 different languages, which breaks down to about one language for every 1,000 people.

The situation in Micronesia falls somewhere in between that for Polynesia and Melanesia. The land areas are quite small, especially the coral atolls found throughout Kiribati, the Marshall Islands, and the Carolines. Some of these scattered atolls are united by a single language with minimal dialect variation (e.g., Marshall Islands, Kiribati), while others are divided by either extreme dialect variation (e.g. Truk), or even separate languages, as is the case in the Central Carolines (e.g. Ngatik, Mokil, Pingelap). Considerable dialect variation can even be found among speakers living on single 'high' islands, such as Ponape, Yap, and Guam.

In short, the Pacific islands, especially Melanesia and Micronesia, are marked by extreme linguistic diversity. A question that has piqued historians, linguists, and other Pacific scholars for years is how it came to be that way. Comparative linguists, working with archaeologists and anthropologists, have worked out some fairly convincing theories

394

of language change, migration patterns, and dates, all of which are still undergoing revision and will not be addressed here.

However, there still remain some tantalizing linguistic puzzles for the comparative linguist, who likes to see clear patterns of language kinship, especially where close proximity of islands would lead one to expect a close relationship between the languages spoken there. In parts of the Pacific the close relationship between languages is quite clear. For example, the major Polynesian languages – Hawaiian, Samoan, Tongan, Tahitian, Maori – all share common vocabulary and grammar. Even though they are not mutually intelligible, one can readily see that they are 'sister' languages.

Such is not the case, however, in western Micronesia, where neighboring island communities speak languages that bear no immediately recognizable kinship to one another. For example, even though the islands of Yap and Ulithi are less than one hundred miles apart, their languages are as different as English is from Russian. A similar situation obtains between Palau and Sonsoral (an island of the Palau group), and between Ngatik and Nukuoro atolls in the Carolines.

In various parts of the Southwest Pacific are found Polynesian 'outliers' surrounded by Melanesian or Micronesian languages (e.g., Takuu, in Papua New Guinea; Kapingamarangi, in Micronnesia; Tikopia, in the Solomon Islands; Mele Village, Efate Island, Vanuatu). Although the question of how and when the Polynesians living on these 'outlier' islands got there has been more or less resolved, many details remain to be answered.

Not to be overlooked in this quick glance at the language mosaic of the Pacific Islands are the aberrant languages, so called because they have no immediate linguistic relatives. (Linguists refer to them as 'isolates'.) Examples are the languages of Nauru, Yap, Palau, and the Mariana Islands. Although obviously members of the Austronesian family, each has no readily apparent sister language.

Specialized Languages

In many oral cultures, in which language is used extensively as a carrier of cultural symbolism and connotations, different forms of the common language are found in some of the Pacific Islands where social stratification, at least until recent times, was rigidly observed. For example, in Truk, Caroline Islands, a form of Trukese known as *Itang* is used among diviners, magicians, and navigators from one end of the Truk continuum to the other, providing a lingua franca known only to this select group of men. (Gladwin 1970, p. 125) On Pohnpei, Caroline Islands, the language known as *Lokaiahn Wahu*, or 'honorific speech', is used when addressing a person of high social rank, particularly in certain ceremonies honoring titled persons (Rehg 1981, pp. 359–375). Inevitably, these specialized languages are rapidly falling victim to the leveling effect of democratization as the societies undergo the process of rapid social change. The young people are simply not learning them.

Pidgin and Creole Languages

In addition to the 800-plus indigenous Austronesian languages of the Pacific Islands, there are the several pidgin and creole languages serving important communication needs. Born (according to some creolists) in the plantations of Hawaii, Queensland, and in parts of Melanesia, where expatriate-owned plantations were developed and worked by Pacific Island laborers, these hybrid languages have grown to serve as much needed linguae francae, particularly in the multilingual populations of Melanesia, where two or more different languages can often be found within an area of ten square kilometers. Born of the basic need to communicate with speakers of other languages, these English-based pidgin languages took firm root, and are now used extensively in Papua New Guinea (*Tok Pisin*), Solomon Islands (*Pijin*), and Vanuatu (*Bislama*), with dialect variations. A second, and until recently widely spoken, pidgin – *Hiri Motu*, or *Police Motu* – based on *Motu* (an Austronesian language of Papua), is still used in the Papuan area of Papua New Guinea, which includes the capital city, Port Moresby, where *Tok Pisin* is now the dominant language of the mostly immigrant population.

Hawaiian Creole English (also called *Pidgin*) is an advanced stage of a pidginized language. It is a pidgin that evolved into a mother tongue. While true pidgin languages are used as a second language between speakers of different native languages, creole languages are those that are learned as the first, and usually only, language by children. Creole languages are, in a sense, pidgin languages that get taken into the home.

These pidgin languages should be viewed as the products, rather than the cause of social change. They now hold an important place in the contemporary Pacific.

It should be noted that the pidgin languages of Melanesia and Hawaiian Creole English are hotly debated political topics. By some they are regarded as 'bastardized' or 'corrupt' forms of standard English, which should be eliminated. However, because these languages evolved naturally from an inherent need to communicate, they are strongly identified as native languages, belonging to the people. They are also viewed by many as the alternative to the introduced colonial language that is identified with colonization. The dilemma facing the young governments in Melanesia is whether to embrace the pidgin languages as a show of national pride, or to insist on proper use of the world language in their education and government systems. The answer is not an easy one.

Intrusion and Early Influence of European Languages

The first-known language contact between Pacific Islanders and 'outsiders' was with the early European explorers of the sixteenth, seventeenth, and eighteenth centuries: Magellan, Drake, Mendaña, de Quiros, Bougainville, La Perouse, and others, including the celebrated Captain James Cook. The linguistic influence the early explorers had on the people of the Pacific was minimal compared with what was to follow in the wake of Cook. As far as is known, they left few marks on the languages of the Pacific. Examples of archaic borrowings that represent the areas of technology introduced to the Pacific Islanders by the early explorers are *amara*, the Tahitian word for 'hammer'; *ainpoat* 'pot', and *kedilahs* 'cutlass' in Ponapean. *Manawa*, from 'man-of-war', is used throughout Micronesia for 'warship'.

The traders, whalers, beachcombers, and missionaries of the early nineteenth century left more visible marks on the societies of the Pacific. In many respects their impact was devastating, particularly the diseases they introduced. However, in the domain of language the secular intruders, whose interests in the Pacific Islands were limited to the exploitation of their natural resources and women, wrought much less change than the missionaries.

The beachcombers and the missionaries learned the languages of the Islanders, each for their own reasons. The beachcombers sought membership in the island societies. The missionaries sought to convert the minds and souls of the people to a new way of thinking, and knew from the onset that the quickest and surest route was through the indigenous languages. Although the Holy Writ was the message, it was preached in the languages of the people the preachers sought to convert. In between their frequent sermons, the missionaries lost no time in developing writing systems for the languages of their new converts, and teaching them how to read and write it. The subject matter of their lessons was almost exclusively biblical.

Following on the heels of the missionaries, the governments of Spain, England, Germany, and France sent armed frigates throughout the Pacific Islands to raise their European flags, thereby claiming possession of the islands, which, from their Eurocentric view, were theirs for the taking. It was an era when the white Europeans were convinced that the entire globe belonged to them, and the missionaries assured them that in the eyes of God they were right.

The great Pacific colonial stakeout was a relatively peaceful, almost uneventful affair. Where resistance did occur, it was minimal. Missionaries, planters, traders, and government functionaries became partners in establishing the Pacific outposts of the expanding Capitalist-Christian world. Within a very short time, colonial governments were established throughout the Pacific Islands, arbitrarily creating new political entities of disparate groups of island people. Outside the church, the languages that ruled were European.

Introduction and Early Uses of Literacy

In addition to the technologies of hammers, iron pots, metal fish hooks, nails and axes – all of which were eagerly sought by the Islanders – there was another technology, introduced by the missionaries, that was to have more profound social impact. That was the technology of literacy – the conversion of language to a silent, visible form, the effect of which was a shift of authority from the spoken words of the traditional chiefs and spiritual diviners to the written words and those who could write them. The notion of the WORD as law, in the tradition of Moses and the Ten Commandments, began to take root in the Pacific Islands. Although this notion began in the church, it was not long before it was extended from God's law to that of the white man. Along with this shift in authority came a dramatic shift in social institutions and world view. According the Harvey Graff, the change would be pervasive:

> The penetration of literacy into an overwhelmingly oral, native culture tends to cause massive social, religious, ideological, political, economic, and cultural changes (Graff 1987, p. 380).

The most obvious of the new social institutions founded on the written word was the church. Beginning on Guam in the seventeenth century, with the establishment of the Catholic Mission by Fr. Diego Luis de Sanvitores, the Christian church, in both Protestant and Catholic forms, spread throughout the Pacific Islands over the course of the next two centuries.

The reception given the missionaries varied from place to place. In parts of Melanesia and Micronesia there was indifference or outright hostility to the bearers of this new belief system. The Chamorros of Guam are reported to have resisted fiercely, as did the New Georgians of the Solomon Islands (Cf. Carano and Sanchez, 1964; Bennett, 1987). In Polynesia, by contrast, there was almost immediate acceptance of the church and its written message. By the middle of the nineteenth century the soldiers of the cross had established themselves firmly in Polynesia, from where they converted and sent Polynesian missionaries, armed with the written word, to other parts of the Pacific to spread the Gospel.

What is of importance here is the effective use and the impact of the written word. The missionaries themselves were the products of a long tradition of religion of and by the book. The written word of the Holy Bible, to their minds, was the divine instrument for the conversion of all mankind. The missionaries set about translating and printing the Holy Word immediately upon arrival in Polynesia, and began to teach the Islanders how to read it.

From the very beginning, the missionaries realized that in literacy they had a very powerful medium. As Clammer (1976) observes:

> Missions throughout the Pacific regarded literacy as an essential element in their struggle for converts, both so that religion amongst the people could be self-sustaining and not forever dependent on the word and presence of the missionary, and so that the training of native pastors could proceed beyond the most rudimentary level. (p. 13).

There were other aspects of the written word that did not go unnoticed. In the beginning, it was not the message of Christianity that motivated the Islanders. Rather, it was literacy itself – the seemingly magic power of marks on paper to speak. Awed by the power to transmit messages silently across distance, and by the comparative wealth of the Europeans, the Islanders were eager to learn to read and write. The price of tuition in literacy was participation in the church and the church schools. The subject matter was God's Holy Word. It was thus through literacy that the first of the western institutions – the churches – became established in the Pacific Islands, where they remain firmly rooted today.

The history and impact of the church in the Pacific Islands have been enormous, and have been described in considerable detail elsewhere (Cf. Latukefu 1974; Schütz 1977; Langmore 1989). What concerns us here is its promulgation of literacy, and some of the implications.

Social Consequences

In addition to the establishment of institutionalized Christianity, there were at least two other immediate social consequences of the new literacy in the islands: the creation of a literate class, and the establishment of institutionalized schooling. Because the European missionaries were few in numbers – the harsh life in distant island outposts had limited appeal – some of the Islanders who showed a flair for literacy were recruited and trained to become men of the cloth. Some stayed at home to preach the gospel, while others were entrusted to take the word to more remote areas in the Pacific, such as New Guinea. A new priestly class emerged, displacing the traditional spiritual leaders, attracted by the rewards that came with literacy, i.e., power, status, and money. The result was the beginning of a drastic realignment of the social strata.

The establishment of mission schools in the island – also an institution founded on literacy – had serious social ramifications. In addition to creating new social classes – students and teachers, literates, and illiterates – new and alien idea were put forth, all supported by the authoritative presence of the book. In keeping with the doctrine of the church, the mission schools reinforced the notion that what was written commanded an authority far greater than the spoken word.

The authoritative power of the written word soon became manifested in several respects. Written prohibitions, the missionaries argued, were much more enforceable than oral taboos, which were already weakened in Polynesia. Agreements, when written, were tangible proof, decidedly more compelling than more spoken words. Entitlements to rank, property, and even sovereignty could all be established incontrovertibly through writing. Thus, the seeds of governments by paper and a new ruling class were sown, and would take firm root during the period of colonization. Chiefs who traditionally governed by voice would ultimately be replaced by certified literate politicians.

This notion of the authority of the written word was probably the most serious consequence of literacy in the oral world of the Pacific Islanders, and was to have its full effect in the colonial and post-colonial years, as we shall see.

Another significant and early consequence of literacy was in the area of monetized commerce, which was virtually unknown to the Islanders before the coming of the Europeans. Since everything in the islands was perishable – houses, canoes, clothing, food – little thought was given to conservation and record-keeping beyond that stored in human memory. This was to change, however, with the development of commercial farms and plantations, which, according to Creighton (1978), was well under way in Hawaii in the early 1830s, shortly after the missions were established.

The shift from subsistence to commercial agriculture required literacy in order to keep records of hours worked, amounts sold, monies owed, labor contracts, and the like. The introduction of money – itself a form of literacy – drew the masses further into the literacy web already developed by the church and schools. The institution of the cash economy would eventually become firmly established throughout the

Pacific, competing with the traditional systems of cooperation and trade.

The social institutions described above – church, school, government, commerce – were introduced throughout the Pacific Islands during the nineteenth century (earlier in some cases, e.g., Guam). They are all the products of literate culture, and require literacy to one degree or another for their very existence. To participate in any of them, one must play by their rules. The impact on the Pacific Islanders has been dramatic, as we shall see when we look at the subsequent eras of change.

The Colonial Period

The colonial period in the Pacific began with the Spanish acquisition of Guam in 1565. Next on Spain's list was Easter Island in 1770. The other European colonial powers – Germany, Great Britain, France – swarmed throughout the Pacific between 1841, when France claimed possession of Tahiti, and 1915, when the Gilbert and Ellice Islands were declared a Colony by Great Britain. 'In 1870, the Pacific Islands belonged to anyone who cared to raise a flag'. (*Pacific Islands Yearbook* 1978, p. 474). The American 'grand slam' was played in 1898–1899, when, at the close of the Spanish-American War, Washington got seriously involved in the colonial business in the Pacific, and took the Philippines, Guam, Hawaii, and American Samoa with a single concentrated effort. By 1915, the entire Pacific area was completely colonized, with the technical exception of the Kingdom of Tonga (which was never colonized because a maverick missionary, the Rev. Shirley Baker, persuaded King Tupou to let him write a constitution establishing Tonga as a sovereign state, which therefore could not be colonized).

In the case of the Spanish annexation of Guam, the mere planting of the flag and a verbal declaration symbolized the change of status. All of the subsequent declarations were legitimized by the written word: 'a deed of cession' in American Samoa; 'a document was drawn up' on Easter Island; 'A formal Deed of Cession was signed' in Fiji; Hawaii 'signed a reciprocity treaty with the U.S.' (All quotations taken from the *Pacific Island Year Book* 1978, passim.) Indeed, literacy played a crucial role in the transformation of the free people of the Pacific Islands to wards of the West in the great nineteenth-century colonial stakeout in the Pacific islands.

From the outset of the colonial era in the Pacific there were problems of communication. The colonial governments, like those that spawned them, were based on the written word. Their forms were those that grew out of the literate cultures of Europe, and they were administered by men whose minds had been thoroughly alphabetized. That is, the colonial governors were steeped in such fundamental western concepts as atomism, logic, linearity, rationality, analytical reasoning, hierarchical structures, individualism, and written statutory laws to govern a society. The governed – the Pacific Islanders – came from a different mindset, based on holism, integration, spiritualism, pluralism, and a communal society governed by a traditional set of laws, known to and understood by all.

The fundamental differences between these two mindsets have been discussed thoroughly by Havelock (1963), Ong (1967), McLuhan (1964), Goody (1977), Logan (1986), and others. They will be briefly reviewed, though not argued here. Some of the claims that are relevant to the present discussion would include the following intellectual by-products of alphabetic literacy: deductive-analytic reasoning, abstract thinking, hierarchical classification, segmentation of time and space. Jack Goody states the case in the following way:

> Literacy is for the most part an enabling rather than a causal factor, making possible the development of complex political structures, syllogistic reasoning, scientific inquiry, linear conceptions of reality, scholarly specialization, artistic elaboration, and perhaps certain kinds of individualism and alienation. (Goody 1968, p. 153)

While some may challenge the validity of some or all of Goody's claims, it is evident that there were (and still are) major differences between the Pacific Islanders and the Europeans in the ways that they viewed themselves and the world around them. These differences are commonly attributed to differences in cultures. If we accept the hypothesis, usually associated with Benjamin Lee Whorf, that language expresses culture, and in turn is shaped by culture, then we can accept the position that the differences in mindset of the Pacific Islanders and their colonial masters were determined, to a large extent, by their different languages, plus the fact that the Pacific Island languages were oral constructs, while those of the white man had been influenced by a long tradition of literacy.

Language Status

It was during this period of colonial rule that a distinction of language status began to emerge. That is, there was the language of the government and the language(s) of the governed. The government language represented status and power, since it was used in its written form to formulate laws, treaties, agreements, certificates, edicts, and regulations, all of which impacted directly on the lives of the governed.

Whereas the missionaries had learned the indigenous languages in both spoken and written forms, and had used them for interaction with their congregation and community, it was extremely rare for a colonial servant to do so. In government affairs, the indigenous languages were clearly subordinate, and the Islanders were very conscious of that fact.

The inequality of languages was most apparent in the area of law and governance. Even though many of the colonial governors made good faith efforts to respect and even incorporate 'custom law' into the system, it was the white man's law that prevailed when conflict occurred. This was so partly because the law was backed by the garrison, but also because it was written law. And, since the time of Moses, written law has prevailed over oral law *because it is written*, and therefore, at least in the minds of men who wrote and enforced it, more authoritative. This imbalance of authority in law lends credence to the claim that, somehow, words carry more weight when you can see them. Without doubt, this was the perception of the Pacific Islanders whose lives and social institutions were being directly impacted by the transition from traditional oral law to the written laws of the men in white suits.

Language in Education

The growing dominance of the European languages can also be traced in the development of education in the islands. The earliest schools were founded and staffed by missionaries and were conducted, for the most part, in the indigenous languages of the islands. The establishment of colonial governments, however, created a need for additional schools for the children of the civil servants and settlers. In most cases, there was a single government-supported school, to which privileged children of island leaders would be admitted to matriculate alongside the children of the colonial servants. There they were trained in the ways of western thought through the medium of print in the colonial languages, after which they could serve as useful clerks in the government offices.

The graduates of these government schools formed a new class in the island communities. They wre literate and trained in the use of numbers. Unlike the products of the earlier missionary schools, these new graduates were literate in the white man's tongue, a skill which enabled many of them to enter into partnership – albeit an unequal one – with the ruling class. Being in such positions set these Islanders apart from their peers.

The shift from mission schools to government schools, with a parallel shift from the indigenous languages to the colonial languages, was a gradual one and was much more effective in some areas than in others. In Hawaii, for example, the shift was complete by 1896, when the Hawaiian language was completely abandoned as a medium of instruction, or even a subject of study. In other areas, for example, Western Samoa, Tonga, mission schools conducted in the indigenous languages dominated until the end of World War II. And in many parts of the Pacific, such as Micronesia and Papua New Guinea, schooling in any form was very limited during the colonial period. (During the Japanese period in Micronesia [1914–1945], for example, mission schools were abandoned, and the Japanese government provided three years of schooling for boys in the district centers).

Although the shift from mission to government schools did not occur simultaneously throughout the Pacific, when it did the pattern was essentially the same. Indigenous languages, if not completely abandoned, as in the case of Hawaii, New Zealand, and Guam, were relegated to the early primary grades. The emphasis shifted to schooling in the colonial languages in both spoken and written forms. The effect was to further the divide between the schooled and unschooled, the old and the young, the literate and the oral, the indigenous and the colonial languages.

Between the establishment of literacy in the Islands and the end of the colonial period, there was a gradual shift in written language use. The indigenous languages continued to be used in the church, for personal letters, and in newspapers (where they existed). However, it was in the major social institutions – government and schools – that the colonial languages began to emerge as the dominant ones, thereby setting the stage for the linguistic imbalance that has developed during the modern post World War II era. It is this shift in languages that lies at the core of all the other aspects of social change that are discussed in this volume.

The Modern Era

The great Pacific war ushered in the 'modern era,' and in the process caused social upheaval throughout the Pacific Islands. In addition to the hundreds of thousands of military personnel that swarmed the islands, in many cases outnumbering the indigenous populations, the war created dislocation of people and destruction on a scale never seen before. The effects on the island people were profound, and somewhat varied. For example, places such as Guam, Saipan, Tarawa, Peleliu, and Guadalcanal experienced nearly total devastation of structures and even vegetation. The Trukese witnessed the annihilation of an entire naval fleet through aerial bombardment. Ni Vanuatu (Melanesians of the former New Hebrides) were aghast at the massive amounts of the materials of war that passed through their islands, and for many of those employed by the military, through their hands. They were even more confounded by the wanton disposal of unused materials after the fighting stopped. Even in islands that were spared from the staging of combat troops and fighting, the lives of the people were touched by the war indirectly. The Pacific would never be the same.

Although war was not new to the Islanders, this type of war, with its massive, impersonal destruction was a traumatic physical as well as psychological experience for them. The wrenching social changes in the islands that were to follow would probably have come about anyway in due course, but not at the same bewildering speed. War in the Pacific served as the catapult to launch the islands into the mainstream of the global society. And one of the key operators in this new society was the global language.

The end of World War II also set the stage for the end to legitimate colonialism in the Pacific, the last stronghold of the colonial world. Independence was slow in coming. It was not until 1962 that the first independent Pacific Island Nation (Western Samoa) raised its own flag. Except for the remaining Chilean, French and US territories, the other island nations reclaimed their sovereignty during the next twenty years, at least in form – written form – if not in substance. Almost without exception, the form that they took was the exact replica of their colonial mentors, with written constitutions, laws, and education systems based upon books from the metropolitan countries from whose political shackles they had sought freedom. Not surprisingly, the new Pacific mini-states had been cloned.

Change in the Institutions

The accelerated changes in the social institutions of the newly emerging Pacific Island countries can be seen in the restructuring of government, law, commerce, and education, in which traditional practices were drastically modified (or eliminated) to fit the westen models. These will be examined separately, with an eye on the role of language.

Government

The prerequisite for the establishment of post-colonial governments was a set of documents, invariably conceived and written in the colonial language, and subsequently translated into the 'national language' in the countries where one could be identified. (The polyglot countries of Melanesia have no indigenous national languages). The

405

ticket for recognition and entry to the global network of sovereign nations was a proper set of documents that served as proof of sovereignty. In a world governed by literate nations, where nothing is legal unless written, traditionally oral people had to become literate in order to join the system.

There were documents establishing the new political status including the special relationship with the mother country (e.g. Commonwealth, Free Association). There were more documents which spelled out the political structures of the newly established nation states (e.g. Constitutions). And from these documents began the flow of the never-ending stream of documents establishing laws, departments, bureaus, portfolios – all of the accoutrements of a western form of government.

Although these documents were prepared separately for each emerging nation, they bore a striking similarity to each other, as did the bureaus that they created. The leaders and clerks of the new island nations were committed to the new world of literacy, most of it in the languages of Europe. Yet, daily life outside the offices was conducted in the indigenous languages, forcing many of the island people into the schizophrenic role of trying to live in two separate cultures governed by different languages and thought patterns. As the govenment ranks grew in number, more and more of the Islanders were affected by this cultural dichotomy.

Perhaps the most significant social change in the new paper-fueled governments was the quiet overthrow of the traditional leaders. The new governments, legitimized by written documents, demanded new leaders who were literate in the metropolitan languages. Traditional leaders were, for the most part, relegated to advisory roles, while the new elite, elected by popular vote, became the voice of authority. It was they who wrote the documents that became the new law of the land, usually in languages barely known by the electorate. It was they who were elected because they could write the documents, having been educated overseas in the colleges and universities of the mother countries. The new leaders throughout the Pacific carried attaché cases, sat behind large desks in air conditioned offices, and had secretaries (usually expatriate) to prepare and file the documents, the major products of government. On rare occasions (e.g. Ratu Sir Kamisese Mara, of Fiji), a traditional chief also served as an elected leader. There were countless others, however, whose role was defined as 'advisory', a role for which knowledge of the new language of the government was not needed.

The Courts

As the new governments issued new laws, courts were established to interpret and administer them. Since court decisions in the islands were always subject to review by higher courts with expatriate judges, the prevailing languages were those of the metropoles. The men who were licensed to argue in the courts – the lawyers – were either expatriates or Islanders who were trained in the metropolitan countries and languages.

It was in the legal system, perhaps more than anywhere else, that language played a dominant role, dictating who could participate. For the relatively simple cases, such as public drunkenness, assault, or larceny, justice could be determined by indigenous judges using indigenous languages. However, for offenses of a higher order – especially where real property was involved – briefs, deliberations, and decisions were processed and documented through the appropriate European language. Indeed, the courts in the new Pacific took on the image as well as the languages of the mother countries.

Commerce

Another significant social change resulting from the increased reliance on literacy was the shift from subsistence to wage economies. The new economic system, based on the exchange of paper money, required a literate workforce to perform the requisite functions of writing, typing, filing, counting, teaching, and so forth. Young men and women who in previous generations would have been performing the traditional roles of farming, fishing, food preparation, and such, found themselves toiling in offices behind desks covered with written documents of one sort or another. This shift in economic structure was a concomitant part of the change from an oral to a literate system.

The effects of the shift in economic systems on family structures and demographic patterns have been dramatic. The erosion of extended family relationships and population shifts from rural areas to the urban centers have both been well documented (Cf. Hezel, n. d.). It is reasonable to claim that the shift from an oral to a literacy-based economic system has played a major role in these changes.

Education

The new governments in the islands soon became awash in the documents of government, most of them written in the metropolitan languages: letters, memos, reports, notices, announcements, and invitations to political functions. In order to build and maintain a cadre of scribes who could keep the flow going, and replace the legions of expatriates who lingered after independence in the employment of the new governments, the training grounds had to be expanded. Hence, the establishment of the largest and most fundamentally literate of the branches of government in all Pacific Island countries: the Department of Education.

Upon entering school, the child's first task is mastery of the ABCs – the key with which to unlock the world of literacy. From the very start, the child is led into a world of abstraction, a world in which silent marks on a piece of paper represent sounds which, when strung together, form pronounceable strings representing words and phrases. Within those silent strings are embedded ideas which are, by some unseen process, to be extracted by the reader. This is the process of literacy, the introduction to the world of abstract reasoning.

As the beginning school child struggles to master the abstract world of literacy, the task is compounded in many parts of the Pacific Islands by the language that is required by the schools. The Hawaii Legislative Assembly mandated English as the sole language of instruction in 1896. Guam followed suit in 1898, an automatic consequence of annexation. The Micronesian islands of the Trust Territory shifted into 'English only' in 1963, three years before the US Peace Corps flooded the islands with instant 'specialists' in Teaching English as a Foreign Language. Thousands of children from the traditionally oral cultures throughout the Pacific Islands have been forced into the multiply-compound task of acquiring literacy – a highly abstract form of conceptualization in itself – in a foreign tongue. This practice has had grave consequences for the children of the islands, which can be seen in the widespread failure and alienation.

With a government in place, and an educational system to supply it with clerks, the newly-emancipated Pacific Island societies are deeply involved in the business of governance in the Western mold. The generation of an endless flood of paper has begun. Governing bodies meet regularly to create new laws and amendments to old laws. Judges and lawyers emerge as the new high priests who interpret the esoteric tomes of the new language. Secretaries, file clerks, typists, court reporters – all required to support the flow of the written world – are trained as the new working class in the changing social order.

The language-driven changes in the social institutions described above are the clearly visible ones. What is perhaps less visible are the broader social and cognitive changes that accompany them.

INFLUENCE OF LITERACY

A number of studies of literacy (Ong, 1967; Goody, 1977; Havelock, 1963; Olson, 1977; Graff. 1987) focus on the cognitive effects on oral societies when they experience the shift from orality to literacy. Their arguments point generally to the same conclusion, which is that the widespread acquisition of alphabetic literacy by people from traditionally oral cultures brings about profound changes in the way they perceive and interpret phenomena. How is this so?

One of the major arguments is that alphabetic writing forces us to view language differently, which in turn affects our perception of all other phenomena. Language, when it is written, becomes visible. Thus, it acquires a new dimension, providing a different experience from spoken language. It is processed through the eyes rather than the ears. Furthermore, unlike speech, language written with alphabetic symbols is linear, segmented into indivisible, sequential units, and goes from left to right (or right to left in the case of Semitic languages). It is this linear sequentiality, inherent in alphabetic literacy, that gave rise to analytic, logical thinking, which is fundamental to syllogistic reasoning and the so-called scientific method.

Another feature of written language that may be of major significance in this discussion is its abstractness, which is claimed to have given rise to abstract thinking. Language itself is an abstract system in which utterances are attempts to represent reality. In written language, visual symbols are used to represent utterances. In alphabetic writing, the system is even more abstract in that the symbols used do not represent utterances directly; rather, they represent individual segments of sound which are then strung together with the eye and assembled by some process of the brain, after which a conclusion as to their meaning is then drawn. Processing a written language is a very different experience from processing a spoken one, and is a highly abstract phenomenon.

If these points are valid, as I believe them to be, it is quite easy to see how they contrast with some of the features associated with traditionally oral societies, such as holism, concretism, and spiritualism.

The early literacy in the Pacific Islands did not penetrate far enough in the population to have brought about immediate conflict in the thinking of the island people. By most accounts, reading was a matter of sounding out words, or rote memory of texts. Most reading was perfunctory, and done in the church or school setting. Writing by indviduals was very restricted, and was generally for the purpose of personal letters. The cognitive effects described by Ong and others were to develop slowly over time, as literacy became more widespread.

Perhaps the most striking feature of literacy, and the one which had the most immediate impact on the oral societies of the Pacific, was the authority of the written word. The fact that literacy was associated with what was perceived as superior western technology and material wealth probably enhanced its authoritativeness. Books, letters, diaries, and official documents were visible artifacts of the Europeans who had come to the Pacific Islands to plant their flags, crops, and ideas. The technology, wealth, and superior weapons of the white men lent support to the authoritative nature of the written word.

In addition, the written word possesses an authoritativeness of its own. Somehow, when words get written down they become more real, powerful, and permanent. They take on an existence of their own. They almost defy challenge. This view was established early on, and with serious consequences.

Some of the concepts introduced to the islands through the authoritative literacy of Christianity posed a clash of mythologies. The new Christian concepts, such as self as individual, sin, salvation, damnation and eternal torture, stood in direct contradiction to established belief systems, which were soon to give way. Throughout the Pacific, and in a relatively short time, the new written mythology prevailed, mainly because it was written. Such is the power of literacy, which lends the stamp of unchallengeable authority.

The concept of God as creator was not, in itself, such an alien one inasmuch as it fitted the pattern of many of the traditional accounts of creation by powerful human-like personae. However, the notion of a single, perfect, omnipresent, omniscient, and vengeful god, who sits in judgment of one's day-to-day actions, required a drastic shift in thinking. The concept of sinful actions that must be atoned for in a life after death struck terror in the hearts of the proselytes, thus providing strong motivation for learning about salvation and eternal bliss.

It was concepts of Christianity such as these that set the pattern for major cognitive shifts that were to develop over time as Christianity, delivered through the written word, took root among the Pacific Islands people.

The literacy taught in the church was further promulgated through the schools. The children of the Pacific Islands, like children in the western world, now spend the best part of their daytime hours in schools. Instead of learning skills, their environment, and their history through observation and interactive participation – the traditional learning styles of their cultures – they are required to adapt to an alien style of learning through the silent, abstract medium of print.

In addition to adapting to a new learning style, the children of the Pacific are confronted with new and often alien types of knowledge.

410

Instead of learning such things as the environmental cycles, around which their own cultures and social systems developed, they focus on reading and writing skills, multiplication tables, the structure of the United Nations, and world geography. Local histories, ecology, kinship systems, and attendant obligations are all but ignored by the schools, largely because the education authorities do not consider them important, and these are not subjects covered by the metropolitan textbooks that are used in the schools. Schools without textbooks, even in the Pacific Islands, are simply unthinkable. Such is the force of literacy in education.

The students who successfully adapt to the learning styles and the subject matter of the schools experience a drastic change in their world view. Their island cultures come to be seen as quaint relics of a former era. Their native languages are overshadowed by the masses of volumes written and studied in the European languages. The fact that very few of the Pacific Island languages are used for any kind of serious writing is in itself a statement about their relative worth. It is a message that is driven home very effectively to the young people of the Pacific, a message which deeply affects their view of themselves and their island world.

Other aspects of world view that are affected by this shift from an oral world to the literate world of school include traditional mythology, history, governance and law, all of which have been displaced by the views from the West.

Although these cognitive changes resulting from the shift from orality to literacy are more difficult to quantify, they are just as real as the social changes, and the effects are deeply rooted.

Summary and Conclusion

This chapter has attempted to give a broad picture of the role of language in the changing Pacific Islands. Simply put, from the time of initial contact with the western world, the Pacific Islands have been changing in ways that clearly reflect western influence. Language has played a prominent role in that change, and continues to do so with increasing intensity and speed. The linguistic profiles of Hawaii, New Zealand, and Guam – where the native languages were robust and played a major social role as recently as 50 years ago – are likely indicators of where many of the Pacific Island languages will be in the not too distant future. Unless drastic steps are taken immediately to turn things around, more of the languages of the Pacific are destined for extinction. The result will be the ultimate social change. When the languages die, the crucial link between the people and their cultural past is broken. To speak of being Hawaiian, or being Maori, or being

Chamorro in a foreign language somehow has a hollow ring. One's cultural identity is hard to find through someone else's tongue.

An unfortunate consequence of the shift from indigenous languages to English and French in the major social institutions has been the emergence of language as a gatekeeping force. It begins in the schools where language proficiency is used to determine who passes and who fails. It continues into adult life, where proficiency and literacy in the European language determines who passes through the gate to certain jobs, political office, and influential positions. While superiority in language performance was always valued throughout the Pacific – witness the importance of orators, oral historians, composers – it was never used as a standard by which to brand someone a failure. The use of language proficiency to define social roles is a by-product of the adoption of western languages and the ideas of correctness that are associated with them. This has been a major factor of social change in the Pacific Islands.

The shift in language use described in this chapter has also established a newly stratified social order throughout the Pacific Islands. There are new lines separating the literate from the illiterate, the traditional from the modern, and the old from the young. A new elite has risen to displace the old, just as new rituals (e.g. high school graduation) have become the accepted rites of passage. Storytellers, navigators, healers and diviners have all been shunted aside in favor of movies, scheduled airlines, allopathic physicians and theologians – all of them products of the literate societies of the western world. Although there are still some areas in the Pacific – usually rural – where the old social hierarchies and practices still prevail, they are relatively insignificant. If the dominant pattern prevails, these pockets of the traditional Pacific way will follow the path that has been set in the urban centers.

Perhaps the most lamentable consequence of the encroachment of the western languages in the Pacific is the resulting language loss and concomitant cultural erosion. Following the pattern of the Polynesians of Hawaii and New Zealand, and the Chamorros of Guam, the island people from one side of the Pacific to the other run the serious risk of losing their languages. The danger signs are already present.

In the schools, playgrounds, and even in the offices, it is more and more common to hear English or French spoken by one Islander to another, even when both are speakers of the same Pacific language. More and more frequently, town-dwelling teenagers speak to each other in the school language. Increasing numbers of young urban parents, educated in the languages of the schools, speak to their children (and, for some strange reason, to their dogs) in the new language. Many young people of the islands are not learning the traditional 'respect' languages, and the adults lament that their children are not learning to speak their mother tongue correctly. The warning signs of rapid language change and possible erosion are apparent in all of the urban centers, from which social changes radiate.

The bilingual education programs found in some of the schools of the Pacific Islands represent an effort to maintain the indigenous languages along with the encroaching world languages. While these programs are laudable in their intent, and may ultimately succeed in fulfilling it, the fact remains that they form a very small part of the school curriculum, and are treated more as addenda to the core of study, getting about the same amount of time and support as art and music. If the goal of bilingual societies in the Pacific is to be achieved – and the odds are probably against it – much more time and conscious effort will be required, not just in the schools, but by the Pacific people at large.

Finally, there is the matter of social changes that stem from changes in the way people structure their thoughts. Belief systems, kinship systems, customs, rituals, courtship, taxonomies, histories – all of the component parts of what we call culture are shaped by language. The variety found in these systems throughout the world reflects the variety in languages. It was through language that they came to exist. As powerful new languages penetrate more and more the lives of the Pacific Islanders, then the systems devised by those new languages are going to prevail. What we are seeing in the Pacific Islands are social changes resulting from the imposition of western rational thought as conceived and developed in the languages of vastly different cultures. It is the growing use of these concepts and languages among the island people that is leading them headlong into rapid social change which the vulnerable cultures of the Pacific may not survive.

15

Children's Survival in the Pacific Islands

George Kent

The infant mortality rate – the number of infants who die before their first birthday for every thousand born alive – is a good indicator of the quality of life in societies. Data for the Pacific islands are provided in Table 1.[1] Infant mortality conditions are worst in the independent Melanesian countries, but they are also serious in parts of Micronesia and Polynesia. The children's mortality rate, sometimes called the under-five mortality rate, is the number of children who die before their fifth birthday for every thousand born alive. In 1987 the children's morality rate for Papua New Guinea, for example, was 85 (Grant 1989, p. 95). Current data for the other Pacific island nations are not available.

The immediate causes of high children's mortality are disease, especially infectious disease, and malnutrition.

Disease

The incidence of infectious disease in the Pacific islands increased sharply following European contact, peaking in the 1700s, and declining steadily since then, largely because of the steady build-up of immunities. Non-communicable diseases such as heart disease, diabetes, and cancer have been increasing over the last few decades.

The change in disease patterns in the islands over time may be following the 'epidemiological transition' pattern observed elsewhere in the world. Epidemiological transition theory is placed within the broader context of demographic transition theory, which is concerned with the way in which countries move from conditions of high birth rates and high death rates to conditions of low birth rates and low death rates (World Bank 1984).

Epidemiological transition theory recognizes three major stages. The first stage is the era of pestilence and famine. Infectious diseases such as typhoid, tuberculosis, cholera, diphtheria, and plague are dominant, as was the case in the middle ages and in Europe through most of the eighteenth century. Mortality rates were high and life expectancies were low.

In the second stage, infectious diseases are gradually controlled, mortality rates decline, and life expectancies begin to increase. In this stage the major causal factors are improved hygiene, sanitation, housing, and nutrition. Medical care does not contribute very much.

Table 1 *Infant Mortality in the Pacific Islands*

Country	Infant Mortality Rate (deaths/1000 live births)	
	I	II
Melanesia		
Papua New Guinea	77	61
Fiji		27
Fijians	30	
Indians	41	
Solomon Islands	53	44
Vanuatu	94	101
New Caledonia		
Melanesians	39	
Europeans	9	
Polynesia		
Western Samoa	33	33
French Polynesia	57	
Tonga	41	
American Samoa	18	
Cook Islands	29	
Wallis & Futuna	49	
Tuvalu	43	
Niue	11	
Tokelau	37	
Micronesia		
Guam	13	
Federal States of Micronesia	45	
Kiribati	93	
Marshall Islands	45	
North Marianas Islands	26	
Palau	28	
Nauru	31	

Sources:
Column I: Taylor, Lewis, and Levy (1989).
Column II: Grant 1989, pp. 95, 108. Column II data are for 1987.

In the third stage of the epidemiological transition, degenerative diseases such as cancer, heart disease, and stroke become dominant. Mental illness becomes more common. Stress, industrial exposure, and environmental factors become major causes of illness and death. Mortality declines and life expectancies become high. The elderly account for a large portion of the population (Omran 1971, pp. 509–538).

This stage is comparable with the South Pacific Commission's account of disease trends in the Pacific:

> Between the World Wars, the mortality from infectious diseases gradually declined. By the end of the Second World War, mortality in many Pacific islands was further reduced by public health programmes that built piped water supply and sanitation systems, by the building of clinics and hospitals, and by the training of medical staff.
>
> The 1950s saw the introduction and use of powerful antibiotics and other drugs that were useful in the widespread control and treatment of infectious diseases. Although infectious diseases were a major cause of death in almost all Pacific Island countries through the 1950s, this pattern began to change in certain islands during the 1960s. For example, in Rarotonga (Cook Islands), Nauru, Fiji, and Guam non-communicable diseases such as heart diseases, diabetes and cancer have emerged as the major causes of death. (1988a, pp. 1–2)

The islands may be grouped according to their current disease patterns, as shown in Table 2. In the group in which infectious diseases are dominant.

> Serious diseases include malaria, diarrhoea and respiratory illness, and tuberculosis. This pattern is generally present in countries where socio-economic development is low, general sanitation is lacking, and health services are under-developed. Infant mortality is high in the countries grouped in this category and life expectancy is generally low (under 60 years) with little difference between the sexes. Approximately 70% of the Pacific's population falls into this category. (South Pacific Commission 1988a, p.3)

Table 2 *Disease Patterns in the Pacific*

Predominantly infectious diseases
 Kiribati
 Papua New Guinea
 Solomon Islands
 Vanuatu

Combination of infectious and non-communicable diseases
 Federated Statesof Micronesa
 Fiji
 Marshall Islands
 Niue
 Tokelau

Tonga
Tuvalu
Wallis and Futuna
Western Samoa

Predominantly non-communcable diseases
American Samoa
French Polynesia
Guam
Nauru
Northern Marianas Islands
Palau

Source: South Pacific Commission (1988a)

Malnutrition

Contrary to the romanticized visions of the Pacific islands long held by outsiders, there is widespread malnutrition in the islands. Most attention has been given to the 'new' malnutrition that seems to result from modernization. A review published by the South Pacific Commission in 1984 concluded that urbanization in the islands

> . . . is accompanied by a decrease in physical exercise, a reliance on imported foods, and an increase in the chronic diseases of western societies.
>
> Diseases such as diabetes, hypertension, gout and dental caries, which were uncommon in traditional Pacific culture occur now, in urban Pacific populations, at rates up to or exceeding the rates found in the affluent industrialised countries. The frequency of coronary heart disease, alcoholism and cancer appears to be rising . . . Infant and child malnutrition and diarrhoea seem to be more frequent and severe in urban centres compared to traditional rural populations. The rise in infant malnutrition is often considered to be due to a decline in breast feeding. (Coyne, Badcock and Taylor 1984, p. 140)[2]

This new malnutrition is a serious cause for concern, but it should be put into context. There is also the 'old' malnutrition of the sort that occurs in remote outer islands and in the Papua New Guinea Highlands, where modernization has not yet had much effect. Sailors in ancient times judged the islanders to be robust and healthy, but they saw only the survivors. Mortality rates were probably high. There may have been abundant food supplies, but there also were times of famine, and much of the available food supply was reserved for the chiefs. Many islanders who still live in traditional ways suffer from serious malnutrition and disease, and have high levels of children's mortality:

Undernutrition in infants and pre-school children appears to have occurred among traditional-living Pacific island groups leading to mild growth retardation in the 1 to 2-year age groups. Infant and second year mortality rates, which can be considered indicators of nutritional status, were probably high in the traditional setting. (South Pacific Commission 1988b, p.2)

The alarm has been raised because of the high level of dependency on imported foods, and extensive consumption of junk foods. Unquestionably, the rising incidence of non-communicable diseases is associated with the new diet patterns. But life expectancies are increasing and children's mortality rates are decreasing. The increased incidence of non-communicable diseases is accompanied by a decreased incidence of infectious diseases, a decrease which is also partly attributable to new diet patterns. There are real nutrition problems in the islands, but it has not been shown that the net result of changes in dietary patterns has been a deterioration of overall health status.

A high level of dependency on imported foods is unwise because it can drain away scarce foreign exchange, put the nation's food security at risk and result in deterioration of the domestic food production sector. But imported food is not necessarily unhealthy. Junk food, consumed in excessive quantities, is unhealthy, whether it is produced locally or abroad. People who now have a steady supply of rice and canned meat may be better off in nutritional terms than when they depended on root crops and erratic supplies of fresh meat and fish. In Papua New Guinea it has been shown that the traditional diet of yam, sago, and sweet potato resulted in widespread malnutrition (Corden 1979, pp. 48–51). In Simbu province, increased consumption of store-bought foods, especially cereals and fish, provided a marked improvement in protein and energy intakes (Harvey and Heywood 1983, pp. 27–35).

One serious cause for concern is the widespread use of breastmilk substitutes for infant feeding:

There has been a reduction in breast-feeding, including length of time breast-fed and an increase in bottle-feeding. According to one summary of 11 island nations in 1986 although approximately 70% of mothers were still breast-feeding for up to 4 to 6 months, numbers were seen to be declining and infants were being breast-fed for a shorter time. The trend away from breast to bottle-feeding is listed by many investigators in Pacific island communities as the major cause of the increase in malnutrition in infants. (South Pacific Commission 1988b, p.5; also see Marshall 1985)

Papua New Guinea has taken decisive action:

To combat the problem of malnutrition associated with bottle-feeding, Papua New Guinea instituted legislation in 1977 to control the sale of baby bottles and teats. Reports from Port Moresby two years after the legislation showed a significant increase in breast-feeding and a decline in the prevalence of severely malnourished children. (South Pacific Commission 1988b, p.6; also see Biddulph 1981; Agyei 1988)

The Marshall Islands

Data for the Pacific islands are often unavailable, and when available they are often unreliable. There are enormous differences among the Pacific islands, so generalizations about them must be tenuous at best. It may be more useful to examine a particular case in depth.

Consider the example of the Republic of the Marshall Islands (RMI), until recently part of the Trust Territory of the Pacific Islands. The RMI is now independent, but associated with the United States under the terms of the Compact of Free Association. Its population in the 1988 census was 43,380 (Republic of the Marshall Islands 1989). The Marshallese are scattered over 24 small islands, with about 40 percent living on outer islands without electricity or plumbing. The average annual rate of population growth from 1980 to 1988 was 4.25 percent, an extremely high rate. The median age is about 15. In 1988 there were 8,651 children under five years of age, or 19.9 percent of the population.

The economy has been highly dependent on infusions of aid from the United States. Copra is the Marshall's major export commodity, yielding an income which has rarely exceeded $100/capita/year. The Marshalls has a consistently negative balance of trade. Food is the major import category; an estimated 75 percent of local food requirements are imported.

According to the *Marshall Islands Statistical Abstract 1988*, in 1986 there were 61 deaths of children under five years of age, accounting for about one-third of all deaths. There were 38 deaths of children under one year of age, yielding an infant mortality rate of 26.1. However, the abstract shows enormous variability in infant mortality rates from 1973 to 1986, swinging erratically from a low of 16.4 to a high of 48.2, suggesting that the data may not be accurate (RMI 1988). A study devoted to making the best possible estimate of the infant mortality rate in the Marshall Islands based on direct interviews and other available data concluded that 'a value of 60 per 1,000 may be taken as the final estimate for 1980–84' (Levy and Booth 1988, pp. 11–20, 31).

Malnutrition is a significant factor contributing to the high levels of children's mortality. According to a 1984 development plan:

A rather disturbing development in recent years is the increasing incidence of diseases related to nutritional imbalance, particularly in the urban areas. Such diseases include hypertension, heart diseases, diabetes and illnesses related to obesity. . . . It is believed that the high incidence of these diseases, which are usually associated with economically advanced countries, is due to the high content of fat, carbohydrates and sugar combined with sedentary type urban life. (Republic of the Marshall Islands 1984)

According to the plan, infant malnutrition 'seems to be the result of two factors: a significant increase in recent years in the consumption of non-nutritional Western foods at the expense of traditional foods and an increasing number of women switching from breast feeding to bottle feeding their infants'.

Malnutrition occurs not only because of inadequate food supplies but also because of diarrhea, parasites, and frequent infections. Diets are poor, with little consumption of fruits and vegetables, and much consumption of junk food. Rice has replaced root crops as the staple. Combined with the sedentary lifestyle of many residents, this has resulted in widespread obesity. The report of the 1984 National Women's Workshop said 'there has been a drastic switch from breast-feeding to bottlefeeding in recent years in the urban areas.' Many cases of undernutrition have been documented, but no national nutrition survey has been undertaken.

Water supply and sanitation facilities are very primitive, especially on the outer islands. As indicated in Table 3, a smaller proportion of Marshallese have access to safe water than any of the other Pacific islands people for which data are available. Similarly, with the exception of 'urban' Tuvalu, a smaller proportion of Marshallese have access to adequate sanitation.

The problems are readily illustrated. In July 1985, on Ebon atoll, a two-year-old child had a fever of 104 degrees, cough, and severe diarrhoea. The child was unable to take oral fluids, and no intravenous solution was available. No aircraft could be obtained, and the child died. In that same month, a one-year-old child on Namdrick atoll had a high fever, and meningitis was suspected. No injectable antibiotics were available on the island, and no aircraft was available. The child died. In September 1985, on Wotje atoll, an 18-month-old boy contracted meningitis. No injectable antibiotics were available on the island. The child died before the ship arrived.

Pregnant women also are very vulnerable. In 1985 a pregnant 19-

year-old on Arno atoll suffered seizures and was diagnosed as having preclampsia. No magnesium sulfate (which would have cost $0.23 for a four gram dose) was available, so an aircraft was sent to medevac her to Majuro, at a cost of $1800. The mother survived but the child was stillborn. There have been many such cases.

Table 3 *Estimated Proportion of Population Covered by Safe Water and Adequate Sanitation Systems*

	Safe Water		Adequate Sanitation	
	Rural (percent)	Urban (percent)	Rural (percent)	Urban (percent)
Polynesia				
Fiji	94	66	83	60
Samoa, Western	100	90	95	75
Tonga	90	80	90	63
Cook Islands	99	40	90	60
Niue		99		99
Tuvalu	95		50	
Micronesia				
Kiribati	93	35	87	80
Rep. of Marshall I.	70	23	72	5
Melanesia				
Vanuatu	88	25	96	30
Solomon Islands	95	24	94	19

Source: WHO-Suva 1985, p. 30.

A Women's Health Survey in 1985 estimated the infant mortality rate at about 35 to 40 per 1,000 live births. Urban rates were estimated to be 30 to 35 per 1,000 live births while rural rates were about 50 per 1,000 live births. The major causes of infant mortality were identified as prematurity, diarrheal diseases, influenza, and pneumonia, and malnutrition.

In December 1985 a government-sponsored Task Force on Health warned that 'the health of the Marshallese people is far below acceptable standards. The number of infants dying is shockingly high, the highest in Micronesia. Malnutrition is affecting a large number of children. Many children are unnecessarily dying due to diarrhea'.

On July 17, 1986 Guam's *Pacific Daily News* headline story, 'Malnutrition Takes Toll in Marshalls,' spoke of a four-year-old girl, weighing only 20 pounds, who had died of malnutrition at Majuro's hospital. The story told of the 'steady stream of malnutrition cases seen in the Marshall Islands'.

In 1987 immunization levels were described by a Ministry of Health Services official as 'distressingly low', with only 3.6 percent of two-year-olds in the outer islands being fully immunized in accordance with standards of the United States Centers for Disease Control. The overall immunization rate for the Marshalls was estimated to be 17 percent. (The immunization levels are somewhat better in terms of the World Health Organization's less demanding standards.) These estimates were based only on children with birth certificates. Since children without birth certificates are less likely to be immunized, the immunization rates are probably lower than indicated by these figures. The conclusion was that 'all of the Marshall Islands, particularly the urban areas, are at extremely high risk for epidemics, particularly polio, pertussis, and measles'.

In the first six months of 1988, seven young children died in the hospital in Majuro of symptoms related to malnutrition (*Marshall Islands Journal* 1988). According to an observer on pediatric grand rounds at the hospital in Majuro, on June 17, 1988, 'there were eight patients in the pediatric ward, and all but one was noted as either suffering from failure to thrive, underweight, or malnourishment'.

In 1988 the Marshall Islands' Secretary of Health Services said that malnutrition in the Marshalls 'is on the rise. The number of malnourished children from outer islands admitted to the main hospital in Majuro is increasing . . . but the majority of malnourished children are from the two urban centers' (*Pacific Islands Monthly* 1988).

A study published by the University of Hawaii in 1984 found that 'nutrition education and services, especially for pregnant women and children, are either missing or inadequate' (University of Hawaii 1984). The study also said that 'the orientation of the RMI health system is inordinately skewed in favor of major attention to acute medical care with the result that vital and basic public health and preventive services have been neglected'. In 1985 a new hospital was opened in Majuro, raising concern that it might draw resources away from primary health care, especially in the outer islands. Beginning in 1986, however, there has been new emphasis on primary health care, as documented in the Ministry of Health Services newsletter, *Ejmour Ebed Ilo Lubiden Beim* (a newsletter of Primary Health Care Action in the Marshall Islands). Immunization rates have increased sharply, and much more attention is being given to child survival in the context of the overall primary health care program.

Powerlessness

Modernization certainly affects the patterns of malnutrition and disease in the Pacific islands, but the nature of the linkage is not entirely clear. Modernization typically involves a complex cluster of phenomena including westernization, urbanization, monetization, increased salaried employment, economic growth, population growth, environmental stress, the decline of community, and the rise of individual wealth accumulation. Which are the critical elements? How do we account for the fact that as early as the 1950s 'bread and sugar were regularly consumed by villagers engaged in subsistence agriculture on Raro-tonga. The average family was consuming 1,535 grams of white bread and 106 grams of sugar per day' (Coyne, Badcock and Taylor 1984).

An analysis of the Pacific islands prepared for the United Nations Children's Fund (UNICEF) observed that 'Most countries show a greater proportion of mortality due to degenerative disease (heart, cancer, endocrine disorders) than would be normally expected in developing countries, but also a higher proportion of deaths due to acute or infectious diseases than would be expected in developed countries' (UNICEF 1985, p.68). The study said these data 'reflect the transitional nature of Pacific society'. But the notion that this is only a passing phase may be overly optimistic. The islanders are increasingly exposed to the health risks of the modern world, but they are not reducing their exposure to more traditional health risks. On the whole, islanders seem to be suffering the worst of both traditional and modern worlds, and there is little indication of change for the better. Indeed, a recent review of the health situation in Fiji, Samoa, Solomon Islands, Tonga, and Vanuatu concluded that they 'are emerging from an era of epidemiological transition', but on the whole the transition has been a deterioration:

> They retain the health characteristics of developing countries – relatively high infant mortality and incidence of communicable diseases – while exhibiting some of the most extreme rates of chronic, degenerative disease. (Bloom 1987, p.43)

The benefits of modernization are passing many of the islanders by because of a complex array of political, economic, and cultural factors, but the essence of it is that many islanders simply have very little control over their own life circumstances. The pattern of loss of control is evident in a way even in that most untypical of Pacific islands, Nauru. The island nation is very wealthy as a result of the mining of its guano deposits over the years. But Nauru has a much higher infant mortality than would be expected on the basis of its high income level. Nauruan men have a life expectancy of only 54 years. In Nauru, 'some of the world's richest people are eating themselves to death'. One in four have diabetes, and many have gout, high blood pressure, and cancer. A doctor who has studied Nauru says 'they have embraced all the worst aspects of Western culture, and none of the good. . . . They get virtually nothing that is fresh. They grow nothing of their own, and no Nauruan knows how to fish' (Field 1988, p.29; also see Taylor 1985, pp.149–155).

Some sense might be made of Nauru's situation in terms of a new fourth stage of the epidemiological transition that has been proposed. In this 'hybristic' stage, deaths are increasingly attributable to individual behaviors and lifestyles:

> Hybris is an excessive self-confidence, a belief that you cannot suffer, that you are invincible. Morbidity and mortality in the hybristic stage are affected by man-made diseases and increasing modernization as well as individual behaviors and potentially destructive life-styles. Increases in physical inactivity, pernicious dietary practices, and excessive drinking and cigarette smoking can contribute to heart disease, diabetes, chronic nephritis, lung cancer, and cirrhosis of the liver and such social pathologies as accidents, suicides, and homicides. (Rogers and Hackenberg 1988, p.240)

Behavior-related problems such as smoking, alcoholism, drug use, AIDS, accidents, and suicide become major factors in this fourth stage. With this formulation, the pattern observed in Nauru is not anomalous. Rather it reflects a recognizable pattern which can be linked to findings such as the extraordinarily low life expectancy of native Hawaiians and the high suicide rate among Micronesian youth (Hezel 1982; 1986; 1989; Hezel, Rubenstein, and White 1985).

This pattern may reflect a kind of arrogance, but perhaps this self-destructive behavior is better understood as indicating a deep and pervasive sense of powerlessness. Alcoholism, drug use, and obesity are most prominent among groups with little control over their own life circumstances. Native Hawaiians, like American Indians and blacks, have extremely high injury mortality rates and high mortality rates from cancer and heart disease. A Hawaiian physician suggests that the real causal factors 'are no control over our lives, language and religion in our own homeland' (*Honolulu Star-Bulletin* 1989, p.A5). High levels of risk-taking behavior, whether in diet, sexual behavior, smoking, or driving, are related to powerlessness.

Poverty alone does not explain the sharp skew of health and other resources in favor of small urban elites. Widespread powerlessness in the Pacific islands, especially of the rural poor, is apparent in the lack of participation in governance in general, and in the health sector in particular. There is a strong urban bias in health expenditures. Often programs are done to, rather than with, local people. Local people are rarely asked for their views, and they are not invited to participate in shaping the programs. Programs are undertaken to respond to issues that are viewed as problematic to outsiders, according to the outsiders' understanding of the problems.

Traditional forms of malnutrition persist in the Pacific islands because many people never had much control over their life situations.

Modern malnutrition is largely due to the fact that many islanders are losing what control they did have. People are losing control over their own diets. The increasing availability of money and thus of store goods, particularly imports, increases people's options, at least for those who can afford the products. But the process also enlarges the islanders' vulnerability to alien influences in the formation of their choices from this rich variety. This outside influence, whether from local store operators or from foreign corporations, is not neutral and it is not designed to promote good nutrition. The primary purpose for which food is imported is to provide profits for the sellers, not to improve the nutrition of consumers.

The unfortunate fact is that junk food tends to be more profitable than wholesome food. Local merchants promote soft drinks rather than coconut milk because selling soft drinks is more profitable. Products such as white bread and doughnuts are promoted because of the profits they can yield, not because of their nutritive value. Imported food does not have to be nutritionally inferior, but there are strong tendencies for it to be that way.

Production also is increasingly oriented toward accommodating outside interests. Some development plans give little attention to the production of food for local consumption. The agriculture budget for local food production (excluding livestock) in a recent Marshall Islands development plan amounted to 0.15 percent of the overall budget for 1985 to 1989 (RMI 1984). The inattention to production for local consumption is indicated by the data in Table 4. Of the nations listed, significant gains in local food production per capita were achieved only in the Solomon Islands.

To the extent that local resources are devoted to production for export, local needs must be fulfilled with imports. The islands now import very large proportions of the food they consume. This yields advantages for some individuals, but with a declining capacity for self-provisioning, the islands' bargaining powers are reduced, and they become compelled to accept unfavorable prices in their international trade. That is clearly the case for their exports.

Outside nations with greater bargaining power derive substantial benefits from trade with the Pacific islands. Thus they promote a level of dependency on trade which goes well beyond what is warranted in terms of the islanders' own interests. The large-scale imports of junk food demonstrate this pattern. Unlike the major continental powers, the Pacific islands are likely to obtain only meager shares of the benefits of international trade because of their low bargaining powers. The poor in the islands are doubly disadvantaged because they are politically weak in countries that are politically weak.

The market orientation introduces pressures into the islands that

Table 4 *Index of Food Production per Capita* (1974–76 = 100)

	1973	1974	1975	1976	1977	1978	1979	1980	1981	1982	1983
Cook Islands	101	83	106	111	112	98	64	90	83	92	104
Fiji	112	102	98	100	110	111	139	121	132	134	92
Papua New Guinea	99	100	101	100	98	97	97	96	95	97	97
Western Samoa	98	98	100	102	105	101	100	101	104	103	100
Solomon Islands	92	100	97	102	113	116	126	120	132	129	138
Tonga	88	93	103	104	98	91	90	96	101	86	86
Vanuatu	81	105	98	97	93	106	107	81	99	82	82

Source: Economic Commission for Asia and the Pacific 1985.

lead to deterioration in the quality of nutrition. This is an empirically observed tendency, not a logical necessity. People can be empowered to regain control over their diets. In Yap, for example, the increasing consumption of Coca Cola was reversed with a campaign based on the slogan, 'Things go better with coconuts'. People gained increasing control over their diets through a form of education which went beyond comparison of the nutrients in the two products. It also helped local people to understand why Coca Cola was promoted so vigorously and whose interests it served (Rody 1978; also see Rody 1988; and Kent 1988).

The kinds of health programs that are needed are well known:

Population-based approaches emphasizing preventive and pro-motive (i.e. life style or behavioural) activities are most likely to influence morbidity and mortality patterns, though certain medical and public health measures (notably childhood immunization, malaria control and increased access to health care, safe water and sanitation) remain priorities. Finally, improvements in health sector management, planning and financing are required to ensure efficient use of meagre resources. (Bloom 1987, p.44)

But technical knowledge of what should be done is not enough; it is also important to address the issue of political will. Little action can be expected until health, particularly the health of the rural poor, is given higher priority. That is not likely to happen unless the rural poor themselves are empowered to make more substantial claims on national resources.

Powerlessness means that individuals and groups cannot make a fair claim on local resources, and it means that they cannot play a strong role in shaping programs. Inadequate opportunities lead to frustration and self-destructive behavior, often in a rapid spiral of decline, sometimes culminating in extreme forms such as suicide. Such behaviors must be recognized as a sign of social pathology, and not simply as problems of individuals and families.

Whether in traditional or modern sectors, whether in relation to food or health, some people's options are narrowed because others with more power make decisions which respond to other priorities. Soft drinks and infant formula are promoted because there is a profit to be made. Urban hospitals absorb health budgets because urban constituencies demand it. The root of high children's mortality rates in the Pacific islands is the fact that many islanders have little influence on the disposition of their nation's resources, and little power to control their own life situations. If they had the means, which of them would not save their children?

16

Some Concluding Thoughts on
Social Change

Cluny Macpherson

This is a remarkable collection of papers on the Pacific. It includes examples from states in Polynesia, Melanesia, and Micronesia, from the northern, western, southern, and eastern reaches of the Pacific. It includes examples of micro-states, mini-states, and larger states, resource-rich and resource-poor atoll and high island ecologies, with very different economic potentials. It includes states which have at various times been colonized by Spain, France, England, Japan, and the United States, and in some cases by more than one of these powers. There are examples of states which have been almost completely penetrated by foreign capital and others in which capital's impact has been much less marked. The collection reflects its editor's desire to assemble a collection of papers which reflects the diversity of the Pacific experience of social change since European contact. It goes beyond the collections about the anglophone or francophone Pacific, the northern or southern Pacific, the eastern or western Pacific, or some other arbitrary basis of division. It provides the reader with a valuable opportunity to consider and to compare the experiences of a range of Pacific island societies, and to arrive at some tentative conclusions about the factors which may have shaped them.

There are two ways to approach conclusions to a collection of papers such as this. The first is to provide a summary of each chapter and a very general statement of where the chapters seem to lead. The second is to raise some more general issues which arise from the chapters as a whole an which are not directly connected to particular points. Since there seems to be some tension between the arguments offered in Robillard's introduction and the approaches which most authors have taken to their accounts of social change in various Pacific states, it seemed useful to examine the basis of this tension and to seek common ground. This seems to be a more productive approach than reviewing and summarizing a series of already well-written papers.

In the introduction to this volume we were invited to ponder some fundamental issues about accounts of social change. We were invited to consider why, when many approaches to the description and explanation of social change are available, one has assumed a privileged status among them. Why have 'scientific' accounts produced by 'social scientists' using 'scientific methods' come to dominate the discussion and explanation of social change? Why are other accounts and explanations, or in Robillard's terms discourses, of social change largely consigned to a broad category of 'non-scientific' or 'nonrational' accounts which can only then be studied in terms and a logic defined by 'science'.

Part of the answer may lie in the status that the study of social change enjoys within the social sciences because of its association with their birth. It was the dramatic changes which accompanied the great revolutions in the societies of Europe which produced some of the first 'scientific' studies of society. The founding fathers of contemporary social science, Auguste Comte, Max Weber, Emile Durkheim, and Karl Marx sought to comprehend the causes of the changes which they were witnessing all around them. Their studies were attempts to move from a social philosophy of the human condition to models of the organization of human society with which they could identify the forces which explained social stability, social statics, and change, social dynamics, respectively. Only, they believed, when the structure and organization of human societies were understood could social scientists hope to comprehend the processes and direction of change and to exert some control over them.

Although their models differ in important respects, the power of these early models is such that the history of social science is a history of attempts to refine these theoretical foundations. Indeed, many contemporary theories of both society and social change are the intellectual descendants of one or another of these early models. They have, in various forms, become part of the early professional education of most social scientists and have become the foundations on which later intellectual models have been built. It is perhaps this inheritance more than any other single factor which explains the centrality which social change occupies in social science, and the domination by social science of formal models and studies of social change.

We were then asked to consider the model, or rather two variants of the model, which currently dominate the study of social change, that is, anthropocentric models formulated in the productivist discourse of classical political economy (Baudrillard 1975). Why has this type of analysis come to dominate our studies and explanations of social change in general and the Pacific in general? The answer to this lies in part in the ways in which many social scientists view the connection between societies' social, political, and economic organization, and in part in Thomas Kuhn's work in the sociology of science which shows how, through a series of 'paradigmatic shifts', certain models assume prominence at given times and command the attention of a discipline (Kuhn 1970). The prominence of models is due in part to their power to relate and explain apparently unrelated ideas and unexplained phenomena, and in part to a series of institutional processes which consolidate and extend their hegemony over disciplines at particular times and places.

It is true, as the introduction noted, a macro-sociological model of social change, rooted in political economy, has won the attention of the discipline and has come to dominate the study and explanation of social change in the Pacific: the chapters in this volume are evidence of this. Some might ask whether this is necessarily a bad thing for the discipline. After all, recognition that the political and economic activities and interests of a group of colonial powers have significantly shaped social organization in the peripheral states of the Pacific represents a significant advance on the early and now apparently naive theories which were displaced. The dominance of this paradigm focuses attention on previously neglected issues and throws them into sharp relief.

But the consequences of the dominance of any paradigm can be potentially limiting where it is so complete that alternative ways of describing and explaining phenomena are forced to the margins of disciplinary attention where they languish for lack of attention. We are asked by Robillard to consider whether the model which has assumed center stage in the study of social change is any more than a 'fictional code [which] serves as an imperialistic ideology, ironicizing and effacing actual and possible alternative cultural worlds'. Does our commitment to this paradigm, in which 'discourses in which the world is projected as the creation of extra-anthropological natural forces become false consciousness, perceptions of an objective world mediated by cultural beliefs', prevent us from seeing other possible realities? The answer is that it probably does. We are social scientists because we share a particular way of thinking about and understanding social phenomena. We share beliefs about procedures for studying phenomena and about the necessary standards of 'evidence'. It is also clear however that the models which we share may also limit our perception. The models at the center of our education allow us to think in certain sets of terms about social life and at the same time prevent us from understanding it in others. Our beliefs about appropriate research strategies, and acceptable evidence, also constrain our ability to approach the problems in other, and possibly more creative, ways. We may, in a word, be the victims of 'trained incapacity' and whether we can, or should, do more than acknowledge the existence of other forms of discourse and the possibility that these may produce alternative forms of explanation is not at all clear.

If we cannot be visionaries we can at least reflect carefully on our models and be aware of their limitations. We can productively turn our attention to what have become 'domain assumptions'; parts of our conceptual equipment which we rarely have occasion to question. We can at least review the premises and procedures on which they rest and the adequacy and limitations of our equipment. Various authors in this volume have pointed to one or more of these and the conclusion seems to be an appropriate place in which to review them. It may be that this may cause us to be a little more circumspect in our claims for our models.

My concerns focus first on our units of analysis, secondly on the quality of the data from which we construct our accounts of social change within these 'units', and finally on the comparisons and generalizations which follow from them. Only when we understand these can we fully appreciate the limits of our generalizations. When we are more aware of the inherent limitations of our models, we will be more open to other ways of understanding the phenomenon.

COUNTRIES AS WESTERN FICTIONS

The first problem concerns the most basic data: our units of analysis. We divide up the region into a series of 'national entities' which were originally the products of decisions made by representatives of states and interests outside the region to 'rationalize their interests' in the Pacific. In the process they combined, as Ben Finney notes, several social formations into units simply because it was convenient for economic, political and/or administrative purposes. The impact of common administration of different social formations has been to produce increasingly homogeneous 'national cultures'. In other cases, as Bob Franco notes, colonial interests split a single social formation such as the Samoas into two, again, for reasons connected with national interests. The by-product of different colonial administrations of what was once a single social formation has been the emergence of distinctly different nation states. In rare cases, as Philibert notes in his discussion of Vanuatu, a series of peoples were combined in a single political entity which was then administered jointly by two colonial powers which produced multiple 'national cultures', a series of indigenous forms, a 'British', and a 'French' form, at least until independence. The problem with such an approach is that it may trap us into aggregating groups which should more correctly be considered individually.

This problem does not arise only in larger, more 'complex' national entities such as Papua New Guinea where, as Ogan and Wesley-Smith note, various distinct modes of production have existed. It is evident in smaller entities in the different ways in which coastal and valley populations in high islands with distinctive forms of social and economic organization exist. In smaller entities, as Lawrence notes, distinct differences existed between the social, political and economic organization of the northern and southern groups of islands which now make up the state of Kiribati. Even in micro-states like the Tokelau islands, with a population of 2,200 on three atolls, the patterns of change on each of these atolls has been driven in quite different directions by missionary styles and by the different technology which became available after contact with Europeans. It may be that, as Gailey notes, the impact of contact transformed the sexual division of labor, and that the experiences and lives of women and men were transformed in quite different ways. It is possible that contact with other populations transformed the lives of indigenous elites and commoners in quite distinctive ways and that it is more useful to discuss the impact of contact on these groups than on some mythical entity called 'society'. The net effect of this failure to consider variance within societies may be a series of generalizations about social changes which are poor representations of the 'reality' of social change and which do little to refine our models. The benefits of recognition of intranational differences is apparent in the fine-grain analyses such as Ogan and Wesley-Smith's account of Papua New Guinea and Philbert's account of Vanuatu. This is not to suggest that we abandon nation states as units of analysis because, as Robillard has indicated, 'nation state discourse increasingly encompasses more and more of Pacific Island population'. It is to suggest that we distinguish between sub-populations' experiences of change and recognize that generalizations about nation states, even small ones, may mask small but significant differences. Where these are recognized and embodied in our accounts we improve both the quality of our depiction and of our conceptual equipment. We will return to this issue in more detail shortly.

ACCOUNTS OF PRE-CONTACT SOCIETIES

Central to all discussions of social change are reconstructions of some prior state, usually the period before contact with Europeans commenced. As Ogan and Wesley-Smith note, these pre-colonial societies frequently receive little attention, which is unfortunate. An underlying assumption, that 'contact' began with the European presence is itself problematical since the archaeological record and linguistic analyses are increasingly confirming Pacific Islanders' claims that contact and even trade with other Pacific Island populations was occurring before Europeans arrived. Finney outlines, for example, the local specialist trade which existed between high island and atoll in French Polynesia; and Gailey outlines the triangular trade which existed to move goods and commodities between Fiji, Tonga, and Samoa. A recent, general account of the extent and possible nature of this movement can be found in Howe's *Where the Waves Fall* (1988). There is little doubt that contact with European populations produced some of the most profound change which has occurred in Pacific Island societies, but the suggestion that it commenced only then is to misrepresent the reality which was that contact and change was a continuous process in many Pacific Islands. There is evidence in oral history, and now archaeology, which suggests that diseases also had profound effects on the organization of these populations before European contact. The decision to start accounts of change at the point of European contact may have more to do with the availability of the records with which social scientists are happiest with than it has to do with the reality of the process itself.

Despite these inherent problems, reconstructions of the 'pre-contact situation' assume considerable importance in the study of social change. As the 'baselines' in accounts of change, they become the bases of comparisons which in turn lead to generalizations which are eventually incorporated in theories. Many accounts proceed from a series of generalizations abut the social and economic organization of 'pre-contact society' as if these were unproblematical. But how reliable are our reconstructions of 'pre-contact states'? The answer must surely be that they are variable.

Many social formations, and indeed all discussed in this volume, were preliterate societies when European contact commenced. There was no indigenous written record of the social and economic organization of those social formations before contact. The record on which we have come to depend for our information abut that period comes from several sources. The first are books, reports, and diaries of the Europeans who lived among the indigenous people in the period immediately after contact. While these have the 'authority' of written texts (cf. Topping above) they are no more than depictions of societies, the best of which are adequate, the worst of them quite unreliable. This is hardly surprising. They were often compiled after a limited range of contact with indigenous people by visitors with an imperfect grasp of indigenous languages and were shaped by the various political, economic, and religious agendas of their authors and the local elites with whom they dealt. Christine Gailey's accounts draws our attention to the ways in which certain powerful and privileged groups whose interests coincided at the time were able to control the social construction of the pre-contact situation in Tonga for their own benefit. One could argue that the Tongan case is an atypical one and that in other situations there are more, and more reliable, accounts on which to base reconstructions of pre-colonial' social formations. It would be difficult to do so with any confidence. All of the contacts on which our accounts are based typically involved meetings with selected groups of indigenous people and it is naive to believe that their accounts of the 'pre-contact' social reality did not reflect, either consciously or unconsciously, a particular set of interests and their associated 'realities'.

The quality of any account that depends solely on this source of data may be related, ironically, to the density of post-contact European settlement. The biases which emerge in one account of a society may be countered in another and, where a number of accounts exist it may be possible to reconstruct a plausible composite image from them. Thus, for Samoa, where relatively large numbers of Europeans from a variety of backgrounds lived, reconstruction of the nature of pre-contact Samoan society can draw on considerable amounts of data, whereas in cases such as the Northern Marianas, as Samuel McPhetres notes, a much more limited set of data exists.

Even then the data are problematical. The observations were rarely intended to be systematic 'scientific' observations of a social formation and were often personal and highly subjective. They were often written to persuade readers with little understanding of their situation of the need for more resources or personnel, to seek support for decisions about to be taken and provide justifications for decisions which had already been taken. The information included was filtered and arranged to ensure that the reader arrived at conclusions chosen and constructed by the writer. Take for instance the missionary accounts which were often written to persuade working-class supporters of the Church Missionary Society of the need to offer themselves or their resources to the mission work. To do this it was necessary to demonstrate the immensity of the challenge which faced the earliest missionaries which in turn required selective attention to the facets of indigenous societies which most graphically depicted the 'heathen state of the wretched natives'.

In attempts to characterize situations which they only imperfectly understood, the writers of these accounts often generalized about social and economic organization in ways which misrepresented and because of limits of the data which are available for reconstruction. Furthermore, it is often difficult to break down these generalizations because there is often little or no indication of the size or character of the population to which a claim relates. Thus, we are asked to guess at the significance of the claim that a phenomenon 'was present in the people in all of the bays at which we called on that journey' without any indication of the number and locations of the bays, the length of the journey or of the stops in the bays. The incidence of certain events is often difficult to establish because accounts are sprinkled with such picturesque, but unhelpful, estimates as 'scarcely a one', 'no more than a few among them', 'in shocking degree' and 'rarely if at all are they seen without this affliction'.

The situation is further complicated because the filters through which these authors' experiences and observations were passed were constructed from European cultural beliefs and prejudices of the time (Howe 1988). European accounts were influenced, some might say dominated, in important ways by various images of the non-European world. Where these images were patterned in ways which are well understood it may be possible to make educated guesses about the ways in which they might have influenced interpretations of events. It is possible, for instance, to read some accounts against the background of such potent ideologies or tropes as the 'noble savage' and later the 'ignoble savage', social Darwinism, and Calvinist Protestantism, and to estimate how these might have shaped the images to which we are heir. But can we understand and take into account the individual, unpatterned preferences and prejudices which lie behind these same accounts? Can we ever understand the personality and state of mind which, on a given day, influenced assertions about reality on which our accounts now rest?

The best historians – and not all of those who write about social change are good historians – can take this material and can produce remarkably detailed, and apparently perceptive, accounts of social organization (see, for example, Howe 1988). These draw on a range of source material from other disciplines to construct more detailed and reliable depictions of pre-contact societies which challenge earlier ones. But even these, as the new Pacific historians (Routledge 1984) note, are not without problems which stem not from the competence or integrity of their authors, but from the nature of the source data. I raise this issue, not to challenge the value of history because, as I will argue later, a closer acquaintance with history is essential to good social science. I raise it because of a recent, personal experience which highlighted for me the problems of such reconstructions and of the social process by which they are 'verified'.

My wife and I set out to reconstruct an account of the pre-contact Samoan medical belief and practice from the records of early Europeans as a baseline for discussion of contemporary indigenous medicine (Macpherson and Macpherson 1990).We found very little evidence of an extensive set of pre-contact belief and practice which we expected to find. We had then to consider three possible explanations of the lack of material. First, that an extensive body of belief had existed but that its owners concealed it, for reasons of their own, from those who produced the accounts of the period. If this was the case, no amount of research in these accounts would produce a reconstructed picture of pre-contact indigenous medicine. Second, that a body of material existed but that some parts of it were likely to have been hidden, again for reasons of their own, from particular types of Europeans. If this were true, an account would only emerge from comprehensive study of the accounts of different groups of Europeans who might have been privy to different types of information. Finally, we had to consider the possibility that the Samoans had not hidden their medicine and that what we had found was a more or less complete record of what existed around the beginning of contact with Europeans. We used these hypotheses to organize our research strategy and constructed, with advice from many skilled and helpful people, what we now believe to be a reasonably solid argument for the third case.

In the process the emerging text assumed a new status: it became an academic research project governed by certain theoretical, methodological, and ethical canons which would, if followed, assure our findings of acceptance among other academics. Findings, and the procedures by which they were derived, were presented in various academic forums and won guarded acceptance of our peers. In the process we became progressively more confident about our reconstruction, partly because our findings appeared to coincide with studies of the same phenomenon elsewhere and partly because peers were unable to find any serious flaws in our procedures or the interpretation of our data.

Yet, if we were challenged today we would have to admit that our reconstruction is limited by the nature of the data on which it rests and that a number of important issues cannot be answered with absolute certainty. How, for instance, do we know how the Samoans thought about their medicine, and whether on the basis of that perception they might have decided to hide it? To whom might they have revealed particular aspects of their medical practice and on what basis? Was there an equal probability in each case that what was revealed would be documented? Might the Samoans, for instance, have shared their medicine with the Tahitian and Cook Island 'native teachers', who did much of the missionary work in the years immediately after contact, because they believed the Polynesian teachers were more like themselves? But did the Samoans think that the Tahitian and Rarotongan teachers were like them? Were they seen to be part of the mission, in which case possession of the religion may have been more significant than ethnicity? Did the Samoans think that the Polynesians were similar? Might they have found more in common with one of the two groups and shared their medicine only with them? Might they have believed that the women of that group might have been more interested and sympathetic to beliefs associated with pre-Christian

Samoan society than their husbands who were more closely associated with the introduced theology? What were the prospects of material revealed to women in these circumstances finding their way into the accounts of their Polynesian husbands and eventually into the mission record which was produced largely by Europeans? All of these fundamental questions, and others, must remain unanswered because we cannot understand how the Samoans saw and understood the events of the day. Their perception of their situation is simply not available in the material from which we attempted to reconstruct our account of medical beliefs and practices. All we can often do is establish a series of logical possibilities using a logic given us by our disciplinary training, and by imputing various understandings, derived again from our own cultural worlds, to actors in another world at another point in time.

In the meanwhile our account, having passed through various tests, is about to be published and to become a text. It will be of interest to academics and others because our findings will coincide with those from similar studies elsewhere and will contribute to a broader picture of the consequences of competitions between indigenous and introduced medical systems. However, any strength which our account derives from this apparent confirmation may be illusory if it was constructed from data which suffer from basic limitations. We may all be reflecting not reality but the nature of shortcomings in the data on which our accounts are based. It will in turn be used to construct further models of pre-contact society which will be verified by the same process in the same way as we have used others' accounts. Thus, in some way we will all compound errors which were present in the early accounts and all of those which have been built on them.

This is not an argument for abandoning history nor is it a rejection of our findings. It is an argument for an openness to the serious limitations of our reconstructions of pre-contact Pacific Island social formations and to the accounts of social change which are built around them. If ever a case for openness to the limitations of our accounts needs to be made, David Stannard's *Before the Horror* is that case. Stannard (1989) used data from archaeology, demography, geography, ethnology, and physiology to argue for a revision of the population of pre-contact Hawai'i from conventional estimates of 200,000–300,000 to 800,000. In doing so he exposed both the limitations and possibilities of reconstructions of Pacific history.

ON THEORIES OF SOCIAL CHANGE

The third issue which I would like to deal with in this conclusion are theories of social change. If social science has a contribution to make to humanity, it may be its ability to comprehend and generalize about the effects of given phenomena on the organization of human societies. Generalizing about aspects of human societies, and indeed about changes in them, is a legitimate goal of our activity. Any refinement of our understanding of human social life is likely to result from the formulating and testing of generalizations which then become the bases of newer, and hopefully better, theories of social organization and social change. While this is the case in theory, it is less often the case in practice. Because of the way in which social science is organized there is often an imperfect correlation between the status which a theory assumes within a discipline at a given time and its power to 'explain' social activity. A given model comes to dominate a discipline only partly because of its explanatory power. The rest has to do with associated social and political processes within the scientific community. Thus paradigmatic shifts, those points at which existing theories are replaced by newer and apparently more powerful ones, occur in part because the evidence of their inadequacy mounts to a point at which they can no longer be sustained. It may also occur because the social and political arrangements which sustained a theory are transformed. This issue is raised only because as a teacher of social science I find myself under constant pressure to generalize, to look for the rules of patterns which seem to underlie and shape the mass of data. I am also painfully aware that the generalizations which I employed and about which I taught are constantly being rethought and reformulated. In some cases it is because data which have existed all along have been visited, or revisited, by those who were not under pressure to generalize and were allowed to think in ways which were not constrained by a dominant paradigm. In other cases it is because a new model threw new light on old facts and allowed us to see new links.

The history of attempts to explain patterns of social change and, more recently, economic and social 'development' in the 'third world' is a case in point. There have been what might be called 'fashions' in the recent study of the causes and directions of social change. (The predominance of particular models at given times has been an imperfect indication of their explanatory power as is now generally acknowledged even by those centrally involved [Hoogevelt 1982].) This argument has been reviewed and documented extensively elsewhere (Higgot 1983) and is only summarized here.

As many colonies sought independence in the period following the Second World War, the colonial powers sought to bring about the types of social change which it was thought would produce economic growth and social development. This required first an understanding of the phenomena which were impeding growth. This involved identifying factors in the 'social systems' of 'third world countries' which contributed to stability and a resistance to the social change considered necessary. Out of this activity came a theory known as modernization which held that many 'traditional' societies were held hostage by beliefs, values, and associated institutions which were responsible for a resistance to change. Once these were changed development, or, more correctly, economic growth, would follow. The euphemism for this process was 'modernization'. That model became a theory which was taught in universities and informed a significant amount of development practice. Its status within both theory and practice owed as much to the social context in which it was born and sustained as it did to the theory's explanatory power. Its association with prominent individuals and institutions and its incorporation in programs intended to produce desirable social goals such as a better more equitable world, all went some considerable way to explaining its importance within the field. It is clear in hindsight that the economic situations of nation states had far less to do with the beliefs, values, and institutions than it had to do with their location within the world capitalist system. This is, for the purposes of this demonstration, less important that the fact that a model can come to dominate academic discourse even when it has serious problems.

The evidence which turned the tide on this model suggested that in fact a state's potential for economic growth was significantly influenced by its past relations with other states in the world capitalist system. A group of Latin American social scientists were able to show that the exchange relations were structured in such a way that economic growth in certain 'satellite' states could not occur because the surpluses were systematically 'sucked out' by metropolitan states with which they were associated. Thus, the states were not undeveloped, they had been systematically underdeveloped by metropolitan states. While the various theories of 'underdevelopment' (Roxborough 1979) represented a corrective to the earlier approach to the explanation of social change, they led in turn to another series of generalizations about the third world which were problematical. Again, while external factors clearly do influence the speed and direction of change in all societies, and especially small peripheral states, they are only part of the explanation. The speed with which dependency theories came to dominate academic and popular thought had much to do with their advocates' skills, their ability to represent a complex problem simply, and the existence of willingness to question conventional wisdom of development in the west. The popularity of the dependency models challenged both modernization theorists and Marxists to respond and generated another set of theories, all of which have their data bases, literature, and institutional bases.

439

It is relatively easy to point to these examples of disciplines lurching from one theoretical position to another and to argue that generalization and theorization ought to be postponed until the necessary data are available. But this is not necessarily helpful. Social scientists sometimes complain that historians amass facts without then attempting to generalize about their meaning. Historians sometimes complain that social scientists are in turn willing to generalize about situations without due attention to detail, which is at least as dangerous. The tension between them is potentially productive for both groups and for Pacific scholarship. How might the models which currently underpin sociological models of change in the Pacific benefit from closer acquaintance with the data of the new Pacific history?

The models which underpin many of the arguments in this collection implicitly, or explicitly, start from the presumption that much of the change which has occurred in the Pacific is a consequence of colonization and other forms of domination of the region by larger and more powerful interests in Europe, and later in the United States. This occurs because larger states are able to dominate small states and to determine the terms on which they become incorporated into the interstate system. There is a presumption that the resultant incorporation into the world system triggers a unilinear process which leads eventually to more or less complete displacement of traditional forms of social, political, and economic organization. Thus, the speed and direction of social change within the Pacific social formations have been defined by the colonial powers and specifically the interests of capital.

These 'models' take two forms and it may be useful to distinguish between these because they are often conflated and cause confusion. The first of these is the popular form in which, as Philibert (in this volume) notes, such metaphors as the 'fall from grace' are central and in which Pacific people lost their 'innocence' at the hands of rapacious colonial powers. This popular image of social change focuses on the consequences of the process and there is no doubt that these were in all cases profound, and in some truly horrific. The image reflects a mythology of the Pacific which, as Philibert notes, has

> found such widespread acceptance that it is now hard to argue for anything else and which is so much part and parcel of Western mythology . . . that it is hard to know whom one is really describing, the Islanders as they see themselves, or an inverted version of ourselves and our own culture.

The by-products of the process are lamented and one wonders whether the sense of loss says more about the liberal disillusion with life under late capitalism than it says about the fate of Pacific Islanders. The impact of the change is often defined in very general terms, such as the elimination of native culture. The islander's view of the elimination of culture may be somewhat different, generated as it is by a different set of concerns. As my brother-in-law once noted, as we rebuilt our family's Samoan guest house, the only people who mourn the passing of the thatched hut are those who have never thatched and re-thatched one, and the only people who lament the displacement of the canoe by the outboard motor boat are those who have never had to maintain a course in one in high seas for hours on end. The intensity of the concern is often apparently related to the amount of guilt about some vicarious connection with those responsible for the situation which is to be explained. But the substitution of a nonspecific guilt about the consequences for a genuine desire to understand why Pacific Islanders decide to do various things is an unproductive response to the comprehension of social change.

The academic model is somewhat more helpful. It replaces general concepts like 'loss of culture' with more specific ones such as the transformation of specified features of social, political, and economic organization and focuses more deliberately on the processes by which produced these. It reconstructs pre-contact modes of production and identifies the technological, economic, and social factors which cause changes in over time. A notion common in several of these models is that this process of transformation is a unilinear one: it cannot be effectively resisted because the power of the joint forces of colonialism and capitalism, which are often reified to an alarming extent, is so great. The consequence of this image is that much attention is paid to identification of these forces and of their operation and rather less is to the Pacific Islanders' role in the process. There is a tendency to focus on and generalize about the process of the interests of representatives of capital and the consequences of their activity throughout the Pacific.

This theoretical foundation provides the basis for comparative studies of the impact of penetration of capital and for generalizations about the impact of the European presence in the Pacific. At one level it is possible to assert that the history of social change in the Pacific is the history of its incorporation into the world capitalist system (Howard 1983). But does this mean anything? An examination of some generic propositions associated with these types of theory reveals such variation that it is difficult to decide whether generalizations based on them have any meaning. One of the principal benefits of a collection such as this is that the chapters may, individually, clarify our understanding of the process of social change in particular settings, and collectively, alert us to the limits of our generalizations.

It is often argued that European contact was precipitated by the interests of capital and specifically the requirement of new sources of cheap raw materials and/or markets for finished manufactured products. Thus states, which had by this time been 'captured' by groups associated with capital, pushed out into the Pacific in search of means of forestalling the crises of capitalism. Yet the papers in this collection show that, in fact, contact was precipitated by a range of factors of which capital was only one.

The initial move into the Pacific in the sixteenth century was driven by the geopolitical aspirations of the Spanish and Portuguese and the desire to save souls with force if it proved necessary. The search for souls was more important than the search for raw materials and markets. Some of the earliest European contact started in 1668, as McPhetres notes, in the Marianas where a Jesuit priest established a mission on Guam with 'a small garrison of Tagalog troops, Spanish officers and a few other priests' only to find a Chinese Confucianist mission already in place. As a start to missionary activity it was singularly inauspicious: the Confucianist, Chaco, persuaded the inhabitants that Christian baptism was a ruse for poisoning children and the people revolted against the priests and their Spanish supporters. In what followed, the Spanish, with the aid of Guamanian supporters, punished those who had refused to convert with force, driving them into the northern islands. The subjugation of the locals owed more to their refusal to accept Catholicism than it did to the need to arrest the falling rate of return on European capital. But if force succeeded in this case it was singularly unsuccessful for the Spanish elsewhere in the Pacific. In Melanesia for instance, as Shineberg notes, the people very quickly found effective ways of countering Spanish weaponry and thus their initial military advantage (Shineberg 1971).

The Dutch followed the Spanish into the Pacific. Expeditions funded by commercial enterprises spread out into the Pacific. They had no soldiers to enforce their will and made only fleeting contact with Pacific populations to reprovision their vessels and to assess possibilities for their sponsors. In the event they found little of interest and, having contributed more to the understanding of the geography of the Pacific than they contributed to their sponsors' profits, were withdrawn.

The English followed but, as Howe notes, they were more interested in intercepting and raiding Spanish galleons which travelled between their outposts in Manila and Acapulco than they were in trading. In the event they spent much time in the north Pacific and had little if any impact on the South Pacific. As a consequence, in a period of some 250 years, between 1513 and 1760, and despite the presence of three colonial powers, the European impact was confined to Guam and the Marianas (Howe 1988, p. 81).

By the mid-eighteenth century, missionary interest in the region was quickening and led to another wave of European activity. It is often asserted that missionaries transformed indigenous values and generated a demand for commodities which swelled the market for capitalist production. The implication in such statements, found typically in macro-sociological studies of social change, is that missionaries moved unimpeded around the Pacific transforming Pacific societies as the witting vanguard of capitalism. There was clearly a widespread commitment in Europe to the saving of souls (Gunson 1978), but beyond this it is difficult to generalize further about their impact. The style, vigor, and success with which missions pursued their activities in the Pacific varied markedly. Not all could depend on the support of a garrison as the early Spanish had done. The English missionariers established their mission in Tahiti and fanned out to the west with varying degrees of success. They had limited power and were frustrated in many places by a complete lack of interest in their message. In Eastern Polynesia they were with time, and the consent of local elites, allowed to destroy the shrines of indigenous religions, to display their moral outrage, while in Samoa the missionaries were warned by Samoan advisers to avoid such gestures which would alienate the Samoans and impede the missionary cause. Thus London Missionary Society (LMS) and Wesleyan missionaries went 'easy on sin' in Samoa rather than risk alienating chiefs on whom they were dependent for patronage.

The dependent relationship in which missionaries found themselves is evident in the numbers of 'martyrdoms' in the missionary record (Gunson 1978). In places the chiefs 'laid claim' to missionaries and were able to determine where and on what terms their work could proceed. Why missionaries were received and allowed to proceed in various ways in the societies to which they went will never really be known. What is clear however is that the Pacific Islanders played an extremely important role in determining the success or otherwise of missionary activities. In Samoa, for instance, they were accepted because a chief, Malietoa Vai'inupo, believed that they were the people who, according to his dream, would come from the sky and increase his power, while elsewhere they were murdered because they were associated with people who did not speak the local language and had on a previous occasion brought disease depended on understanding the situations into which they went.

The same is true of the progress of the various missions. In some cases animosities, produced by the religious equivalent of imperialism, hindered their activities. Tension developed in places between the London Missionary Society, the Church Missionary Society, and the Wesleyan Missionary Society, and they had to divide up the territory in much the same way as imperial powers establish spheres of influence, before they could proceed without hostility. In Kiribati, for instance, tension developed between the Roman Catholics and the London Missionary Society (Lawrence in this volume) and between the 'formal' missionary activity, which consisted of Samoan pastors representing the LMS and Europeans representing Roman Catholicism, and the 'informal' activity of I Kiribati returning from elsewhere in the Pacific and attempting to convert their families to faiths with which they had come into contact in their travels which extended to New South Wales, Samoa, Hawai'i, and South America.

But other things further limited their power. They faced opposition from beachcombers who had established themselves and had no intention of letting go of any power which they might have held and were in a position to influence both the reception and progress of the missionaries. Members of the mission suffered from poor health and stress and were constantly leaving the mission field in search of 'gentler climes'. Some who postponed their departure died in the field. Others deserted to become warriors as in the case of George Vason in Tonga, or 'succumbed to the fleshly temptations,' took wives and joined local communities. The missions' problems were not confined to the conduct of their European members.

'Native teachers' who were, for various mostly economic reasons, the mainstay of the mission became increasingly aware that while their burdens were heavier often than those of their European brethren, they were not accorded the same rewards and privileges. Increasing dissatisfaction with their conditions of service led eventually to the teachers' 'revolts.' In Samoa in 1850 for instance,

> teachers demanded books, clothing and that the missionary salaries be paid by British donors so freeing local Samoan contributions for paying the teachers' salaries. LMS missionaries in Samoa had no option but to give in, for without the teachers there could be no mission, especially now that the LMS headquarters in London began reducing the number of missionaries in Samoa (and elsewhere) as a cost-cutting measure. (Howe 1988, p. 242)

All of these diverted their energy and attention for varying periods of time and limited the effectiveness of the missions.

The missions operated in very different ways. Some reserved the power to interpret and disseminate their theology for Europeans, while others opened their hierarchies to Pacific Islanders with profound consequences for the shape and progress of their activity. The ways in which missions were able to operate were defined to a large extent by the characteristics of local social organization. In Hawai'i, Tahiti, and Tonga where centralized states existed it was possible to establish and retain control of a particular interpretation of religious doctrine and form of practice with the aid, as Howe notes, of those who controlled the state. In others, such as Samoa, where no central state had existed, Samoans increasingly gained control of both interpretation and practice of religion, to the European missionaries' despair.

This is not to underestimate the significance of their activities but to acknowledge that it varied greatly. Some promoted literacy more vigorously than others which depended more, for various reasons, on rote learning of scripture. Again, as Topping notes (in this volume) the consequences even of limited literacy in these societies were profound. Some missions with access to suitable land introduced new plant species and became involved in agricultural training which had a marked impact on local economy. Others with access only to atoll environments had much more limited possibilities and did little to change the structure of the local economy.[1]

It is true that in places the missionaries had sought to train people in the 'habits of productive labor' as an antidote to what they considered to be unproductive and wasteful lives. It is also true that some missionaries encouraged their flocks to produce certain commodities, such as coconut oil, for sale in a market to fund the extension of missionary activity. It is fashionable to argue, on the basis of these two facts, that the missionaries were the vanguard for capitalism and that they set out deliberately to prepare a way for capitalism. But again, it is not at all clear that this was the case. Whether they set out to advance the cause of capitalism is open to question. The desire to establish the habits of productive labor owed as much to their beliefs about the means to the expiation of sin as it did to the conscious establishment of the preconditions for capitalism. Similarly, their encouragement to their congregations to produce commodities which could be sold owed as much to their commitment to extending their missionary activity into new areas as it did to destroying the precapitalist mode of production to prepare the ground for a capitalist mode. There were undoubtedly missionaries who abused their position to obtain land and who exploited the labor power of their people for their own economic and political purposes. But to impute to all missionary activity a conscious desire to further the advance of capitalism is to go beyond conclusions suggested by the data. There are also examples of considerable hostility which developed as missionaries attempted by various means to frustrate traders' attempts to acquire land and to limit their activities and profits. The record is a fascinating one but does little to confirm some popular generalizations abut the unimpeded impact of missionary activity. What we do know is that there was considerable variety in what we know as religious incursion and that we should generalize with caution.

In the period following the missionary activity the interests of European capitalism started to move into the Pacific. It is true that the European capital followed the missions into the Pacific. The argument follows that interests of capital penetrated Pacific societies and led to their incorporation into the world capitalist economy. But it is useful to look carefully at this proposition. Penetration was, as these chapters show, very uneven and was determined by a range of factors beside the interests of capital.

First, physical and environmental factors determined the probability of penetration which meant that in many areas the activities and influences of Europeans were initially at least confined to more access-ible usually coastal regions. This meant that the extent of penetration differed from, say, atolls in which all people were encouraged to produce a surplus for sale, to higher islands in which smaller pro-portions of populations were involved in any form of production of surplus. The prospects of capitalism having any appreciable impact on local economic organization differed with size of the unit. In atolls and low islands a single trader could have a marked impact on the organization of production, while in larger islands a greater number of traders might have only a limited impact on production in the immediate vicinity.

An even more significant influence on the impact of capitalism was the Pacific Islanders' attitude to trade and willingness to become involved. The earliest trade out of New South Wales was into Tahiti to buy pigs, the Marquesas and Fiji for sandalwood, and the Tuamotus for pearlshell. These were, as Howe notes, all maritime ventures and did not involve the establishment of bases on shore They did, how-ever, depend completely on Pacific Islanders harvesting and selling the commodities they sought. It is popular to portray the Pacific Island-ers as naive passive reactors but as Howe notes

> Without the voluntary participation of Pacific Islanders as labor-ers, most European commercial ventures would have been impossible. This active, cooperative role of island communities has usually been overlooked by those writers who have seen early culture contact as a clash between dominant, initiative-taking Europeans, and passive, helpless Islanders reeling back-ward from the 'invasion'. But culture contact throughout much of the Pacific was not a one-way process so much as a more subtle and complex interaction (1988, p. 95).

Furthermore, the record shows that the Pacific Islanders were not merely the laborers but that they were also clever entrepreneurs and enjoyed certain advantages in the trade. They were able to set price and volume for the traders who had no other means of obtaining the commodities. This early trade was effectively dominated by Pacific

Islanders because the commodities which they produced for trade were not perishable and because while they were not dependent on this trade for their livelihood, the traders and their crews were. In Kiribati, for instance, Lawrence explains how the coconut palm which can be both a subsistence staple and a commercial crop offered a range of options to I Kiribati and limited their dependence on the market (in this volume). The record shows that Pacific Islanders were also astute traders and would turn away commodities which were of little interest, of poor quality, or out of fashion.

With time the Pacific Islanders themselves became increasingly involved in trade. They reorganized economic production to increase their wealth. In some cases, such as in Tahiti and Hawai'i they capitalized on traditional authority and power to organize production of commodities for sale in the international market. The aristocrats of Hawai'i were at least as oppressive as their European counterparts in their supervision of the sandalwood extraction. In Fiji chiefs became involved, if not as oppressively, in the organization of commoners in the collection and sale of *bêche-de-mer* and sandalwood. In other islands, such as New Zealand, local social organization prevented the ruthless exploitation of the commoners that was possible where the monarchies existed. Yet even there the Maoris reorganized production and were soon competing successfully with Europeans in shipping, flour production, and production of vegetables to the New South Wales market. In places such as the Samoas local social organization was so fragmented that people could not be easily reorganized into other forms of production and the same dramatic reorientation to the international market did not occur. In each of these cases, as history suggests, the Pacific Islanders played active and quite different roles in determining the fate of early capitalism.

It is fashionable in some writing to depict the Pacific Islanders' early involvement in capitalism as the product of naivete and a product of their innocence of its social consequences. Such a view is difficult to sustain. Pacific Islanders had sought in various ways to limit the risks which came with certain forms of social and economic organization within a particular ecological zone. They planted reserve crops of yams which would survive storms and hurricanes, they stored seed between seasons, and practiced various forms of resource conservation. There is evidence that in places such as Hawai'i, powerful chiefs had at various times already found means of generating surpluses which permitted the creation of major capital works such as the fish ponds which made food supplies more stable and their exploitation more efficient. They also engaged in strategic visiting, trading, marriage, and adoption practices to create linkages with other groups and to minimize the possibility of avoidable, destructive conflict. A people as concerned with inherent physical and social instability of their situation, may well have perceived the early opportunities to participate in the world economy as means of further minimizing these risks. Thus as Lawrence notes (in this volume), flotillas of up to forty canoes are reported to have travelled up to twenty kilometres offshore to trade with whaling vessels. By this mean they extended the range of crops available to them to include pumpkin, squash, and watermelon; their sources of meat to include pigs and goats; and their technology to include various metals. In the process they diversified their food sources, increased the efficiency with which they exploited their physical environment and ensured that they would no longer be dependent on the limited ecology which was periodically devastated by storms and droughts. These are not the actions of a naive, and gullible people.

Pacific Islanders had also, like human groups everywhere, sought means of gaining, consolidating, and extending their power by the same means and where necessary by force. A people as concerned with power would quickly have seen the possibilities which trade offered them. The Hawaiian, Tahitian and Tongan elites clearly seized the significance of the opportunities with which they were presented. The Ngapuhi tribe in New Zealand saw this possibility and began to trade for weapons which were subsequently used to conquer other groups. This is of course speculation. It does however draw attention to the possibility that Pacific Islanders' may have played a rather more active and deliberate role in this process than we are typically asked to believe by some of the models which draw principally from political economy and particularly the more determinist forms of that argument.

What Pacific Islanders thought and why they chose to begin to participate in the world economy we will never know. We get tantalizing glimpses in Topping's chapter of how Pacific Islanders might have seen the significance of writing and how this might have led to an eagerness to participate in the activities of a group which controlled such a potent system of signs. We can begin to understand why a group might have cooperated to gain control of that system of symbols and the power which was apparently associated with it. We see the same concern to understand the Pacific Islanders' world view in Philibert's attempt to explain why the people of Efate chose to become involved in the polity and the economy of Vanuatu in the way they have.

Later capitalism in the Pacific Islands is a somewhat different phenomenon. By then, Europeans and other had taken advantage of Pacific Islanders' willingness to become involved in various ways to establish themselves and pursue their interests in the Pacific without, in many cases, much regard for the Pacific Islanders rights or interests.

448

Vijay Naidu's paper shows the power and the skill with which the representatives of capital manipulated events to establish, extend, and then maintain their control of the Fijian political economy. Ogan and Wesley-Smith's paper shows the same phenomenon at work in parts of Papua New Guinea. But even during this phase of capitalist penetration, against far more powerful and better organized interests, Pacific Islanders were not simply rolling over and submitting. The Pacific Islanders' reluctance to become involved in plantation labor on the terms on which employment was offered forced plantation owners to recruit indentured labor from elsewhere in the Pacific and Asia to provide labor power.

It is possible to argue that the Pacific Islanders miscalculated the consequences of their involvement with the strangers. They could not have foreseen the devastation which would result from the introduction of diseases to which they had no immunity for instance. It is, however, not possible to sustain the argument that they did not attempt to calculate the consequences. Our ignorance of their perception should not be an excuse. It should be a warning to us about the limits of our comprehension.

Even as I write this, a small group on Bougainville is successfully frustrating the attempts of the one of the world's largest mineral multinationals and the Papua New Guinea state from extracting copper from mines which provide a significant amount of wealth for both the company and the state. I am not suggesting that by focusing on a few heroic examples of resistance to capitalism in the Pacific Islands will make the phenomenon go away. It is undoubtedly established as a central feature of the organization of the economies of the Pacific. I am arguing that the Pacific Islanders' role in decisions is given more weight in arguments about the Pacific.

Thus we find the argument that uneven development on the Pacific rim has generated a demand for labor which has resulted in the migration of Pacific Islanders from relatively low wage zones to relatively higher wage zones. This growth and associated migration has had negative consequences for national economies of the Pacific States which are still recovering from falls in commodities, prices on the world market. This weakness has consequences for the economic viability and political sovereignty of the Pacific States. At one level this is all true. I have written this sort argument myself. What is missing from this is the Pacific Islanders' explanation of why what is happening is happening. It is probably easier to examine behavior and drawing from our own economistic models of human behavior impute motives to Pacific Island migrants than it is to attempt to understand why Pacific Islanders do what they do and how they see it. This is what Robillard is arguing when he asks us to abandon out simple models centered around Homoeconomicus and to admit other ways of understanding and other motives for doing things.

In some recent research I was looking at Samoan migration from village society in which productive relations are closely connected with the organization of kinship and the polity. In that society and others like it, productive relations are part of a more general set of social relations. Much agricultural production is consumed by those who produce it by combining factors of production which they control. Co-workers are often also members of the same kin-group, coresidents in the same village in which they have land rights by virtue of kinship, and spend considerable amount of their leisure time together in similar activities. People leave this setting for one in which productive relations are largely separated from other sorts of relations. In these settings, their labor power which is in effect their only asset is purchased at the lowest possible price and applied to means of production which are owned by the purchaser to create surplus value which is then expropriated for the purchaser's use. At a point at which the purchaser determines that the activity is insufficiently profitable they are free to dismiss the labor force and to invest their capital in some other form of activity without concern for the social consequences for the labor force.

Now using our models we are able to construct an explanation. The migrants become aware of the wage differentials and are persuaded to abandon a society in which they enjoy security by virtue of their control of the means of production for one in which they must sell labor in a market over which they have not control. Their migration is encouraged by states which have been captured by the interests of capital to swell the pool of available labor power, to reduce its price and thus improve the rate of return on invested capital. All of this is quite correct, at least in its own terms. What it fails to do is to attempt to understand the experiences and meanings which the migrants attach to this movement. It imputes motives to the migrants' activities which reflect the Homoeconomicus at the center of the model. Their meanings and experiences, which are incidentally, rich, colorful and have as much to do with a desire for freedom from social control as they have to do with wage differentials, are consigned to a category called false consciousness which requires little study because it is, after all, only a by-product of the ideological hegemony of capitalism.

Thus, our models of Pacific society in particular and human society in general are dominated by a particular conception of humanity and human activity which allows us to consign all other forms of discourse to a category where it languishes because it defined as irrelevant. In doing so we exclude from our accounts the discourse of those who were, and are, perhaps most centrally involved in the activities which we seek to explain. This is not an argument against theory in general or against a particular theory, but rather for a fuller theory which draws on the widest range of data from multiple fields of discourse, to produce a fuller picture of the complexity of human activity.

Notes

CHAPTER 1: INTRODUCTION: SOCIAL CHANGE AS THE PROJECTION
OF DISCOURSE

Notes
1 For a very available review of the postmodern novel see Edmundson,
  Mark. 1989. 'Prophet of a New Postmodernism: The Greater Challenge of
  Salman Rushdie', p. 62–71 in *Harper's* December.
2 For a comprehensive treatment of contemporary Fiji politics, see the special
  edition of *The Contemporary Pacific*, 1990, Volume 2, Number 1, Spring.
3 For an instructive account of the early Spanish galleon trade with the West
  Indies see Carla Rahn Phillips (1986).

CHAPTER 2: PAPUA NEW GUINEA: CHANGING RELATIONS OF
PRODUCTION

Notes
1 Ogan is grateful for the support of a Bush Sabbatical Fellowship from the
  University of Minnesota, and the facilities provided by a Visiting Fellowship, Department of Anthropology, Research School of Pacific Studies,
  Australian National University, in preparing his portion of this article.
  Don Gardner and Robin Hide made helpful comments on earlier versions;
  however, the authors take full responsibility for any deficiencies in the
  finished product. The author's names are listed alphabetically.
2 The most notable exceptions were those dependent on gathering sago as
  a dietary staple.
3 See de Lepervanche (1973) for an approach which, though different from
  ours, also seeks common features amidst the diversity of Papua New
  Guinean societies.
4 However, an ethnography may talk of 'cognatic descent' when what is
  meant is what we call 'ego-centered kinship'.
5 Cf. Amarshi, et al (1979, p. 3): 'ownership of the means of production and
  the distribution of the social product within the group were governed by
  the rules of the kinship organization'; Modjeska (1982, p. 52); 'To pursue
  the relations of production to their heart only to find structures of production to their heart only to find structures of kinship is by now predictable, if disappointing'.
6 A useful analogy is with a board game like Monopoly, in which the 'player'
  in the Papua New Guinea situation described could, for example, activate

an affinal relationship to obtain land rights when his or her descent group found itself short of land: cf. Crocombe and Hide (1971, pp. 304–307).

7 However, if possible to regard pigs as a form of surplus production embodying the labor of women producers, and to argue that men appropriated this surplus since women were seldom involved in determining how it was to be used (Modjeska 1982).

8 The question of gender inequality in the modern history of Papua New Guinea is a key point of Denoon's (1987); article and we gladly acknowledge the intellectual stimulation thus provided for our own analysis.

9 Even in relatively profitable New Guinea the administration consisted of less than 400 people. The total length of vehicular roads was less than 500 miles (New Guinea Annual Reports).

10 An individual was allowed to spend a maximum of two three-year periods under indenture, with at least three months spent in the home village in between stints. The 'twenty-five percent rule' required that a maxiumum of 25 percent of adults males could be absent from an area at any one time. If village life was threatened by absenteeism the area was closed to recruiters. The specter of a 'landless proletariat' was often raised in the official reports of the time.

11 Production for the market, rather than for use or gift exchange, was not a part of the traditional economy. Its destructive potential is particularly evident in the pressures it brought to bear on existing land tenure systems. The semi-permanent nature of the crops grown for the market required more clearly-defined usufruct land rights than were usually available to the individual. Furthermore, cash-cropping gave the land itself a new reified value that made its tranformation into a commodity a distinct future possibility.

12 Missionaries often preceded those who pursued economic interests in Papua New Guinea, and certainly had effects on relations of production, especially with regard to gender relations (e.g., Denoon [1987, p. 55] notes their 'pungent patriarchalism . . . on the subject of proper family relations.'), but their impact cannot be treated at length here.

13 Indeed, Sack (1986, p. 126) suggests that, in the early days, 'The point was to be the master rather than to be rich.' This would not have been true during the days of high copra prices in the 1920s.

14 For example, Connell (1978, pp. 27–28) estimates wartime population losses from both deaths and constraints on fertility ranging from 20–30 percent in South Bouganville.

15 Although only what had been the Mandated Territory of New Guinea now fell under the United Nations Trusteeship, wartime administration of that Territory and Papua as a single unit continued after the war.

16 Pacification of the highlands, traditionally characterized by endemic intergroup fighting, made women's lives considerably easier since they could carry out gardening and other activities without what must have been pervasive fear of rape and murder.

17 The Australian subsidy for 1946–1947 was nearly 40 times that for 1940, and doubled approximately every five years thereafter until the mid-1960s.

18 Denoon (1987, p. 57) provides the reminder that most schools, aid posts and clinics were in the hands of Christian missions until the 1960s.

19 One innovation, the Ex-servicemen's Credit Ordinance 1958–1961, was initially directed toward expatriates, providing both land leases and a loan of A$50,000 to develop new plantations of coconuts and cocoa, a crop

which attracted as much postwar attention in the islands as coffee did in the highlands (Crocombe and Hide 1971, pp. 318, 327).

20 There were nearly five times as many positions in the Department of Agriculture in 1968 as there had been in 1963 (Fitzpatrick 1979, p. 95). State officials were emphasizing individual effort from an early date. For example, a patrol officer in Bougainville reported in 1963 that he had 'strongly discouraged' communal planting in the South Nasioi 'wherein work and returns are shared' and advocated instead of 'individual owner-ship' of plots (Patrol Report #5, 1962–1963).

21 The combined overtures of the Bank and agricultural extension officers leads Fitzpatrick (1979, p. 97) to observe that 'The overall picture . . . may approximate less to ventures set up and run by big men assisted by the state and more to ventures set up and run by the state assisted by big men . . .'

22 Vagrancy regulations making it illegal for Papua New Guinea to remain in town for more than four days without either employment or official per-mission were not repealed until 1969.

23 The number of expatriates in Papua New Guinea nearly doubled in the decade to 1971, when there were 53,123 (Downs 1980, p. 283).

24 Nearly 6,000 teade store licenses had been issued to Papua New Guineans by 1966 (Epstein 1972, p. 312). Other popular business ventures included road haulage, taxis, building and some services.

25 Although the capitalist economy grew rapidly after about 1968, the part of it owned and controlled by Papua New Guineans remained extremely small, and their relative share of the total generated income actually decreased (Donaldson and Turner 1978).

26 The total number of Papua New Guineans employed by the administration increased from 19,219 in 1962 to 28,972 in 1969 (Parker 1972, p. 91).

27 The discovery of the Bougainville copper deposit in 1964 precipitated an exploration boom that has involved a host of mining companies, and identified at least ten major mineral prospects.

28 Under Australian mining law imposed upon Papua New Guinea in 1922, all sub-surface mineral rights were vested in the colonial state. This prin-ciple was retained after independence.

29 Others, of course, could claim subsidiary land rights under the traditional system; cf. Ogan (1971).

30 The agreement was very favorable for the company. Indeed, the only revenue allotted to the government when operations began was a 1.25 percent royalty on the value of minerals produced.

31 For example, occupation fees paid to individuals in the tailings area ranged in 1988 from K2 to K27,960 (Connell 1989, p. 19). At this writing K1 = US$1.15.

32 At the end of 1984, about 16 percent of the national employees at Bougain-ville Copper had been with the company since production commenced in 1972. About 50 percent had been there for at least seven years (information supplied by Bougainville Copper Limited).

33 Denoon (1987, p. 54) claims that the resentment of emerging local bureau-crats toward colonialism in the decade before independence 'was largely the outcome of personal sights rather than objections to the structure of a colonial regime and the export-led economic strategy.'

34 The Bougainville District became the North Solomons Province shortly after independence.

35 More than 90 percent of all positions, and most senior level ones, are filled by Papua New Guineans. There were nearly 5,000 more public servants in 1981 than there had been in 1976 (Goodman, Lepani, and Morawetz 1985, p. 54).

36 For example, retail liquor licenses in Port Moresby were available to non-citizens until 1988.

37 Not all Papua New Guinean-owned urban enterprise is of a petty nature. The Bougainville Development Corporation, for example, now runs, among other things, a limestone mining and processing industrial complex, and a cement works.

38 In theoretical terms, this can be explained as a product of the articulation of indigenous and introduced modes of production. Thus traditional relations of production have not been generally replaced by capitalist relations of production, but by combined forms that, arguably, offer few of the advantages of either progenitor.

CHAPTER 3: NEW CALEDONIA: SOCIAL CHANGE, POLITICAL CHANGE AND TRADITION IN A SETTLER COUNTRY

Notes

1 The demographic and economic history of New Caledonia is discussed in detail in Connell (1987a).

2 Tijiboau himself has commented 'we have already tried the other method-taking action to make people leave' (*Sydney Morning Herald*, 3 September 1988) and *Les Nouvelles Caléniennes* (26 July 1988) has documented much forced migration from the countryside (e.g., 'Un broussard quitt sa terre').

3 This, and other quotations, are from Connell (1988).

4 For example, J-P. Doumenge, 'La Nouvelle-Calédonie: la recherche de son équilibre économiquet et de son unité sociale', *Les Cahiers d'Outre-Mer*, 39, 1986, 221–48. This is reviewed in L. Wacquant, 'Land and History in New Caledonia: The Politics of Academic Writing', *New Zealand Sociology*, 3, (May 1988), pp. 49–55.

CHAPTER 4: SOCIAL CHANGE IN VANUATU

Notes

I wish to thank SSHRCC for providing a leave fellowship which allowed me to spend six months in Vanuatu in 1983 and later seven months at the Australian National University as a Research Fellow. Many thanks to the Australian National University for the offer of a Visiting Fellowship and to Professor Anthony Forge, Department of Prehistory and Anthropology, Faculty of Arts, for his generous hospitality. I am as usual grateful to the Erakor people for tolerating my curiosity about their life and for making research so pleasant. I am particularly indebted to Jane Philibert for translation and editorial advice.

Section 2 has already appeared in print in French in 'Vers une Symbolique de la Modernisation au Vanuatu', 1982 *Anthropologie et Sociétés* 6(1), 69–98. This material has not until now been easily accessible to English-speaking readers.

1 Not all island groups were in a position to maintain their political and cultural independence. In some cases, depopulation or strong missionary activity forced them to come to the coast. In other instances, Christian

coastal groups were now able to assume political control over inland groups. This happened during a period of Tannese history known as the *Tanna Law*. Needless to say, missionaries were rarely aware of the political dimension of conversion to Christ.

2 Anthropologists have in general either exaggerated the integration of traditional knowledge or introduced an antiquarian view of the authentic as age-old traditions quite foreign to the views of many indigenous groups. Traditional knowledge was being created daily, the way 'traditional artifacts' are produced for ritual purposes. Larcom (1982) shows how Malekulan traditional knowledge involved a great deal of cultural borrowing from neighboring groups. She goes so far as to say that traditional culture – rituals, knowledge, objects – were commoditized in the sense that they could be bought, exchanged, or sold at will. 'The kind of *kastom* which Mewun informants described in their history that had little to do with the past in the sense of traditional, continuous or authentic'. (1982, p. 333).

3 Bonnemaison, a geographer, focused his research on the spatial dimension of Tannese social life, something he was better equipped to do than anthropologists. Fortunately for him, on Tanna space is much more than a metaphor with which to conceptualize social ife; people display what Bonnemaison in a pun called a form of 'territorial fixation'. This approach gave him a special insight into Tannese culture which he turned into two old fashioned monographs (1986; 1987), the sort of books anthropologists have stopped writing ever since they started feeling unsure of the concept of culture and being unable to leave themselves out of the narrative. Bonnemaison provides a totalizing view of Tannese culture or an ideal model of it which may well be – probably is – oversystematic, too tightly integrated. However, in a place as varied and complex as Tanna, it often is easier to subtract than to add to the large picture.

4 An old villager from North Efate once told me that inland people regrouped into coastal villages by missionaries in the nineteenth century were never able to adjust and quickly died of sorrow, their hearts broken by having to leave their place of birth.

5 There are two national parties in Vanuatu, the Vanuaaku Pati and the Moderate Party. The former has a Western-style party organization and a single program of action. The party was founded in the early 1970s by educated English-speaking ni-Vanuatu who were university students, administrators working for the British Residency, or pastors of the Anglican or Presbyterian churches. The Moderate Party is a loose national alliance of various island-based opposition groups ranging from French-educated, Catholic civil servants working for the French administration, to the French *colons*, cargo cultists, traditional groups, etc. It could never mobilize its members the way the Vanuaaku Pati could. Its name came from its weaker anti-colonial position.

Tanna is a very complex island; for instance, five different languages are spoken on the island. However, the rebellion on Tanna is only partially explained by national debates. Minorities, be they religious, cultural, or linguistic, regrouped themselves in larger units such as the *Kapiel* group of Middle Bush traditionalist, *Kastom,* another group of traditionalists, and the John Frum people, the Prince Phillip cultists, the *forcona* movement, etc. These made up the Tan Union political group in opposition to Tannese affiliated to the Vanuaaku Pati. Nationally, the Tan Union belonged to the Moderate Party.

6 The British administration, which considered that independence should be achieved quickly, unofficially supported the Vanuaaku Pati as the one party with the national organization required to foster national unity and create a nation-state. The French, because of their numerous nationals in Vanuatu, some of whom had been there as *colons* for a few generations, favored a slower road to independence and a decentralized form of government with local devolution of powers. This position was also the result of a French domino theory in the Pacific whereby Vanuatu's independence would lead to a stronger pro-independence movement in New Caledonia and then Tahiti, strategically the most important possession in the Pacific. This had led the French Residency in the recent past to suport 'minority' groups on Tanna, such as the traditionalists from the Middle Bush and the John Frum movement.

7 It should be pointed out that the Vanuaaku Pati refused to take part in the 1977 national election. This political imbroglio was subsequently resolved when the Moderates gave 5 of the 10 ministerial posts to Vanuaaku politicians and formed what became known as the Government of National Unity. When the Vanuaaku Pati won the 1979 election, with a sizable majority, it did not reciprocate and refused to give even a few ministries to the Moderate Party. Whether or not a second Government of National Unity would have staved off the Santo and Tanno rebellions is hard to say. I personally feel it is doubtful.

8 This is first of all, as Forman notes (1972), because the usual relationship between missions and colonial powers was reversed: the missions were already firmly established before the imperial powers arrived, and were generally more influential, and the colonial governments weaker, in the South Pacific than elsewhere. On top of this, there were proportionally far more missionaries in the South Pacific than almost anywhere else. At the beginning of the century there was one missionary for every thousand inhabitants in the Pacific, whereas in Africa, for example, there was only one for every 50,000 inhabitats at the same period (Forman 1978, p. 39).

9 The reasons which led the Melanesians to convert themselves to Christianity were of course many. Some groups used the missionaries, unbeknown to the latter, in traditional political conflicts, as Guiart emphasizes in the case of Vanuatu (Guiart 1962). Others were looking for European protectors to support their cause in interracial disputes where official justice was often on the side of the strongest. After all is said and done, the missionaries were often the only ones to concern themselves with the ni-Vanuatu, live among them – often for very long periods – learn their languages, and advise them.

10 As Harwood (1978, p. 240) points out, 'Islanders were amazed by what they regarded as an indiscriminate and profligate offer by the missionaries to dispense knowledge'.

11 A Canadian missionary, the Reverend J. W. MacKenzie, played a leading part in transforming the village. Between 1872 and 1912 when he left the New Hebrides, MacKenzie put a stop to warfare between the villages of south-west Efate, and to cannibalism; he abolished polygamy, the ancestor cult, the drinking of kava (*Piper methysticum*), inter-village feasts and dancing, the institution of the men's house and the use of sorcery (magic stones). In view of the depopulation among the local people, he regrouped them in coastal villages under a single chief, chosen for his amenable attitude. Local kinship solidarities, already undermined by mortality and

an influx of outsiders, were further weakened when the mission altered the rules of land tenure in favor of personal ownership, abandoning the customary rights of clans.

12 Some of the early Presbyterian missionaries belonged to the Reformed Presbyterian Church of Scotland whose model for church-state relations was the Second Reformation of 1638. Indeed, according to Adams (1988, p. 84), 'They tenaciously, at times frantically, clung to the ancient Presbyterian goal of a covenanting nation with ministry and magistracy united in yielding obedience to God. Until 1863 they refused to acknowledge the jurisdiction of the British Constitution, to vote for Parliament, to take an oath of office, even to use a postage stamp or buy a rail ticket.' This quasi-millenarian attempt to reconcile the temporal and the spiritual, state and church, fell on good ground in Vanuatu.

13 During the Second World War, Vanuatu was one of the important allied bases in the Pacific with large garrisons on the islands of Efate and Santo. Over 100,000 men were permanently stationed on Santo alone.

14 'Before, anyone who made a mistake [committed an offense] received his punishment the very same way: an accident or illness. This is why people became Christian so quickly. Otherwise today people would all belong to the John Frum movement [a messianic movement on Tanna]. God acted in this way to show his power. Before the law was swift; anyone who disobeyed got punished the same day. If he did not repent he could expect to die. That was good because it made people afraid.' An informant.

15 Joe Garanger, both an archaeologist and an ethnohistorian, has been able to verify the substance of a myth cycle concerning the adventures of legendary hero, Roy Mat, by means of archaeological excavation. According to Garanger, Roy Mata

. . . appears to have been one of the first chiefs who came up from the south and landed on the shores of Efate 'long before the cataclysmic eruption of Kuwae'. He and his people settled first at the spot where they had landed, Maniura Point. There, feasts were organized at which new titles were handed out. The chiefs instituted in this way were sent out in all directions to impose the supremacy of these 'people from the south' . . . Roy Mata gradually succeeded in this way in bringing under his sway all the inhabitants of Efate's coastal regions . . . In particular, he instituted a peace festival held every five years. During this festival the entire population had to lay down their arms, and personal symbols were attributed. These symbols were plants or animals, transmitted matrilineally, and had a part to play in the control of marriages. They also enabled anyone holding such a symbol to find refuge with anyone else who had been attributed the same symbol. It is said that this institution reduced the extent and the number of intestinal wars, and so enabled the population to increase considerably (Garanger 1972, pp. 238–40, my translation).

16 As I have published elsewhere on the present-day image of modernity in Erakor and the threat it poses to the villages' successful adaptation to the modern world, I shall be brief here. Interested readers should consult Philibert (1984; 1986; 1989).

17 There are numerous definitions of ideology. I consider ideology to be a way 'to handle the gap between actuality and aspiration. It incorporates an analysis of present dissatisfaction, an outline of the social desideratum and an analysis of the means by which this gap may be closed' (Peel 1973, p. 300).

18 The material for this discussion of *kastom* is taken from a 1982 special number of *Mankind* (Vol. 13, No. 4) edited by R. M. Keesing and T. Tonkinson devoted to *kastom* in Vanuatu and the Solomon Islands. A second source is Philibert (1986).

19 For the invention of tradition in nineteenth century Europe, see Hobsbawm and Ranger (eds.) (1983).

CHAPTER 5: SOCIAL CHANGE AND THE SURVIVAL OF NEO-TRADITION IN FIJI

Notes

1 Warfare, the major method of overturning land ownership and control, was eliminated altogether (Crocombe 1971).

2 'The Fijian administration acquired control over almost all aspects of Fijian social life, from child rearing to laws against polygamy, assembly and production. Able-bodied Fijians between the ages of 14 to 60 years could be called upon to perform a large number of "social services", including road and house-building, planting and upkeep of food crops, supplying Fijian visitors with food and so on. In addition many of the personal services to a chief were made mandatory unless legal exemption was obtained' (Samy 1977, p. 44; cf. Nayacakalou 1971, p. 220).

3 Such lands were obtained with difficulty and at great costs (MacNought 1982, p. 122).

4 Gujeratis penetrated the countryside as hawkers, trademen and store-keepers, and they set up shops in the towns as jewellers, tailors, grocers, drapers, laundrymen, barbers, and bootmakers, many of these according to their caste-occupations. 'In sharp contradiction to ex-indentured Indians and Indo-Fijians, Guieratis consciously maintained a separate cultural identity *from the very beginning*, speaking regional (Indian) Gujerati language, not marrying with Indo-Fijians, and living in closely-knit but segregated communities. The relations between this ethnic category and the Indo-Fijians was predominantly commercial. Gujeratis did not engage in farming or in wage labour outside the family or exclusively Gujerati enterprises. To label them under the blanket term "Indian" distorted the real situation' (Samy 1977, pp. 79–80, emphasis in the original). See Ali (1976, pp. 18–21) for Gujerati and Punjabi exclusiveness and predominance in commerce.

5 Prior to the 1860s few settlers in Fiji and Samoa showed little, if any, social distance from the islanders. Indeed many Europeans married or had liaisons with island women and raised mixed race children (Ralston 1977, p. 156). However, with the arrival of the cotton planters in the 1860s a transformation occurred in race relations which was to be reinforced under colonial rule.

6 For an unsympathetic account of various anticolonial movements among Ethnic Fijians (largely because of the lack of a broader horizon) but which fails to grasp the extent of popular discontent with the 'administrative society' see MacNaught (1982, Chs. 6 and 7).

7 More or less detailed accounts of these industrial disputes and economic protests have been given by other scholars (see Mayer 1963; Gillion 1977; Ali 1978; Mamak and Ali 1979; Narsey 1979; Sutherland 1984; Hampenstall and Rutherford 1984). These accounts together with the reports of the commissions of enquiry that invariably followed the protestations show

certain patterns that provide insights on both the behavior of the antagonists in the disputes as well as the part played by the state.

8 The name of the person was Modgiran (Immigration Pass Number, 54985, *Fiji Times*, February 20, 1920).

9 In 1940 the United Kingdom Colonial Development and Welfare Act insisted upon trade legislations when support for schemes involving local labor where embarked upon.

10 In his periodic visits to London, Knox, the general manager of CSR raised issues of concern to the company with the Colonial Office which sought to placate him whenever the issue of concern was CSR relations with its cane producers (see Gillion 1977).

11 Both of whom were India-born, Patel was lawyer and Rudrananda was priest of the Ramakrisham Mission to Fiji.

12 Ratu Sukuna took the view that the cane farmers and their union leaders had been impeding the war effort for five months by hindering the production of an essential commodity (Gillion 1977, p. 184).

13 According to one authority sugar production in Fiji decreased by £2 million (Mayer 1963, p. 74).

14 In 1943, Alport Barker, member of the Legislative Council proposed an unofficial majority in the Legislature with six elected members from each of the three ethnic categories. This proposal was defeated and in the ensuing debate, the Indo-Fijian war record was dragged in. Again in the 1946, A. A. Ragg, MLC, noted the increasing population of Indo-Fijians, their growing economic strength and their calls for common roll and moved a motion to guarantee that Fiji '. . . be preserved and kept as a Fijian country' for all time. In the cathartic discussion that followed, both European and Fijian members, raised the spectre of Indo-Fijian dominance, their poor war record, the possibility of conflict over land and the harm done by Indian political agitators (Gillion 1977, p. 196). Sutherland has shown that the racial outburst by Ragg and other European MLCs was supported by the Fijian representation. He maintains that white capital manipulated Ethnic Fijian concerns to offset challenge by Indo-Fijians (1984, p. 175).

With respect to the war effort, in contrast to Indians/Indo-Fijians, 2,071 Ethnic Fijians fought for the Allies, with 42 being killed in action. The war also partially proletarianized Fijians:

The war was such a disruption of village life that many feared the dislocation would be permanent. Neglected womenfolk were unwilling or unable to repair their houses and some *tikinas* had less than a score of men available for the program of work. In April 1944, a year past the peak of the war effort, there were still 9503 Fijians working for wages, or 36.5 per cent of able-bodied males between 16 and eighty. Nearly 7000 were in uniform or directly employed by the military (MacNaught 1982, p. 151).

At the end of the war, efforts by the colonial state were directed to ensure that Ethnic Fijian ex-servicemen were smoothly re-absorbed into their village environments. About 7,000 men were involved and it was considered important to re-settle them quickly in their villages (CP 8/1946). With new widened social horizons, the returned servicemen were considered to be a potential cause of instability in the colonial society. A Fiji Servicemen's After-care Fund of more than £45,000 was established to rehabilitate the demobilized soldiers and labor corps personnel. Grants were made towards the purchase of tools and clothing as well as for the education of the ex-servicemen's children (CP 19 1946).

15 The CSR had made a threat it would withdraw from Fiji if the commission's report was not favorable to it. It became a priority for the colonial state to ensure that the commission would do its best to fulfill CSR's requirements (Moynagh 1981, p. 116).

16 Elections were on a separate role and literacy and property qualifications limited franchise to only 1,404 registered voters (Gillion 1977, p. 132).

17 The establishment of the Native Land Trust Board and the separate Fijian Administration in 1944 (the latter was a form of separate development or apartheid) ensured the basis of their reproduction as a social class.

18 In 1916, the Education Ordinance established a Department of Education and a Board of Education. This legislation was repealed in 1929 and replaced by the Education Ordinance No. 1 of 1929 (Narayan 1984, p. 22).

19 The refusal of Indo-Fijians to fight within an army based on discriminatory conditions, coupled with their antagonism to the dominant white and Fijian chiefly-class, ensured that the recruitment policies excluded Indo-Fijian participation.

20 Nayacakalou concurred with many of these critics but pointed out the threats of 'Indians' made the Ethnic Fijians seek protection in their own institutions and chiefs (1975, p. 138–139). Racial rhetoric and racism had taken effect and became a support of chiefly reproduction.

21 See Meller and Anthony (1967) for a detailed study of the 1963 Fiji General Elections.

22 In 1956, 85 percent of the adult Indian racial category was Fijian-born, as was nearly 100 percent of the non-adult population (Ali 1977, p. 49).

23 The Federation Party had accepted the Constitution under protest because the British Secretary of State had entreated it to give it a try and said it was not the final constitution.

24 As a response the Alliance Party formed the Indian Alliance as a constituent body. Mandatory racial membership was established and direct membership by individuals (irrespective of race) was abolished (Ali 1977, p. 70).

25 This is an after the event justification because only the gross over-representation of the General Electors was diminished to mere over-representation. Ali makes out that secrecy was the key in the dialogue because any dramatic change from the status quo was going to be decried by the *Fiji Times*.

26 Britain had never tried to create a multiracial society in Fiji (see Ali 1980). 'As elsewhere', Ali wrote, 'colonial rule in Fiji bequeathed unresolved problems to its heirs to grapple with and seek their own solution in the era of independence' (1977, p. 74). This voluntarist resolution is hardly likely given the objective conditions that throw up racial politics.

27 Since the coups, Japanese investors have purchased all three of these resorts.

CHAPTER 7: THE MILITARIZATION OF GUAMANIAN SOCIETY

Notes

1 The Annual Report of the Governor of Guam for 1934 states that after 36 years of American occupation in Guam, use of English was not in general use among Chamorros. Children learned English in school, but forgot it due to lack of practice in the home where parents did not understand the language (Governor of Guam 1934, p. 4).

CHAPTER 8: ELEMENTS OF SOCIAL CHANGE IN THE CONTEMPORARY
NORTHERN MARIANA ISLANDS

Notes and References
1 On 13 September 1988, Judge Ramon Villagomez of the Commonwealth
Trial Court issued a decision calling into question millions of dollars in
land transactions over the past ten years. The case involved the Chamorro
wife of an American mainlander who used her husband's money and that
of his business colleague to purchase land from another Chamorro. The
buyer then leased the land to a Japanese firm at a healthy profit. The
original owner challenged the deal, saying that it was the source of the
money that was significant and that the husband and his colleague were
actual owners since they furnished the money. The original owner con-
tended that the deal was unconstitutional. VIllagomez agreed with the
plaintiff. We will see appeals all the way to the US Supreme Court. Many
Chamorros who were go-betweens in land deals will be scrambling to
keep their commissions. The American brokers and foreign leasees may be
affected as well.

A comprehensive bibliography can be found in the series of annual Trust
Territory reports to the Trusteeship Council, United Nations, published by
the U.S. State Department. The last edition of the Report covering the
Marianas was that of 1986.
There are two key works on the social anthropology of the Marianas.

1. Fritz, Georg, *The Chamorros*, 1984. This paper has been translated and is
available through the Historic Preservation Office, Commonwealth of the
Northern Marianas 96950.

2. Joseph and Murray: *Chamorros and Carolinians of Saipan*, Personality Studies of Saipanese Children and Adults, including a report on Psychopathology.
A report published by the Coordinated Investigations in Micronesian Anthropology (CIMA).

There is not much on the early history of the Marianas but the following
publications contain the basic essence:

1. Pigafetta. Antonio: *The Voyage of Magellan*. A translation by Paula Spurlin
Paige from the edition in the William L. Clements Library, University of
Michigan, Ann Arbor. Prentice-Hall, Inc. Englewood Cliffs, N.J. 1969.

2. Risco, Alberto S. J., *The Apostle of the Marianas, The Life, Labors and Martyrdom of Ven. Diego Luis De San Vitores, 1627–1672*. Translated from the Spanish
by Juan Ledesma, S. J. Published by the Diocese of Agana, Guam 1970.

3. Carano and Sanchez: *A Complete History of Guam*. Published by Charles
E. Tuttle Co. Inc. Tokyo 1968.

For miscellaneous background materials and political developments relating
to the social environment, the following publications are useful:

1. Craig and King, ed., *Historical Dictionary of Oceania*, Greenwood Press,
Westport, Conn. 1980.

2. Peattie, Mark R. *Nan'yo – The Rise and Fall of the Japanese in Micronesia,
1885–1945*, Pacific Islands Monograph Series No. 4, Honolulu, University of
Hawaii press, 1988.

Much of the recent development in the Marianas has no precedent or substantial rooting in the distant past. To understand what is going on it is necessary to be familiar with the politics of the Commonwealth.

1. McHenery, Donald: *Micronesia: Trust Betrayed, Altruism v. Self-Interest in American Foreign Policy*. Carnegie Endowment for International Peace, N.Y. 1976.

2. US State Dept: *Annual Report of the Administering Authority of the Trust Territory of the Pacific Islands to the United Nations Trusteeship Council*. Published annually by the Trust Territory. The last edition to contain materials relevant to the Northern Marianas is the 1986 edition. This edition contains the following materials:

A. The 1947 Trusteeship Agreement between the UN Security Council and the United States Government.

B. The Proclamation of President Reagan of 3 November 1986 making the Northern Marianas covenant fully applicable without unilaterally terminating the Trusteeship Agreement.

C. Statistical information concerning the political, social, economic, educational, and legal aspects of the Marianas for the fiscal year 1986.

D. Bibliography of publications prepared by the Curator, Pacific Collection, Hamilton Library, University of Hawaii at Manoa, Honolulu.

3. *Covenant to Establish a Commonwealth of the Northern Marianas in Union with the United States of America*. Negotiated between the US and the Northern Marianas and signed on 15 February 1975, approved by the people of the NMI on 17 June 1975 and by the U.S. Congress on 1 April 1976.

4. *Constitution of the Commonwealth of the Northern Marianas*. Became effective on 9 January 1978 with the first constitutional government being sworn in on Saipan.

CHAPTER 9: KIRIBATI: CHANGE AND CONTEXT IN AN ATOLL WORLD

Notes
1 In fact even as independence was approaching in 1979 no widely accepted indigenous name identifying the islands as a group could be agreed upon. This resulted in the choice of *Kiribati* which is an indigenization of *Gilberts* the name given the group by Von Krusenstern in the 1820s in recognition of the sightings of the archipelago by Captains John Marshall and Thomas Gilbert in 1788 (Macdonald 1982, p. 15).
2 From the Samoan term *faipule* or councillors.
3 Limited resettlement of Tabuaeran (Fanning Island) and Teraina (Washington Island) began in October 1989.

CHAPTER 11: STATE FORMATION, DEVELOPMENT, AND SOCIAL CHANGE IN TONGA

Notes
1 Stanley Diamond (1974) has provided a thoroughgoing critique of the notion of progress through civilization.
2 Social reproduction is a concept in the Marxist tradition which describes the material and ideological means through which social continuity occurs. This includes recruitment or replacement of the labor force and nonproductive groups, the rituals and other cultural means of socialization and status acquisition, the replication of the division of labor, conservation and continuity of other ecological relations, belief structures and values which

reinforce the status quo, and – often ignored – reproduction of the tensions and conflicts inherent in certain social relations and structures.

3 On cultural forms of resistance to domination, see James Scott (1985); Diamond has discussed the opposition of law and custom (1974).

4 For a recent restatement of this view, see Eric Wolf (1982). This is a view adopted by modernization theorist and in sectors of the left influenced by the Second International (Gailey 1985b). There is much to criticize about its Eurocentric view of history, which privileges capitalism as socially dynamic and, thus, implicitly or explicitly portrays colonialism as progressive, even if destructive of precapitalist modes.

5 These orders have been termed by those from monarchical societies as 'nobles' and 'commoners,' a conflation which, unfortunately, has persisted in the literature. They become nobles and commoners with state formation, but it is unwise to gloss nobles with chiefs or commoners with nonchiefly people, as relative political relations and labor claims are quite different (Gailey 1987a, p. 85; see also Marcus 1977).

6 Until the establishment of the Wesleyan Methodist mission in the late 1820s, few of the men remained for more than four years: They either left eventually on passing ships, were caught between factions (and, so, without protection) in interchiefly wars and killed, or were executed by their sponsors for outraging Tongan mores.

7 The Europeans also introduced a partially realized potential for chattel slavery using war captives in trade. But the interests of missionaries and, from 1830 on, their prominence in mercantile activities truncated this development. But the missionaries did encourage the use of 'criminal' labor on state and mission estates.

8 This neutrality – a refusal to reject United States or French military policies – has impeded efforts by most other countries in the region to create a nuclear-free Pacific (see Alcalay 1988; S. Firth 1987).

9 Families deliberate carefully about which members would be most likely to do well overseas and be reliable in sending remittances. Rank (including age) and kin role generally determined who received remittances for how long (Gailey 1987b).

10 This was reported by a number of returned migrants, both women and men.

11 The 1986 survey and in-depth interviews were designed by myself and administered by two then-unemployed high school graduates from the community. The sample included adult members of all households in two urban neighborhoods in Vava'u, and all returned migrants in those neighborhoods (Gailey, in press). I also interviewed a snowball sample of returned migrants in the capital.

CHAPTER 12: FRENCH POLYNESIA: A NUCLEAR DEPENDENCY

Notes

1 This paragraph enormously simplifies the diverse, if related, systems of social organization found in these islands. See Oliver (1974) for Tahitian social organization; Ottino (1972) for Rangiroa in the Eastern Tuamotus; Hansen (1970) for Rapa in the Australs; and Dening (1980) for the Marquesas.

2 Accounts of various aspects of Tahitian and French Polynesian history during this era include Arii Taimai (1964); Beaglehole (1955); Davies (1961);

Dening (1974); Ellis (1829); Gunson (1977); Newbury (1980); and Oliver (1974, Vol. 3).

3 The imprecision of the *Demi* label, complicated by a recent switch in census categories, make it impossible to be precise about this proportion. The 1983 census (Recensement 1983) indicated that 11.6 per cent of the population was of European (mostly French) descent, 8 per cent was of Chinese of mixed Chinese descent, and 10 per cent was 'Polynesian-European', a new census category which replaces the old *Demi* category. Except for a few foreigners who did not fit in any of the above categories, the rest of the population was classified as Polynesian. In that 17 per cent of the people identified themselves as *Demi* in the 1977 census, it is likely that many of these people opted for the Polynesian category in the 1983 census.

4 See photos of the slums and interviews with tahitian social work in Le Mintier (1985).

5 Le Minter (1985); See Crocombe and Hereniko (1985) for a collection of interviews with poor Polynesian living on Tahiti and analyses of their social problems by Polynesian and French officials.

6 This account compresses and simplifies a complicated tale of French and Polynesian political mascinations. For more details see Daniellsson (1983); Danielsson and Danilesson 1986b; and Tagupa (1976; 1983).

7 The phrase, 'confetti of empire', is from Guilliband (1976).

CHAPTER 13: WHERE IS SOCIAL CHANGE IN HAWAI'I?

Notes

1 This paper does not address the issue of native Hawaiians, the indigenous population. Nor does it focus on an overview of all the ethnic groups of Hawai'i, like *People and Cultures of Hawaii* (McDermott, Tseng, and Maretzki 1980). The paper avoids realistic accounts of natural history, choosing instead to describe the ideological structure of social change. This structure is what makes change sensible.

2 I obtained the idea of incommensurate social structures in conversation with Harold Garfinkel.

3 I am using structuration in the sense of Anthony Giddens (1981).

4 The reader should refer to the work of Jean Baudrillard (1975) for a notion of possible alternative discourses.

5 This book was privately financed by the authors. I am indebted to Neghin Modavi for this information. That the book could not obtain publisher support in the US because of a limited market and the fear of lawsuit is a partial indication of the limited generality or equivalence of the discourse of the book.

6 By introduced I mean that this was the first time a shirt called the aloha shirt was produced. They were anticedent shirts. It must be remembred that the shirt is not indigenous to Hawaii. It is a European item.

7 There is an entire genre of aloha attire called the picture shirt.

8 I want to emphasize the word said. That people are seen and reported to choose aloha shirts by class is a production of the natural evidenice of discourse.

9 The side of the material on which the print is made is worn inside the garment.

10 I am thankful to Alfred Fortin for pointing out the importance of the pants worn with the Reyn's aloha shirt.

11 I am told by Alfred Fortin that in large bureaucracies in Hawaii that the knowledge of the Reyn's aloha shirt is explicitly prescriptive.

12 The first shirts in Hawaii were dyed with vegetable dyes. They faded quickly, creating a muted pattern. The Reyn's aloha shirt is worn with the pattern on the inside, producing the same effect. I cannot help but wonder about the intention of the anitqueing of the Reyn's shirt.

13 I am indebted to Cluny Macpherson who interviewed the sales staff at the Ala Moana Reyn's store.

14 Baudrillard makes brief mention of fashion. Roland Barthes devotes an entire book to the subject, *The Fashion System* (1983).

15 I am indebted to Fred Fortin for his descripton of working in a Honolulu based insurance company and how he came to learn about the Reyn's aloha shirt code.

16 The Reyn's shirt is an exclusively male code. There are Reyn's shirts made for women. However, the women who wear the shirt in offices do not have the same effect as the male wearer. Women are aware of the male exclusivity of the shirt code and attribute it to male hegemony of commerce. I am thankful for this observation by Meredith Burns.

17 I recognize that the shirt is not a Hawaiian garment. It only became an item in Hawaii with Western contact. It has always been a commodity, produced for exchange, at least insofar as the growing, weaving, sewing, advertising, distribution and sales of the shirt.

18 I must acknowledge that the production and exchange of sign values is not new or post-modern. The production of signs can be seen in the potlach and the kula circle and in many other contexts. I contrast the production of signs with the idealization of production for use value only to point to the relatively increased proportion of production of signs in post-modern capitalism.

19 *Hapa haole* literally means half a foreigner. However, haole in practice has come to mean Caucasian. Hapa haole refers to anyone of mixed parentage between a Caucasian and other ethnic groups in Hawaii, mostly Asian and Pacific Islanders.

20 I am thankful to Cluny Macpherson for this observation.

21 These observations are from seven and a half years of living in a rural Cagayano family in Hawaii.

CHAPTER 15: CHILDREN'S SURVIVAL IN THE PACIFIC ISLANDS

Notes

1 The Taylor, Lewis, and Levy data may also be found in Lewis (1988). Mortality data in the Pacific islands are generally unreliable because of the weakness of their vital registration and health information systems.

2 Comparable accounts may be found in Bruss (1987).

CHAPTER 16: SOME CONCLUDING THOUGHTS ON SOCIAL CHANGE

Notes

1 I realize this is an argument of geographic determinism. This is a western ecological discourse.

References

Althusser L and Balibar E 1970, *Reading Capital*. London: New Left Books.

Alcalay Glen 1988, 'The Ethnography of Destabilization: Pacific Islanders in the Nuclear Age', *Dialectical Anthropology* 13 (3): 243–251.

Adams Henry 1930, *Letters of Henry Adams, 1859–1891*. Edited by Ford W C. Boston: Houghton Mifflin.

Adams R W 1988, *In the Land of Strangers*. Pacific Research Monograph Number 9. Canberra: The Australia National University.

Agyei William K A 1988, *Fertility and Family Planning in the Third World: A Case Study of Papua New Guinea*. London: Croom Helm.

Aldrich Robert, Connell John 1990 (in press), *France's Overseas Frontier. The Departements et Territories de l'Outre-mer*. Sydney: Cambridge University Press.

Ali A 1976, *Society in Transition, Aspects of Fiji-Indian History 1879–1939*. Suva: University of the South Pacific.

Ali A 1977, *Fiji: From Colony to Independence 1874–1970*. Suva: University of the South Pacific.

Ali A 1978, 'The Indians of Fiji: Poverty, Prosperity and Security', *Economic and Political Weekly* 8:1655–1660 (September).

Ali A 1979, *The Indenture Experience in Fiji*. Suva: Bulletin of the Fiji Museum No. 5.

Ali A 1980, *Plantation to Politics*. Suva: University of the South Pacific

Alkire William H 1965, *Lamotrek Atoll and Inter-Island Socioeconomic Ties*. Illinois Studies in Anthropology, No. 5. Urbana: University of Illinois Press.

Alkire William H 1977, *An Introduction to the Peoples and Cultures of Micronesia*. 2nd ed. Menlo Park: Cummings.

Alkire William H 1978, *Coral Islanders*. Arlington Heights, Illinois: AHM.

Allen J, Golson J, and Jones R (eds) 1977, *Sunda and Sahul: Prehistoric Studies in Southeast Asia, Melanesia and Australia*. London: Academic Press.

Amarshi, Azeem, Good Kenneth, Mortimer Red 1979, *Development and Dependency: The Political Economy of Papua New Guinea*. Melbourne: Oxford University Press.

American Samoa Department of Education 1974, *Think Children*. Department of Education Annual Report. Pago Pago, American Samoa.

Anderson Benedict 1983, *Imagined Communities: Reflections on the Origin and Spread of Nationalism*. New York: Verso.

Annear, P. 1973. 'Foreign Private Investment in Fiji', pp. 39–54 in *Fiji, A Developing Australian Colony*, edited by A. Rokotuivuna and International Development Action. North Fitzroy, Victoria: IDA.

Anova-Ataba Apollinaire 1984, *D'Atai à l'indépendance*. Noumea: Edipop.

Apple, Russell 1980, *Guam: Two Invasions and Three Military Occupations. A Historical Summary of War in the Pacific National Historical Park, Guam*. Guam: Micronesian Area Research Center.

Arii Taimai 1964, *Mémoires d'Arii Taimai*. Publications de a Société des Océanistes, no. 12. Paris: Musée de l'Homme.

Athens John Stephen. 1980. *Archaeological Investigations at Nan Madol: Islet Maps and Surface Artifacts*. Agana, Guam: Pacific Studies Institute.

Babha H K 1983, 'The Other Question: The Stereotype and Colonial Discourse', *Screen* 24 (6): 18–37.

Bain A 1986, 'Labour Protests and Control in the Goldmining Industry of Fiji, 1930–1970', *South Pacific Forum* 3 (1): 37–59.

Bain K R 1954, *Official Record of the Royal Visit to Tonga, 1953*. Nuku'alofa, Tonga: Government Printer.

Baker John 1977, 'Contemporary Tonga: An Economic Survey', pp. 228–246 in *Friendly Islands*, edited by N. Rutherford. Melbourne: Oxford University Press.

Baker L, Baker Shirley 1951, *Memoirs of the Rev. Dr Shirley Waldeman Baker*. Dunedin: Coulls Somerville Wilke.

Balasuriya T 1973, 'Developing the Poor by Civilising the Rich', *Pacific Perspective* 2 (1): 5–15.

Baldauf R B 1981, 'Educational Television, Enculturation, and Acculturation: A Study of Change in American Samoa', *International Review of Education* 27: 227–245.

Barnes H 1982, 'Population Growth and Status of Women' pp. 255–263 in *Population of Papua New Guinea*, ESCAP/SPC.

Barnett, T. 'Politics and Planning Rhetoric in Papua New Guinea', *Economic Development and Cultural Change* 27: 776.

Barrington J M 1973, 'United Nations and Educational Development in Western Samoa During the Trusteeship Period, 1946–1961', *International Review of Education* 19:255–261.

Barthes Roland 1977, *Leçon*. Paris: Seuil.

Barthes Roland 1983, *The Fashion System*. New York: Hill and Wang.

Baudrillard Jean 1968, *Les Système des Objets*. Paris: Denoel-Gonthier.

Baudrillard Jean 1975, *The Mirror of Production*. Translated by Mark Poster. St. Louis: Telos Press.

Baudrillard Jean 1981, *For a Critique of the Political Economy of the Sign*. Translated by Charles Levin. St. Louis: Telos Press.

Baudrillard Jean, 1983, 'The Ecstasy of Communication', pp. 126–134 in *The Anti-Aesthetic: Essays On Postmodern Culture*, edited by Hal Foster. Port Townsend, Washington: Bay Press.

Baudrillard Jean 1985, 'The Child in the Bubble', *Impulse*, Winter: 12–13.

Bayliss Smith T P 1978, 'Batiki in the 1970's: Satellite of Suva', *UNESCO/UNFPA Fiji Island Reports No. 4, The Small Islands and Reefs*. Canberra: Development Studies Centre, Australian National University.

Bayliss-Smith T P 1978, 'Batiki in the 1970's: Satellite of Suva', *UNESCO/UNFPA Fiji Island Report No. 4, The Small Islands and Reefs*. Canberra: Development Studies Centre, Australian National University.

Beaglehole John C (ed) 1955, *The Journals of Captain James Cook on His Voyages of Discovery*. Vol. 1. Hakluyt Society, Extra Series, no. 35. London: Cambridge University Press.

Beaglehole, Ernest and Beaglehole Pearl, 1941. *Pangai*. Wellington: The Polynesian Society.

Beauchamp Edward 1975, 'Educational Policy in Eastern Samoa: An American Colonial Outpost', *Comparative Education* 11: 23–30.

Bedford R D 1978, 'Kabara in the 1970's: Home in Spite of Hazards, Economy and Population', *UNESCO/UNFPA Fiji Island Reports No. 4, The Small Islands and Reefs*. Canberra: Development Studies Centre, Australian National University.

Bedford R D and Mamak A F, 1977. *Compensating for Development: The Bougainville Case*. Christchurch: Bougainville Special Publication £2.

Bedford R D, Macdonald B and Munro D 1980, 'Population Estimates for Kiribati and Tuvalu, 1850–1900: Review and Speculation', *Journal of the Polynesian Society* 89 (2): 199–246.

Beebe C E 1954, *Report on Education in Western Samoa*. Wellington, New Zealand: Government Printer.

Beechert Edward D 1985, *Working in Hawai'i: A Labor History*. Honolulu. University of Hawai'i Press.

Beers Henry P 1944, *American Naval Occupation and Government of Guam, 1898–1902*. Washington, D. C.: Navy Department.

Bell Leonard 1980, *The Maori in European Art: A Survey of the Representation of the Maori by European Artists from the Time of Captain Cook to the Present Day*. Wellington: Reed.

Bellwood Peter 1978, *Man's Conquest of the Pacific*. New York: Oxford University Press.

Belshaw S W 1964, *Under the Ivi Tree*. Berkeley: University of California Press.

Bennett J A 1976, 'Immigration, "Blackbirding" and Labour Recruiting: The Hawaiian Experience 1877–1887', *Journal of Pacific History* 11: 3–27.

Bennett Judith A 1987, *Wealth of the Solomons*. Honolulu: University of Hawaii Press.

Bergesen Albert 1983, *Crises in the World-System: Volume 6, Political Economy of the World-System Annuals*. Beverly Hills: Sage Publications.

Bertram Geoffrey 1986, 'Sustainable Development in Pacific Micro-Economies', *World Development* 14: 808–822.

Bertram I G and Watters R F 1985, 'The MIRAB Economy in South Pacific Microstates', *Pacific Viewpoint* 26 (3): 497–519.

Bertram I G and Watters R F 1986, 'The MIRAB Process: Earlier Analyses in Context', *Pacific Viewpoint* 27 (1): 47–59.

Biddulph John 1981, 'Promotion of Breast-feeding: Experience in Papua New Guinea', pp. 169–174 in *Advances in International Maternal and Child Health*, Vol. 1, edited by D. B. Jelliffe and E. F. Patrice Jelliffe. New York: Oxford University Press.

Blanchet Gilles 1984. *L'Economie de la Polynésie Francaise de 1960 à 1980 – Une Apercu de son Evolution*. Notes et Documents, no. 10. Pape'ete, Tahiti: Office de la Recherche Scientifique et Technique Outre-Mer.

Blanchet, Gilles 1987, *Croissance Induite et Développement Autocentre en Polynésie Francaise*. Colletion: Travaux et Documents Microédités. Paris: Office de la Recherche Scientifique et Technique Outre-Mer.

Bloch Maurice (ed) 1975, *Marxist Analyses and Social Anthropology*. New York: Wiley.

Bloch Maurice (ed) 1983, *Marxism and Anthropology: The History of a Relationship*. Oxford: Clarendon Books.

471

Bloom Abby L 1987, 'A Review of Health and Nutrition Issues in the Pacific', *Asia-Pacific Population Journal* 1 (4): 17–48.

Boggs Stephen T 1985, *Speaking, Relating, and Learning: A Study of Hawaiian Children at Home and at School*. Norwood, NJ: Ablex.

Bonnemaison J 1979, 'Les voyages et le'enracinement: formes de fixation et de mobilité dans les sociétés traditionnelles des Nouvelles-Hébrides', *L'Espace Géographique* 4: 303–318.

Bonnemaison J 1980, 'Espace géographique et identité culturelle en Vanuatu', *Journal de la Société des Océanistes* 68: 181–188.

Bonnemaison J 1984, 'The Tree and the Canoe: Roots and Mobility in Vanuata Societies', *Pacific Viewpoint* 25 (2): 117–151.

Bonnemaison J 1985a, 'Un certain refus de l'Etat', *International Political Science Review* 6 (2): 230–247.

Bonnemaison J 1985b, *La Dernère île*. Paris: ARLEA/ORSTOM.

Bonnemaison J 1986, *L'arbre et la Pirogue*. Paris: Ed de l'ORSTOM.

Bonnemaison J 1987, *Tanna: Les Hommes Lieux*. Paris: Ed de l'ORSTOM.

Bougainville Copper Limited. *Annual Reports*. Melbourne.

Boyd R 1911, *Report by the Commissioner Appointed to Superintend the Taking of the Census of the Colony*. Suva: Government Printer.

Braudel, Fernand 1967, *Capitalism and Material Life 1400–1800*. Translated by Miriam Kochan. New York: Harper and Row.

Brewer A 1980, *Marxist Theories of Imperialism: A Critical Survey*. London: Routledge and Kegan Paul.

Brewster A B 1922, *The Hill Tribes of Fiji*. Philadelphia: Lippincott.

Britton S G 1980, 'Tourism and Economic Vulnerability in Small Pacific Island State: The Case of Fiji', pp. 239–263 in *The Island States of the Pacific and Indian Oceans: Anatomy of Development*, edited by R. T. Shand. Canberra: Australian National University Press.

Britton S G 1982, 'The Political Economy of Tourism in the Third World', *Annals of Tourism Research* 9: 31–58.

Bronson Vernon 1968, *A System Manual for the Staff and Faculty of the Department of Education*. Utulei: Government of American Samoa.

Brookfield H C 1972, *Colonialism Development and Independence the Case of the Melanesian Islands in the South Pacific*. Cambridge: Cambridge University Press.

Brookfield H C 1987, 'Export or Perish: Commercial Agriculture in Fiji', in *Fiji, Future Imperfect*, edited by M. Taylor. Sydney: Allen and Unwin.

Brookfield H C with Hart D 1971, *Melanesia: A Geographic Interpretation of an Island-world*, London: Methuen.

Brookfield H C, United Nations Fund for Population Activities, and UNESCO. 1978. *Taveuni, Land, Population and Production*. Canberra: Development UNESCO/UNFPA Island Reports.

Brower Kenneth 1983, *A Song for Satawal*. New York: Harper and Row.

Brunton R 1979, 'Kava and the Daily Dissolution of Society on Tanna, New Hebrides', *Mankind* 12 (2): 93–103.

Bruss, Mojdeh (ed) 1987, *The Pacific Conference: Nutrition Challenges in a Changing World*. Proceedings of a Conference held July 13–14, 1987 at the University Of Hawaii. Honolulu: Hawaii Nutrition Council/Society for Nutrition Education.

Bryant Coralie and White Louise G 1982, *Managing Development in the Third World*. Boulder: Westview Press.

REFERENCES

Buck Peter 1930, *Samoan Material Culture*, Honolulu: Bishop Museum Bulletin No. 75.

Burns Report 1960, *Report of the Commission of Enquiry into the National Resources and Population Trends of the Colony of Fiji, 1959*. Published as Legislative Council Paper No. 1 of 1960.

Burns Alan 1963, *Fiji*. London: HMSO.

Caces Maria Fe F 1985, *Personal Networks and the Material Adaption of Recent Immigrants: A Study of Filipinos in Hawai'i*. Unpublished doctoral dissertation, Department of Sociology. University of Hawai'i.

Cameron John 1983, *An Overview of Tourism and its Employment Generating Potential in Fiji in the 1990s*. Discussion paper No. 42. Norwich: School of Development Studies, University of East Anglia (September).

Carano Paul and Sanchez Pedero C 1964, *A Complete History of Guam*. Tokyo: Charles E. Tuttle Co.

Carstairs R T and Prasad R D 1981, *Impact of Foreign Direct Investment on the Fiji Economy*. Suva: Centre for Applied Studies and Development, University of the South Pacific.

Carter, John (ed) 1981, *Pacific Islands Yearbook*. Fourteenth edition. Sydney: Pacific Publications.

Central Planning Office 1975, *Fiji's Seventh Development Plan, 1976–1980. (DP7)*. Suva: Government of Fiji.

Chandra R 1980, *Maro Rural Indians of Fiji*. Suva: South Pacific Social Services Association.

Chapman, Murray and Mansell R Prothero 1985, *Circulation in Population Movement: Substance and Concepts from the Melanesian Case*. London: Routledge & Kegan Paul.

Chapple W A 1921, *Fiji: Its Problems and Resources*. Auckland: Whitcombe and Tombs.

Chowning A 1977, *An Introduction to the Peoples and Cultures of Melanesia*. Menlo Park: Cummings.

Christnacht Alain 1987, *La Nouvelle-Calédonie*. Paris: La Documentation Francaise.

Cicourel Aaron Victor 1964, *Method and Measurement in Sociology*. New York: Free Press of Glencoe.

Clammer J R 1975, 'Colonialism and the Perception of Tradition in Fiji', in *Anthropology and the Colonial Encounter*, edited by T. Asad. London: Ithaca Press.

Clammer J R 1976, *Literacy and Social Change: A Case Study of Fiji*. Leiden: E. J. Brill.

Collocott E E V 1923, 'An Experiment in Tongan History', *Journal of the Polynesian Society* 32: 166–184.

Collocott E E V and Havea John 1922, *Proverbial Sayings of the Tongans*. Honolulu: Bernice Pauahi Bishop Museum Press.

Connell John 1978. *Taim Bilong Mani: The Evolution of Agriculture in a Solomon Island Society*. Canberra Development Studies Centre, Monograph £23, Australian National University.

Connell John 1979, 'The Emergence of a Peasantry in Papua New Guinea', *Peasant Studies* 8 (2): 103–137.

Connell John 1982, 'Development and Dependency: Divergent Approaches to the Political Economy of Papua New Guinea', in *Melanesia: Beyond Diversity*, edited by R. J. May and H. Nelson. Canberra: The Australian National University.

Connell John 1987a, *New Caledonia or Kanaky? The Political History of a French Colony*. Pacific Research Monography No. 16. Canberra: Australian National University National Centre for Development Studies.

Connell John 1987b, ' "Trouble in Paradise". The Perception of New Caledonia in the Australian Press', *Australian Geographical Studies* 25:54–65.

Connell John 1988, *New Caledonia. The Matignon Accord and the Colonial Future*. RIAP Occasional Paper No. 6. Sydney: RIAP.

Connell John 1989, 'Compensation and Development: Contemporary Struggles at the Bougainville Copper Mine, Papua New Guinea', Paper presented to the Institute of Australian Geographers Conference, Adelaide, February 1989.

Connell John and Aldrich Robert 1988, 'Remnants of Empire: France's Overseas Departments and Territories', pp. 148–169 in *France in World Politics*, edited by R. Aldrich and J. Connell. London: Routledge.

Constitution of the Commonwealth of the Northern Marianas. Saipan.

Contemporary Pacific, The: A Journal of Island Affairs. 1990. 2 (1) (Spring).

Cook Captain James 1777, *A Voyage to the Pacific Ocean* 2 vols. London: W. and A. Strahan.

Cooper George and Daws Gavan 1985, *Land and Power in Hawai'i*. Honolulu: Benchmark Books.

Coppenrath, Gérald 1967, *Les Chinois de Tahiti: De l'Aversion à l'Assimilation 1865–1966*. Publications de la Société des Océanistes, no. 21. Paris: Musée de l'Homme.

Corden Margaret W 1979, 'Subsistence Diet Patterns in Papua New Guinea', *Food and Nutrition Notes and Review* 36 (2): 48–51 (April/June).

Coulter J W 1942, *Fiji: Little India of the Pacific*. Chicago: University of Chicago Press.

Covenant to Establish a Commonwealth of the Northern Marianas in Union with the Unites States of America. 1975. Saipan.

Cox Leonard M 1926, *The Island of Guam*. Washington, D. C.: Government Printing Office.

Coyne Terry, Badcock Jacqui, and Taylor Richard 1984, *The Effect of Urbanisation and Western Diet on the Health of Pacific Island Populations*. Noumea, New Caledonia: South Pacific Commission.

Craig Robert D and King F P (eds) 1980, *Historical Dictionary of Oceania*. Westport, Conn.: Greenwood Press.

Creighton Thomas H 1978, *The Lands of Hawaii: Their Use and Misuse*. Honolulu: The University Press of Hawaii.

Crocombe R (ed) 1971, *Land Tenure in the Pacific*. Melbourne: Oxford University Press.

Crocombe Ron and Hide Robin 1971, 'New Guinea: Unity in Diversity', pp. 292–333 in *Land Tenure in the Pacific*, edited by R. Crocombe. Melbourne: Oxford University Press.

Crocombe Ronald G and Hereniko Patricia G (eds) 1985, *The Other Side*. Translated by Kushnma Patel. Suva, Fiji: University of the South Pacific.

Cumming Constance Gordon F 1882, *A Lady's Cruise in a French Man of War*. Vol. 1. Edinburgh: William Blackwood and Sons.

Cummings H B 1977b, 'Tongan Society at the Time of Contact', pp. 63–89, in *Friendly Islands*, edited by N. Rutherford. Melbourne: Oxford University Press.

Cummings H G 1977a, 'Holy War: Peter Dillon and the 1837 Massacres in Tonga', *Journal of Pacific History* 12 (1): 25–39.

Curtain R L 1981, 'Migration in Papua New Guinea: The Role of the Peasant Household in a Strategy of Survival', in *Population, Mobility and Development: Southeast Asia and the Pacific*, edited by G. W. Jones and H. V. Richter. Development Studies Centre Monograph No. 27. Canberra: Australian National University.

Curtain R L 1984, 'The Migrant Labour System and Class Formation in Papua New Guinea', *South Pacific Forum* 1 (2): 117–141.

Daniel P and Sims R 1986, *Foreign Investment in Papua New Guinea: Policies and Practices*. Pacific Research Monograph £2. Canberra: National Centre for Development Studies, Australian National University.

Danielsoon, Bengt and Marie-Thérèse 1986a, *Poisoned Reign: French Nuclear Colonization in the Pacific*, Ringwood, Victoria, Australia: Penguin.

Danielsson, Bengt and Marie-Thérèse 1986b, 'Flosse Settles into the Seats of Power', *Pacific Islands Monthly* 57: 20–21 (July).

Danielsson Bengt 1983, 'French Polynesia: Nuclear Colony', pp. 193–228 in *Politics in Polynesia*, edited by R. Crocombe and A. Ali. Suva, Fiji: Institute of Pacific Studies and the University of the South Pacific.

Davidson J W 1967, *Samoa mo Samoa: The Emergence of the Independent State of Western Samoa*. Melbourne: Oxford University Press.

Davies John 1961, *The History of the Tahitian Mission 1799–1830*. Edited by C. W. Newbury. Hakluyt Society, Second Series, no. 116. London: Cambridge University Press.

Daws Gavan 1974, *Shoal of Time: A History of the Hawaiian Islands*. Honolulu: University of Hawai'i Press.

de Lepervanche Marie 1973. 'Social Structure', pp. 1–60 in *Anthropology in Papua New Guinea: Readings from the Encyclopaedia of Papua and New Guinea*, edited by I. Hogbin. Melbourne: Melbourne University Press.

De Rego Jaime 1930, 'Nuevo aspecto va tomando la Mision de Carolinas bajo in dominacion japonesa', in *El Angel de Carolinas*, 23: 2–3 (March); 24: 3–5 (April); 26: 2–3 (June).

Deleuze Gilles and Guattari Felix 1972, *Anti-Oedipus: Capitalism and Schizophrenia*. New York: Viking Press.

Dening Greg 1980, *Islands and Beaches, Discourses on a Silent Land: Marquesas 1774–1880*. Honolulu: University of Hawai'i Press.

Denning Rt Hon Lord 1970, *The Award of the Rt. Hon. Lord Denning in the Fiji Sugar Cane Contract Dispute 1969*. Suva: Fiji Royal Gazette.

Denoon Donald 1985, 'Capitalism in Papua New Guinea: Development or Underdevelopment', *Journal of Pacific History* 20 (3): 119–134.

Denoon Donald 1987, 'An Agenda for the Social History of Papua New Guinea', *Canberra Anthropology* 10 (2): 51–64.

Derrick R A 1950. *A History of Fiji*. Suva: Government Printer.

Derrida Jacques 1973, *Speech and Phenomena: And Other Essays on Husserl's Theory of Sign*. Translated by David B. Allison, Evanston: Northwestern University Press.

Derrida, Jacques 1982, *Margins of Philosophy*. Translated by Alan Bass. Chicago: University of Chicago Press.

Development Bank. Various years. *Annual Report and Financial Statements*. Port Moresby: Papua New Guinea Development Bank.

Diamond Stanley 1974, *In Search of the Primitive: A Critique of Civilization*. New Brunswick: E. P. Dutton/Transaction Books.

Dogan Mattei and Pelassy Dominique 1984, *How to Compare Nations: Strategies in Comparative Politics*. Chatham, New Jersey: Chatham House Publishers.

475

Dommen E 1980, 'External Trade Problems of Small Island States in the Pacific and Indian Ocean', pp. 170–199 in *The Island States of the Pacific and the Indian Oceans: Anatomy of Development*, edited by R. T. Shand. Canberra: Australian National University Press.

Donaldson Mike 1982, 'Contradiction, Meditation and Hegemony in Pre-capitalist Papua New Guinea: Warfare, Production and Sexual Antagonism in the Eastern Highlands', in *Melanesia: Beyond Diversity*. Vol. II, edited by R. J. May and H. Nelson. Canberra: Research School of Pacific Studies, Australian National University.

Donaldson Mike and Turner D 1978, *The Foreign Control of Papua New Guinea's Economy and the Reaction of the Independent State*. Port Moresby: University of Papua New Guinea.

Donaldson Mike and Good Kenneth 1988, *Articulated Agricultural Development: Traditional and Capitalist Agriculture in Papua New Guinea*. Aldershot, UK: Avebury, Gower Publishing Company Ltd.

Douglas Bronwen 1980, 'Conflict and Alliance in a Colonial Context: Case Studies in New Caledonia', *Journal of Pacific History* 15: 22.

Douglas Bronwen 1985, 'Ritual and Politics in the Inaugural Meeting of High Chiefs from New Caledonia and the Loyalty Islands', *Social Analysis* 18: 73.

Doumenge Jean-Paul 1986, 'La Nouvelle-Calédonie: la récherche de son equilibre économique et de son unité sociale', *Les Cahiers d'Outre-Mer* 39: 221–248.

Downs I 1980, *The Australian Trusteeship: Papua New Guinea 1945–75*. Canberra: Australian Government Publishing Service.

Dunn John 1987, 'Pape'ete Erupts', *Pacific Islands Monthly* 58: 11 (December).

Durutalo S 1982, 'The Logging Industry in Fiji: An Introduction', *USP Sociological Society Newsletter* 2: 6–9 (August).

Durutalo S 1986, *The Paramountcy of Fijian Interest and the Politicization of Ethnicity*. South Pacific Forum, Working Paper No. 6. Suva: USP Sociological Society.

Economic Commission for Asia and the Pacific 1985. *Statistical Yearbook for Asia and the Pacific*. Bangkok, Thailand: Economic Commission for Asia and the Pacific.

Edmundson Mark 1989, 'Prophet of a New Postmodernism: The Greater Challenge of Salman Rushide', *Harper's*, December: 62–71.

Ehrenreich Barbara 1989, *Fear of Falling: The Inner Life of the Middle Class*. New York: Pantheon.

Eichner Lorie and Wright Judi 1983, 'Business: Guam', *Pacific Magazine* 8 (5): 21.

Ellis F 1983, 'An Overview of Unemployment in Agriculture, Forestry and Fisheries: Past Trends and Current Policy Issues', *Discussion paper No. 147*. Norwich: School of Development Studies, University of East Anglia. (October).

Ellis William 1829, *Polynesian Researches*. 2 Vols. London: Fisher, Son, and Jackson.

Epstein T S 1968, *Capitalism Primitive and Modern: Some Aspects of Tolai Economic Growth*. East Lansing: Michigan State University Press.

Epstein T S 1972, 'Economy, Indigenous', pp. 306–314 in *Encyclopaedia of Papua and New Guinea*, Vol. I, edited by P. Ryan. Carlton, Victoria: Melbourne University Press.

Erskine Cpt John E 1853/1967, *Journal of a Cruise Among the Islands of the Western Pacific*. London: Dawsons.

REFERENCES

Evans Peter B, Rueschemeyer Dietrich, and Skocpol Theda (eds) 1985, *Bringing the State Back In*. New York: Cambridge University Press.

Fairburn T 1985, *Island Economies*. Suva: Institute of Pacific Studies, University of the South Pacific.

Field Michael 1988, 'Nauruans Dying of Wealth', *Pacific Islands Monthly* 59 (10): 29.

Fiji Blue Book 1877, Suva: Government Printer.

Fiji Employment and Development Mission (FEDM). 1984. *Interim Report to the Government of Fiji*. Parliamentary Paper (PP) No. 5 of 1984.

Fiji Royal Gazette. 1884. Vol. 10. Suva: Government Printer.

Fiji Times. February 20, 1920.

Finlay Marike 1988, *The Romantic Irony of Semiotics: Friedrich Schlegel and the Crisis of Representation*. New York: Mouton de Gruyter.

Finney Ben 1973, *Big-men and Business: Entrepreneurship and Economic Growth in the New Guinea Highlands*. Honolulu: University of Hawai'i Press.

Finney Ben 1973, *Polynesian Peasants and Proletarians*. Cambridge, Mass.: Shenkman.

Finney Ben 1979, 'Tahiti et Mama France', pp. 19–26 in *The Emerging Pacific Island States*, edited by J. N. Hurd. Honolulu: Pacific Islands Studies Program, University of Hawai'i.

Finney Ben 1987, *Business Development in the Highlands of Papua New Guinea*. Honolulu: Pacific Islands Development Program, Research Report £6, East-West Center.

Firth Raymond 1968, 'A Note on Decent Groups in Polynesia', pp. 213–224, in *Kinship and Social Organization*, edited by P. Bohannan and J. Middleton. Garden City: Natural History Press,

Firth Stewart 1977, 'German Firms in the Pacific Islands, 1857–1914', pp. 3–25 in *Germany in the Pacific and Far East, 1870–1914*, edited by John A. Moses and Paul M. Kennedy. St. Lucia: University of Queensland Press.

Firth Stewart 1987, *Nuclear Playground: Fight for an Independent and Nuclear-Free Pacific*. Honolulu: University of Hawaii Press.

Firth Stewart 1989, 'Sovereignty and Independence in the Contemporary Pacific', *the Contemporary Pacific* 1 (1–2): 75–96.

Fischer John L and Fischer Ann 1957, *The Eastern Carolines*. New Haven: Human Relations Area Files Press.

Fisk E K 1962. 'Planning in a Primitive Economy: Special Problems in Papua New Guinea', *Economic Record* 38: 462–78.

Fisk E K 1974, 'Rural Development', *New Guinea and Australia, The Pacific and Southeast Asia* 9 (1): 51–60.

Fisk E K 1975, 'The Response of Nonmonetary Production Units to Contact with the Exchange Economy', pp. 53–83 in *Agriculture in Development Theory*, edited by L. G. Reynolds. New Haven: Yale University Press.

Fisk E K 1982, 'Subsistence Affluence and Development Policy', In *Regional Development Dialogue*. United Nations Centre for Regional Development, Nagoya, Japan. Special Issue.

Fisk E K and Shand R T 1969, 'The Early Stages of Development in a Primitive Economy: The Evolution from Subsistence to Trade and Specialisation', In *Subsistence Agriculture and Economic Development*, edited by C. R. Wharton Jr. Chicago: Aldine.

Fitzpatrick P 1978, 'Really Rather Like Slavery: Law and Labour in the Colonial Economy in Papua New Guinea', In *Essays in the Political Economy of Austral-*

ian Capitalism, Vol. 3, edited by E. L. Wheelright and K. Buckley. Sydney: Australia and New Zealand Bookshop.

Fitzpatrick P 1979, 'The Creation and Containment of the Papua New Guinea Peasantry', pp. 85–121 in *Essays in the Political Economy of Australian Capitalism*, Vol. 4, edited by E. L. Wheelright and K. Buckley. Sydney: Australia and New Zealand Bookshop.

Fitzpatrick P 1980, *Law and State in Papua New Guinea*. London: Academic Press.

Fitzpatrick P 1985, 'The Making and Unmaking of the Eight Aims', pp. 22–31 in *From Rhetoric to Reality: Papua New Guinea's Eight Point Plan and National Goals After a Decade*. edited by P. King, W. Lee and V. Warakai. Port Moresby: University of Papua New Guinea Press.

Fodor Eugene 1968, *Fodor's Hawaii*. New York: D. McKay Co.

Fonua Pesi and Fonua Mary 1981, *A Walking Tour of Neiafu, Vava'u*. Neiafu, Tonga: Vava'u Press, Ltd.

Force Roland W 1960, *Leadership and Cultural Change in Palua*. Fieldiana: Anthropology, Vol. 50. Chicago: Natural History Museum.

Forman C W 1972, 'Missionaries and Colonialism: The Case of the New Hebrides in the Twentieth Century', *Journal of Church and State* 14: 75–92.

Forman C W 1978, 'foreign Missionaries in the Pacific Islands During the Twentieth Century', In *Mission, Church, and Sect in Oceania*, edited by J. A. Boutilier, D. T. Hughes, and S. W. Tiffany. Ann Arbor: University of Michegan Press.

Foster-Carter A 1978, 'The Modes of Production Controversy', *New Left Review* 107: 47–77.

Foucault Michel 1977, *Discipline and Punish: The Birth of the Prison*. Translated by Alan Sheridan. New York: Vintage Books.

Foucault Michel 1980, *Power/Knowledge: Selected Interviews and Other Writings 1972–1977*. Translated by Colin Gordon, Leo Marshall, John Mepham, Kate Soper. New York: Pantheon Books.

France P 1969, *The Charter of the Land*. Melbourne: Oxford University Press.

Frank Andre Gunder 1979, *Dependant Accumulation and Underdevelopment*. New York: Monthly Review Press.

Fraser Helen 1987, *New Caledonia: Anti-Colonialism in a Pacific Territory*. Canberra: Department of the Parliamentary Library.

Fried Jacob and Molnar Paul 1978, *Technological and Social Change: A Transdisciplinary Model*. New York: Petrocelli Books.

Fritz George 1904, *The Chamorros*. Saipan: Historic Preservation Office, Commonwealth of the Northern Marianas.

Fuchs Laurence H 1961, *Hawai'i Pono: A Social History*. New York: Harcourt Brace Jovanovich.

Fusitu'a, 'Eseta and Rutherford Noel 1977, 'George Tupou II and the British Protectorate', pp. 173–189 in *Friendly Islands*, edited by N. Rutherford. Melbourne: Oxford University Press.

G.I.E.C. F34/4/15 *Reports and Returns Travelling Diary Tamana Islands*, Vols I and II. Unpublished. WPHC Archives, Suva, Fiji.

Gailey Christine Ward 1980, 'Putting Down Sisters and Wives: Tongan Women and Colonization', pp. 294–322 in *Women and Colonization*, edited by M. Etienne and E. B. Leacock. New York: Praeger.

Gailey Christine Ward 1985a, 'The Kindness of Strangers: Transformations of Kinship in Precapitalist Class and State Formation', *Culture* V (2): 3–16.

Gailey Christine Ward 1985b, 'The State of the State in Anthropology', *Dialectical Anthropology* 9 (1–4): 65–90.

Gailey Christine Ward 1987a, *Kinship to Kingship: Gender Hierarchy and State Formation in Tonga*. Austin: University of Texas Press.

Gailey Christine Ward 1987b, 'State, Class and Conversion in Commodity Production: Gender and Changing Value in the Tongan Islands', *Journal of the Polynesian Society* 96 (1): 67–79.

Gailey Christine Ward 1987c, 'Culture Wars: Resistance to State Formation', pp. 35–56 in *Power Relations and State Formation*, edited by T. C. Patterson and C. W. Gailey. Washington, D. C.: American Anthropological Association.

Gailey Christine Ward. Forthcoming. 'Overseas Migration and the Changing Family in Tonga', In *The Predicaments of the Family*, edited by A. Ong and C. Szanton. Berkeley: University of California Press.

Gailey Christine Ward and Patterson Thomas C 1987, 'Power Relations and State Formation', pp. 1–26 in *Power Relations and State Formation*, edited by T. C. Patterson and C. W. Gailey. Washington, D. C.: American Anthropological Association.

Gailey Christine Ward and Patterson Thomas C 1988, 'State Formation and Uneven Development', pp. 77–90 in *State and Society: The Emergence and Development of Social Hierarchy and Political Centralisation*, edited by B. Bender, J. Gledhill, and M. Larsen. London: George Allen and Unwin.

Galeano, Eduardo 1985, *Memory of Fire: Genesis*. Translated by Ed McCaughan. New York: Pantheon Books.

Galeano Eduardo 1990. 'Lanauge, Lies and Latin Democracy', *Harper's*, February: 19–22.

Garanger J 1972, 'Mythes et archéologie en Océanie', *La Recherche* 3 (21): 233–242.

Gardner Don 1985, 'Introduction', pp. 81–86 in *Recent Studies in the Political Economy of Papua New Guinea Societies*, edited by D. Gardner and N. Modjeska. Sydney: Mankind Special Issue 4.

Garfinkel Harold 1967, *Studies in Ethnomethodology*. Englewood Cliffs, New Jersey: Prentice Hall.

Garnaut R and P Baxter 1984, *Exchange Rate and Macroeconomic Policy in Independent Papua New Guinea*. Pacific Research Monograph £10. Canberra: Development Studies Centre, Australian National University.

Garraud O, et al 1986, 'Le rhumatisme articulaire aigu en Nouvelle-Calédonie', *La Presse Medicale* 15: 2047–2050.

Geddes W H 1983, 'North Tabiteuea', *Atoll Economy: Social Change in Kiribati and Tuvalu, No. 2*. Canberra: Development Studies Centre, Australian National University.

Geddes W H, Chambers Anne, Sewell Betsy, Lawrence R and Watters R 1982, 'Islands on the Line, Team Report.' *Atoll Economy: Social Change in Kiribati and Tuvalu, No. 1*. Canberra: Development Studies Centre, Australian National University.

Geddes W R 1945, *Deuba: A Study of a Fijian Village*. Wellington: Polynesian Society Memoir No. 22.

Geertz Clifford 1973, *The Interpretation of Cultures*. New York: Hutchinson.

Geertz, Clifford 1983, *Local Knowledge: Further Essays in Interpretive Anthropology*. New York: Basic Books.

Gellner Ernest 1983, *Nations and Nationalism*. Oxford: Blackwell.

Gerritsen R 1975, 'Aspects of the Political Evolution of Rural Papua New

Guinea: Towards a Political Economy of the Terminal Peasantry', Paper presented to the Canberra Marxist Discussion Group Seminar, 26 October.

Giddens Anthony 1981, *A Contemporary Critique of Historical Materialism: Vol. 1 Power Property and the State*. Berkeley: University of California Press.

Giddens Anthony 1987, *The Nation-State and Violence: Volume Two of A Contemporary Critique of Historical Materialism*. Berkeley: University of California Press.

Gifford Edward Winslow 1923, *Tongan Place Names*. Honolulu: Bernice Pauahi Bishop Museum Press.

Gifford Edward Winslow 1929, *Tongan Society*. Honolulu: Bernice Pauahi Bishop Museum.

Gillion K L 1962, *Fiji's Indian Migrants: A History to the End of Indenture in 1920*. Melbourne: Oxford University Press.

Gillion K L 1977, *The Fiji Indians: Challenge to European Dominance 1920–1946*. Canberra: Australian National University Press.

Gladwsin Thomas 1970, *East is a Big Bird*. Cambridge: Harvard University Press.

Glick Clarence Elmer 1980, *Sojourners and Settlers, Chinese Migrants in Hawai'i*. Honolulu: University Press of Hawai'i.

Godelier Maurice 1986, *The Making of Great Men: Male Domination and Power Among the New Guinea Baruya*. Cambridge: Cambridge University Press.

Goldman Irving 1970, *Ancient Polynesian Society*. Chicago: University of Chicago Press.

Golson Jack 1982, 'The Ipomoean Revolution Revisited: Society and the Sweet Potato in the Upper Wahgi Valley', pp. 109–136 in *Inequality in New Guinea Highland Societies*, edited by A. Strathern. Cambridge: Cambridge University Press.

Goodenough W H 1953, *Native Astronomy in the Central Carolines*. Philadelphia: University Museum, University of Pennsylvania.

Goodman R, Lepani C, and Morawetz D 1985, *The Economy of Papua New Guinea: An Independent Review*. Canberra: Development Studies, Australian National University.

Goody Jack 1977, *The Domestication of the Savage Mind*. Cambridge: Cambridge University Press.

Government of Guam, Department of Commerce 1977. *Guam Annual Economic Review 1976: Towards a Sound Economic Direction*. Guam: Government of Guam.

Government of Guam, Department of Commerce 1980. *Guam Annual Economic Review 1980: Guam '80. Challenging the New Decade*. Guam: Government of Guam.

Government of Guam, Department of Commerce 1984. *Guam Annual Economic Review for 1984*. Guam: Government of Guam.

Government of Guam 1951. *Guam. Information on Guam . . .* Transmitted by the United States to the Secretary General of the United Nations Pursuant to Article 73(e) of the Charter. Prepared by the Government of Guam in Cooperation with the Department of the Interior.

Government of Tonga 1907. *The Law of the Government of Tonga*. Auckland: Brett Printing Co.

Government of Tonga 1967. *The Law of Tonga, Revised Edition*. 3 vols. London: Sweet and Maxwell; Nuku'alofa: Government Printer.

Government of Tonga 1973. *Lands and Survey Report*. Nuku'alofa: Government Printer.

REFERENCES

Governor of Guam 1934. *Annual Report of the Governor of Guam, Naval Government of Guam*. Guam: Naval Government of Guam.

Goward E W 1902, 'Report of Work in the Tokelau, Ellice and Gilbert Groups, LMS September 1900 to September 1902', Unpublished. South Seas Odds. London Missionary Society Records. School of Oriental and African Studies, University of London.

Graff Harvey J 1987, *The Legacies of Literacy: Continuities and Contradictions in Western Culture and Society*. Bloomington: Indiana University Press.

Grant James P 1989, *The State of the World's Children 1989*. New York: Oxford University Press.

Grathen C H 1963, *The Southwest Pacific to 1900: A Modern History*. Ann Arbor: University of Michigan Press.

Graves A 1984, 'The Nature and Origin of Pacific Islands Labour Migration to Queensland 1863–1906', In *International Labour Migration*, edited by S. Mark and B. Richardson. London: Oxford University Press.

Green L, Bukhari M S, and Lawrence R 1979, 'Decentralisation in the Gilbert Islands', Unpublished. Report submitted to the Gilbert Islands Government, Development Planning Unit, University College, London.

Gregory C A 1979, 'The Emergence of Commodity Production in Papua New Guinea', *Journal of Contemporary Asia* 9 (4): 389–409.

Gregory C A 1982, *Gifts and Commodities*. London: Academic Press Ltd.

Gregory J W and Piché V 1978, 'African Migration and Peripheral Capitalism', *African Perspectives* 1:37–50.

Griffin James, Nelson Hank, and Firth Stewart 1979, *Papua New Guinea: A Political History*. Richmond, Vic.: Heileman Educational Australia Pty. Ltd.

Grimble A F 1933–34, 'The Migrations of a Pandanus People; as Traced from a Preliminary Study of Food, Food Traditions and Food-Rituals in the Gilbert Islands', *Polynesian Society Memoir 12*. Supplement to the *Journal of the Polynesian Society* 42: 1–84; 43: 85–112.

Groube L 1971, 'Tonga, Lapita Pottery, and Polynesian Origins', *Journal of the Polynesian Society* 80: 278–316.

Guiart Jean 1956, 'Un Siècle et Demi de Contacts Culturels à Tanna', Paris: *Journal de la Société des Océanistes*.

Guiart Jean 1962, 'The Millenarian Aspects of Conversion to Christianity in the South Pacific', In *Millenarian Dreams in Action*, edited by S. Thrupp. The Hague: Mouton & Co.

Guiart Jean 1988, 'The Turth about Jean-Marie Tjibaou', *Island Business*, September: 23–29.

Guiart Jean. 'Forerunners of Melanesian Nationalism', *Oceania* 22: 81–90.

Guilliband Jean-Claude 1976, *Les Confettis de l'Empire*. Paris: Editions du Seuil.

Gunson Neil 1977, *Messengers of Grace: Evangelical Missionaries in the South Seas 1797–1860*. Melbourne: Oxford University Press.

Habermas Jurgen 1973, *Theory and Practice*. Translated by John Viertel. Boston: Beacon Press.

Habermas Jurgen 1981, *The Theory of Communicative Action: Volume 1 Reason and the Rationalization of Society*. Translated by Thomas McCarthy. Boston: Beacon Press.

Habermas Jurgen 1988, *On the Logic of the Social Sciences*. Cambridge, Massachusetts: MIT Press.

Hager Carl 1886, *Die Marshall-Inseln in Erd-und Völkerkunde, Handel und Mission*. Leipzig: G. Lingke.

Hall C W 1969, 'Samoa, the Nonconformist', *Educational Screen and Audiovisual Guide* 48:18.

Hanlon David L 1988, *Upon A Stone Altar: A History of the Island of Pohnpei to 1890*. Honolulu: University of Hawai'i Press.

Hanson F Allan 1970, *Rapan Lifeways, Society and History on a Polynesian Island*. Boston: Little, Brown.

Harada, Wayne. 1989. 'Review of the Frank DeLima Show', page 1 of the Living Section. *The Honolulu Advertiser*, December 6.

Harvey, P. W. and P. F. Heywood. 1983. 'Twenty-five Years of Dietary Change in Simbu Province, Papua New Guinea', *Ecology of Food and Nutrition* 13(1):27–35.

Harwood, F. 1978. 'Intercultural Communication in the Western Solomons: The Methodist Mission and the Emergence of the Christian Fellowship Church', In *Mission, Church, and Sect in Oceania*, edited by J. A. Boutilier, D. T. Hughes, and S. W. Tiffany. Ann Arbor: University of Michigan Press.

Hasluck, P. 1976. *A Time for Building: Australian Administration in Papua New Guinea 1951–1973*. Carlton, Victoria: Melbourne University Press.

Havelock, Eric A. 1963. *Preface to Plato*. Cambridge, Mass.: Belknap Press.

Haverlandt, R. Otto. 1975. 'The Guamanian Economic Experience', pp. 92–122 in *The Social-Economic Impact of Modern Technology Upon a Developing Insular Region: Guam*, Vol. 3, pt. 6, coordinated by J. Vail. Guam: University of Guam Press.

Healey, A. M. 1967. *Bulolo: A History of the Development of the Bulolo Region, New Guinea*. Canberra: Australian National University Press.

Healey, Christopher J. 1985. 'New Guinea Inland Trade: Transformation and Resilience in the Context of Capitalist Penetration', pp. 127–144 in *Recent Studies in the Political Economy of Papua New Guinea Societies*, edited by D. Gardner and N. Modjeska. Sydney: *Mankind* Special Issue 4.

Heidegger, Martin. 1962. *Being and Time*. Translated by John Macquarrie and Edward Robinson. New York: Harper.

Hempenstall, P. and N. Rutherford. 1984. *Protest and Dissent in the Colonial Pacific*. Suva: University of the South Pacific.

Henningham, Stephen. 1989. 'Keeping the Tricolor Flying: The French Pacific into the 1990s', *The Contemporary Pacific* 1(1–2):97–132.

Heritage, John. 1984. *Garfinkel and Ethnomethodology*. New York: Polity Press.

Hervier, R. P. Jean. 1902. 'L'archipel des Iles Tonga', pp. 125–48 in *Les Missions Catholiques Francaises Aux XIXe Siecle*, edited by J. Pere and B. Piolet. Paris: Armand Colin.

Hess, Michael. 1983. 'In the long run . . .': Australian Colonial Labour Policy in the Territory of Papua New Guinea', *The Journal of Industrial Relations* 25:51–67.

Heyerdahl, Thor. 1950. *The Kon-Tiki Expedition*. London: Allen & Unwin.

Hezel, Francis X. 1974. 'Reflections of Micronesia's Economy', *Friends of Micronesia Newsletter* 4.1:15–19.

Hezel, Francis X. 1978. 'The Role of the Beachcomber in the Carolines', pp. 251–272 in *The Changing Pacific*, edited by N. Gunson. Melbourne: Oxford University Press.

Hezel, Francis X. 1982. 'Micronesia's Hanging Spree', in *Reflections on Micronesia: The Collected Papers of Father Francis X. Hezel, S. J.* Honolulu: Pacific Island Studies Program, University of Hawaii.

Hezel, Francis X. 1983. *First Taint of Civilization: A History of the Caroline and*

Marshall Islands in Pre-Colonial Days, 1521–1885. Honolulu: University of Hawaii Press.

Hezel, Francis X. 1984. 'A Brief Economic History of Micronesia', in *Past Achievemets and Future Possibilities.* Report on a conference on Economic Development held in Kolonia, Pohnpei. May 22–25, 1984. Micronesian Seminar, Majuro.

Hezel, Francis X. 1986. *The Dilemmas of Development: The Effects of Modernization on Three Areas of Island Life.* Truk, Caroline Islands: Micronesian Seminar.

Hezel, Francis X. 1989. 'Suicide and the Micronesian Family', *The Contemporary Pacific* 1(1–2):43–74.

Hezel, Francis X. and Michael Levin. 1989. 'Micronesian Emigration: The Brain Drain in Palau, Marshalls and the Federated States', Forthcoming In *Migration and Development in the South Pacific,* edited by J. Connell.

Hezel, Francis X. and Thomas B. McGrath. 1989. 'The Great Flight Northward: FSM Migration to Guam and the Northern Marianas Islands', *Pacific Studies* 13:1 (Nov).

Hezel, Francis X., Donald H. Rubenstein, and Geoffrey M. White. 1985. *Culture, Youth and Suicide in the Pacific.* Honolulu: University of Hawaii and Institute of Culture and Communication, East-West Center.

Higgott, R. A. 1983. *Political Development Theory.* London: Croom-Helm.

Hince, K. W. 1971. 'Trade Unionism in Fiji', *Journal of Industrial Relations* 13(4):368–389.

Hindness, B. and P. Q. Hirst. 1975. *Pre-capitalist Modes of Production.* London: Routledge and Kegan Paul.

Hjarno, Jan. 1979. 'Social Reproduction: Towards an Understanding of Aboriginal Samoa', *Folk* 21:73–123.

Hobsbawm, E. and T. Ranger (eds). 1983. *The Invention of Tradition.* Cambridge: Cambridge University Press.

Hocart, A. M. 1915. 'Chieftainship and the Sister's son in the Pacific', *American Anthropologist* 17:631–646.

Honolulu Star-Bulletin. 1989. 'Study on Hawaiian Violent Death Called "Unsurprising" ', January 9:A–5.

Hoogevelt, A. M. M. *The Third World in Global Development.* London: Macmillan Press.

Houbert, Jean. 1985. 'Settlers and Natives in Decolonisation. The Case of New Caledonia', *The Round Table* 295:217.

Howard, Michael C. 1982. 'The South Pacific and the Capitalist World Economy', *USP Sociological Society Newletter* 3:2–3 (August). Suva.

Howard, Michael C. 1983. *The Political Economy of the South Pacific: An Introduction.* Suva, Fiji: University of the South Pacific.

Howard, Michael C. 1986. 'Ethnicity in the South Pacific', Paper presented to the United Nations University International Conference on Ethnicity and Nation Building, held at the University of the South Pacific, Suva.

Howard, Michael C. 1987. 'The Trade Union Movement in Fiji', In *Fiji, Future Imperfect,* edited by M. Taylor. Sydney: Allen and Unwin.

Howard, Michael C., etal. 1983. *The Political Economy of the South Pacific.* South East Asian Monograph Series No. 13. Townsville: James Cook University.

Howe, K. R. 1977. *The Loyalty Islands: A History of Culture Contacts 1840–1900.* Canberra: Australian National University Press.

Howe, K. R. 1984. *Where the Waves Fall: A New South Sea Islands History from First Settlement to Colonel Rule.* Pacific Islands Monograph Series, No. 2

Honolulu: University of Hawai'i Press and the Pacific Islands Studies Program, Center for Asian and Pacific Studies, University of Hawaii.

Howlett, D. 1973. 'Terminal Devleopment: From Tribalism to Peasantry', in *The Pacific in Transition: Geographical Perspectives on Adaptation and Change,* edited by H. C. Brookfield. New York: St. Martin's Press.

Hughes, A. 1973. 'What is Development?', *Pacific Perspectives* 1(2):8–19.

Im Thurn, Everard. 1905. *Report on Tongan Affairs (December 1904-January 1905).* Suva: Edward John March, Government Printer.

Inder, Stuart, (general ed.). 1978. *Pacific Islands Yearbook, 13th Edition.* Sydney: Pacific Publications.

International Bank for Reconstruction and Development (IBRD). 1965. *The Economic Development of the Territory of Papua and New Guinea.* Port Moresby: Law Reform Commission Monograph £5. Kieta Patrol Reports Held in North Solomons Provincial Government Library.

Jameson, Fredric. 1981. *The Political Unconscious: Narrative as a Socially Symbolic Act.* Ithaca, New York: Cornell University Press.

Jarman, R. 1838. *Journal of a Voyage to the South Seas in the "Japan" Employed in the Sperm Whale Fishery under the Command of Captain John May.* London: Longmans.

Jayawardena, C. 1971. 'The Disintegration of Caste in Fiji Indian Rural Family', pp. 9–20 in *Anthropology in Oceania,* edited by C. R. Hiatt and C. Jayawardena. Sydney: Angus and Robertson.

Johnson, Giff. 1984. *Collision Course at Kwajalein: Marshall Islanders in the Shadow of the Bomb.* Honolulu: Pacific Concerns Resource Center.

Joralenon, Victoria L. 1988. 'Collective Land Tenure and Agricultural Development', pp. 154–180 in *French Polynesia,* edited by N. Pollock and R. Crocombe. Suva, Fiji: Institute of Pacific Studies of the University of the South Pacific.

Joseph and Murray. n.d. *Chamorros and Carolinians of Saipan.* Saipan: Coordinated Investigation in Micronesian Anthropology (CIMA).

Kaiser, T. 1965. 'Classroom TV Comes to Samoa', *Saturday Reveiw* 48:58–59 (June 19).

Kanaky Update. 1988. 10(3):3–4.

Keesing, Felix M. and Marie M. Keesing. 1956. *Elite Communication in Samoa; A Study of Leadership.* Stanford Anthropological Series, No. 3. Palo Alto: Stanford University Press.

Keesing, Roger M. 1982. 'Kastom in Melanesia: An Overview', In *Reinventing Traditional Culture: The Politics of Kastom in Island Melanesia,* edited by R. M. Keesing and R. Tonkinson. *Mankind* 13(4):297–301.

Keesing, Roger M. 1982. 'Traditionalist Enclaves in Melanesia', pp. 39–54 in *Melanesia: Beyond Diversity,* edited by R. J. May and H. Nelson. Canberra: Australian National University.

Keesing, Roger M. 1989. 'Creating the Past: Custom and Identity in the Contemporary Pacific', *The Contemporary Pacific: Journal of Island Affairs* 1(1–2):19–42.

Kent, George . 1988. 'Nutrition Education as an Instrument of Empowerment', *Journal of Nutrition Education* 20(4):193–195.

Kent, Noel. 1983 . *Hawai'i, Islands Under The Influence.* New York: Monthly Review Press.

King, Pauline. 1987. 'Structural Changes in Hawaiian History: Changes in the Mental Health of a People', pp. 32–44 in *Contemporary Issues in Mental Health*

REFERENCES

Research in the Pacific Islands, edited by A. B. Robillard and A. J. Marsella. Honolulu: University of Hawai'i Press.

Kirch, Patrick V. 1984. 'Tonga', pp. 217–242 in *The Evolution of the Polynesian Chiefdoms*. Cambridge: Cambridge University Press.

Kirchhoff, Paul. 1955. 'The Principles of Clanship in Human Society', *Journal of Anthropology* 1:1–10.

Kirbati National Planning Office, Ministry of Finance. 1988. *Sixth National Development Plan 1987–1991*. Tarawa, Kiribati.

Knapman, B. 1983. 'Capitalism and Colonial Development: Studies in the Economic History of Fiji, 1874–1939', Ph.D. Thesis, Department of Pacific and Southeast Asian History, Australian National University.

Knapman, B. 1988. 'The Economic Consequences of the Coups', in *Fiji Shattered Coups*, edited by R. T. Robertson and A. Tamanisau. Leichhardt: Pluto Press.

Koch, G. 1965. *Materielle Kulture der Gilbert-Inseln*. Berlin: Museums für Völkerkunde.

Kolig, E. 1986. 'Andreas Reischek and the Maori: Villainy or the Nineteenth Century Scientific ethos', *Pacific Studies* 10(1):55ff.

Kooijman, Simon. 1972. 'Tonga', pp. 297–341 in *Tapa in Polynesia*. Honolulu: Bernice Pauahi Bishop Museum Press.

Korn, Shulamit Decktor. 1978. 'After the Missionaries Came: Denominational Diversity in the Tongan Islands', pp. 395–422 in *Mission, Church and Sect in Oceania*, edited by J. Boutillier, D. H. Hughes, and S. Tiffany. Ann Arbor: University of Michigan Press.

Korn, Shulamit Dector. 1974. 'Tongan Kin Groups: The Noble and the
- Common View', *Journal of the Polynesian Society* 83:5–13.

Koskinen, A. A. 1953. 'Missionary Influence as a Political Factor in the Pacific Islands', Thesis, Helsingfors Universitet.

Kotzebue, Otto von. 1821. *A Voyage of Discovery into the South Sea and Beering's Straits . . . in the Years 1815–1818*. 3 vols. London: Longman and Brown.

Krämer, Augustin. 1902. *Die Samoa-Inseln*. 2 vols. Stuttgart: E. Nagele.

Kuhn, T. S. 1970. *The Structure of Scientific Revolutions*. Chicago: University of Chicago Press.

Kuper, Leo. 1974. *Class and Power: Ideology and Revolutionary Change in Plural Societies*. London: Duckworth.

Kuruduadua, S. 1979. *The Fiji Employers' Consultative Association: Its Development and Role in Industrial Relations*. Research Paper. Industrial Relations Centre. Wellington: Victoria University.

Kuva, A. 1974. *The Solomons Community in Fiji*. Suva: South Pacific Social Sciences Association.

Lacan, Jacques. 1968. *The Language of the Self; The Function of Language in Psychoanalysis*. Translated by Anthony Wilder. Baltimore: John Hopkins Press.

Lacan, Jacques. 1977. *Ecrits: A Selection*. Translated by Alan Sheridan. New York: Norton.

Lacey, R. 1983. *Our Young Men Snatched Away: Labourers in Papua New Guinea's Colonial Economy, 1884–1942*. Occasiional Paper in Economic History #3. Port Moresby: University of Papua New Guinea.

Laclau, E. 1971. 'Feudalism and Capitalism in Latin America', *New Left Review* 67: 19–38.

Lal, B. V. 1983. *Girmityas*. Canberra: The Journal of Pacific History.

485

Langmore, Diane. 1989. *Missionary Lives: Papua, 1874–1914*. Honolulu: University of Hawaii Press.

Larcom, J. 1982. 'The Invention of Convention', *Mankind* 13(4): 330–337.

Lasaqa, I. Q. 1984. *The Fijian People Before and After Independence*. Canberra: Australian National University Press.

Latukefu, Sione. 1968. 'Oral Traditions', *Journal of Pacific History* 3: 135–143.

Latukefu, Sione. 1974. *Church and State in Tonga: The Wesleyan Methodist Missionaries and Political Development*. Canberra. Australia National University Press.

Latukefu, Sione. 1977. 'The Wesleyan Mission', p. 114–135 in *Friendly Islands*, edited by N. Rutherford. Melbourne: Oxford University Press.

Lawrence, P. 1964. *Road Belong Cargo*. Manchester: Manchester University Press.

Lawrence, R. 1983. 'Tamana', In *Atoll Economy: Social Change in Kiribati and Tuvalu, No. 4*. Canberra: Development Studies Centre, Australian National University.

Lawrence, R. 1984. 'Tamana: A Study of a Reef Island Community', Unpublished Ph.D. thesis in Geography, Victoria University Wellington.

Le Mintier, A. 1985. 'La Misère au Paradis!', *La Depeche*, 16 January: 15–17. Pape'ete, Tahiti.

Leenhardt, Maurice. 1979. *Do Kamo. Person and Myth in the Melanesian World*. Chicago and London: University of Chicago Press.

Legge, J. D. 1958. *Britain in Fiji 1858–1880*. London: Macmillan.

Les Nouvelles Calédoniennes. 1988. 26 July.

Les Temps-Modernes. 1985. 'Entretien avec Jean-Marie Tjibaou.' 464: 1599, 1601.

Levi-Strauss, Claude. 1966. *The Savage Mind*. Chicago: University of Chicago Press.

Levi-Strauss, Claude. 1969. *The Raw and the Cooked: Introduction to a Science of Mythology: I*. Translated by John and Doreen Weightman. New York: Harper and Row.

Levy, Robert I. 1973. *Tahitians: Mind and Experience in the Society Islands*. Chicago: University of Chicago Press.

Levy, Susan J. and Booth, Heather. 1988. 'Infant Mortality in the Marshall Islands', *Asia and Pacific Population Forum* 2(3–4): 11–20, 31.

Lewis, David. 1972. *We, the Navigators: The Ancient Art of Landfinding in the Pacific*. Honolulu: University Press of Hawai'i.

Lewis, James L. 1967. *Kusaiean Acculturation 1824–1948*. Saipan: Division of Land Management, Resources and Development, Trust Territory Government.

Lewis, Nancy Davis. 1988. *Modernization, Morbidity and Mortality: Noncommunicable Disease in the Pacific Islands*. Honolulu: Pacific Islands Development Program, East-West Center.

Lindstrom, L. 1982. 'Leftemap Kastom: The Political History of Tradition on Tanna (Vanuatu)', *Mankind* 13(4): 316–329.

Lindstrom, L. 1984. 'Doctor, Lawyer, Wise Man, Priest: Big-men and Knowledge in Melanesia', *Mankind* 19: 291–309.

Lockwood, Victoria Joralemon. 1988. 'Capitalist Development and the Socioeconomic Position of Tahitian Peasant Women', *Journal of Anthropological Research* 44: 263–286.

Logan, Robert K. 1986. *The Alphabet Effect*. New York: St. Martin's Press.

Low Richard Barrett. 1967. *Problems in Paradise: The View from Government House*. New York: Pageant Press, Inc.

Lukes, Steven. 1973. *Individualism*. Oxford: Basil Blackwell.

Lütke, Frederic. 1835. *Voyage autour du monde, exècutè par ordre de Sa Majestè l'empereur Nicolas Ier, sur la corvette le Sèniavine, dans les annèes 1826, 1827, 1828 et 1829.* Part 2: Partie Historique, 2 vols. Paris: Firmin Didot. (Reprinted 1971. Bibliotheca Australiana No.s 58, 59. New York: Da Capo/Amsterdam: N. Israel.)

MacClancy, J. 1980. *To Kill a Bird with Two Stones: A Short History of Vanuatu.* Port Vila: Vanuatu Cultural Centre Publications.

MacDonald, Barrie. 1982. *Cinderellas of Empire: Towards a History of Kiribati and Tuvalu.* Canberra: Australian National University Press.

MacDonald, D. 1892. 'Efate, New Hebrides', *Australian Association for the Advancement of Science* 4: 720–735.

MacNaught, J. J. 1982. *The Fijian Colonial Experience.* Canberra: Australian National University Press.

MacPherson, C. B. 1962. *The Political Theory of Possessive Individualism.* Oxford: Clarendon Press.

Macpherson, Cluny and Macpherson, Cecelia. 1990. *Traditional Samoan Medicine.* Auckland: University of Auckland Press.

MacWilliam, Scott. 1988. 'Smallholdings, Land Law and the Politics of Land Tenure in Papua New Guinea', *The Journal of Peasant Studies* 16(1): 77–109.

Malik, Michael. 1989. 'A Wealth of Disputes', *Far Eastern Economic Review*, 3 August: 18–23.

Mamak, A. and Ali, A. 1979. *Race and Class and Rebellion in the South Pacific.* Sydney: George Allen and Unwin.

Mann, Michael. 1986. *The Sources of Social Power: Volume I A History of Power from the Beginning to A.D. 1760.* New York: Cambridge University Press.

Marcus, George. 1977. 'Contemporary Tonga – The Background of Social and Cultural Change', Pp. 210–227 in *Friendly Islands*, edited by N. Rutherford. Melbourne: Oxford University Press.

Marcus, George. 1980. *The Nobility and the Chiefly Tradition in Modern Tonga.* Honolulu: University of Hawaii Press.

Mariner, Will. 1827. *An Account of the Natives of the Tongan Islands.* 2 vols. Edinburgh: John Constable.

Marshall Islands Journal. 1988. 'Seven Die of Malnutrition', September 15, 19(29).

Marshall, Leslie, (ed.). 1985. *Infant Care and Feeding in the South Pacific.* New York: Gordon and Breach.

Marx, K. 1987. 'Primitive Accumulation', In *The Political Economy of Law: A Third World Reader*, edited by Y. Ghai, R. Luckham, and F. Snyder. Delhi: Oxford University Press.

Maude, Alaric. 1971. 'Tonga: Equality Overtaking Privilege', pp. 106–127 in *Land Tenure in the Pacific*, edited by R. G. Crocombe. Melbourne: Oxford University Press.

Maude, H. E. with Leeson, Ida. 1968. 'The Coconut Oil Trade in the Gilbert Islands', In *Of Islands and Men*, edited by H. E. Maude. Melbourne: Oxford University Press.

Maudslay, Alfred. 1878/1930. *Life in the Pacific Fifty Years Ago.* London: G. Routledge and Sons.

Mayer, Adrian C. 1961. *Peasants in the Pacific: A Study of Fiji Indian Rural Society.* London: Routledge and Kegan Paul.

Mayer, Adrian C. 1963. *Indians in Fiji.* London: Oxford University Press.

Mayer, Adrian C. 1973. *Peasants in the Pacific: A Study of Fiji Indian Rural Society.* 2 ed. London: Routledge and Kegan Paul.

Mayo, Larry W. 1984. 'Occupations and Chamorro Social Status: A Study of

Urbanization in Guam', Unpublished doctoral dissertation, University of California-Berkeley.

Mazrui, A. 1970. 'An African's New Guinea', *New Guinea* 5(3).

McArthur, Norma. 1967. *Island Populations of the Pacific*. Canberra: Australian National University Press.

McDermott, John F., Jr., Wen-Shing Tseng, and Thomas W. Maretzki. 1980. *People and Cultures of Hawaii*. Honolulu: University of Hawaii Press.

McDougall, R. S. *Fijian Administration Finance: A Report, 1957*.

McHenery, Donald. 1976. *Micronesia: Trust Betrayed, Altruism vs. Self-Interest in American Foreign Policy*. New York: Carnegie Endowment for International Peace.

McKillop, R. and Firth, S. G. 1981. 'Foreign Intrusion: The First Fifty Years.' Pp. 85–103 in *A Time to Plant and a Time to Uproot: A History of Agriculture in Papua New Guinea*, edited by D. Denoon and C. Snowden. Port Moresby: Institute of Papua New Guinea Studies.

McLellan, David. 1977. *Karl Marx: Selected Writings*. Oxford: Oxford University Press.

McLuhan, Marshall. 1964. *Understanding Media: The Extensions of Man*. New York: McGraw-Hill.

Mead, Margaret. 1930. *Social Organization of Manu'a*. Honolulu: Bishop Museum Bulletin No. 76.

Meller, N. and Anthony, J. 1967. *Fiji Goes to the Polls; The Crucial Legislative Council Elections of 1963*. Honolulu: East West Center Press.

Melville, Herman. 1982. *Typee: A Peep at Polynesian Life During a Four Month's Residence in a Valley of the Marquesas*. New York: New American Library.

Michener, James A. 1947. *Tales of the South Pacific*. New York: The Macmillan Company.

Michener, James A. 1959. *Hawaii*. New York: Random House.

Mikesell, R. F. 1975. *Foreign Investment in Copper Mining: Case Studies of Mines in Peru and Papua New Guinea*. Baltimore and London: The Johns Hopkins University Press.

Mishra, V. 1979. *Rama's Banishment: A Centenary Tribute to the Fiji Indians, 1879–1979*. Auckland: Heinemann Educational Books.

Modjeska, Nicholas. 1982. 'Production and Inequality: Perspectives from Central New Guinea', pp. 50–108 in *Inequality in New Guinea Highland Societies*, edited by A. Strathern. Cambridge: Cambridge University Press.

Moench, Richard A. 1963. 'Economic Relations of the Chinese in the Society Islands', Ph.D. thesis, Harvard University, Cambridge, Massachusetts.

Momis, Fr. John and Ogan, Eugene. 1971. 'Bouganville '71: Not Discovered by CRA', *New Guinea* 6(2): 32–40.

Moorehead, Alan. 1966. *The Fatal Impact*. New York: Harper and Row.

Moorehead, Alan. 1971. *The Fatal Impact: The Invasion of the South Pacific 1767–1840*. London: Hamilton.

Morrell, W. P. 1946. 'The Transition to Christianity in the South Pacific', *Transactions of the Royal Historical Society* 28: 101–120.

Morrell, W. P. 1960. *Britain in the Pacific Islands*. Oxford: Clarendon.

Moynagh, M. 1981. *Brown or White: A History of the Fiji Sugar Industry, 1873–1973*. Pacific Research Monograph No. 5. Canberra: Australian National University Press.

Murphy, Joe. 1986. 'Advent of Tourism on Guam Remarkable', *Pacific Daily News (Guam)*, Special Supplement, 30 March: B–2.

Murphy, Joe. 1986. 'Diverse Blends Make Guam Unique', *Pacific Daily News (Guam)*, Special Supplement, 30 March: D–4.

Naidu, V. 1980a. *The Violence of Indenture in Fiji*. Suva: World University Service and the University of the South Pacific.

Naidu. V. 1980b. 'The Plural Society Thesis and Its Relevance to Fiji', MA Thesis, School of Social and Economic Development, University of the South Pacific.

Nairn, Ian. 1981. *The Break-Up of Britain. Crisis and Neo-Nationalism*. London: Verso.

Narayan, J. *The Political Economy of Fiji*. Suva: South Pacific Review Press.

Narokobi, Bernard. 1980. *The Melanesian Way*. Port Moresby: Institute of Papua New Guinea Studies.

Narsey, W. L. 1979. 'Monopoly Capital, White Racism and Superprofits in Fiji: A Case Study of CSR', *The Journal of Pacific Studies* 3: 66–146.

Nash, Jill. 1981. 'Sex, Money and the Status of Women in Aboriginal South Bougainville', *American Ethnologist* 8: 107–26.

Nash, Jill and Ogan, Eugene. 1990. *The Red and The Black: Bougainvillean Perceptions of Other Papua New Guineans*. Pacific Studies, (March).

Nayacakalou, R. R. 1971. 'Fiji Manipulating the System', In *Land Tenure in the Pacific*, edited by R. Crocombe. Melbourne: Oxford University Press.

Nayacakalou, R. R. 1975. *Leadership in Fiji*. Melbourne: Oxford University Press.

Néchéro-Joredié, Marie-Adèle. 1988. 'An Ecole Populaire Kanake (EPK): The Canala Experiment', pp. 198–218 in *New Caledonia: Essays in Nationalism and Dependency*, edited by M. Spencer, A. Ward, and J. Connell. Brisbane: University of Queensland Press.

Nelson, Hank. 1976. *Black, White and Gold: Goldmining in Papua New Guinea 1878–1930*. Canberra: Australian National University Press.

Nelson, Lyle. 1970. *Report of the Educational Television Taskforce*. Honolulu: Hawaii, p. 16. *New Guinea Annual Reports*. Melbourne: Government Printers.

Newbury, C. 1975. 'Colour Bar and Labour Conflict on the New Guinea Goldfields 1935–1941', *Australian Journal of Politics and History* 21(3): 25–38.

Newbury, Colin. 1980. *Tahiti Nui: Change and Survival in French Polynesia 1767–1945*. Honolulu: University of Hawai'i Press.

Nipperdey, T. 1983. 'In Search of Identity: Romantic Nationalism, Its Intellectual, Political and Social Background, in Romantic Nationalism', p. 15 in *Europe*, edited by J. C. Eade. Canberra: Australian National University Humanities Research Centre Monograph No. 2.

Norton, R. 1977. *Race and Politics in Fiji*. St. Lucia: University of Queensland Press.

Nozikov, Nikolai. 1946. *Russian Voyages Round the World*. London: Hutchinson.

O'Callaghan, Mary Louise. 1988. 'Tjibaou's Quiet Optimism', *Sydney Morning Herald*, 16 May.

O'Connor, Verna. 1986. 'ROTC Continues Soldier's Tradition', *Pacific Daily News (Guam)*, Special Supplement, 30 March:D–22.

O'Neill, John. 1983. 'Power and the Splitting (Spaltung) of Language' in *New Literary History: A Journal of Theory and Interpretation* 14:695–710.

O'Neill, John. 1985. *Five Bodies: The Human Shape of Modern Society*. Ithaca: Cornell University Press.

Ogan, Eugene. 1971. 'Nasioi Land Tenure: An Extended Case Study', *Oceania* XLII:81–93.

Okamura, Jonathan Y. 1982. 'Immigrant Filipino Ethnicity in Honolulu,

Hawai'i', unpublished doctoral dissertation, Department of Anthropology, University of London.

Oliver, Douglas L. 1961. *The Pacific Islands*. 2nd ed. The Natural History Library. New York: Harvard University Press.

Oliver, Douglas L. 1974. *Ancient Tahitian Society*. 3 Vols. Honolulu: University of Hawai'i Press.

Oliver, Douglas L. 1989. *The Pacific Islands*. Honolulu: University of Hawaii Press.

Olson, David R. 1977. 'From Utterance to Text: The Bias of Language in Speech and Writing', *Harvard Educational Review* 47:257–281.

Omran, Abdel R. 1971. 'The Epidemiologic Transition: A Theory of the Epidemiology of Population Change', *Milbank Memorial Fund Quarterly* 49(4):509–538 (Part I).

Ong, Walter J. 1967. *The Presence of the Word*. New Haven: Yale University Press.

Ormerod, Beverley. 1981. 'Discourse and Dispossession: Edouard Glissant's Image of Contemporary Martinique', *Carribbean Quarterly* 27:1–12.

Osborne, D. 1966. 'The Archaeology of Palau Island', *Bishop Museum Bulletin* 230. Honolulu: Bishop Museum Press.

Ottino, Paul. 1972. *Rangiroa: Parenté Etundue, Résidence et Terres dans un Atoll Polynésien*. Paris: Editions Cujas.

Pacific Courts and Justice. 1977. Suva: Institute of Pacific Studies, University of the South Pacific.

Pacific Daily News. 1987. 'Squadron Helps Village With Water Shortage', 19 April:41.

Pacific Islands Monthly. 1963. 'Western Samoa', September:18.

Pacific Islands Monthly. 1964. 'Much Ado about UN Experts Bombshell on Education', August:45–47.

Pacific Islands Monthly. 1967. February:38.

Pacific Islands Monthly. 1988. 'War Declared on Junk Food', August:44.

Pacific Magazine. 1989. 'Guam Land Sales Make Clan Rich', 14(4):18.

Palomo, Tony. 1984. *An Island in Agony*. Guam.

Panholzer, Tom. 1985. 'Guam Land Settlement Gets Hearing in Cuba', *Pacific Magazine* 10(4):14.

Papua Annual Reports. Melbourne: Government Printers.

Papua New Guinea. 1973. *Commission of Enquiry into Land Matters*. Report. Port Moresby.

Parker, R. S. 1972. 'Public Administration', pp. 982–993 in *Encyclopaedia of Papua and New Guinea*, edited by P. Ryan. Carlton, Victoria: Melbourne University Press.

Parnaby, O. W. 1964. *Britain and the Labour Trade in the Southwest Pacific*. Durham, N.C.: Duke University Press.

Parsons, Talcott. 1937. *The Structure of Social Action*. New York: The Free Press.

Parsons, Talcott. 1954. *The Social System*. New York: The Free Press.

Pawley, A. 1972. 'On the Internal Relationships of Eastern Oceanic Languages', in *Studies in Oceanic Culture History, Pacific Anthropological Records No. 13*, edited by R. C. Green and M. Kelly. Honolulu: Bishop Museum Press.

Peattie, Mark R. 1988. *Nan'yo – The Rice and Fall of the Japanese in Micronesia*. Honolulu: University of Hawaii Press.

Peel, J. D. Y. 1973. 'Cultural Factors in the Contemporary Theory of Development', *Archives Européennes de Sociologie* 14(2):283–303.

Peel, J. D. Y. 1978. 'Olaju: A Yoruba Concept of Development', *The Journal of Development Studies* 14(2):139–165.

Philibert, Jean-Marc. 1981. 'Living Under Two Flags: Selective Modernization in Erakor Village', in *Vanuatu: Politics, Economics and Ritual in Island Melanesia*, edited by M. Allen. Sydney: Academic Press.

Philibert, Jean-Marc. 1984. 'Affluence, Commodity Consumption, and Self-image in Vanuatu', in *Affluence and Cultural Survival*, edited by R. F. Salisbury and E. Tooker. Washington: The American Ethnological Society.

Philibert, Jean-Marc. 1986. 'The Politics of Tradition: Toward a Generic Culture in Vanuatu', *Mankind* 16(1):1–12.

Philibert, Jean-Marc. 1988. 'Women's Work: A Case Study of Proletarianization of Peri-urban Villagers in Vanuatu', *Oceania* 58(3):161–175.

Philibert, Jean-Marc. 1989. 'Consuming Culture: A Study of Simple Commodity Consumption', in *The Social Economy of Consumption*, edited by H. J. Rutz and B. S. Orlove. Lanham: University Press of America.

Phillips, Carla Rahn. 1986. *Six Galleons for the King of Spain: Imperial Defense in the Early Seventeenth Century*. Baltimore: The Johns Hopkins University Press.

Pigafetta, Antonio. 1969. *The Voyage of Magellan*. Englewood Cliffs, N.J.: Prentice-Hall.

Pisani, Edgard. 1985. *Interview: face a l'outre-mer*. Radio RFO Noumea, 24 September.

Plange, N. K. 1985. 'Coming in from the Cold: Gold Mining and Proletarianization of Fijians, 1920–1985', *Capital and Society* 18(1):88–127 (April).

Pollner, Melvin. 1987. *Mundane Reason: Reality in Everyday and Sociological Discourse*. New York: Cambridge University Press.

Powell, T. 1879. 'Samoa Upolu-Visit to N.W. Outstations', 2 Oct – 18 Nov. Unpublished. South Seas Journals. London Missionary Society Records. School of Oriental and Afican Studies, University of London.

Prasad, A. 1962. *Kisan Sangh Ka Itihas*. Vols I–II. Rajket.

Prasad, S. (ed.). 1988. *Coups and Crisis: Fiji a Year Later*. North Carlton, Victoria, Australia: Arena Publications.

Pratt, G. 1872. 'Samoa – Visit to Tokelau, Ellice and Gilbert Groups', 15 July – 16 Aug. Unpublished. South Seas Journals. London Missionary Society Records, School of Oriental and African Studies, University of London.

Pugh, Paul E. Radm. 1971. 'Economic Impact of the Military on Guam', pp. 66–68 in *Proceedings of the 1971 Guam Economic Conference*. Agana, Guam: Information Service Inc.

Purcell, David. 1967. 'Japanese Expansion in the South Pacific, 1890–1935', PhD dissertation, University of Pennsylvania.

Qualo, R. 1982. 'The Need to Reform Local Government in the Fiji Islands', MSocSc thesis, Faculty of Commerce and Social Science, University of Birmingham.

Quiggin, H. A. 1949. *A Survey of Primitive Money*. London: Methuen.

Ralston, C. 1977. *Grass Huts and Warehouses: Pacific Beach Communities of the Nineteenth Century*. Canberra: Australia National University Press.

Ranger, Terence. 1968. 'Connexions between "Primary Resistance" Movements and Modern Mass Nationalism in East and Central Africa', *Journal of African History* 9:437.

Rapson, Richard L. 1980. *Fairly Lucky You Live Hawai'i*. Washington, D.C.: University Press of America.

Réalités. 1987. 'Pape'ete en Feu', *Réalités du Pacifique* November-December, No. 5:3–6.

Recensement. 1983. *Résultats du Recensement di la Population de la Polynésie Francaise, 15 Octobre, 1983*. Paris: INSEE.

Reddy, J. 1974. 'Labour and Trade Unions in Fiji', MA thesis, Department of Economics, University of Otago.

Rehg, Kenneth. 1981. *Ponapean Reference Grammar*. Honolulu: The University Press of Hawaii.

Republic of the Marshall Islands. 1984. *First Five-Year Development Plan, 1985–1989*. Majuro, Marshall Islands: Nitijela Paper No. 1.

Republic of the Marshall Islands. 1988. *Marshall Islands Statistical Abstract 1988*. Majuro: Office of Planning and Statistics.

Republic of the Marshall Islands. 1989. *1988 Census of Population and Housing: Interim Report*. Majuro: Office of Planning and Statistics.

Rex, John. 1980. 'The Theory of Race Relations – A Weberian Approach', pp. 117–142 in *Sociological Theories: Race and Colonialism*. Paris: UNESCO.

Rey, P. P. 1973. *Les Alliances de Classes*. Paris: Maspero.

Ribeiro, Darcy. 1968. *The Civilizational Process*. Washington, D.C.: Smithsonian Institute.

Richard, Dorothy E. 1957. *United States Naval Administration of the Trust Territory of the Pacific Islands*. 3 vols. Washington: Office of Chief of Naval Operations.

Ringon, Gérard. 1971. *Une Commune de Tahiti à l'Heure de Centre d'Experimentation du Pacifique: Faaa*. Paris: Office de la Recherche Scientifique et Technique Outre-Mer.

Risco, Alberto S. J. 1970. *The Apostle of the Marianas, The Life, Labors and Martydom of Ven. Diego Luis De San Vitores, 1627–1672*. Translated from the Spanish by Juan Ledesma, S. J. Guam: Published by the Diocese of Agana.

Ritter, Philip L. 1978. 'The Repopulation of Kosrae: Population and Social Organization on a Micronesian High Island', PhD dissertation, Stanford University.

Robie, David. 1988. 'Mururoa Moves Shock Paris', *Pacific Islands Monthly* 59:33 (September).

Robillard, Albert B. 1987. 'Introduction', pp. v – xix in *Contemporary Issues in Mental Health Research in the Pacific Islands*, edited by A. B. Robillard and A. J. Marsella. Honolulu: University of Hawai'i Press.

Rodman, M. 1986. 'Enracinement de l'identité: Analyse sémiotique de la tenure foncière au Vanuatu', *Culture* 6(2):3–13.

Rody, Nancy. 1978. 'Things Go Better With Coconuts: Program Strategies in Micronesia', *Journal of Nutrition Education* 10:19–22.

Rody, Nancy. 1988. 'Empowerment as Organizational Policy in Nutrition Intervention Programs: A Case Study from the Pacific Islands', *Journal of Nutrition Education* 20(3):133–141.

Rogers, Richard G. and Robert Hackenberg. 1988. 'Extending Epidemiologic Transition Theory: A New State', *Social Biology* 34(3–4):234–243.

Rokotuivuna, Amelia and International Development Action. 1973. *Fiji, A Developing Australian Colony*. North Fitzroy, Victoria: IDA.

Rollat, Alain. 1987. 'Problems come to Paradise', *Pacific Islands Monthly* 58:32–33 (October).

Rollin, Louis. 1929. *Les Iles Marquises*. Paris: Société d'Editions.

Roth, G. K. 1973. *Fijian Way of Life*. Melbourne: Oxford University Press.

Rousseau, Jérome. 1978. 'On Estates and Castes', *Dialectical Anthropology* 3(1):85–96.

Routledge, D. 1984. 'Pacific History as Seen from the Pacific Islands', *Pacific Studies* 8(2):81ff.

Roxborough, I. 1979. *Theories of Underdevelopment*. London: Macmillan Education Ltd.

Rubinstein, Donald. 1982. Personal Communication.

Ruhen, Olaf. 1966. *Harpoon in My Hand*. Sydney: Angus and Robertson.

Runciman, W. G. 1969. 'The Three Dimensions of Social Inequality', pp. 46–63 in *Social Inequality*, edited by A. Beteille. Harmondsworth, Middlesex: Penguin Books, Inc.

Rutherford, Noel. 1971. *Shirley Baker and the King of Tonga*. Melbourne: Oxford University Press.

Rutherford, Noel. 1977. 'George Tupou I and Shirley Baker', pp. 154–172 in *Friendly Islands*, edited by N. Rutherford. Melbourne: Oxford University Press.

Sabatier, E. 1977. *Astride the Equator: An Account of the Gilbert Islands*. Translated by Ursula Nixon. Melbourne: Oxford University Press.

Sack, Peter. 1986. 'German New Guinea: A Reluctant Plantation Colony?' *Journal de Societe des Oceanistes* XL:82–83, 109–127.

Sahlin, M. 1972. *Stone Age Economics*. Chicago: Aldine Publishing Co.

Sahlins, Marshall. 1958. *Social Stratification in Polynesia*. Seattle: University of Washington Press.

Said, Edward W. 1979. *Orientalism*. New York: Vintage Books.

Said, Edward W. 1985. 'Orientalism Reconsidered', *Cultural Critique* I:89–107.

Salibsury, Richard F. 1962. *From Stone to Steel: Economic Consequences of a Technological Change in New Guinea*. Melbourne: Melbourne University Press.

Samy, J. 1977. 'Some Aspects of Ethnic Politics and Class in Fiji', unpublished MPhil thesis, Institute of Development Studies, University of Sussex.

Sanchez, P. C. 1955. *Education in American Samoa*. Unpublished doctoral dissertation. Palo Alto: Stanford University.

Scarr, Deryck. 1967. *Fragments of Empire: A History of the Western Pacific High Commission, 1877–1914*. Canberra: Australian National University Press.

Scarr, Deryck. 1970. 'Recruits and Recruiters: A Portrait of the Labour Trade', pp. 225–251 in *Pacific Islands Portraits*, edited by J. W. Davidson and D. Scarr. Canberra: Australian National University Press.

Scarr, Deryck. 1984. *Fiji: A Short History*. Sydney: George Allen and Unwin.

Scarr, Deryck. (ed.) 1983. *The Three-Legged Stool: Selected Writings of Ratu Sir Lala Sukuna*. London: Macmillan Education Ltd.

Schramm, W. L., L. M. Nelson, and J. M. Betham. 1981. *Bold Experiment: Story of Educational TV in American Samoa*. Palo Alto: Stanford University Press.

Schutz, Albert J. 1977. *The Diaries and Correspondence of David Cargill, 1832–1343*. Canberra: Australia National University Press.

Schutz, Alfred. 1962–1966. *Collected Papers*. The Hague: M. Nijhoff.

Schutz, Alfred. 1967. *The Phenomenology of the Social World*. Evanston, Illinois: North-Western University Press.

Schutz, B. and R. Tenten. 1979. 'Adjustment', in *Kiribati: Aspects of History*. Kiribati: Institute of Pacific Studies and Extension Services, University of the South Pacific and Kiribati Ministry of Education, Training and Culture.

Scott, James. 1985. *Weapons of the Weak: Everyday Forms of Peasant Resistance*. New Haven: Yale University Press.

Seddon, David (ed.). 1978. *Relations of Production: Marxist Approaches to Economic Anthropology*. London: F. Cass.

Sewell, Betsy. 1983. 'Butaritari', *Atoll Economy: Social Change in Kiribati and Tuvalu, No. 3*. Canberra: Development Studies Centre, Australian National University.

493

Shapiro, Michael J. 1981. *Language and Political Understanding*. New Haven: Yale University Press.

Shapiro, Michael J. 1988. *The Politics of Representation: Writing Practices in Biography, Photography, and Policy Analysis*. Madison: University of Wisconsin Press.

Sheperd Report. 1945. *The Sugar Industry of Fiji*. London: H.M.S.O.

Shineberg, Barry. 1986. 'The Images of France: Recent Developments in French Polynesia', *Journal of Pacific History*. 21:153–168.

Shineberg, Dorothy L. 1967. *They Come for Sandalwood: A Study in the Sandalwood Trade in the South West Pacific 1830–1865*. Melbourne: Melbourne University Press.

Shineberg, Dorothy L. 1971. 'Guns and Men in Melanesia', *Journal of Pacific History* 6:61–82.

Shineberg, Dorothy L. 1971. *The Trading Voyages of Andrew Cheyne 1841–1844*. Pacific History Series, No 3. Canberra: Australian National University Press.

Shore, Bradd. 1982. *Sala'ilua: A Samoan Mystery*. New York: Columbia University Press.

Shutler, R., and M. E. Shutler. 1975. *Oceanic Prehistory*. Menlo Park, California: Cumming Publishing Co.

Simmons, R. T., J. J. Graydon, D. C. Gadjusek and P. Brown. 1965. 'Blood Group Genetic Variations in Natives of the Caroline Islands and in Other Parts of Micronesia', *Oceania* 36(2):132–70.

Smith, Anthony. 1979. *Nationalism in the Twentieth Century*. Canberra: Australian National University Press.

Somerville, Terry Sra. 1987. 'Andersen Wins SAC Award', *Pacific Daily News (Guam)*, 26 April:31.

South Pacific Commission. 1988a. *Patterns in Diseases and Causes of Deaths in the Pacific Islands*. Noumea, New Caledonia: South Pacific Commission.

South Pacific Commission: 1988b. *Food and Nutrition Issues in the Pacific*. Noumea, New Caledonia: South Pacific Commission.

Sapte Report. 1959. *The Fijian People: Economic Problems and Prospect*. CP13/59. Suva: Government Printer.

Spoehr, A. 1957. 'Marianas Prehistory', *Fieldiana Anthropology* 48.

St. Julian, Charles. 1857. *Official Report on Central Polynesia*. Sydney: Fairfax and Sons.

Stanmore, Arthur Hamilton Gordon, Baron. 1880. *Fiji, Records of Private and Public Life 1875–1880*. Vol. 1. Edinburgh: R. and R. Clark.

Stannard, D. E. 1989. *Before the Horror. The Population of Hawaii on the Eve of Western Contact*. Honolulu: Social Science Research Institute and University of Hawaii Press.

Stavenuiter, S. 1983 (April). *Income Distribution in Fiji*. Suva: Central Planning Office and ILS.

Stavrianos, L. S. 1981. *Global Rift: The Third World Comes of Age*. New York: William Morrow.

Steele, H. Thomas. 1984. *The Hawaiian Shirt: Its Art and History*. New York: Abbeville Press.

Steinberg, David Joel (ed.). 1987. *In Search of Southeast Asia*. Honolulu: University of Hawai'i Press.

Sterndale, H. B. 1874. 'Memoranda on Some of the South Sea Islands', p. 1–6 in *Papers Relating to the South Sea Islands, part 3*. Wellington: George Didsbury, Government Printer.

Stevens, Russell L. 1953. *Guam, USA: Birth of a Territory*. Honolulu: Tongg Publishing Co., Ltd.

Stevenson, Robert Louis. 1948. *Island Nights' Entertainment*. Paris: Les Deux Sirenes.

Strethern, Andrew, 1982a. 'Two Waves of African Models in the New Guinea Highlands', pp. 35–49 in *Inequality in New Guinea Highland Societies*, edited by A. Strathern. Cambridge: Cambridge University Press.

Strathern, Andrew. 1982b. 'Tribesmen or Peasants?', pp. 137–157 in *Inequality in New Guinea Highland Societies*, edited by A. Strathern. Cambridge: Cambridge University Press.

Strethern, Andrew. 1985. Lineages and Big-men: Comments on an Ancient Paradox. Pp. 101–109 in *Recent Studies in the Political Economy of Papua New Guinea Societies*, edited by D. Gardner and N. Modjeska. Sydney: Mankind Special Issue 4.

Sunday Morning Herald. 1988. 3 September.

Sundrum, R. M. 1983. *Development Economics: A Framework for Analysis and Policy*. New York: John Wiley & Sons.

Sutherland, W. M. 1984. *The State and Capitalist Development in Fuji*. Ph.D. Thesis. Canterbury: University of Canterbury.

Tagupa, William. 1976. *Politics in French Polynesia 1945–1975*. Wellington: New Zealand Institute of International Affairs.

Tagupa, William. 1983. 'Electoral Behavior in French Polynesia, 1977–1982', *Political Science* 35:38–57.

Takayama, J., H. Takasugi and K. Kaiyama. 1988. 'The 1988 Archaeological Expedition to Kiribati: A Preliminary Report', unpublished report to Kiribati Ministry of Home Affairs and Decentralisation. Tarawa, Kiribati.

Taylor, M. (ed.). 1987. *Fiji, Future Imperfect*. Sydney: Allen and Unwin.

Taylor, R., N. Lewis, and S. Levy. Forthcoming. *Mortality in Pacific Islands Countries – A Review Circa 1980*. Noumea: South Pacific Commission.

Taylor, Richard. 1985. 'Mortality Patterns in the Modernized Pacific Island Nation of Nauru', *American Journal of Public Health* 75(2):149–155.

Thiele, B. 1976. 'Ethnic Conflict and the "Plural Society" Ideology in Fiji', unpublished BA honours thesis, Department of Sociology, Flinders University.

Thomas, R. Murray. 1981. 'Evaluation Consequences of Unreasonable Goals: The Plight of Education in American Samoa', *Education Evaluation and Policy Analysis* 3(2):41–49 (March-April).

Thomas, Stephen D. 1987. *The Last Navigator*. New York: Henry Holt and Company.

Thompson, Laura. 1940. *Southern Lau, Fiji: an Ethnography*. Honolulu: Bishop Museum Bulletin No. 162.

Thompson, Laura. 1969. *Guam and Its People*. Third Edition. New York: Greenwood Press.

Thompson, Laura. 1972. *Fijian Frontier*. New York: Octagon Books.

Thompson, R. C. 1980. *Australian Imperialism in the Pacific: The Expansionist Era 1820–1920*. Melbourne: Melbourne University Press.

Thompson, Virginia and Richard Adloff. 1971. *The French Pacific Islands: French Polynesia and New Caledonia*. Berkeley: University of California Press.

Thomson, Basil. 1904. *Diversions of a Prime Minister*. London: William Blackwood and Sons.

Tinker, H. 1974. *A New System of Slavery: The Export of Indians Overseas 1830–1920*. London: Oxford University Press for the Institute of Race Relations.

495

Tjibaou, Jean-Marie. 1976. 'Recherche d'identité Mélenesienne et société tradi-tionelle', *Journal de la Société des Océanistes* 53:285.

Tjibaou, Jean-Marie. 1981. 'Entre Melanésien Aujourd'hui', *Esprit* 57:87.

Tjibaou, Jean-Marie. 1983. 'Mon idée de développement', *Trente Jours* 19:16 (October).

Tonkinson, R. 1982. 'National Identity and the Problem of Kastom in Vanuatu', *Mankind* 13(4):306–315.

Tryon, D. T. 1971. *The Contemporary Language Situation in the New Hebrides*. Canberra: Australia National University Press.

Tukuaho, J. U. and Basil Thomson. 1891. *Criminal and Civil Code of the Kingdom of Tonga*. Auckland: Brett, for the Tongan Government.

Turner, Mark M. 1984. 'Class Analysis in Sociological Explanation: The Case of Papua New Guinea', *Yagl-Ambu* 11(4):1–19.

Turner, Mark M. 1987. 'Reducing Inequality in Papua New Guinea: Gains, Losses and Prospects', *Manchester Papers on Development III* 3:25–36.

U.S. Bureau of the Census. 1941. *Sixteenth Census of the United States: 1940. Guam: Population and Agriculture*. Washington, D. C.: Government Printing Office.

U.S. Bureau of the Census. 1953. *United States Census of Population: 1950. Vol. 2, Characteristics of the Population, parts 51–54, 'Territories and Possessions.'* Washington, D.C.: Government Printing Office.

U.S. Bureau of the Census. 1983. *General Social and Economic Characteristics: Hawaii*. Washington, D. C.: Government Printing Office.

U.S. Bureau of the Census. 1984. *Detailed Social and Economic Characteristics: American Samoa*. Washington, D.C.: Government Printing Office.

U.S. Department of State. 1981. *34th Annual Report on the Trust Territory of the Pacific Islands to the United Nations*. Saipan: Trust Territory of the Pacific.

U.S. Navy Department. 1948. *Information on Guam Transmitted by the United States to the Secretary-General of the United Nations Pursuant to Article 73(e) of the Charter*. Washington, D.C.: U.S. Navy Department.

U.S. Navy Department. 1949. *Information on Guam Transmitted by the United States to the Secretary-General of the United Nations Pursuant to Article 73(e) of the Charter*. Washington, D.C.: U.S. Navy Department.

U.S. State Department. 1986. *Annual Report of the Administering Authority of the Trust Territory of the Pacific Islands to the United Nations Trusteeship Council*. Saipan: Trust Territory.

UNESCO/UNFPA. 1976. *Project on Population and Environment in the Eastern Islands of Fiji. Draft General Report No. 1*. Canberra.

United Nations Children's Fund. 1985. *The Pacific Child: A Situation Analysis*. Manila: UNICEF Programme of Cooperation for the Pacific Region.

University of Hawaii. 1984. *An Evaluation of Health Systems in the Pacific Insular Jurisdictions of the United States*. Honolulu: Schools of Medicine, Nursing, and Public Health.

Urbanowitz, Charles. 1979. 'Change in Rank and Status in the Polynesian Kingdom of Tonga', p. 225–242 in *Political Anthropology: The State of the Art*, edited by S. L. Seaton and H. J. M. Claessen. The Hague: Mouton.

Vasil, R. K. 'Communalism and Constitution Making in Fiji', *Pacific Affairs* (45(1):21–41.

Vason, George. 1910. *An Authentic Narrative of Four Years of Tongatapu*. London: Longman, Hurst, Rees, and Orme.

Ve'ehala and Topou Posesi Fanua. 1977. 'Oral Tradition and Prehistory', p.

REFERENCES

27–38 in *Friendly Islands*, edited by N. Rutherford. Melbourne: Oxford University Press.

Volti, Rudi. 1988. *Society and Technological Change*. New York: St Martin's Press.

Wacquant, L. 'Land and History in New Caledonia: The Politics of Academic Writing', *New Zealand Society* 3:49–55 (May).

Wallerstein, Immanuel. 1979. *The Capitalist World Economy: Essays*. New York: Cambridge University Press.

Wallerstein, Immanuel. 1984. *The Politics of the World-Economy: The States, the Movements, and the Civilizations*. New York: Cambridge University Press.

Wallerstein, Immanuel. 1989. *The Modern World-System III: The Second Era of The Capitalist World Economy, 1730s–1840s*. San Diego: Academic Press.

Walsh, A. C. 1982. 'Migration, Urbanisation and Development in South Pacific Countries', *Country Report VI Comparative Study on Migration, Urbanisation and Development in the ESCAP Region*. Thailand: ESCAP.

Ward, Alan. 1982. *Land and Politics in New Caledonia*. Canberra: Australian National University.

Ward, R. G. 1972. 'The Pacific Beche-de-mer Trade with Special Reference to Fiji', p. 91–119 in *Man in the Pacific Islands*, edited by R. G. Ward. London: Oxford University Press.

Ward, R. G. 1980. 'Plus ça Change . . . Plantations, Tenants, Proletarians of Peasants in Fiji', p. 134–152 in *Of Time and Place: Essays in Hounour of O. H. K. Spate*, edited by J. N. Jennings and G. J. R. Linge. Canberra: Australian National University Press.

Ward, R. Gerard and J. A. Ballard. 1976. 'In Their Own Image: Australia's Impact on Papua New Guinea and Lessons for future Aid', *Australian Outlook* 30:439–453.

Ward, R. Gerard. n.d. *Contract Labour Recruitment from the Highlands of Papua New Guinea, 1950–1974*. Forthcoming in International Migration Review.

Watson, James B. 1965. 'From Hunting to Horticulture in the New Guinea Highlands', *Ethnology* 4:295–309.

Watters, R. F. 1969. *Koro – Economic Development and Social Change in Fiji*. Oxford: Clarendon.

Watters, R. F. with K. Banibati. 1984. 'Abemama', *Atoll Economy: Social Change in Kiribati and Tuvalu, No. 5*. Canberra: Development Studies Centre, Australian National University.

Wendt, Albert. 1987. 'Novelists and Historians and the Art of Remembering', p. 78–91 in *Class and Culture in the South Pacific*, edited by A. Hooper, S. Britton, R. Crocombe, J. Huntsman, and C. Macpherson. Suva: University of the South Pacific Press.

Wesley-Smith, Terence A. 1988. 'Melanesians and Modes of Production: Underdevelopment in Papua New Guinea with Particular Reference to the Role of Minding Capital', unpublished dissertation, University of Hawaii.

Wessman, James W. 1981. *Anthropology and Marxism*. Cambridge, MA: Schenkman Publishing Co.

West, F. J. 1960. 'Background to the Fijian Riots', *Australian Quarterly* 32(1) (March):46–53.

West, Rev. Thomas. 1846/1865. *Ten Years in South-Central Polynesia*. London: James Nisbet.

Western Samoa Department of Education. 1973. *Annual Report*. Apia: Government Printing.

Western Samoa, Government of. 1975. *Western Samoa's Economic Development Plan*. Apia: Government Printing.

497

Western Samoa, Government of. 1976. *Western Samoa's Census of Population*. Apia: Government Printing.

Western Samoa, Government of. 1984. *Western Samoa's Fifth Development Plan*. Apia: Government Printing.

Whiting, Alfred F. 1954. *Nan Madol*. Ponape District: Education Department.

Whorf, Benjamin Lee. 1956. *Language, Thought and Reality*. Cambridge: MIT Press.

Wilkes, C. 1845. *Narrative of the United States Exploring Expedition During the Years 1838, 1839, 1840, 1841, 1842*. London: Lea and Blanchard.

Wilkes, Charles. 1852. *The United States Exploring Expedition During the Years 1838, 1839, 1840–1842*. 2 vols. London: Ingram, Cooke and Co.

Williams, John. 1984. *The Samoa Journals of John Williams*. Edited by R. A. Moyle. Canberra: Australian National University Press.

Wilson, James. 1799/1968. *A Missionary Voyage to the Southern Pacific Ocean, 1796–1798*. New York: Praeger.

Wilson, Walter Scott. 1975. 'Historical Summary of Cultural Influences on the People of Guam', p. 91–96 in *The Social-Economic Impact of Modern Technology Upon a Developing Insular Region: Guam*, Vol. 3, pt. 6, coordinated by J. Vail. Guam: University of Guam Press.

Winkler, J. E. 1982. *Losing Control . . . Towards an Understanding of Transnational Corporations in the Pacific Islands Context*. Pacific Conference of Churches. Suva: Lotu Pasfika Productions.

Wittaker, Elvi W. 1986. *The Mainland Haole: The White Experience in Hawaii*. New York: Columbia University Press.

Wolf, Eric. 1982. *Europe and the People without History*. Berkeley and Los Angeles: University of California Press.

Wolfe, Tom. 1987. *The Bonfire of the Vanities*. New York: Farrar, Straus Giroux.

Wollin, J. and J. Roberts. 1973. 'Fiji's Sugar Industry', p. 74–77 in *Fiji, A Developing Australian Colony*, edited by A. Rokotuivuna and International Development Action. North Fitzroy, Victoria: IDA.

Wolpe, Harold (ed.). 1980. *The Articulation of Modes of Production: Essays from Economy and Society*. London: Routledge and Kegan Paul.

World Bank. 1984. *World Development Report 1984*. Washington: D.C.: World Bank.World Bank. 1985. *World Development Report 1985*. New York: Oxford University Press.

World Bank. 1986. *World Development Report*. New York: IBRD and Oxford University Press.

World Bank. 1987. *The World Bank Atlas 1987*. Washington: World Bank.

World Health Organization – Suva. 1985. 'Report on the International Drinking Water and Sanitation Decade', presented in United Nations Children's Fund-/Manila, *The Pacific Child: A Situation Analysis*. Manila: UNICEF Programme of Cooperation for the Pacific Region.

Worsley, D. 1967. *The Third World*. London: Weidenfeld and Nicholson.

Worsley, P. 1957. *The Trumpet Shall Sound: A Study of 'Cargo' Cults in Melanesia*. London: MacGibbon and Kee.

Worsley, P. 1981. 'Social Class and Development', in *Social Inequality*, edited by G. Berreman. New York: Academic Press.

Wright, L. B. and M. I. Fry. 1936. *Puritans in the South Seas*. New York: Henry Holt.

Wu, David Y. H. 1982. *The Chinese in Papua New Guinea, 1880–1980*. Hong Kong: Chinese University Press.

Young, J. 1970. 'Evanescent Ascendancy – The Planter Community in Fiji', in

REFERENCES

Pacific Island Portraits, edited by J. W. Davidson and D. Scarr. Canberra: Australian National University Press.

Primary Sources on Fiji

Constitution of Fiji, 1970. Suva: Government of Fiji.

Fiji Constitutional Discussions. Report on Lord Shepard's Visit to Fiji, January 1970. (CP 1/1970).

Fiji Legislative Council Debates, 1961. 1961. Suva: Government of Fiji.

Fiji Legislative Council Paper No. 8 of 1946 (CP 8/1946). Suva: Government of Fiji.

Fiji Legislative Council Paper No. 10 of 1960 (CP 10/1960). Lowe Report. Report of Commission of Inquiry into the Disturbances in Suva, December 1959. Suva: Government of Fiji.

Fiji Legislative Council Paper No. 19 of 1946 (CP 19/1946). Suva: Government of Fiji.

Fiji Legislative Council Paper No. 20 of 1961 (CP 20/1961). Eve Report. Report of the Fiji Sugar Inquiry Commission. Suva: Government of Fiji.

Fiji Legislative Council Paper No. 25 of 1946 (CP 25/1946). Wages and Salaries Report. Suva: Government of Fiji.

Fiji Legislative Council Paper No. 39 of 1914 (CP 39/1914). Suva: Government of Fiji.

Fiji Legislative Council Paper No. 46 of 1920 (CP 46/1920). Suva: Government of Fiji.

Fiji Legislative Council Paper No. 67 of 1920 (CP 67/1920). Suva: Government of Fiji.

Fiji Legislative Council Sessions, 1967. 1967. Suva: Government of Fiji.

Index

Absenteeism, from villages, 165
Acculturation, in Micronesia, 218
Agriculture, 211; in French Polynesia,
   360; in Melanesia, 69, 71, 72; in
   Samoa, 305; in Tonga, 330
Akhil Fiji Krishnak Maha Singh (A
   Fiji Farmer's Union), 173ff
Alienation, 88; of land, 141
Aloha Friday, 381
Aloha Week, 381
Aloha shirt, as social order, 30
American Samoa, see Samoa
Anglo-French Protocol, 111
Atai, leader of revolt in Melanesia, 66
Atomic testing program, 30, 367
Atomic weapons, 359; see also Nuclear
   weapons
Austronesian language, 394

Beachcombers, 397
Before the Horror, 437
Belau, 16, 202; German
   administration in, 211
Bethan, Mere, director of education
   in Samoa, 317
Biculturalism, 315
Big-men, in Papua New Guinea, 42
Bi-planar theory of language, 4
Bigotry, in Fiji, 134
Bikini, nuclear tests on, 214
Bilingualism, education programs in,
   410; in Samoa, 314, 315
Blood of the Ancestors, see *Te Toto
   Tupuna*
Blood of the Lamb movement, 166
Brigham Young University – Hawai'i,
   389
British Colombia Sugar Refining Co.,
   151

Bulolo Gold Dredging Company
   Limited, 49

Caldoches, French settlers in New
   Caledonia, 75, 79, 80
Canoe, imagery of, 105; war canoe of
   Fiji, 137
Capitalism, 24; in Fiji, 140, 144; as
   world system, 36
Cargo Cults, 107; in Vanuatu, 101
Caroline and Marshall Islands, 26,
   203–20; sanitation in, 420; Spanish
   annexation of, 208, 419
Caroline Islands, see Caroline and
   Marshall Islands
Catholic missions, 309; see also
   Missionaries
Catholicism, see Christianity; Religion
Censorship, of press, 79
Census, in Samoa, 319
Centre d'Experimentation du Pacifique,
   30
Chaco (on Guam), 243, 442
Chamorros, 27, 220, 411; in Northern
   Marianas, 27
Children, 31; and diet, 31; survival of,
   31, 414–27
China trade, 206
Chinese: in Fiji, 160; in French
   Polynesia, 362; in Guam, 27
Christianity, 88, 89, 113, 350; concept
   of, 410; introduction of, 25; see also
   Religion
Church Missionary Society, 434ff
Clan organization, 122
Class, formation of: in Fiji, 183; in
   Tonga, 334
Clothing, 372–92
Club Med, 70
Coca Cola, 427

501

Coconut oil, 138, 349
Colonial Sugar Refining Company (CSR), *see* CSR
Colonial period, and language, 402; in the Pacific, 401; in Vanuatu, 99, 101, 106
Commonwealth of the Northern Marianas Islands, *see* Northern Marianas Islands
Compact of Free Association, 218; with the US, 26
Condominium of the New Hebrides, 111, 112, 126
Constancy principle, 5
Cooperatives, 89
Copra, 26, 163, 212, 340, 345, 354ff; in Kiribati, 28; in Tonga, 332, 335; trade in, 207, 209
Cotton, 140, 354
Coups, 26; in Fiji, 134
Coutume (tradition), 88
Creation myths: and prehistory, 13; of Vanuatu, 13, 25; *see also* Myths
CSR (Colonial Sugar Refining Company), 150ff
Cults, ancestor, 116; religious, in Fiji, 166
Cultural beliefs, of Europeans, 435
Culture, traditional, 106

Dance forms, 210
De Gaulle, Charles, 358, 364
Death rates, *see* Mortality
Debt bondage, 337
DeLima, Frank [entertainer in Hawai'i], 390
Demis, locals in French Polynesia, 362
Depopulation, 354, 437
Depression (1880s and 1890s), 150
Descent, partilineal: in Fiji, 137; principle of, in Papua New Guinea, 41
Disease, 350; infectious, 414ff; non-communicable, 416
Dravidian kinship terminology, 103
Drugs, 363

Education, 407ff; and bilingualism, 412; and churches, 399; in Guam, 226ff; in Pacific Islands, 399; in Samoa, 303, 304, 308, 309, 314; in Tonga, 341; in Vanuatu, 112; under

Japan, 211; under US Navy, 214; uses of TV in, 316
Emigration, 218
Emperor Gold Mines, 162
Enewetak, nuclear tests on, 214
Epidemics, 142; *see also* Disease
Epidemiological transition theory, 414ff
Erakor, village in Vanuatu, 112, 123
European contact: in French Polynesia, 348; in Northern Marianas, 242
Eve Commission, 181–2
Exploration, 442; in Micronesia, 205

Fa'a Samoa, 309ff
Federated States of Micronesia, 203, 217
Fédération pour une Nouvelle Société Calédonienne, *see* FNSC
Fertility, 320
Fiji, 26, 134–99; bigotry in, 134; canoes of, 137; capitalism in, 140; Chinese in, 160; class formation of, 183; constitution, provisions of, 189–90; cults in, 166; demography of, 161; descent in, 137; economy of, 198; farmers' union in, 173ff; and France, 145; governments in, 167; lumber industry in, 194; militarism in, 134; political parties in, 187; population in, 186; protests in, 167; racism in, 134; religion in, 139; social order in, 134; sugar industry in, 197; tourism in, 192
Fiji Industrial Workers Congress, 175
Fiji Sugar Co., 151
Fijian Affairs Ordinance, 146
Filipinos, in Guam, 235
Fine mats, 320
Fishing, 355
FLNKS, 75ff
FNSC, 75
Food: consumption of, 425; imported from France, 419; junk, 425; *see also* Nutrition
Free association, 202
Free migrants, 158
French Polynesia, 25, 29, 65, 346–71; agriculture in, 360; banning of language in, 366; changes in, 360ff; Chinese in, 362; copra production

in, 354; European contact in, 348; and Greenpeace Affair, 370; political parties in, 357; population decline in, 353; preservation of Tahitian in, 370; protests in, 368; trade in, 348; women in, 347
Front de Libération Nationale Kanake et Socialiste, see FLNKS
Funerals, 338

Garment industry, 27
Gaullists, 365
Germany, 26; administration of Samoa by, 310; and copra, 26; and New Guinea, 44
God, concept of, 410
Godeffroy and Son, 141, 336
Gold: fields in Papua New Guinea, 46; mine at Vatukoula, 149; rush, in New Caledonia, 68
Gordon, Sir Arthur, 142
Government, post-colonial, 405; see also Colonialism
GovGuam, see Guam, government of
Grande Terre, 72
Great Revolt of Atai, see Atai
Green Revolution, 78
Greenpeace Affair, 370
Guam, 27, 220–40, 394; Chinese in, 27; community relations in, 237; economy of, 229ff; education in, 226ff; ethnic diversity of, 234; Filipino workers in, 235; government of, 27, 222ff; and Japanese occupation, 223–4; linguistic changes in, 27; Spanish acquisition of, 401
Gujeratis, as shopkeepers and craftsmen, 160

Harada, Wayne [entertainment critic], 390
Hawai'i, 30, 381–92; holidays in, 381; political economy in, 377
Health programs, 427
Hienghene, massacre at, 87
Holidays, in Hawai'i, 381

Immigrants, and clothing, 384
Immigration, French, 366; from Philippines, 384
Indenture system: in Papua New Guinea, 48; workers in, 156ff, 169

Independence, 95; in New Caledonia, 75–6
India, laborers from, 155
Individualism ideology, 386
Infant mortality, see Mortality
Insurrection: in the Pacific islands, 25; in Papua New Guinea, 40; in Pohnpei, 208; see also Rebellion
Internment camps, in Guam, 224
Ipomoean Revolution, in Papua New Guinea, 40
Islands of Thieves, 241
Isle of Pines, 70

Jaluit Company, 209
Japan: colonization by, 210–12; immigration, 212; language usage in Guam, 224; mandate in Marshall and Caroline Islands, 26; Occupation of Guam, 223; in World War I, 210
John Frum movement, 127
Juvenile delinquency, 363

Kamehameha, family of, 15
Kanak rebellion (1878), 24, 25
Kanaks (Melanesian), 95; political party, 73–4
Kanaky, 90; nationalism, 91; Republic of, 75
Kastom, 25, 108, 119, 127ff, 252; creation of, 25; and national culture, 7, 91
Kinship, 41, 345; ego-oriented networks of, 41; ties, in Fiji, 136; systems, 6
Kiribati, 28, 30; copra trade in, 28; missionization of, 28
Kisan Sangh (Farmers' Association), 171ff
Kolonia, 18
Kosrae, 203, 205
Kwajalein, military installation on, 217

Labor, 27, 318; division of, 329, 338; immigrant, 154; in Fiji, 141; from India, 155; from Japan, 27; in Melanesia, 67, 71; migration of, 342; recruitment of, 142, 154; in Tonga, 336; in Vanuatu, 112
Land and Power in Hawai'i, 376ff
Land: access to, 42, 43, 341; alienation

of, 66, 73, 75, 141, 229; claims for, 86, 140; development of, 375; government ownership of, 211; ownership of, 86; pressures of, 73; shortages, 164; struggle for, 66; tenure, 6, 90, 208, 330, 340; use-rights to, 336

Land Tenure Conversion Act (1973) in Papua New Guinea, 54

Lands Office, New Caledonia, 79

Languages; aberrant, 395; banning of, 366; bi-planar theory of, 4; changes in, 393; differences in, 395; European influence on, 397, 403; in Fiji, 143; of indigenous Pacific, 31, 403; introduced, 393; preservation of, 370; and social change, 393–413; and social science, 3; specialized, 395; status of, 402; written, 404

League of Nations, and Japanese mandate, 211

Life expectancy, 423

Lineage, 38, 326

Literacy, 399, 402; influence of, 408; and religion, 398; see also Education

London Missionary Society, 309, 333, 349, 443

Loyalty Islands, 66

Mandated Territory of New Guinea, 48

Marriage, 117; in Papua New Guinea, 41; strategies of, 307; in Samoa, 306; in Tonga, 325

Marshall Islands, see Caroline and Marshall Islands

Marxism, 19, 23, 26, 439; as conceptual framework, 35; theory of, 11

Matignon Accord, 84, 85

Medicine, 32, 435

Melanesia, 23, 24; agriculture in, 69, 71–2; penal settlement in, 65, 66; political party in, 74; revolt in, 66

Micronesia, 23, 203–19; acculturation in, 218; exploration in, 205; oral history of, 204; Peace Corps in, 216

Migration: to New Caledonia, 65, 67, 78; to the Pacific, 13; patterns of, 395; from Samoa, 318

Militarization, in Guam, 239

Military government, 225

Mining, in Papua New Guinea, 48

Mission schools, 309, 310, 400, 403

Missionaries, 136, 138, 170, 337–8, 397, 442; conflict with, 140; in Fiji, 140; in French Polynesia, 349; in Hawai'i, 373ff; in Kiribati, 28; in Samoa, 309; in Tonga, 323ff, 332–3; in Vanuatu, 101

Mitterand, François, 369

Moderates (political party in Vanuatu), 107

Money, introduction of, 400

Mormons, 340, 390; see also Religion

Morobe gold fields, 48; see also Gold

Mortality, 350; child, 31; infant, 414; rates, 141, 155, 414

Myths, 1, 13, 440; Christian, 118; and social change, 18; in Vanuatu, 102, 104

Nan Madol, 18, 19, 205

Nanyo Boeki Kaisha (NBK), 212

Nanyo Kohatsu Kaisha (NKK), 212

Nation-state: organization of, 21; perspective of, 22

National culture, and kastom, 7; in Vanuatu, 7

Nationalism, 86, 92ff; in Melanesia, 87

Native Land Trust Board (NLTB), 188

Native Lands Commission, 145

Native Regulation (1877), 146

Navigation, 5, 8

Nekabesuam, 65

Neo-Marxism, 11, 23; see also Marxism

New Guinea, western capitalism in, 35

New Guinea Goldfields Limited, 48; see also Gold

New Caledonia, 24, 65–97; migration to, 65, 67, 78; political unrest in, 70

Ni-Vanuatu, 25, 105; see also Vanuatu

Nickel, mining of, in New Caledonia, 69

Nimitz, Admiral Chester, 224

Northern Marianas Islands, 27, 241–3; Chamorros in, 27; pre-contact history of, 240ff; tourist industry in, 27

Noumea, 66, 68, 70, 71, 74

Nuclear tests, 359; protests against, 364

Nuclear weapons, 346; future of, 358; see also Atomic Testing Program

Nutrition, problems of, 417ff; see also Food

Occupations, divisions of, 163

Oral history: in Micronesia, 204; in Tonga, 325

Organic Act, 220, 226, 236

Pacific Islanders, as health workers, 6

Pacific Islands, exploration of, 442; language in, 393–413; under Spanish rule, 2, 202

Palau, see Belau

Papua New Guinea, 16, 24, 29, 35–64, 418; access to land in, 42, 43; big-men in, 42; colonial penetration in, 43; descent principle in, 41; German claims to, 44; goldfields in, 46; indenture system in, 48; Ipomoean Revolution in, 40; kinship in, 41; land tenure in, 54; marriage in, 41; partilineal descent in, 39; plantations in, 45–6; postwar colonialism in, 50; strikes in, 49

Parkinson, Richard, German scientist, 47

Peace Corps, in Micronesia, 216

Pearls, 347, 353

Piailug, master traditional navigator, 8

Pisani, Edgard, High Commissioner to New Caledonia, 75

Pohnpei, 7, 18, 203, 204

Political economy, 23; in Hawaii, 377

Political institutions, 104, 139

Political movements, 171

Political parties, 75, 76, 357; in Fiji, 139, 187; in French Polynesia, 357; in Guam, 187; in Melanesia, 73–4; in New Caledonia, 73; in Vanuatu, 107, 121

Polynesia, 23, 24, 29, 300–413

Polynesian Culture Center, 389

Pomare, family of, 15

Population: decline of, 207, 353; increase in, 73; of Europeans, 66; in Fiji, 186; in Tonga, 323

Postmodern critique, purpose of, 2

Powerlessness, 422

Pre-colonial societies, 432

Prehistory, and ceation myths, 13; see also Myths

Protestant American Board, 207

Protests, 368; economic, 167ff; industrial, 167ff; in Fiji, 167, see also Rebellion

Pula, Nikolao, director of education in Samoa, 317

Racial division, 161, 162, 168

Rank, 328

Rassemblement pour la Calédonie dans la République, see RPCR

Ratu Sir Lala Sukuna, see Sukuna, Ratu Sir Lala

RDPT party, 365

Rebellion, 368; in Fiji, 167; in Vanuatu, 108; see also Protests; Insurrection

Religion, 89, 116, 139, 350; Catholicism in Fiji, 139; as education, 123; and literacy, 398; see also Christianity

Religious movements, 166

Revolutions, see Insurrection

Reyn Spooner, 377

Reyn's aloha shirt, 30, 372ff

Roy Mata, legend of, 119

RPCR (Rassemblement pour la Calédonie dans la République), RPCR, Sahul, 23

Sahul, 23

Salote, Queen, 339ff

Samoa, 29, 301–19; administration by Germany of, 310; administration by New Zealand of, 310; after World War II, 303; agriculture in, 304, 307; bilingualism in, 314–15; census in, 319; currency in, 308; education in, 303, 304, 308, 309, 314, 316; emigration from, 311; fine mats in, 320; marriage in, 306; medical beliefs in, 435; missionaries in, 309; occupations in, 303; social status in, 304; status of women in, 305ff; talking chiefs in, 304, 307; tattooing in, 305

Sandalwood trade, 16

San Vitores, 243

Sanvitores, Fr Diego Luis de, 398

Satawal atoll, 8, 10

Schools, see Education

Scottish Moral Philosophers, 19
Shame, 385
Ships' logs, 5
Shirt, Reyn's aloha, 372–92
SLN, 69
Social change: alternative analysis of,
    1; consensus of, 3; political
    economies of, 2; theories of, 2, 437ff
Social science, consensus on, 3
Sociétal Le Nickel (SLN), 69
Sorcery, 19
South Pacific Sugar Mills Ltd (SPSM),
    159
Spain: annexation of Caroline and
    Marshall Islands by, 208, 419;
    exploration by, 205; and settlement
    of Guam, 220
Spanish-American War, 221
Strikes, 169ff; in Papua New Guinea,
    49
Styles, in aloha wear, 383
Subsistence economy, 407
Suffrage, 76
Sugar industry, 148; in Fiji, 135, 197
Suicide, of indentured workers, 157
Sukuna, Ratu Sir Lala, 162
Supernatural, belief in, 117

Talking chiefs, in Samoa, 304, 307
Tan Union, 108, 131
Tarawa, 28
Taro, 305
Tattooing, 210, 305
Te Toto Tupuna, 368
Television, programs filmed in
    Hawai'i, 383; use of in education,
    316ff
Tenure, see Land
Tjibaou, Jean-Marie, president of
    Republic of Kanaky, 75, 79
Tolai, 47
Tonga, 24, 322–44, 401; agriculture in,
    330; class formation in, 334; copra
    trade in, 332, 335; education, 332,
    335, 340; labor in, 329, 336; land in,
    330, 340, 341; marriage in, 325;
    missionaries in, 323ff, 332; oral
    history in, 325; population in, 323;
    prehistory of, 323ff; and tourism,
    343; trade in, 331, 332; and western
    contact, 327; women in, 338
Tourism, 237; and apparel, 377; in

Fiji, 135, 192; in Guam, 27; in
    Melanesia, 70; in Tonga, 343
Trade unions, 165, 171, 175ff
Trade: in Fiji, 137; in French
    Polynesia, 348; in Micronesia, 204;
    in Tonga, 331
Traders, 397ff
Traditionalism, in Vanuatu, 100
Treaty of Paris, 221
Truk, 203
Trusteeship Council Mission, in
    Western Samoa, 310

UC (Union Calédonienne), 73; see
    also Political parties
Unemployment, in Tonga, 344
Union Calédonienne (UC), 73
Unions, see Trade unions
United Nations, 213; agencies of, 24;
    International Trusteeship System,
    310
US Navy, education under, 214; in
    Guam, 222

Vanuaaku Pati (political party in
    Vanuatu), 107, 110
Vanuatu, 10, 17, 25, 97–133; cargo
    cults in, 101; colonialism in, 99,
    101, 106; conspicuous consumption
    in, 124; creation myths of, 13, 25;
    Erakor village in, 112, 123; labor in,
    112; modernization of, 121;
    national culture of, 7, 9; political
    divisions in, 103; political parties
    in, 107; rebellion in, 108; rituals in,
    103; social organization in, 102,
    103; symbolization in, 99;
    traditionalism in, 100
Venereal disease, 207, 350

Wages, disparity in, 230; rates, 156
Waianae, HI, social problems in, 8
Warfare, 332
Wesleyan Methodist mission, 332
Western contact, 204, 331, 433; and
    language, 393
Western Samoa, see Samoa
Whalers, 206, 350, 397
Wholesale and Retail General
    Workers' Union (WRGWU), 176
Women: as chiefs, 327; as domestic
    servants, 153; exchange of, 118; in
    French Polynesia, 347; in Fiji, 137;

as laborers, 156, 318; in Micronesia, 212; in Papua New Guinea, 62; in Samoa, 29, 305ff; in Tonga, 323, 327, 338
World Bank, 24
World system theory, 11, 26; in Vanuatu, 98
World War I, 210

World War II, 341, 357, 404ff; in Guam, 223, 438; in Micronesia, 213; and mortality, 416; in New Caledonia, 67–8; in Papua New Guinea, 45

Yap, 202